Geophysical Monograph Series

Including

IUGG Volumes
Maurice Ewing Volumes
Mineral Physics Volumes

Geophysical Monograph Series

76 Relating Geophysical Structures and Processes: The Jeffreys Volume (IUGG Volume 16) K. Aki and R. Dmowska (Eds.)

77 The Mesozoic Pacific: Geology, Tectonics, and Volcanism Malcolm S. Pringle, William W. Sager, William V. Sliter, and Seth Stein (Eds.)

78 Climate Change in Continental Isotopic Records P. K. Swart, K. C. Lohmann, J. McKenzie, and S. Savin (Eds.)

79 The Tornado: Its Structure, Dynamics, Prediction, and Hazards C. Church, D. Burgess, C. Doswell, R. Davies-Jones (Eds.)

80 Auroral Plasma Dynamics R. L. Lysak (Ed.)

81 Solar Wind Sources of Magnetospheric Ultra-Low Frequency Waves M. J. Engebretson, K. Takahashi, and M. Scholer (Eds.)

82 Gravimetry and Space Techniques Applied to Geodynamics and Ocean Dynamics (IUGG Volume 17) Bob E. Schutz, Allen Anderson, Claude Froidevaux, and Michael Parke (Eds.)

83 Nonlinear Dynamics and Predictability of Geophysical Phenomena (IUGG Volume 18) William I. Newman, Andrei Gabrielov, and Donald L. Turcotte (Eds.)

84 Solar System Plasmas in Space and Time J. Burch, J. H. Waite, Jr. (Eds.)

85 The Polar Oceans and Their Role in Shaping the Global Environment O. M. Johannessen, R. D. Muench, and J. E. Overland (Eds.)

86 Space Plasmas: Coupling Between Small and Medium Scale Processes Maha Ashour-Abdalla, Tom Chang, and Paul Dusenbery (Eds.)

87 The Upper Mesosphere and Lower Thermosphere: A Review of Experiment and Theory R. M. Johnson and T. L. Killeen (Eds.)

88 Active Margins and Marginal Basins of the Western Pacific Brian Taylor and James Natland (Eds.)

89 Natural and Anthropogenic Influences in Fluvial Geomorphology John E. Costa, Andrew J. Miller, Kenneth W. Potter, and Peter R. Wilcock (Eds.)

90 Physics of the Magnetopause Paul Song, B.U.Ö. Sonnerup, and M.F. Thomsen (Eds.)

91 Seafloor Hydrothermal Systems: Physical, Chemical, Biological, and Geological Interactions Susan E. Humphris, Robert A. Zierenberg, Lauren S. Mullineaux, and Richard E. Thomson (Eds.)

92 Mauna Loa Revealed: Structure, Composition, History, and Hazards J. M. Rhodes and John P. Lockwood (Eds.)

93 Cross-Scale Coupling in Space Plasmas James L. Horwitz, Nagendra Singh, and James L. Burch (Eds.)

94 Double-Diffusive Convection Alan Brandt and H. J. S. Fernando (Eds.)

95 Earth Processes: Reading the Isotopic Code Asish Basu and Stan Hart (Eds.)

96 Subduction Top to Bottom Gray E. Bebout, David Scholl, Stephen Kirby, and John Platt (Eds.)

97 Radiation Belts: Models and Standards J. F. Lemaire, D. Heynderickx, and D. N. Baker (Eds.)

98 Magnetic Storms Bruce T. Tsurutani, Walter D. Gonzalez, Yohsuke Kamide, and John K. Arballo (Eds.)

99 Coronal Mass Ejections Nancy Crooker, Jo Ann Joselyn, and Joan Feynman (Eds.)

100 Large Igneous Provinces John J. Mahoney and Millard F. Coffin (Eds.)

101 Properties of Earth and Planetary Materials at High Pressure and Temperature Murli Manghnani and Takehiki Yagi (Eds.)

102 Measurement Techniques in Space Plasmas: Particles Robert F. Pfaff, Joseph E. Borovsky, and David T. Young (Eds.)

103 Measurement Techniques in Space Plasmas: Fields Robert F. Pfaff, Joseph E. Borovsky, and David T. Young (Eds.)

104 Geospace Mass and Energy Flow: Results From the International Solar-Terrestrial Physics Program James L. Horwitz, Dennis L. Gallagher, and William K. Peterson (Eds.)

105 New Perspectives on the Earth's Magnetotail A. Nishida, D. N. Baker, and S. W. H. Cowley (Eds.)

106 Faulting and Magmatism at Mid-Ocean Ridges W. Roger Buck, Paul T. Delaney, Jeffrey A. Karson, and Yves Lagabrielle (Eds.)

107 Rivers Over Rock: Fluvial Processes in Bedrock Channels Keith J. Tinkler and Ellen E. Wohl (Eds.)

108 Assessment of Non-Point Source Pollution in the Vadose Zone Dennis L. Corwin, Keith Loague, and Timothy R. Ellsworth (Eds.)

Geophysical Monograph 109

Sun-Earth Plasma Connections

James L. Burch
Robert L. Carovillano
Spiro K. Antiochos
Editors

American Geophysical Union
Washington, DC

Published under the aegis of the AGU Books Board

Library of Congress Cataloging-in-Publication Data

Sun-earth plasma connections / James L. Burch, Robert L. Carovillano,
    Spiro K. Antiochos, editors.
        p. cm. -- (Geophysical monograph ; 109)
      Includes bibliographical references.
      ISBN 0-87590-092-5
        1. International Solar-Terrestrial Physics Program Congresses.
2. Sun Congresses. 3. Solar magnetic fields Congresses. 4. Space
plasmas Congresses. 5. Magnetosphere Congresses. 6. Astrophysics
Congresses. I. Burch, J. L., 1942- . II. Carovillano, Robert
L., 1932- . III., Antiochos, Spiro K. IV. Series.
    QB520.S86 1999
    523.7'2--dc21   99-28915
                                                                        CIP

ISBN 0-87590-092-5
ISSN 0065-8448

Cover
Top: an eruptive prominence on the Sun, imaged at 304 angstroms (He II) with the Extreme-ultraviolet Imaging Telescope (EIT) on the SOHO spacecraft. (Photo courtesy of the SOHO-EIT consortium. SOHO is an ESA/NASA project of international cooperation.)

Middle: an energetic neutral atom image showing the enhancement of the Earth's ring current during the geomagnetic storm of April 11, 1997. The image was produced from data acquired with the Comprehensive Energetic Particle Pitch Angle Distribution (CEPPAD) instrument on board the Polar spacecraft. (Image courtesy of the NASA/Polar project and the CEPPAD instrument team.)

Bottom: an intensification of the Earth's x-ray aurora during the April 11, 1997 magnetic storm, as imaged with the Polar Ionosphere X-ray Imaging Experiment (PIXIE) on the Polar spacecraft. (Courtesy of the NASA/Polar project and the PIXIE team.)

Copyright 1999 by the American Geophysical Union
2000 Florida Avenue, N.W.
Washington, DC 20009

Figures, tables, and short excerpts may be reprinted in scientific books and journals if the source is properly cited.

Authorization to photocopy items for internal or personal use, or the internal or personal use of specific clients, is granted by the American Geophysical Union for libraries and other users registered with the Copyright Clearance Center (CCC) Transactional Reporting Service, provided that the base fee of $1.50 per copy plus $0.35 per page is paid directly to CCC, 222 Rosewood Dr., Danvers, MA 01923. 0065-8448/99/$01.50+0.35.

This consent does not extend to other kinds of copying, such as copying for creating new collective works or for resale. The reproduction of multiple copies and the use of full articles or the use of extracts, including figures and tables, for commercial purposes requires permission from the American Geophysical Union.

Printed in the United States of America.

CONTENTS

Preface
James L. Burch, Robert L. Carovillano, and Spiro K. Antiochos vii

The International Solar Terrestrial Physics Program: The Great Observatory for the Sun-Earth Connection

ISTP and Beyond: A Solar-System Telescope and a Cosmic Microscope
D. N. Baker and M. J. Carlowicz 1

The SOHO Mission
A. I. Poland 11

Geotail Mission: Accomplishments and Prospects
A. Nishida 19

The Correspondence of EUV and White Light Observations of Coronal Mass Ejections with SOHO EIT and LASCO
B. J. Thompson, O. C. St. Cyr, S. P. Plunkett, J. B. Gurman, N. Gopalswamy, H. S. Hudson, R. A. Howard, D. J. Michels, and J.-P. Delaboudinière 31

GEOTAIL Substorm/Storm Studies
Rumi Nakamura 47

New Results on the Polar Cap and PSBL Dynamics
G. Parks, M. Brittnacher, L. J. Chen, M. McCarthy, D. Larson, R. P. Lin, G. Germany, J. Spann, H. Reme, and T. Sanderson 57

Global Energy-Resolved X-Ray Images of Northern Aurora and Their Mappings to the Equatorial Magnetosphere
David L. Chenette, William L. Imhof, Steven M. Petrinec, Michael Schulz, Joseph Mobilia, John G. Pronko, Michael A. Rinaldi, John B. Cladis, Frances Fenrich, Nikolai Østgaard, and Michael C. McNab 65

Numerical Cavity Mode Simulation and Polar Data From the January 1997 Magnetic Cloud Event
J. Goldstein, R. E. Denton, M. K. Hudson, W. Lotko, and J. G. Lyon 77

Polar/TIDE Results on Polar Ion Outflows
T. E. Moore, M. O. Chandler, C. R. Chappell, R. H. Comfort, P. D. Craven, D. C. Delcourt, H. A. Elliott, B. L. Giles, J. L. Horwitz, C. J. Pollock, and Y.-J. Su 87

The Low-Latitude Boundary Layer: Application of ISTP Advances to Past Data
M. Lockwood and M. A. Hapgood 103

Major Space Plasmas and Field Processes of the Sun-Earth System

The Role of Magnetic Reconnection in Solar Activity
Spiro K. Antiochos and C. Richard DeVore 113

Unresolved Questions About the Structure and Dynamics of the Extended Solar Corona
William C. Feldman 121

CONTENTS

Models for Coronal and Interplanetary Magnetic Fields: A Critical Commentary
Kenneth H. Schatten 129

A Multi-Spacecraft Study of Solar Wind Structure at 1 AU
K. I. Paularena, J. D. Richardson, F. Dashevskiy, G. N. Zastenker, and P. A. Dalin 143

Plasma Entry, Transport, and Loss in the Magnetosphere and Ionosphere
Patricia H. Reiff 149

Cusp Ion Composition as an Indicator of Non-Steady Reconnection
S. A. Fuselier and K. J. Trattner 161

Simulation of Radiation Belt Dynamics Driven by Solar Wind Variations
M. K. Hudson, S. R. Elkington, J. G. Lyon, C. C. Goodrich, and T. J. Rosenberg 171

Origins and Transport of Ions During Magnetospheric Substorms
M. Ashour-Abdalla, M. El-Alaoui, V. Peroomian, J. Raeder, R. J. Walker, L. A. Frank, and W. R. Paterson 183

Ionospheric Outflow
R. W. Schunk 195

Future Missions and Scientific Objectives:
Evolution of the International Solar Terrestrial Physics Program

The Science of Solar-B
Spiro K. Antiochos 207

The Solar Stereo Mission
D. M. Rust 213

Magnetospheric Multiscale and Global Electrodynamics Missions
Barry H. Mauk, Richard W. McEntire, Roderick A. Heelis, and Robert F. Pfaff, Jr. 225

Solar Probe: A Mission to the Sun and the Inner Core of the Heliosphere
G. Gloeckler, S. T. Suess, S. R. Habbal, R. L. McNutt, J. E. Randolph, A. M. Title, and B. T. Tsurutani 237

Magnetospheric Constellation: Past, Present, and Future
V. Angelopoulos and H. E. Spence 247

A Mercury Orbiter Mission
D. N. Baker 263

PREFACE

The Sun and the terrestrial magnetosphere have been the subjects of active research since the dawn of the space age. The capabilities of observing both systems with greater and greater detail evolved separately until the 1980s, when it was realized that definitive results on the connection between the Earth and the Sun would require a concerted and joint effort. It was also realized that sophisticated solar-terrestrial research communities existed within all the space-faring nations of the world and that no one of them could launch such an effort by itself. This realization led to the creation of the International Solar-Terrestrial Physics (ISTP) program, which now comprises at least 12 spacecraft and includes extensive ground-based observations and theory and modeling efforts.

ISTP research has succeeded beyond anyone's wildest dreams. For example, before ISTP, we knew that coronal mass ejections (CMEs) were associated with geomagnetic storms. With ISTP, we can image CMEs as they leave the Sun, track them through interplanetary space by receiving the Type II radio bursts emitted at their leading shock waves, measure the details of the solar wind and interplanetary magnetic field as they pass by the L1 libration point (one hour upstream from the Earth), measure the intermingling of solar and terrestrial plasmas at the magnetopause, sense the injection of particles into the inner magnetosphere, and image the global light output of the aurora in the polar upper atmosphere.

The stage for the magnetospheric portion of ISTP has been set by the preceding and ongoing Geospace Environment Modeling (GEM) program of the National Science Foundation. Recognizing the need for close interaction between the ISTP and GEM communities and the special opportunity afforded by the rapidly approaching maximum of the current solar cycle, a workshop titled "Toward Solar Max 2000: The Present Achievements and Future Opportunities of ISTP and GEM" was held on February 10–13, 1998 at Yosemite National Park, California. The discussions that took place at the workshop formed the basis for this monograph, which will be a resource for both new and veteran members of these research communities as they prepare for research on interactions between the Sun and the Earth's space environment during its most active epoch.

In addition to papers documenting the accomplishments of the ISTP and GEM era, this monograph includes a set of papers describing planned or proposed Solar Terrestrial Probes and other future missions. These papers set forth some of the key science questions that will be addressed during the early twenty-first century as we seek to raise our knowledge of the physics of the Sun-Earth connection to a new level of insight and understanding, and they outline the novel mission strategies and technologies that will be employed to answer those questions.

The editors thank both the authors and the reviewers for finding the time and energy in the press of many other activities—research, teaching, proposal writing, serving on review panels and various professional committees, etc.—to make their respective contributions to this volume. We dedicate this monograph to the memory of our late friend and colleague, Tom Potemra, whose work contributed in no small way to our knowledge of the Sun-Earth connection.

James L. Burch
Southwest Research Institute
San Antonio, Texas

Robert L. Carovillano
Boston College
Chestnut Hill, Massachusetts

Spiro K. Antiochos
Naval Research Laboratory
Washington, DC

ISTP and Beyond: A Solar-System Telescope and a Cosmic Microscope

D.N. Baker

LASP, University of Colorado, Boulder

M.J. Carlowicz

NASA/GSFC, Greenbelt, MD

The International Solar Terrestrial Physics (ISTP) program has coordinated the activities of an armada of spacecraft, ground-based observatories, and theoretical modeling centers. This assemblage represents the first great observatory for space physics, providing both a global, telescopic view of the Sun-Earth system, as well as a magnificent microscopic view of its physical processes. In just two years of coordinated observations, ISTP has already reaped a rich scientific harvest. However, the future holds even more opportunities to pursue a new, profound understanding of the Sun and the Earth and the interplanetary space between them. Continuing the mission through solar maximum (2000-2001) and beyond offers potentially prodigious rewards for a remarkably low cost.

1. INTRODUCTION

Humans have always been fascinated by the Sun and its relationship to Earth. Every civilization has speculated about the place of our planet in the realm of the stars and about our relationship to our own star. Stonehenge and sundials and the folklore of eclipses are testimonials to that fascination. Ancient Chinese and Greek observers saw sunspots centuries before telescopes proved they were there. However, it has only been within the past few hundred years – since Copernicus and Galileo – that we have closely examined the changing face and place of the Sun in our skies. The role of the Sun in driving magnetic disturbances at Earth has only been appreciated in the last 150 years. It has really been in the 20th century – and primarily the last 40 years – that we have arrived at a relatively clear picture of solar activity and its effect on Earth. Moreover, what we have found through astronomical observations is that our Sun is rather common and ordinary, akin to the many main sequence stars in the universe. In essence, our Sun-Earth system is the physical prototype for stellar systems throughout the cosmos. It is also the only one we can study up close.

As our appreciation of the Sun-Earth system has grown more sophisticated, so too has our technology. Today, a tangled web of electrical and communication links has been woven across Earth's surface, while fleets of spacecraft work in the electric space above us. By using electromagnetic techniques to enhance communication, navigation, reconnaissance, and weather prediction that generally make the world safer, we have also put ourselves in harms way. Every tool and gadget that relies on radio waves, conducting wires, and sensitive transistors and processing chips can be affected by disturbances in the solar-terrestrial system. Furthermore, many more such disturbances lie in our immediate future.

The Sun reaches a maximum of activity every 11 years or so, and as it reaches the peak it expels with increasing frequency huge magnetic clouds of material (called coronal mass ejections, or CMEs) which can move outward at speeds sometimes approaching 2000 km/s. The shock waves preceding such clouds can accelerate particles to tremendous energies—sometimes more than 100 million electron volts (MeV). If the CMEs and the shock waves they produce strike Earth's magnetosphere, they can cause violent geomagnetic storms that can disturb power systems, communication links, and the constellations of spacecraft on which society increasingly relies. The appreciation of CMEs as the primary drivers of such disturbances has only come about in the past few years, and this paradigm shift has had a far-reaching impact on how we perceive solar-terrestrial relationships [e.g., Gosling, 1993].

Given the many thousands of years we have waited to achieve our current view of the Sun, the Earth, and the space in between, we now enjoy a most remarkable situation. The International Solar-Terrestrial Physics (ISTP) program has put into place an amazing array of spacecraft and ground facilities for studying the Sun-Earth environment [Baker and Carovillano, 1997]. Sensitive telescopes in space examine the Sun's many layers in unprecedented detail. Other spacecraft sample the hot, high-speed plasmas flowing past the Earth from the expanding solar corona. Still more satellites continuously monitor the plasmas which ebb and flow within the magnetosphere as it is buffeted by the solar wind. There is even an international network of ground stations recording the magnetospheric and ionospheric signatures of interaction between Sun and Earth. Merged as they are into a unified, comprehensive mission, the many components of ISTP afford an opportunity to extend greatly our understanding of the physics of solar-terrestrial processes.

2. THE ASTROPHYSICAL CONNECTION: THE SUN AS A STAR

In the past century, massive ground-based telescopes expanded and refined our view of the solar system and then began to reveal the origins of the galaxy and universe. Improvements in the spatial and spectral resolution of ground-based telescopes brought astronomy to a point where the mysteries of the distant cosmos -- such as black holes, quasars, active galaxies, and developing planetary systems -- are being discovered and explored at an exciting pace.

Astronomical studies have reached a crescendo with the space-based "Great Observatories": the Hubble Space Telescope (HST), Compton Gamma-Ray Observatory (GRO), Advanced X-ray Astronomy Facility (AXAF), and the Space Infrared Telescope Facility (SIRTF). Each has pushed back the cosmic frontiers (or soon will) in their respective wavelength regimes. Plate 1a, for example, shows the distant galaxies of the "Hubble Deep Field" study [Williams et al., 1996]. In this project, long observations of small regions of the sky have brought human viewing to the edge of the expanding universe.

But the image from HST also illustrates the limitations of astronomy. In studies with astronomical telescopes, researchers can gather grand views of large segments of the cosmos. Yet the details of the physical processes in the far reaches of space may never be directly visible. Immense modeling efforts and subtle detective work are needed to tease out the physics of distant, obscure objects. It is only in our own "cosmic backyard" that can we study directly, and in detail, the physical processes that drive much of the observable universe.

Within our own solar system, we can explore a typical main sequence star and a wide variety of planets in remarkable detail. Plate 1b, for example, shows the exquisite structure of the active regions around the Sun's equatorial belt. This SOHO (Solar and Heliospheric Observatory) image reveals the small-scale magnetic structures that can persist for long periods. Such images show intense, dynamic energy-conversion events that even the largest astrophysical observatory will never be able to observe on another star with such spatial (and temporal) resolution. Therefore, it is likely that only by detailed examination of our Sun can we ever hope to understand the physical processes operating in other stellar atmospheres.

3. OBSERVING LIFE CYCLES: EVENTS FROM CRADLE TO GRAVE

While it is useful to consider the relevance of solar-terrestrial observations to astrophysical research, the greatest impact of ISTP lies in something far more familiar and relevant: the chronicling of solar-terrestrial events from start to finish. From the origin of disturbances deep in the solar interior to collisions with the magnetosphere to the dissipation of energy in aurora and plasmoids, the ISTP constellation can analyze Sun-Earth connections from a global perspective.

A series of events in May 1998 provide a useful case study of how ISTP is changing and improving our understanding of the solar-terrestrial system. The sequence began with observations from the SOHO spacecraft at ~1340 UT on 2 May 1998. Figure 1 shows observations taken by the EIT (Extreme Ultraviolet Imaging Telescope) experiment on SOHO (courtesy J. Gurman and J.-P. Delaboudiniere). The image is recorded in the FeXII line (195Å) and

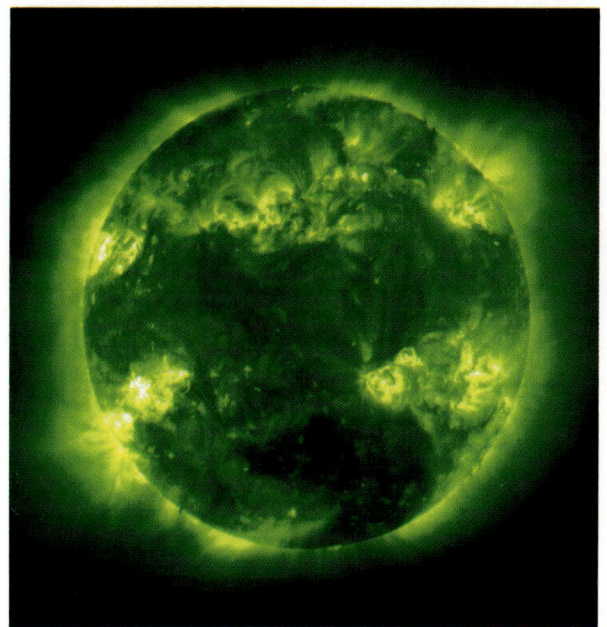

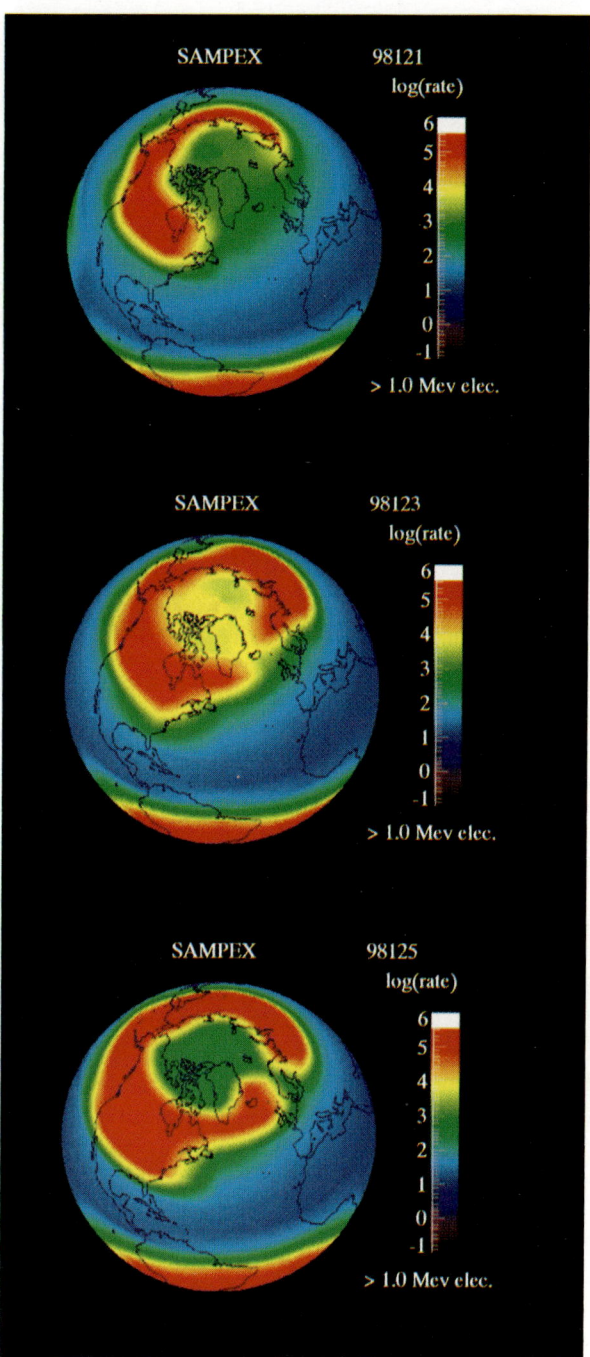

Plate 1. (a) A portion of the Hubble Deep Field showing distant galaxies [*Williams et al.*, 1996]. (b) A SOHO image of an active region on the Sun (courtesy of the SOHO-EIT Consortium).

Plate 2. A sequence of global maps of >1 MeV electrons measured by SAMPEX (data courtesy B. Klecker) for Days 121, 123, and 125 (1, 3, and 5 May) 1998.

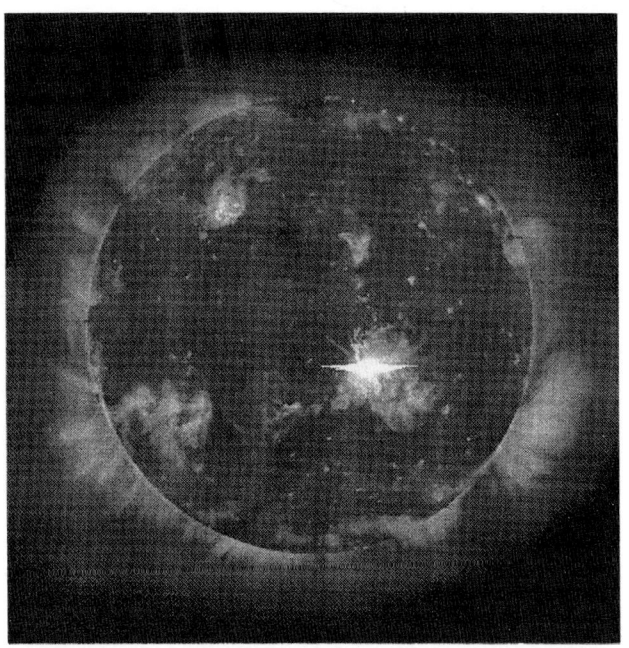

Figure 1. A SOHO EIT image (at 195Å) of the Sun at 1340 UT on 2 May 1998. A solar flare shows up as the bright region in the lower right quadrant (courtesy of J. Gurman and the SOHO-EIT Consortium).

shows a bright, intense flare on the Sun (lower right quadrant). This was an X-class event that produced copious quantities of energetic particles at Earth [see *Baker et al.*, 1998].

The coronal disturbance seen in the SOHO/EIT data rapidly spread outward from the solar surface into interplanetary space. Figure 2 (courtesy of R. Howard) shows a SOHO coronagraph (LASCO) image of material being expelled from the Sun: the image (taken at 1503 UT on 2 May) is from the LASCO C2 coronagraph and shows a huge, bright CME moving outward slightly to the upper right of the image, but directed toward the Earth. The solar disturbances on 2 May 1998 were clearly of great power and immense physical scale (R. Howard, private communication, 1998).

The active solar event on 2 May (Day 122 of 1998) discussed above produced powerful streams of solar wind plasma that were detected upstream of the Earth a few days later. Figure 3a shows the solar wind speed (V_{SW}), the total interplanetary magnetic field (IMF) strength (B_{IMF}), and the IMF north-south component (B_z) for 20 April to 20 May 1998 (DOY 110 to 140), as recorded by the WIND spacecraft (data courtesy K. Ogilvie and R. Lepping). In the period from DOY 121 through DOY 140, there were four separate streams in which V_{SW} reached peak values $\geq$ 600 km/s. Such streams are very effective at producing highly relativistic electron (HRE) events in the magnetosphere [*Baker et al.*, 1994, 1998]. A particularly notable solar wind stream occurred on Day 124 (4 May), when V_{SW} went to ~850 km/s. This is the highest solar wind speed that has been measured in near-Earth space for several years (A. Lazarus, private communication, 1998).

When high solar wind speed occurs in combination with large B_{IMF}, and especially when B_z is strongly negative, then we expect significant electron acceleration [e.g., *Blake et al.*, 1997] and intense geomagnetic activity. Indeed, the planetary magnetic index Kp reached 9 on Day 124. The Dst index on that day reached -218 nT (a major geomagnetic storm) and the provisional auroral electrojet (AE) briefly exceeded 2500 nT (WDC-C2, Kyoto University). All of this information is indicative of powerful, global geospace disturbances on 4 May.

The powerful solar wind streams and southward IMF illustrated in Figure 3a produced disturbed auroral conditions as seen by the POLAR auroral imaging system, VIS [Visible Imaging System; *Frank et al.*, 1995]. Figure 3b shows an auroral image (courtesy L.A. Frank and J. Sigwarth) taken at 0731 UT on 4 May 1998. The image shows an auroral oval that was greatly expanded equator-

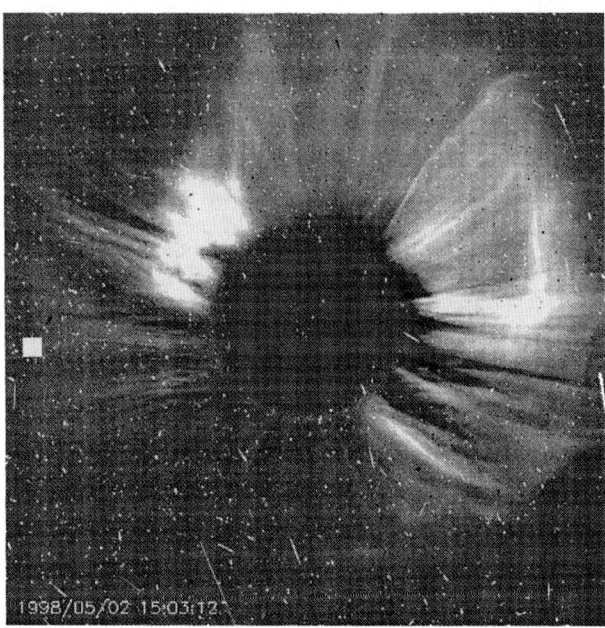

Figure 2. A SOHO LASCO C2 coronagraph image taken at 1503 UT on 2 May 1998 showing a large halo CME event (courtesy R. Howard).

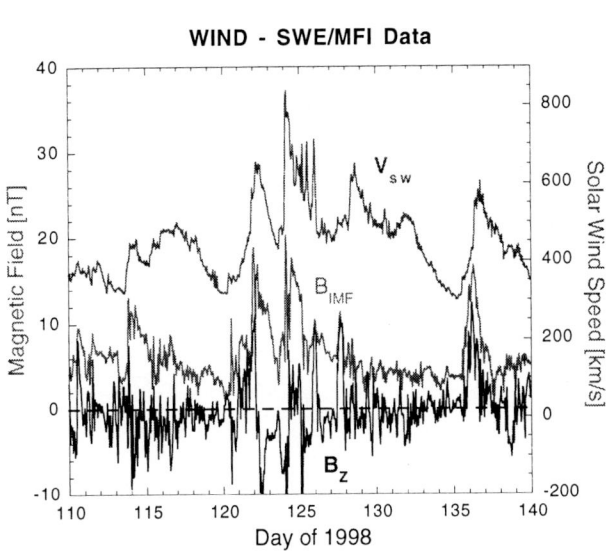

Figure 3. (a) Solar wind speed and IMF data from the WIND spacecraft (courtesy K. Ogilvie and R. Lepping) for Days 110-140 (20 April – 20 May) 1998. (b) A UV image of the Earth taken by the VIS camera onboard POLAR at 0731 UT on Day 124 (4 May) 1998 (courtesy L. Frank and J. Sigwarth).

ward. Moreover, auroras were particularly active and intense that day. As noted above, the AE index was >2500 nT on 4 May, indicating an unusually powerful auroral electrojet.

Energetic particle data revealed a strong acceleration of relativistic electrons quite deep in the magnetosphere. This is particularly clear using particle fluxes measured by an array of solid-state detectors (Heavy-Ion Large Telescope, HILT) on the Solar, Anomalous, and Magnetospheric Particle Explorer (SAMPEX) spacecraft. Flying in a high-inclination (82°), low-altitude (~600 km) orbit, SAMPEX samples magnetic field lines across nearly the entire magnetosphere every 100 min [see Baker et al., 1994 and references therein]. It carries sensors capable of measuring very energetic ions and electrons of both solar and magnetospheric origin. The HILT channel has an electron energy threshold of 1 MeV. This channel also has sensitivity to >4 MeV protons (which is especially important during solar particle events and during surveys of the inner radiation belt).

Plate 2a shows particle flux measured by HILT plotted on a global map according to the color bar to the right of the plate. The data show rather quiet conditions on Day 121 (1 May), with a modest, variable flux of electrons with E>1 MeV in the outer radiation zone (which mapped to rather typical latitudinal positions during that interval).

Plate 2b shows the global map for Day 123 (3 May), when a large solar proton event filled the polar cap with energetic particles [see Baker et al., 1998].

The most striking and notable event in the interval occurred on DOY 124 (4 May). On that day, SAMPEX observed a huge increase of the flux of HREs (highly relativistic electrons) very deep in the magnetosphere ($L \lesssim 3$). The "slot" region between the inner and outer radiation zones was filled, and a new radiation belt feature appeared at $L \simeq 2.2 \pm 0.2$ [Baker et al., 1998]. The relativistic electrons remained high throughout the outer zone for at least two weeks. Electrons filled a broad region from $L \simeq 2$ to beyond $L \simeq 7$ over the next several-day interval. The relativistic electron enhancement, demonstrated for Day 126 (6 May) in Plate 2c, ultimately was as intense, long-lasting, and spectrally hard as any event seen in the magnetosphere over the past several years (cf., Baker et al., 1998).

Figure 4a shows the daily average flux (electrons/cm²-sr-day) of E>2 MeV electrons from 21 April to 20 May 1998, as recorded by National Oceanic and Atmospheric Administration's GOES satellites. Electron fluxes were low (10^4/cm²-sr-day) on 21 April, but the flux then rose progressively over the subsequent week or so, reaching a maximum on 29 April. The electron intensities then were

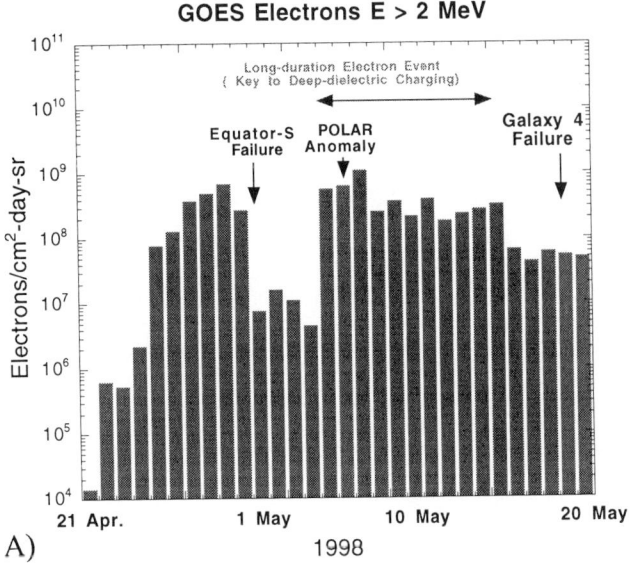

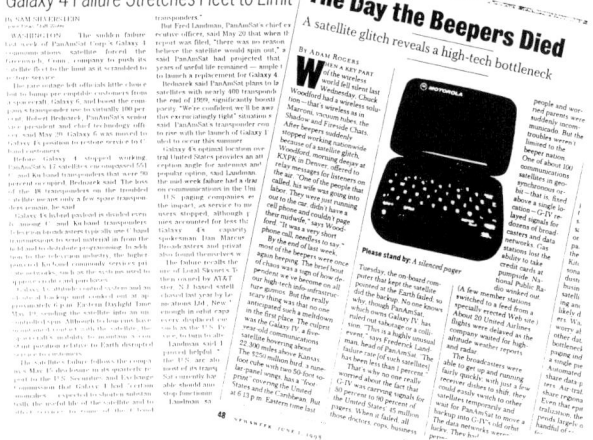

lower for several days (1-4 May). The average electron intensity jumped by two orders of magnitude on 5 May and stayed high for the subsequent 10 days. On 16 May the electron flux diminished by a factor of 2-3, but it remained well above 10^7 until the end of the plotting sequence.

4. THE HUMAN CONSEQUENCES OF SOLAR-TERRESTRIAL DISTURBANCES

As reported in the newspaper headlines shown in Figure 4b, at approximately 2200 UT on 19 May 1998, PanAmSat Corporation's Galaxy 4 spacecraft experienced a catastrophic failure in its attitude control system. Unfortunately, the backup system also had failed -- either at that same time or earlier -- so the operators were unable to maintain stable Earth-link [*Silverstein*, 1998]. PanAmSat and Hughes have been working to determine the exact cause of the Galaxy 4 failure [*Silverstein*, 1998]; in August 1998, company officials proclaimed the failure "an isolated incident" and "a one-time, random event" (Cable News Network, 11 August 1998). In the past, however, long-duration HRE enhancements such as those observed in May 1998 have been convincingly associated with spacecraft failures [e.g., *Baker et al.*, 1994, 1996].

Regardless of whether the space environment played a role in the failure of Galaxy 4, the event puts into sharp focus just how much society has come to rely on satellites and space-based communications. Galaxy 4 was a heavily subscribed communications satellite at geostationary orbit; its sudden failure caused the loss of pager service to some 45 million customers, as well as numerous other communications outages [*Rosenbush*, 1998]. Doctors and fire-fighters could not be reached for emergencies. Stock brokers, business executives, and radio announcers were suddenly incommunicado. In an ironic twist, even the illicit drug trade was slowed by the loss of paging relay systems [*Rogers*, 1998]. In essence, the Galaxy 4 failure provides a hint of what we might see during the enhanced solar-terrestrial interactions that come at solar maximum [*Baker et al.*, 1998].

With modern technological society so reliant on space-based communications -- systems that are vulnerable to the plasmas and radiation environments studied by ISTP -- we need more reliable information about the dynamics of the

Figure 4. (a) Daily flux values of E>2 MeV electrons measured by GOES from 21 April to 20 May 1998 (courtesy H. Singer). (b) Newspaper and magazine accounts of the PanAmSat Galaxy 4 spacecraft failure.

space environment. Whenever there is an operational problem with space hardware, it is advisable to examine broadly the space environmental conditions prior to and at the time of the problem. Using the wide array of space data sets from ISTP, we are able to do just that. In addition, many current models of the radiation belts are based on information that is 30 years old. New observations from ISTP can help scientists and engineers create more accurate models of Earth's radiation environment, and thereby build heartier spacecraft.

5. THE INTERNATIONAL SPACE PHYSICS PROGRAM: A TELESCOPE AND A MICROSCOPE

Earth's space environment traditionally has been explored as a set of independent parts – the interplanetary region, the magnetosphere, the ionosphere, and the upper atmosphere. Consequently, past science missions have advanced the understanding of these geospace components individually. Yet even from the earliest studies, we have known that geospace is composed of highly interactive elements. To understand the system as a whole, we needed to plan a program of simultaneous space and ground-based observations and theoretical studies. It would require that we assess the production, transfer, storage, and dissipation of energy across the entire solar-terrestrial system. In essence, we needed a comprehensive, quantitative study of the energy chain from the Sun's interior to Earth's magnetic tail [Baker and Carovillano, 1997].

The ISTP Program was created out of this need to obtain a comprehensive, global understanding of the generation and transfer of energy from Sun to Earth. ISTP was (and is) coordinated by the InterAgency Consultative Group (IACG), which includes representatives from the U.S., European, Japanese, and Russian space agencies. The stated goal of ISTP has been to establish cause-and-effect relationships between key regions and processes within the solar-terrestrial system.

Figure 5 shows many of the space-based assets now operating to study solar-terrestrial coupling. This program represents a multi-billion dollar investment toward understanding the Sun-Earth system in unprecedented scope and detail. When these spacecraft are linked with the ground-based elements and the theoretical modeling tools of ISTP, the result is a true "great observatory" for space physics. The ISTP observatory provides not only a telescopic view (like its astronomical counterparts), but also a microscopic view. Figure 6 shows the sort of global solar image that ISTP can provide, as well as the exquisite detail that can be obtained from particular active regions. This coupling of microscopic and telescopic views is crucial to understanding the physical processes -- such as magnetic reconnection and particle acceleration -- that drive our solar system.

But this dynamic and complex system of interacting plasmas, magnetic fields, and electrical currents also might serve as an astrophysical prototype. Plasma physics determines the behavior of matter in the solar-terrestrial system on spatial and temporal scales, and with particle densities vastly different from those produced in earthbound laboratories. Thus, solar-terrestrial space is a unique and readily accessible laboratory for investigating the natural plasma processes of astrophysics.

6. FUTURE DIRECTIONS

As measured by sunspot number, this next solar maximum will most likely be a large one, perhaps among the most active of the modern era [Joselyn et al., 1997]. Figure 7 suggests that the next peak of solar activity is likely to occur in the years 2000-2001, and as with past maxima, this one should bring solar disturbances of great power and geoeffective potential. Thus, it is an historic occurrence that the ISTP constellation should be operating as the solar maximum approaches.

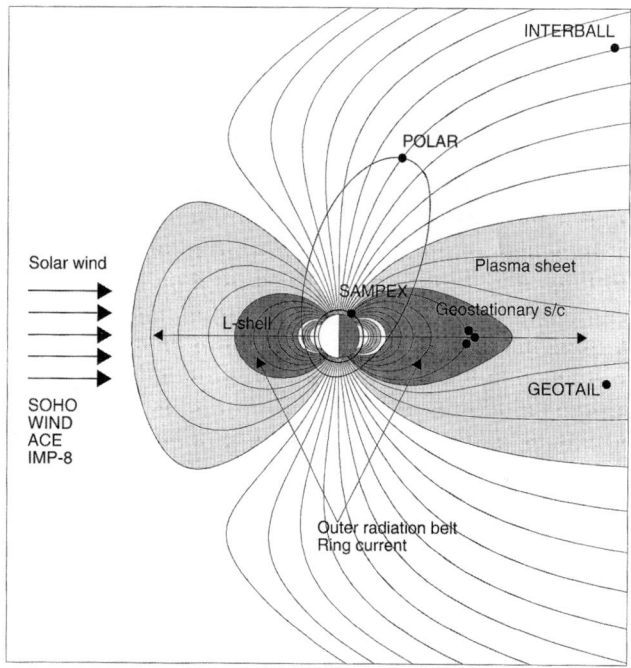

Figure 5. A diagram showing key ISTP spacecraft locations in, and near, the Earth's magnetosphere.

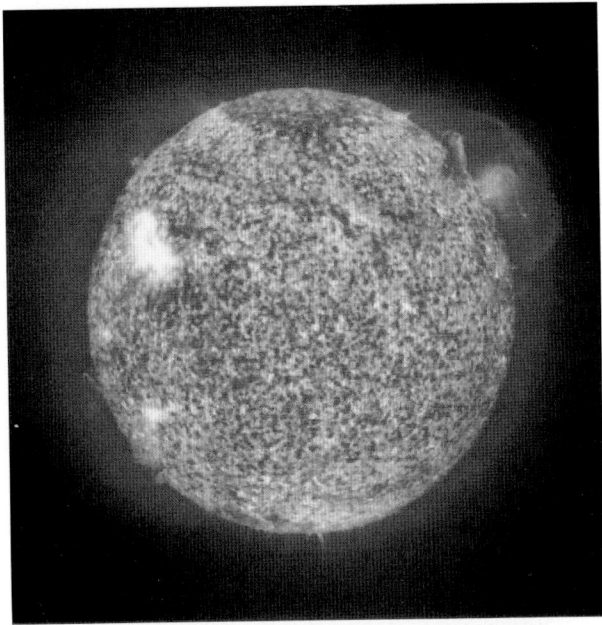

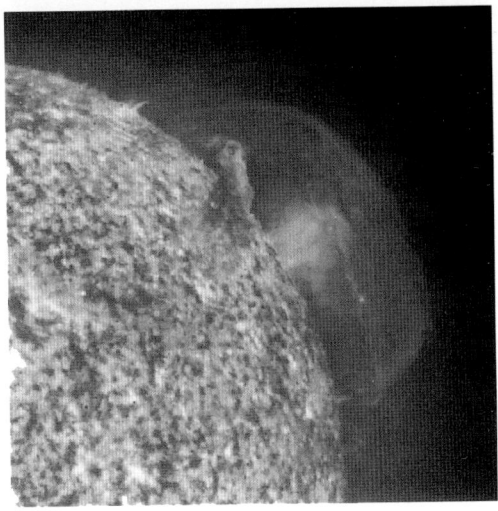

Figure 6. (a) A global image of the Sun from SOHO taken at 1607 UT on 26 August 1997 in He II (304Å); (b) a detailed view of the eruptive prominence feature shown in (a) (courtesy the SOHO-EIT Consortium).

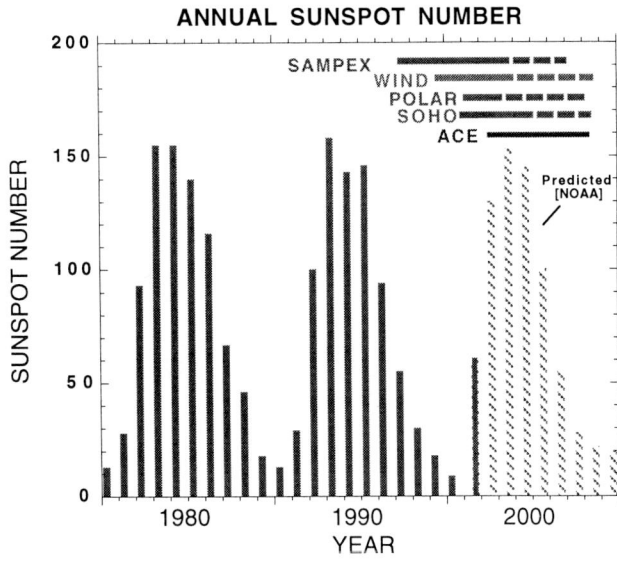

Figure 7. Sunspot number from the 1970s and projected values into early next century. A sunspot maximum is predicted in the years 2000-2001.

ISTP and its affiliated spacecraft can give us the perfect vantage point to finally understand the disturbed Sun and how it disrupts the geospace environment.

Responsible stewardship of the Sun-Earth investment demands operation of ISTP for as long as possible.

Acknowledgments. The authors thank numerous colleagues from the ISTP and related programs for their unselfish support. Nicky Fox, Shri Kanekal, and Tuija Pulkkinen are particularly acknowledged for their help in preparing illustrations. This work was supported by NASA.

Never before have we had such a complete set of tools with which to study the beginning of a new solar cycle (number 23). And never before have we had tools of such power and precision to study our most important star – the Sun – and our most important planet – the Earth. We have a chance to study all aspects of the solar maximum and its consequent effects on near-Earth space, and we can do it for modest costs. An investment in extended operations of

REFERENCES

Baker, D.N., R.D. Belian, P.R. Higbie, R.W. Klebesadel, and J.B. Blake, Deep dielectric charging effects due to high energy electrons in Earth's outer magnetosphere, *J. Electrostat., 20,* 3, 1987.

Baker, D.N., J. B. Blake, L.B. Callis, J.R. Cummings, D. Hovestadt, S. Kanekal, B. Klecker, R.A. Mewaldt, and R.D. Zwickl, Relativistic electron acceleration and decay time scales in the inner and outer radiation belts: SAMPEX, *Geophys. Res. Letters, 21,* 409, 1994.

Baker, D.N., J.H. Allen, R.D. Belian, J.B. Blake, S.G. Kanekal, B. Klecker, R.P. Lepping, X. Li, R.A. Mewaldt, K. Ogilvie, T. Onsager, G.D. Reeves, G. Rostoker, R.B. Sheldon, H.J. Singer, H.E. Spence, and N. Turner, An assessment of space environmental conditions during the recent Anik E1 spacecraft operational failure, *ISTP Newsletter, 6,* No. 2, p. 8, 1996.

Baker, D.N., and R. Carovillano, IASTP and Solar-Terrestrial Physics, *Adv. Space Res., 20*, 531-528, 1997.

Baker, D.N., J.H. Allen, S.G. Kanekal, and G.D. Reeves, Disturbed space environment may have been related to pager satellite failure, *Eos, Trans., AGU, 79*, 477, 1998.

Blake, J.B., D.N. Baker, N. Turner, K.W. Ogilvie, and R.P. Lepping, Correlation of changes in the outer-zone relativistic electron population with upstream solar wind and magnetic field measurements, *Geophys. Res. Lett., 24*, 927-929, 1997.

Frank, L.A., et al., The Visible Imaging System (VIS) for the POLAR spacecraft, *Space Sci. Rev., 71*, 297, 1995.

Gosling, J.T., The solar flare myth, *J. Geophys. Res., 98*, 18,937, 1993.

Joselyn, J.A., et al., Panel achieves consensus prediction of solar cycle 23, *Eos, Trans., AGU, 78*, 205, 1997.

Rogers, A., "The Day the Beepers Died", *Newsweek*, p. 48, June 1, 1998

Rosenbush, S., Loss Shows Key Role of Satellites, *USA Today*, "Money" Section (Section B), p. 1B-3B, 21 May 1998.

Silverstein, S., PanAmSat Scrambles to Restore Service - Galaxy 4 Failure Stretches Fleet to Limit, *Space News*, p. 3 + p. 18, 25-31 May 1998.

The SOHO Mission, *Scientific and Technical Aspects of the Instruments, ESA SP-1104*, ESTEC, 1988.

Vampola, A.L., The aerospace environment at high altitudes and its implications for spacecraft charging and communications, *J. Electrostat., 20*, 21, 1987.

Williams, R., et al., The Hubble Deep Field survey, *STScI*, 26 June, 1996.

D.N. Baker, LASP/University of Colorado, 1234 Innovation Drive, Boulder, CO 80309-0590.

M.J. Carlowicz, Mail Code 695, NASA/Goddard Space Flight Center, Greenbelt, MD 20771.

The SOHO Mission

A. I. Poland

NASA Goddard Space Flight Center, Greenbelt, Maryland

The Solar and Heliospheric Observatory (SOHO) is primarily the solar observing portion of the International Solar Terrestrial Physics (ISTP) mission but includes some particle detecting experiments. This paper describes some selected results from the interior studies, the transition region observations, and the outer coronal observations.

1. INTRODUCTION

SOHO, the solar part of the ISTP mission, was launched in December 1995. Since that time it has provided a wealth of new information about the Sun, from the structure of its interior through the particle composition of the solar wind. Two books, which are duplicates of series in *Solar Physics*, have been published describing the instruments [*Fleck et al.*, 1995], and the first results [*Fleck and Svestka*, 1997]. Together with the other ISTP mission spacecraft, SOHO has helped breathe new life into the study of space physics. In this paper I will present a few of the highlights from the solar studies, including the interior, the outer atmosphere, and the corona.

When this paper was originally written SOHO had been apparently lost permanently. The loss occurred in June 1998. Many of us had very little hope that it would ever be recovered. Thus, some comments in the text are based on a lost SOHO. SOHO has since been recovered through a tremendous effort on the parts of the European Space Agency (ESA) and NASA and through a lot of luck. All of the instruments are again functioning, most with no degradation. SOHO is thus now obtaining impressive data on the Sun as it approaches its maximum of activity.

2. INTERIOR

To date, the interior studies have given us a clear view from the surface to just below the convection zone (approximately one-third the way into the Sun). A longer time base will be necessary before we can probe accurately to deeper levels. Many of the results were obtained in conjunction with studies from the Global Oscillations Network Group (GONG) program [*Harvey et al.*, 1996]. The areas that will be discussed here include: the structure of the convection zone; velocity turbulence that may be the source of the dynamo that drives sunspots; coherent subsurface velocity flows that are comparable to the jet streams on Earth; and the small-scale surface magnetic field that turns over completely in about 40 hours indicating a small-scale dynamo for its generation.

The Michelson Doppler Imager (MDI) [*Scherrer et al.*, 1995] instrument on SOHO measures velocity as a function of position on the Sun to determine its interior density and rotational velocity structure. It uses a 1024 · 1024 CCD camera and a Michelson interferometer which together make images and line profiles of the solar surface. These provide a line-of-sight velocity map of the solar surface once every minute. The analysis of the velocity wave pattern provides the density and material velocity as functions of depth and position in the solar interior.

An early result from the helioseismology studies is a first look at the structure of the convection zone. Previously there had been no observational information on the shape of the convective cells. It was generally assumed that the height to width ratio of these cells was on the order of 1. Observations from MDI indicate that the cells are shaped more like pancakes with a ratio of 0.1. This result has strong implications for the mixing length theory of stellar convection since convective efficiency is related to this ratio. These observational results are discussed in *Duvall et al.* [1997].

One of the significant results from this analysis has been the measurement of a zone of velocity turbulence at the bottom of the convection zone. In Plate 1 we present a map of the interior velocity as a function of depth and latitude. The inner two-thirds of the Sun rotates as a solid body. However, there is a differential rotation above this, and it varies with depth. In its outer one-third the Sun rotates faster near the equator and slower at the poles. These differential rotations appear to move as "rivers" or jet streams under the solar surface to the same depth as the convection zone. Over the entire core there is a sheer at the bottom of the convection zone. It is thought that this shear zone is where the large scale dynamo is generated. This dynamo would be responsible for the large scale solar magnetic field and the sunspot field. A more complete discussion of these results can be found in *Schou et al.* [1998].

MDI also provides high quality magnetograms of the solar surface. These have a good spatial resolution (~4" and ~1" in high resolution mode) with no interference from the Earth's atmosphere. They are made with a time cadence of approximately 30 s. Thus MDI provides a clear view of how the solar magnetic field changes with time. The primary result of these studies is that the surface magnetic field turns over completely in approximately 40 hours. Points of rapid change are correlated with brightenings in transition region emission lines. These results are discussed in *Schrijver et al.* [1998].

If the SOHO mission had continued through the solar cycle we would have expected to be able to answer questions such as: how do the "rivers" change with the solar cycle? how do they relate to sunspot changes? how deep into the interior do sunspots reach? how does the surface magnetic field change with the cycle?

3. TRANSITION REGION

Despite years of spectroscopic observations of the chromosphere and transition region, material flowing out from these parts of the solar atmosphere has still not been observed to any significant extent. Except for dynamic events and the network in coronal holes, material is observed to be moving down, not up. The answer to the question is still problematic - where in the lower solar atmosphere does the low speed solar wind arise? Much of the current observational work has been directed toward observing coronal holes and dynamic upward moving material to determine the upward mass fluxes.

3.1. Quiet Sun

Recent SOHO observations of the chromosphere, transition region, and corona using SOHO have again shown a predominance of downward flows. In papers by *Chae et al.* [1998] and *Brekke et al.* [1997] observations have been extended to 10^6 K (see Figure 1). Previous observations were reliable to approximately 100,000 K to 200,000 K but errors between these values and $1 \cdot 10^6$ K were too large to make the measurements reliable. Both of the above referenced papers show downward flows in quiet Sun regions of a few km/s near 10^4 K increasing to ~10km/s near $2 \cdot 10$ K and decreasing again to a few km/s at 10^6 K. The only significant differences between the works of *Chae et al.* [1998] and *Brekke et al.* [1997] are the measurements near 250,000 K. *Brekke et al.* [1997] show a continued rise to 250,000 K and a rapid fall above that while *Chae et al.* [1998] show a smooth fall above 200,000 K. These differences seem to be primarily due to the choice of laboratory rest wavelengths for the lines used. I do not believe that these differences are significant for understanding transition region structure. The measurements presented in this graph are averages over large areas of the quiet Sun. There are a few points along network boundaries that show upward velocity, but in the quiet Sun these points are very few.

Observations in coronal holes show a blue shift in some transition region lines. The absolute wavelength calibration of previous instruments has made this a relative result. In recent work by *Hassler* [1998] it is shown that there is definitely a blue shift in coronal holes. This would confirm the concept of the high speed wind arising from coronal holes.

A theoretical explanation of why we do not see upward flows in the quiet Sun is presented in a paper by *Chae et al.* [1997]. They use computer models to determine the temperature structure in both up flowing and down flowing transition region material. The temperature gradients are dominated by enthalpy and conduction. The model calculations show that the interplay between these two terms in up flows yields a very steep temperature gradient. The mass at any given temperature in the transition region is very small in this case. For down flows of the same velocity at transition region temperatures the gradient is much less steep. There is thus more mass at each temperature and the lines are visible. Thus, *Chae et al.* [1997] show that in the case of the same mass flux going up and down, only the downward flow would be observable. This is consistent with observations.

3.2. Dynamic Events

There are several studies using SOHO data that involve the observation of dynamic events that may inject material into the solar wind. One type of event observed at approximately 80,000 K has been defined as "jets", the other has been called "blinkers" which are observed at approximately 200,000 K.

Plasma jets in the solar network are reported in a paper by *Innes et al.* [1997]. The Solar Ultraviolet Measurements of Emitted Radiation (SUMER) instrument was used to make high time and spatial resolution spectra and images of the Sun in the Si IV lines, which are formed at approximately 80,000 K. The combination of images and spectra shows the Doppler shift associated with brightenings. Time sequences of these image-spectra show the movement of the brightening and line shift, indicating that one is observing a moving jet of material. The data used consist of 5 s exposures at 6 or 8 scan positions separated by 4.4". Explosive events were observed to last about 4 minutes with velocities of approximately 100 km s^{-1}. The transverse velocity observed was on the order of 20 km s^{-1}. From the observations, the length of the jet was calculated to be approximately $2 \cdot 10^4$ km. Thus these jets are probably reaching into the lower corona. These jets also seem to be bidirectional in nature; thus material is being ejected into the

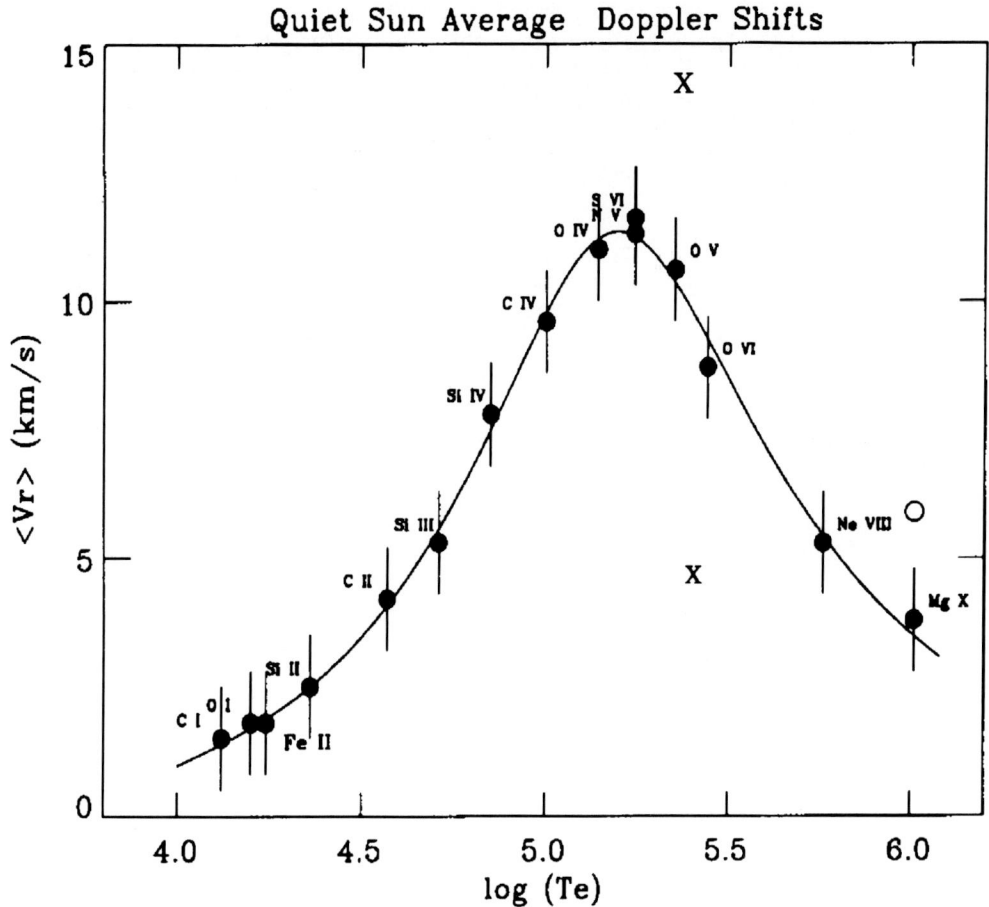

Figure 1. Downward velocity as a function of T_e as measured by *Chae et al.* [1997]. The crosses show the points where *Brekke et al.* [1997] differ significantly from *Chae et al.* [1997].

corona and toward the solar surface. However, they are only seen at transition region temperatures, so the question remains as to their relationship to the corona and the solar wind.

Another dynamic feature observed by SOHO are the "blinkers" observed with the Coronal Diagnostics Spectrometer (CDS) [*Harrison*, 1997]. CDS is able to view a fairly wide spectral range in the EUV (150-800 Å) simultaneously. With this spectral range it can make images and low resolution spectra over a wide temperature range. In a search for fast brightenings in the solar atmosphere the CDS instrument was used to observe sequences of He I, OIV, Mg IX, and Fe XIV or He I, O V, and Mg IX. What was observed in these measurements were enhancements of a factor of 2 to 3 in the transition region lines at network junctions. But, there were no related brightenings in the lower chromosphere or in the corona. Six hours of observation yielded 5 brightenings, called "blinkers", which ranged in lifetime from 1 to 30 minutes. The average lifetime was 13 minutes. It is significant that the brightenings were seen in transition region lines (OI V and O V) but not in the Mg IX coronal line. Again, we have dynamic events in the transition region with no coronal signature.

I have discussed only two of the many observations of dynamic phenomena made by the SOHO spectrographs. These observations have revealed new information about the relations between energetic phenomena at different temperatures in the solar atmosphere. We have seen many cases of various types of brightenings in the transition region. However, we still have not seen dynamic brightenings in the corona. Thus, the heating in the corona is most likely distributed over a large area and thus does not lead to local brightenings.

4. THE CORONA

The solar corona is primarily observed using three SOHO instruments, Extreme ultraviolet Imaging Telescope (EIT), Large Angle and Spectrometric Coronagraph (LASCO), and Ultraviolet Coronagraph Spectrometer (UVCS). EIT makes

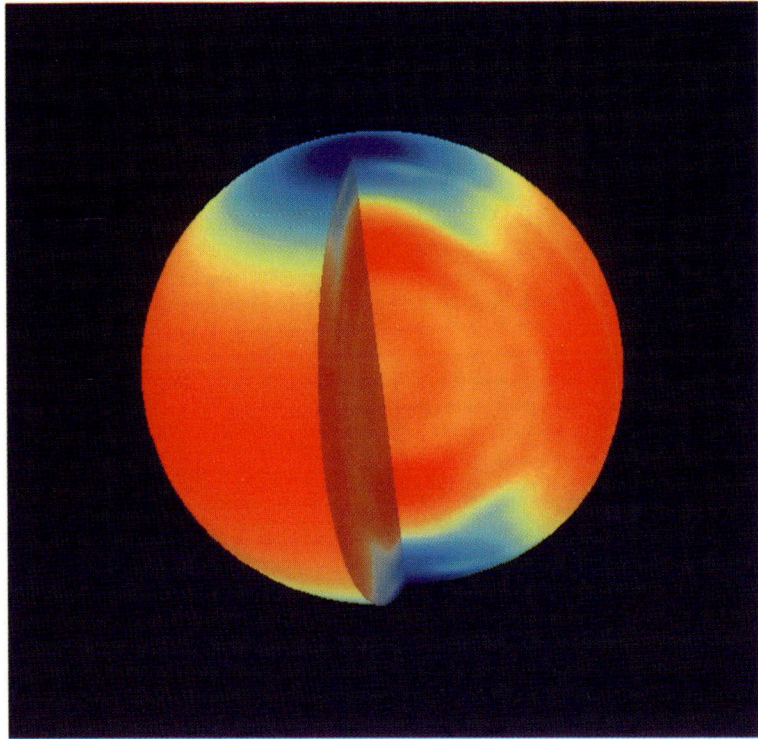

Plate 1. Solar interior rotation velocities. Red is fast, blue is slow.

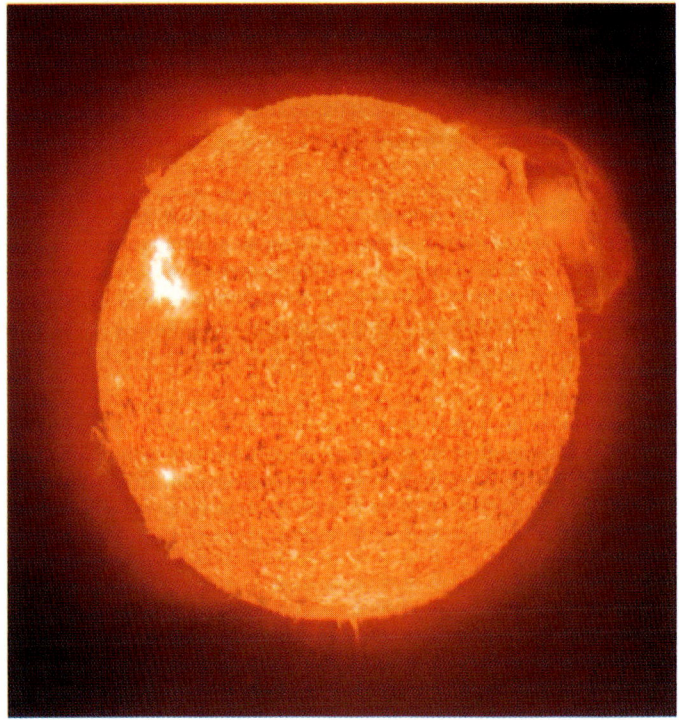

Plate 2. Image of the Sun in He II 304 Å. Notice the large prominence in the NW (upper right), and coronal hole in the N.

Plate 3. EIT image of the Sun in Fe IX/X at 171 Å. Note the active region magnetic loops and the polar hole in the N.

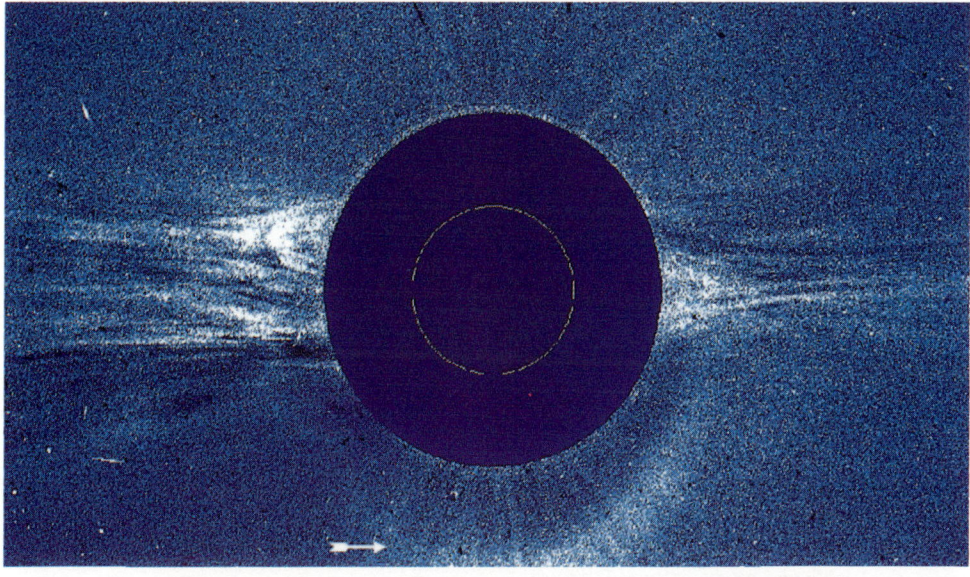

Plate 4. Halo CME observed using a difference image from LASCO C2. The arrow at the bottom points to the Halo ring.

images of the whole Sun in He II (304Å), Fe IX/X (171 Å), Fe XIII(195 Å), and Fe XIV(284 Å). LASCO consists of three telescopes (C1, C2, and C3), which allow one to view the outer solar corona from very near the surface out to 30R☉. UVCS is a spectrograph yielding spectra (line profiles) in Lyα, O VI, and some other minor ions from near the solar surface out to 10R☉.

EIT is designed to observe the whole Sun using 4 multi-layer filters, each tuned to a different wavelength [see *Delaboudiniere et al.*, 1995]. The 304Å line of He is formed near 80,000 K; 171 Å lines of Fe IX and X are formed near 1×10^6 K; the 195 Å line of Fe XII is formed near $1.5 \cdot 10^6$ K; and the 284 Å line of Fe XV is formed near 2-$2.5 \cdot 10^6$ K. Thus the EIT provides images of the transition region, the normal corona, and the hot corona. An example of a He II image is shown in Plate 2. Particularly interesting features include the large erupting prominence observed in the NW(upper right); an active region just inside the NE limb; the dimmer north polar coronal hole; and some large "macrospicules" in the south. The mottled appearance of the image is the super-granular network. The large erupting prominence is the type of phenomena that are part of coronal mass ejections (CMEs).

EIT observations of the low corona are shown in Plate 3. In this image the most prominent features are the large active region magnetic loops. One can also see the north polar coronal hole and polar plumes sticking up from the pole. Images made in the 195 Å line of Fe XII looks similar to the 171 Å line and have been used to observe the magneto-acoustic waves (frequently referred to as coronal Morton waves) that have been observed after some solar flares. The observations indicate that these waves seem to be highly associated with CMEs.

Outer coronal observations by LASCO have greatly enhanced our understanding of the solar wind and coronal mass ejections. An important new result is the observation of eddies in coronal streamers indicative of material flowing out along the axis of streamers. In a paper by *Sheeley et al.* [1997] this outflow is reported as "a continuous outflow of material in the streamer belt". It is identified by eddies in the flow which originate at about 3-4R☉ from Sun center. The speed seems to accelerate from ~150 km s^{-1} near 5R☉ to 300 km s^{-1} near 25R☉. The source of the slow speed solar wind in the corona may thus now have been identified.

While "halo" CMEs had previously been observed by the NRL P78 coronagraph, the LASCO observations have made a significant improvement in the observation of these events. Because of its high sensitivity and low noise the C3 coronagraph has made a significant improvement in our ability to predict events that will impact the Earth. In Plate 4 we present an image of a "halo" CME made with the C2 coronagraph. This image is produced by differencing the pre-event image with the event image. The CME in this case is headed toward the Earth and thus appears in the image as a ring or in this case a partial ring around the Sun. Thus, the observation of "halo" CMEs and coronal Morton waves have improved our ability to predict geomagnetic storms.

The UVCS spectrograph has greatly increased our understanding of the structure of the corona in terms of temperature, density, velocity, and abundance. In a definitive paper on coronal abundances by *Raymond et al.* [1997] it is determined that the He abundance in the corona is lower than in the photosphere; in the core of streamers, O and other high first ionization potential (FIP - defined as ~10ev) elements are depleted by an order of magnitude compared to the photosphere; near the edges of streamers this is only a factor of ~3; and finally, abundances along the edges of streamers resemble elemental abundances in the slow solar wind.

The papers by *Sheeley et al.* [1997] and by *Raymond et al.* [1997] generate a partial answer to the problem of the origin of the slow speed solar wind: coronal streamers seem to be the source. However, the question now becomes: Does the slow solar wind originate from the axis or the edge? The UVCS observations are from 1.4-4R☉ while the LASCO observations start at 3-4R☉. Thus, they are not clearly in disagreement; further analysis is needed.

5. CONCLUSION

Clearly, SOHO is solving many of the problems it was designed to solve, but is opening interesting new questions. I have focused on a small variety of results to convey a general view of what is available from SOHO. There are clearly other important results in these areas and in particle experiment measurements that have not been discussed in this paper. A comprehensive review of the early SOHO results can be found the "The First Results from SOHO" by *Fleck and Svestka*, [1997].

REFERENCES

Brekke, P., D. M. Hassler, and K. Wilhelm, Doppler shifts in the quiet-sun transition region and corona observed with Sumer on SOHO, *Sol. Phys. 175*, 349, 1997.

Chae, J., H. S. Yun, and A. I. Poland, Effects of Non-LTE radiative loss and partial ionization on the structure of the transition Region, *Astrophys. J., 480*, 817, 1997.

Chae, J., H. S. Yun, and A. I. Poland, Temperature dependence of UV line average Doppler shifts in the quiet Sun, *Astrophys. J. Suppl. Ser., 114*, 151, 1998.

Delaboudiniere, J. P., et al., EIT: Extreme-Ultraviolet Imaging Telescope for the SOHO Mission, *Sol. Phys. 162*, 291, 1995

Duvall, T. L., et al., Time-distance helioseismology with the MDI instrument: Initial results, *Sol. Phys. 170*, 63, 1997.

Fleck, B., V. Domingo, and A. I. Poland, (Eds.), *The SOHO Mission*, Kluwer Acad., Norwell, Mass., 1995.

Fleck, B., Z. Svestka, (Eds.), *The First Results from SOHO*, Kluwer Acad., Norwell, Mass., 1997.

Harrison, R. A., EUV Blinkers: The significance of variations in the extreme ultraviolet quiet Sun, *Sol. Phys. 175*, 467, 1997.

Harvey, J. W, F. Hill, and R. P. Hubbard, The Global Oscillation Network Group (GONG) Project, *Science, 272*, 1284, 1996.

Hassler, D. M., Paper presented at Toward Solar Max 2000: The Present Achievements and Future Opportunities of ISTP and GEM, sponsored by National Aeronautics and Space Administration, National Science Foundation, held at Yosemite National Park, CA, Feb. 10-13, 1998.

Innes, D. E., B. Inhester, W. I. Axford, and K. Wilhelm, Bi-directional plasma jets produced by magnetic reconnection on the Sun, *Nature, 386*, 811, 1997.

Raymond, J. C. , et al., Composition of coronal streamers from the SOHO Ultraviolet Coronagraph Spectrometer, *Sol. Phys. 175*, 645, 1997.

Scherrer, P. H., et al., The solar oscillations investigation - Michelson Doppler Imager, *Sol. Phys. 162*, 129, 1995.

Schou, J., et al., *Astrophysical J.*, in press, 1998.

Schrijver, C. J., A. M. Title, K. L. Harvey, N. R. Sheeley, Jr., Y.-M. Wang, G. H. J. vanden Oord, R. A. Shine, T. D. Tarbell, and N. E. Hurburt, Large-scale coronal heating by the small-scale magnetic field of the Sun, *Nature, 394*, 152, 1998.

Sheeley, Jr., N., et al., Measurements of flow speeds in the corona between 2 and 30 Rs, *Astrophys. J. 484*, 472, 1997.

Arthur I. Poland, Code 682, NASA Goddard Space Flight Center, Greenbelt MD 20771.

Geotail Mission: Accomplishments and Prospects

A. Nishida

Institute of Space and Astronautical Science3-1-1 Yoshinodai, Sagamihara 229-8510, Japan

By virtue of the orbit design, which is optimized for studying the plasma sheet over the distances of 10 to 220 R_E, the GEOTAIL mission has been able to significantly advance our understanding of magnetotail physics. We briefly review the progress made on such subjects as IMF control of magnetotail structure and convection, initiation of substorms by the near-Earth reconnection, kinetic structure of the foreshock region of the slow shock, cold dense ion flows in the tail lobe and at the magnetopause, and distribution and nature of plasma waves in geospace. Use of advanced computer simulations has contributed much to the quantitative interpretation of the observations. Further studies of the kinetic properties of the tail plasma at finer time scales below the electron gyroperiod are suggested as a future target.

INTRODUCTION

The GEOTAIL satellite, which was launched on July 24, 1992, has been functioning most successfully. It has already fulfilled the prime objective of clarifying the structure and dynamics of the magnetotail and their dependence on IMF conditions, and is still producing a wealth of new information on the role of the near-Earth region of the magnetotail in the global dynamics of the magnetosphere. Since the magnetotail plasma is collisionless, its kinetic properties as well as the generation of plasma waves are important elements of its physics. In this paper we shall present a brief overview of the results obtained and discuss future missions that will be needed to further elucidate the physics of the magnetotail. An earlier overview was given at the 1996 COSPAR Assembly [Nishida et al., 1997].

The orbit of GEOTAIL has been optimized for accomplishing this mission (Figure 1). In the first two years the apogees were kept on the nightside of the Earth and ranged from x of -80 R_E to about -220 R_E in order to study the distant tail. During this period the orbit was in the lunar orbital plane and the lunar double swingby maneuvers were performed. Later, after November 1994, the apogee was lowered first to 50 R_E and then to 30 R_E in order to study substorm-related processes in the near-Earth tail region. The inclination has been set at -8° so that the spacecraft be continually sunlit at the December solstice when it is in the neutral sheet at the apogee. The perigee has been set at 10 R_E. This orbit strategy has worked highly satisfactorily and we have been able to survey the tail extensively from 10 R_E to 220 R_E.

Because of its high perigee GEOTAIL skims along the magnetopause when the perigee is on the dayside. This has allowed an extensive study of the dayside magnetopause region to be conducted as well. Last year we reduced the perigee to 9 - 9.5 R_E in order to further increase the probability that the spacecraft will be just inside of the dayside magnetopause.

As for the status of the instrument complement, the Low Energy Plasma Analyzer, which had once been paralyzed due to electric arcing, was revived in September 1993. There has been no loss of any instruments since then. In early March of this year GEOTAIL survived an eclipse lasting for about 3 hours which was longer than the maximum eclipse of 2 hours that the satellite had been designed to endure, and thus longevity of the mission seems to be assured.

20 GEOTAIL MISSION: ACCOMPLISHMENTS AND PROSPECTS

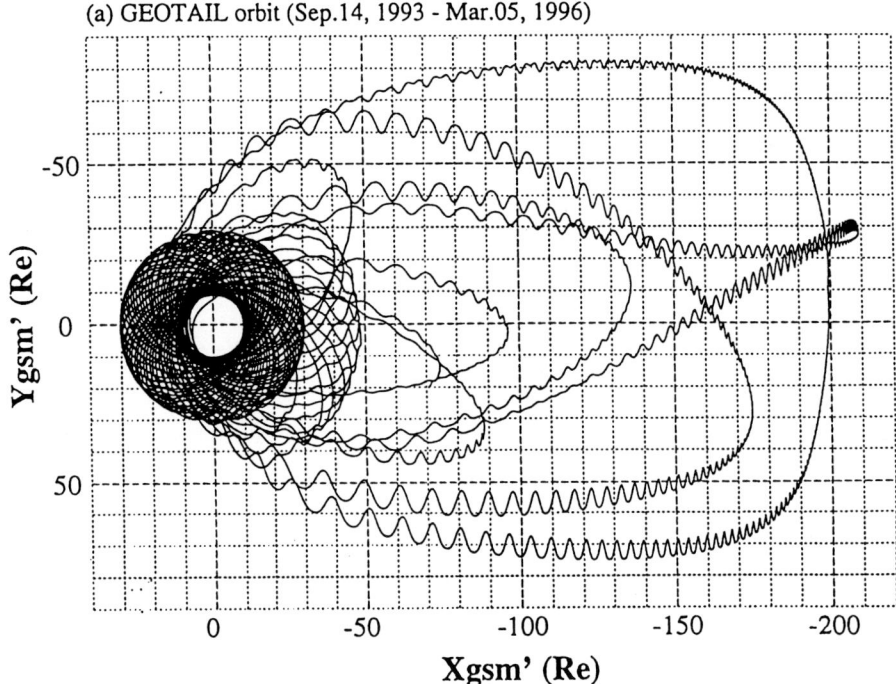

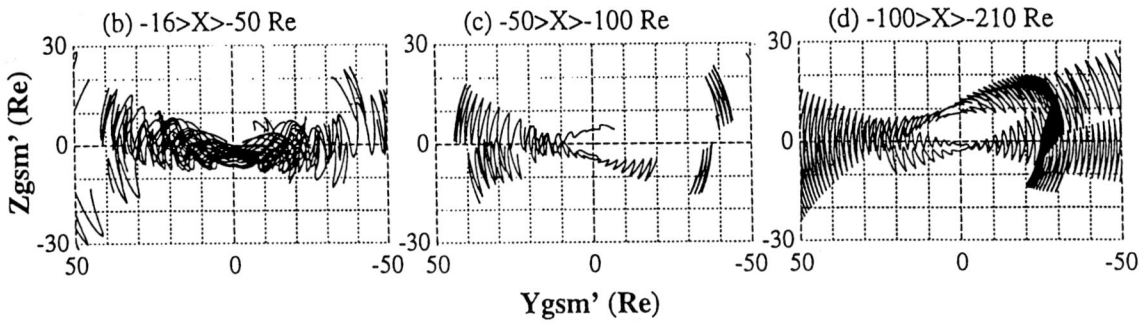

Figure 1. Orbit of GEOTAIL satellite in the modified Geocentric Solar Magnetospheric coordinate system where the aberration angle in the xy plane is 4° and the hinging distance of the neutral sheet is 10 R_E [Ieda et al., 1998].

IMF CONTROL ON MAGNETOTAIL STRUCTURE AND CONVECTION

GEOTAIL observations have demonstrated that magnetic reconnection governs the structure and dynamics of the magnetotail under both southward and northward IMF conditions [Nishida et al., 1998]. However, diffusion of magnetosheath ions onto the closed field lines may also occur when the IMF is northward.

Figure 2 shows the relation between the density versus the x component of the flow velocity observed in the tail at distances beyond 150 Re [Maezawa and Hori, 1998]. Both density and velocity have been normalized by the corresponding values in the upstream solar wind. The occurrence frequency suggests that there are two populations of the tail ions. The first population starts from the same density and velocity as those of the solar wind and tends smoothly toward lower densities and velocities. This part can be interpreted to represent the entry of the solar wind plasma into the magnetotail. The observed decrease in density and velocity agrees very well with the deceleration and rarefaction that are expected for the slow expansion fan generated in the entrant magnetosheath plasma [Siscoe and Sanchez, 1987]. The second population has densities lower

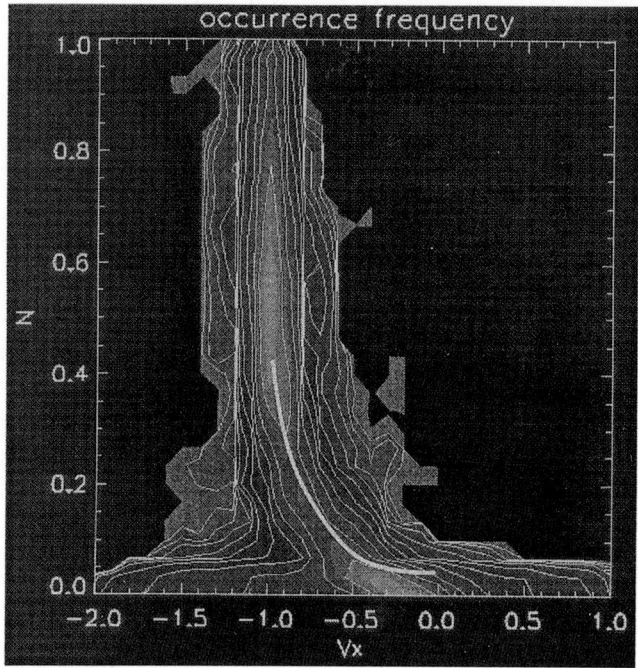

Figure 2. Occurrence frequency of density (N) and x-component of the flow velocity (V_x) of the tail plasma at x < -150 R_E [Maezawa and Hori, 1998].

than one tenth of the upstream solar wind density and shows a wide spread in the velocity. This part represents the plasma accelerated in the magnetotail, and ion temperature tends to be higher than 400 eV and thus substantially hotter than the magnetosheath plasma.

The acceleration and heating in the tail is due to reconnection. In geomagnetically active times, which occur under conditions of southward IMF, the distant neutral line is located at the distance of about 140 R_E [Nishida et al., 1996]. The occurrence of reconnection is not immediately obvious under the conditions of northward IMF since the northward magnetic field lines appear to be convected tailward under such conditions. However, the neutral sheet in the distant magnetotail is twisted under the influence of the B_y component of the IMF and more severely so when the IMF is northward [Maezawa et al., 1997]. Figure 3 illustrates reconnection that occurs in such a twisted neutral sheet. The open field lines move toward the neutral sheet (a) and their reconnection results in closed field lines (b) and IMF-type field lines (c). Note in Figure 3c that the IMF-type field lines that are produced on the far side of the reconnection line can have the northward B_z component although they cross the neutral sheet from its northern side to the southern side [Nishida et al., 1998]. Correspondingly, the electric field parallel to the neutral sheet is positive, that is, directed from dawn to dusk for both southward and northward IMF conditions. This is shown in Figure 4, where the projection of the electric field to the twisted neutral sheet is plotted versus IMF B_z [Maezawa and Hori, 1998].

While the features discussed above are interpreted in the framework of the open model of the magnetosphere where reconnection plays a pivotal role, observations of the low latitude boundary layer could suggest the operation of an interaction process of a different kind [Fujimoto et al., 1998a and b]. In the flank region of the magnetosphere GEOTAIL has observed cold and dense ions that are almost stagnant or flow very slowly sunward. These ions seem to be on closed field lines as they are associated with bi-directional thermal electrons (< 300 eV), and they make a clear contrast to hot ions in the plasma sheet, which are considered to be convected from the distant tail. When the IMF is northward, the layer of cold, dense ions sometimes has substantial thickness and continues to be seen well inside the magnetotail. This intriguing feature is being investigated further, and a mechanism for the diffusive entry of plasma from the magnetosheath onto closed field lines is being developed.

Density and temperature of the near-Earth plasma sheet (at x > -50 R_E) show quantitative differences between northward and southward IMF conditions. It is seen that the temperature is lower and the density is higher when the IMF B_z is northward than when it is southward [Terasawa et al., 1997]. These differences pose a problem to be explained in terms of reconnection or other models.

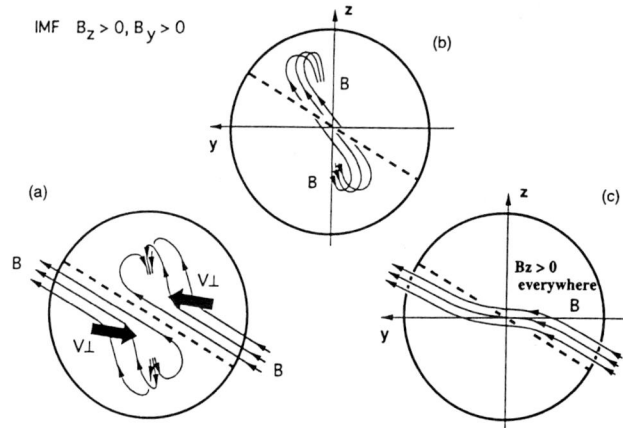

Figure 3. Projection of three types of tail field lines under the northward IMF as viewed toward the Earth. (a) Open field lines whose reconnection at the neutral sheet (dashed line) results in (b) closed field lines and (c) IMF-type field lines which have the northward polarity [Nishida et al., 1998].

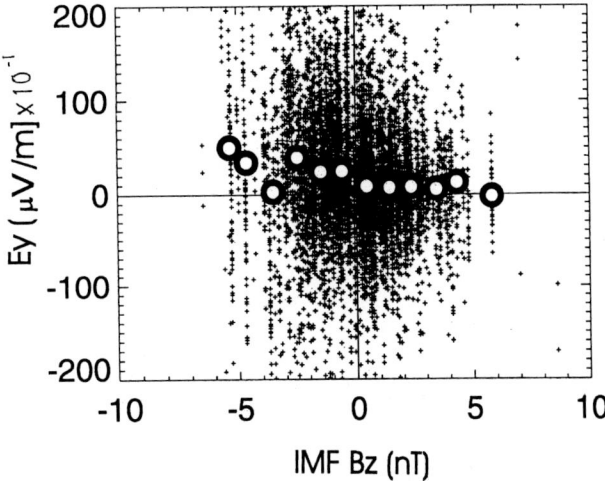

Figure 4. Component E_y of the electric field parallel to the neutral sheet is plotted versus IMF B_z. The twisting angle of the neutral sheet has been determined empirically to be 32° for the northward IMF and 15° for the southward IMF [Maezawa and Hori, 1998].

INITIATION OF SUBSTORM BY THE NEAR-EARTH RECONNECTION

The GEOTAIL mission has given a clear answer to the role of near-Earth reconnection in the magnetospheric substorm. Since the orbit is designed for observations of the plasma sheet in the crucial range of 20 to 30 Re, a substantial body of data has been obtained on the state of the near-Earth plasma sheet at the times of substorms.

Figure 5 shows the onset times of the flows in the plasma sheet relative to the substorm onset time determined by ground and geosynchronous-orbit signatures [Nagai et al., 1998; Nagai and Machida, 1998]. The upper and lower panels are for the tailward flows and for earthward flows, respectively, and there is a demarcation between the tailward and earthward flows in the distance region of 20-30 R_E. Beyond this distance tailward flows tend to be associated with substorm onsets, while earthward of this distance earthward flows do. This demarcation delineates the position of the near-Earth neutral line. It is also seen in the left two panels that the flows that signify the onset of reconnection precede the substorm onset in the midnight-premidnight sector. This suggests that the near-Earth reconnection starts before the expansion phase starts on the ground, and that this neutral line is initially formed in the midnight-premidnight region, which coincides with the local time range where the auroral and geomagnetic signatures of the expansion phase onset have been known to be observed first.

The observations on the propagation of plasmoids tailward of the near-Earth neutral line are summarized in Figure 6 [Ieda et al., 1998]. In the near-Earth tail inside 50 R_E plasmoids expand longitudinally with a speed of about one hundred km/s. Beyond this distance there is little longitudinal expansion and plasmoids propagate tailward with high speeds. The energy flux is on the order of 10^9 W R_E^{-2} and is dominated by the flow of thermal energy [Ieda et al., 1998]. This flux is comparable to the energy flux that is deposited on the ionosphere during substorms.

Velocity distribution functions show characteristic signatures at the passage of plasmoids. Figure 7 is an example of such passage at distance of about 70 R_E. The spacecraft was engulfed in the plasma sheet and observed the northward-to-southward turning of the magnetic field and the high-speed tailward ion flow (see the central panel). At the top and bottom corners of this figure we show four typical ion distribution functions during this plasmoid passage. Immediately before the northward turning of the B_z magnetic field, a high-speed tailward ion flow of > 2500 km s^{-1} is observed in the plasma sheet boundary layer (PSBL). In the phase space density distribution at this time (top left), where vertical and horizontal axes are aligned to the directions of the magnetic field and the convection velocity, respectively, the small dark area located near the center corresponds to the cold lobe ions, and the high-speed component flowing in the direction opposite that of the magnetic field is the PSBL ion beam. The ion

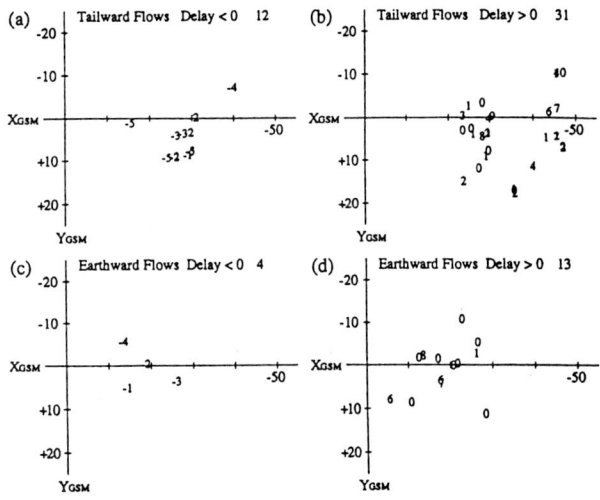

Figure 5. Onset times of tailward (top) and earthward (bottom) flows relative to the substorm onset time are plotted at the (x,y) coordinates of the satellite at the times of the observation [Nagai and Machida, 1998].

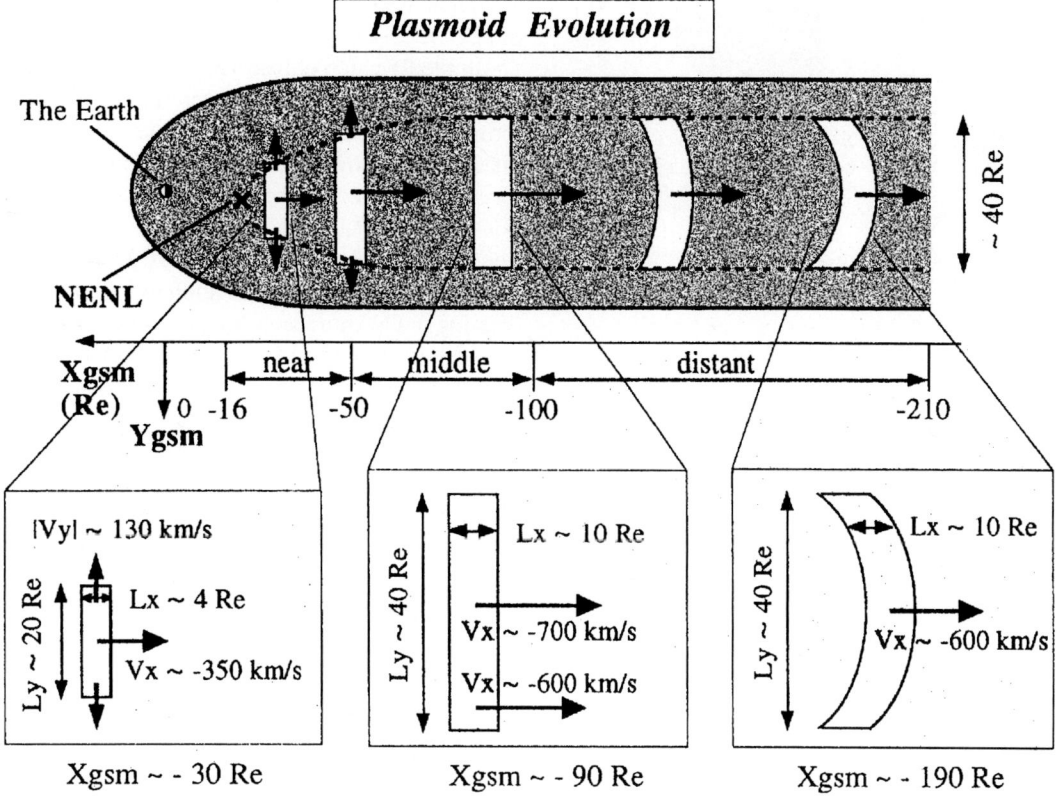

Figure 6. Evolution of plasmoids. Average size and speed of plasmoids are given for three distance ranges: the near tail (left), middle tail (middle) and distant tail (right) [Ieda et al., 1998].

distribution inside the plasmoid (bottom left) is characterized by two cold ion components parallel to the magnetic field [Mukai et al., 1966]. Around the turning point of the B_z polarity (bottom right), an almost thermal ion distribution function is observed. Just after the passage of the plasmoid (top right), we observe two main ion components bunched perpendicular to the magnetic field. Together with another slice of the three dimensional distribution function, we find that the ion distribution function is characterized by non-gyrotropic behavior related to crossings of the ions across the neutral sheet [Hoshino et al., 1998]. These observations have established the kinetic picture of plasmoid evolution, and the computer simulations have confirmed that a wide variety of distributions is produced by collisionless magnetic reconnection in a thin plasma sheet, where the characteristic spatial scale length is not necessarily shorter than the thickness of the plasma sheet [Hoshino et al, 1998; Hoshino, 1998].

Characteristic signatures of the plasmoid passage are seen also in electron velocity distribution functions. In the postplasmoid plasma sheet (PPPS) the low energy (0.1-1 keV) electrons flow earthward into the PPPS while more energetic components (ions and electrons) leak from the PPPS. It is suggested that the earthward-flowing electrons sustain the field-aligned current away from the neutral line and constitute part of the Hall current loop built up in the course of the magnetic reconnection [Fujimoto et al., 1997].

KINETIC PROPERTIES OF THE FORESHOCK REGION OF THE SLOW SHOCK

Acceleration of ions and electrons in the essentially collisionless regime of the tail plasma produces non-gyrotropic distributions in the phase space density. Characteristic phase-space signatures are found in such regions as plasmoids, neutral sheet, and plasma-sheet boundary, and rapid development of the kinetic picture of the magnetotail dynamics is at the heart of the new findings by GEOTAIL. An observation of a plasmoid passagge has been shown in the previous section. In this section we shall show

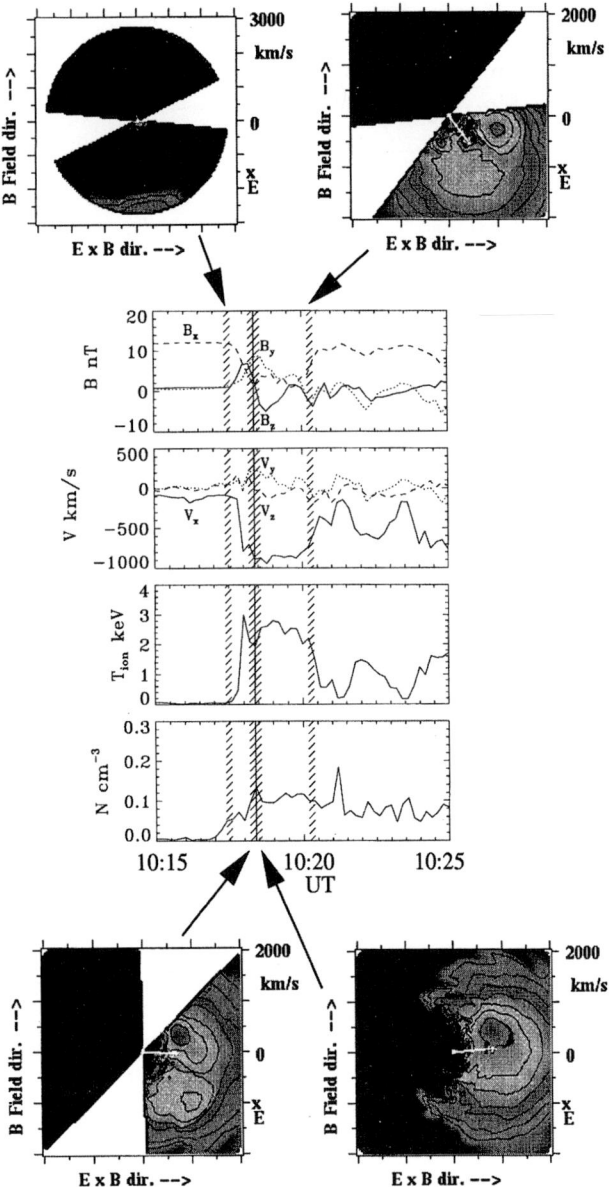

toward understanding energy dissipation process at the slow shock.

In the first step of research in this direction, the structure of the foreshock region has been studied [Saito et al., 1996]. Immediately upstream of slow-mode shocks there is a region that is characterized by counterstreaming ions: cold lobe ions that are incident from the upstream side and hot backstreaming ions that leak out of the plasma sheet. As shown in Figure 8, the perpendicular component of the cold ion temperature is enhanced in the foreshock region to 3 - 20 % of the total ion heating in the entire slow-shock system. Backstreaming ion density is about 1 -15 % of the ion density in the plasma sheet. Since the relative velocity between the cold ions and backstreaming ions is between 1.6 and 2.2 V_A where V_A is the local Alfvén speed, electromagnetic mode can be generated in the foreshock region by the ion cyclotron instability at times when the relative velocity exceeds the threshold of 2 V_A. For both cold and backstreaming ions the perpendicular component of the temperature is higher than

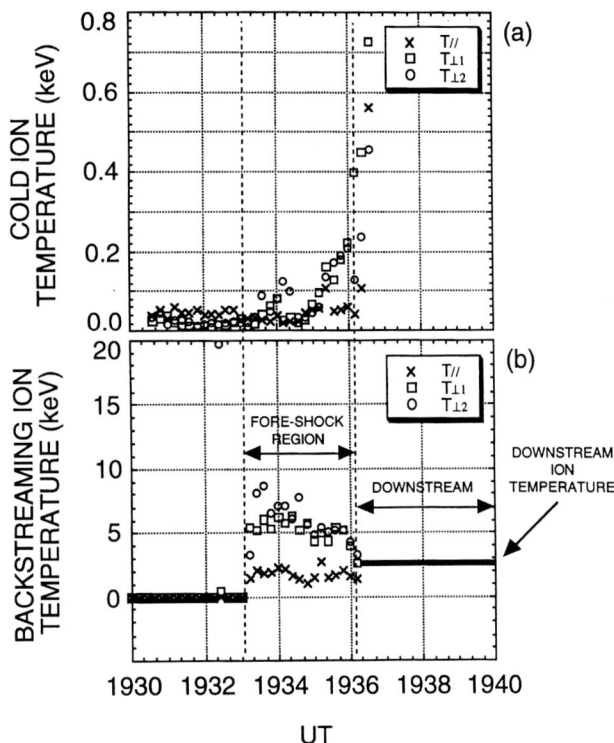

Figure 7. Contours of velocity distribution functions that are observed in four representative regions at the passage of a plasmoid [Hoshino et al., 1998].

GEOTAIL observations of the ion acceleration in the foreshock region of the slow-mode wave.

It was suggested earlier by Petschek [1964] that the reconnection rate is enhanced substantially due to presence of slow shocks that extend from the reconnection line. These slow shocks have been identified by GEOTAIL with use of a full complement of the plasma instrumentation, including the ion analyzer [Saito et al., 1995]. Our interest is now directed

Figure 8. Precursor changes in ion temperatures upstream of the slow shock (second dashed line). Gradual heating of cold ions starts in the foreshock region where backstreaming hot ions are present [Saito et al., 1996].

the parallel component, and it increases toward the direction of the flow of respective components. This suggests that ions are heated by ion cyclotron waves in the foreshock region. The thickness of the foreshock region is about 10 times the ion inertial length.

COLD DENSE ION FLOWS IN THE TAIL LOBE AND AT THE MAGNETOPAUSE

As seen earlier in Figure 2 there are cold dense ions in the magnetotail that have entered from the magnetosheath and are flowing tailward. When observations are made at the boundary between the tail lobe/mantle and the magnetosheath, two types of energy variations are seen. These are illustrated on the lower sides of panels (a) and (b) of Figure 9 [Hirahara et al., 1998]. In type (a) the ion energy simply decreases from the magnetosheath to the lobe/mantle, but in type (b) the ion band has two branches; one of these is characterized by an energy increase toward the lobe/mantle while the other is like type (a). The type (a) structure has been interpreted as representing entry of the magnetosheath plasma along open field lines as illustrated in the first panel, where the entrant plasma is decelerated by the Lorentz force as it crosses the magnetopause. Type (b), on the other hand is attributed to observations on the former lobe/mantle field lines which have been re-reconnected with the IMF on the tail surface as illustrated in the third panel. The Lorentz force operating at the local magnetopause produces the accelerated ion branch in this case, while the decelerated branch comprises ions that have entered the field lines on the downstream side.

The ions that are observed in the magnetotail are not entirely of solar wind origin. In addition to protons, tailward flowing helium and oxygen ions are sometimes observed in the tail lobe. These ions obviously originate in the ionosphere, but their occurrence shows a clear dependence on the IMF sector. As seen in the lower panel of Figure 10, the cold oxygen beams are observed preferentially in the tail sector that contains the recently opened field lines [Seki et al., 1998]. Since it is not likely that the acceleration rate of heavy ions at low altitudes is strongly dependent on the IMF B_y, this dependence suggests that field lines on which the accelerated heavy ions have been supplied are convected to the tail lobe in an IMF-dependent manner. The accelerated ions are probably supplied first on closed field lines and then transported to the tail lobe as these closed field lines are opened on the dayside magnetopause by reconnection. Whatever the actual mechanism of the transport is, it is noteworthy that the governing influence of the IMF on the magnetospheric structure extends to the supply of the ionospheric ions to the magnetotail.

As a related subject, entry of the hot component of the solar wind electrons along open field lines across the magnetopause has been clearly observed [Shirai et al., 1998]. Such electrons reverse their direction from tailward to earthward upon entry into the magnetotail and flow toward the ionosphere to be observed as the polar rain.

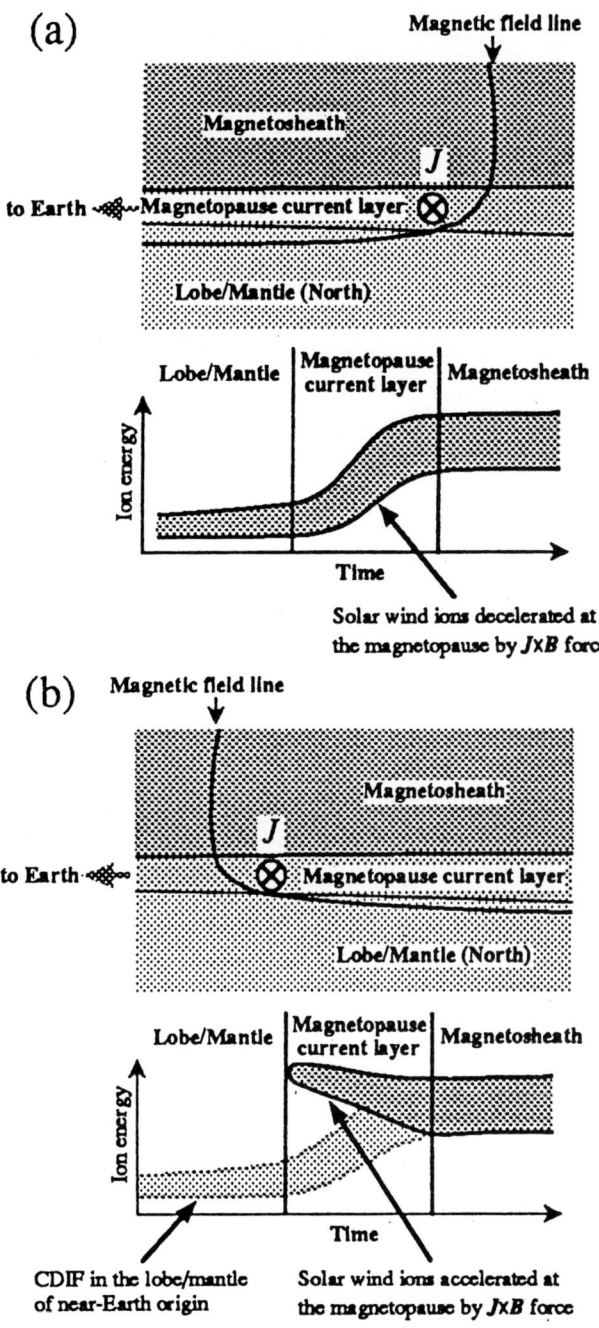

Figure 9. Field line configuration (upper panel) and associated E-t diagram of ions (lower panel) for (a) the open field line configuration and (b) re-reconnected IMF-type field line configuration [Hirahara et al., 1998].

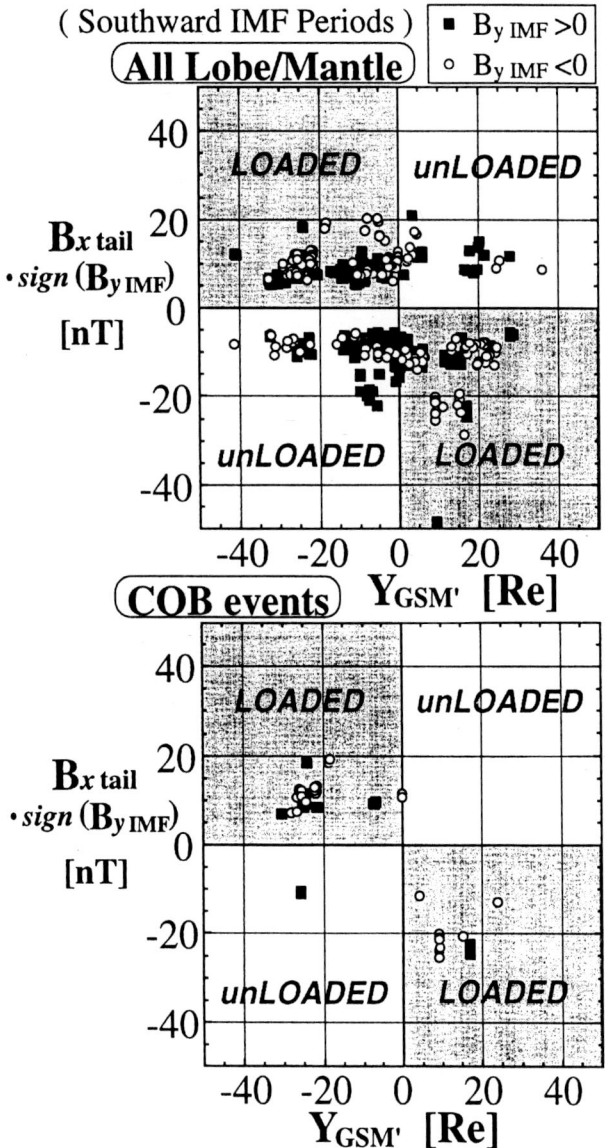

Figure 10. While lobe/mantle observations are made almost evenly in both loaded and unloaded quadrants, the cold oxygen beam (COB) events are observed exclusively in the loaded quadrant (shaded area) where the freshly reconnected IMF field lines are expected to be present. The vertical axis is B_x multiplied by the sign of IMF B_y and the horizontal axis is the aberrated y coordinate [Seki et al., 1998].

DISTRIBUTION AND ORIGIN OF THE PLASMA WAVES IN GEOSPACE

Modes of plasma waves in the magnetotail, magnetopause, magnetosheath, and bow shock upstream regions are mapped extensively and related to the characteristic features of the plasma in respective regions. Non-linear developments of plasma waves are observed and reasonably interpreted with extensive use of numerical simulations [Kojima et al., 1997; Matsumoto et al., 1998].

A prime example of the new findings concerns the nature of the broadband electrostatic noise (BEN) and narrowband electrostatic noise (NEN) which are commonly observed in the plasma sheet boundary layer and the tail lobe regions, respectively (Figure 11). Similar wave emissions are observed in the magnetosheath region. The BEN-type emissions observed in the plasma sheet boundary layer and the magnetosheath consist of isolated bipolar pulses. On the other hand, the waveforms of the NEN-type emissions are quasi-monochromatic. One of the common features of these waves is their burstiness, which means that their amplitudes or frequencies rapidly change with time scales of the order of a few milliseconds to a few hundreds of milliseconds. These waves are electrostatic waves propagating parallel to the ambient magnetic field.

The observed waveform of the BEN-type emissions can be reproduced in non-linear computer simulation of BGK mode waves where an electron beam is used as a free energy source. Since the BGK mode is very stable it can propagate far from the source region and result in the waveform of the BEN-type emission in the plasma sheet boundary layer [Omura et al., 1996].

Intense BEN emissions are seen in the electric component in plasmoids. At the same time the magnetic noise burst (MNB) is often observed in the magnetic field component. At the leading edge of the plasmoid the electron plasma waves are observed that are consistent with the presence of the electron beam, while at the trailing edge rather intense but scattered waves are seen in addition to the electron plasma waves. Differences in the electron spectra that can be related to the different wave features at the opposite edges of the plasmoid have not been identified so far [Matsumoto et al., 1998].

GEOTAIL IN THE CONTEXT OF ISTP

Among the other members of the ISTP fleet of spacecraft, WIND has been the most closely related to GEOTAIL. The information on the upstream solar wind and IMF provided by WIND has been heavily used to place the GEOTAIL observations in the proper context. The data from the old IMP-8 satellite have been valuable too; although IMP-8 has not always been in the upstream solar wind and its data acquisition has been less than full, it has an advantage of monitoring the solar wind and IMF just upstream of the bow shock.

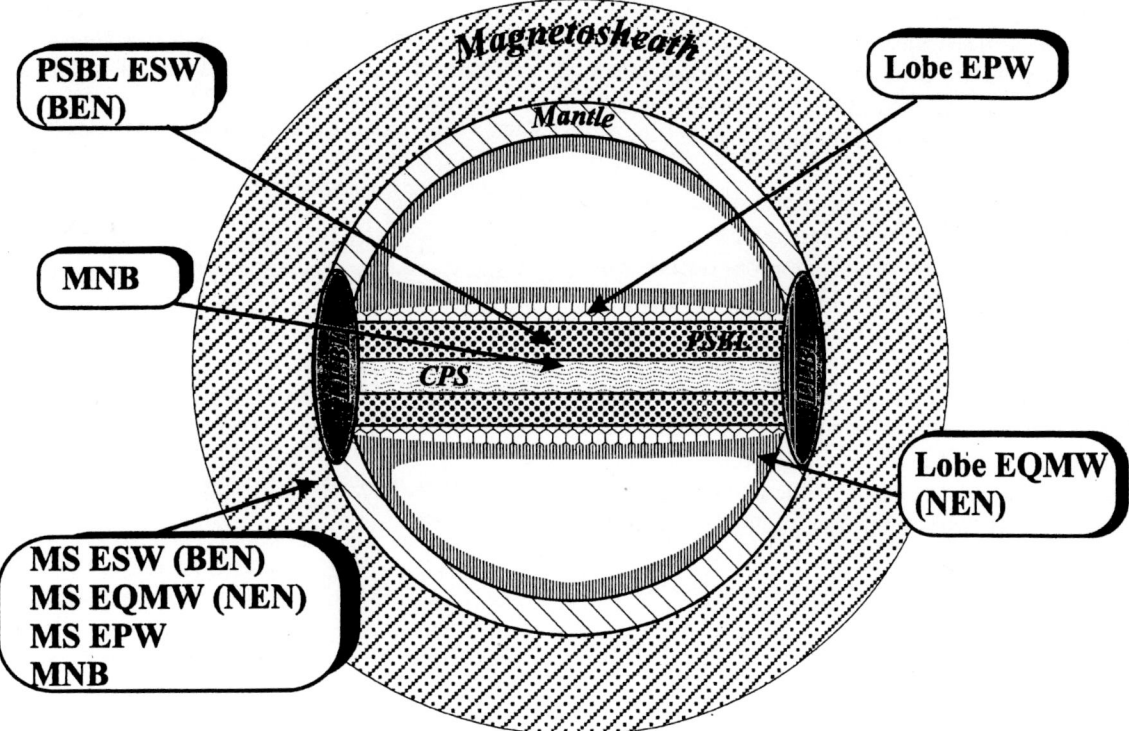

Figure 11. Characteristic modes of plasma waves observed in the magnetotail [Matsumoto et al., 1998]. PSBL ESW stands for plasma sheet boundary electrostatic solitary wave, MNB for magnetic noise burst, lobe EPW for lobe electron plasma waves, and lobe EQMW for lobe quasi-monochromatic wave. BEN and NEN stands for broadband electrostatic noise and narrowband electrostatic noise, respectively. Waves modes in the magnetosheath (MS) are also given.

INTERBALL-TAIL is a companion satellite that explores the relatively high latitude region of the near-Earth tail. Simultaneous observations with this satellite are providing information on the latitudinal, azimuthal, and radial structure of the tail dynamics. Petrukovich et al.[1998] have studied a case where the earthward plasma streaming and dipolarization were observed at 12 R_E at Interball while the tailward energetic ion beam and then the tailward propagating plasmoid were observed at 28 R_E at GEOTAIL. Initial brightening of the aurora delayed the commencement of the reconnection pulse by about 1 min. No significant magnetic disturbance was seen on the ground and the reconnection did not proceed to the open field lines in this pseudobreakup event.

On the downstream side of the energy flow in geospace, POLAR provides information on the energy that is precipitated into the ionosphere. Comparison of auroral activity with the tail condition is producing very interesting results, and will be vigorously pursued. The launch of EQUATOR-S has made possible to study the influence of the tail dynamics on the structure of the inner magnetosphere. We intend to promote a coordinated data analysis between GEOTAIL and EQUATOR-S.

The GEOTAIL scientific data base for public use is being constructed, and Internet access will be made possible for the MGF and LEP data. These data sets will include the 3-s average magnetic field, 12-s moments of the ion distribution function, and E-t diagrams for ions and electrons.

Since the 10 x 30 R_E orbit of GEOTAIL regularly visits such key regions of the magnetosphere as the dayside magnetopause, the earthward boundary region (i.e., Alfvén layer) of the plasma sheet, and the generation region of plasmoids/flux ropes, GEOTAIL observation should be instrumental for clarifying the origin of high latitude observations of the aurora, ionospheric dynamics, and geomagnetic variations. It is hoped that the GEOTAIL data base will be widely used, since the wealth of observations by GEOTAIL is far from having been exploited and is still being enriched as GEOTAIL continues to operate.

SUGGESTED FUTURE MISSION

By virtue of the three-dimensional phase space measurement and time resolution of 12 s, GEOTAIL has been able to study the kinetic properties of the tail ions and has revealed the non-gyrotropic ion velocity distributions in such regions as the neutral sheet and the plasma sheet boundary. These distributions are produced by non-adiabatic ion trajectories in these regions and have provided valuable information on the energization and dissipation processes that operate on the tail ions.

Similar studies of the kinetic properties of the electrons should be important. Such properties could be even more important in the case of electrons than for ions since it is the electron population in which the magnetic field lines are frozen. The departure from the frozen-in relation that is required for reconnection to proceed is the consequence of the non-gyrotropic motions of the electrons. A reasonable projection of the GEOTAIL mission toward the future would therefore be observations of magnetotail physics with an instrument complement that is capable of measuring the electron kinetics. The electron inertia length of about 10 km is traversed in 0.1 s when the velocity is 100 km s^{-1} and the gyroperiod is about 0.01 s. The required time resolution is already within the achievable range.

REFERENCES

Fujimoto, M., M.S. Nakamura, I. Shinohara, T. Nagai, T. Mukai, Y. Saito, T. Yamamoto, and S. Kokubun, Observations of earthward streaming electrons at the trailing boundary of a plasmoid, *Geophys. Res. Lett., 24*, 2893, 1997.

Fujimoto, M., T. Mukai, H. Kawano, M. Nakamura, A. Nishida, Y. Saito, T. Yamamoto, and S. Kokubun, Structure of the low-latitude boundary layer: A case study with Geotail data, *J. Geophys. Res., 103*, 2297, 1998a.

Fujimoto, M., T. Terasawa, T. Mukai, Y. Saito, T. Yamamoto, and S. Kokubun, Plasma entry from the flanks of the near-Earth magnetotail: Geotail observations, *J. Geophys. Res., 103*, 4391, 1998b.

Hirahara, M., K. Seki, and T. Mukai, Cold dense ion flows in the distant magnetotail: The Geotail results, in *New Perspectives of the Earth's Magnetotail*, ed. A. Nishida, D.N. Baker and S.W.H. Cowley, American Geophysical Union, in press, 1998.

Hoshino, M., Kinetic ion behavior in magnetic reconnection region, in *New Perspectives of the Earth's Magnetotail*, edited by A. Nishida, D.N. Baker and S.W.H. Cowley, American Geophysical Union, in press, 1998.

Hoshino, M., T. Mukai, T. Yamamoto, and S. Kokubun, Ion dynamics in the magnetic reconnection: Comparison between numerical simulation and GEOTAIL observations, *J. Geophys. Res., 103*, 4509, 1998.

Ieda, A., S. Machida, T. Mukai, Y. Saito, T. Yamamoto, A. Nishida, T. Terasawa, and S. Kokubun, Statistical analysis of the plasmoid evolution with GEOTAIL observations, *J. Geophys. Res., 103*, 4453, 1998.

Kojima, H., H. Matsumoto, S. Chikuba, S. Horiyama, M. Ashour-Abdalla, and R.R. Anderson, Geotail waveform observations of broadband / narrowband electrostatic noise in the distant tail, *J. Geophys. Res., 102*, 14439, 1997.

Maezawa, K., and T. Hori, The distant magnetotail: Its structure, IMF dependence and thermal properties, in *New Perspectives of the Earth's Magnetotail*, edited by A. Nishida, D.N. Baker, and S.W.H. Cowley, American Geophysical Union, in press, 1998.

Maezawa, K., T. Hori, T. Mukai, Y. Saito, T. Yamamoto, S. Kokubun, and A. Nishida, Structure of the distant magnetotail and its dependence on the IMF By component: GEOTAIL observations, *Adv. Space Res., 20*, 949, 1997.

Matsumoto, H., H. Kojima, Y. Omura and I. Nagano, Plasma waves in geospace: GEOTAIL observations, in *New Perspectives of the Earth's magnetotail*, edited by A. Nishida, D.N. Baker and S.W.H. Cowley, American Geophysical Union, in press, 1998.

Mukai, T., M. Fujimoto, M. Hoshino, S. Kokubun, S. Machida, K. Maezawa, A. Nishida, Y. Saito, T. Terasawa, and T. Yamamoto, Structure and kinetic properties of plasmoids and their boundary regions, *J. Geomag. Geoelectr., 48*, 541, 1996.

Nagai, T., and S. Machida, Magnetic reconnection in the near-Earth magnetotail, in *New Perspectives of the Earth's magnetotail*, edited by A. Nishdia, D.N. Baker, and S.W.H. Cowley, American Geophysical Union, in press, 1998.

Nagai, T., M. Fujimoto, Y. Saito, S. Machida, T. Terasawa, R. Nakamura, T. Yamamoto, T. Mukai, A. Nishida, and S. Kokubun, Structure and dynamics of magnetic reconnection for substorms onsets with GEOTAIL observations, *J. Geophys. Res., 102*, 4419, 1997.

Nishida, A., T. Mukai, T. Yamamoto, Y. Saito, and S. Kokubun, Magnetotail convection in geomagnetically active times, 1. Distance to neutral lines, *J. Geomag. Geoelectr., 48*, 489, 1996.

Nishida, A., T. Yamamoto, and T. Mukai, The GEOTAIL mission: Principal characteristics and scientific results, *Adv. Space Res., 20*, 539, 1997.

Nishida, A., T. Mukai, T. Yamamoto, S. Kokubun, and K. Maezawa, A unified model of the magnetotail convection in geomagnetically quiet and active times, *J. Geophys. Res., 103*, 4409, 1998.

Omura, Y., H. Matsumoto, T. Miyake, and H. Kojima, Electron beam instabilities as generation mechanism of electrostatic solitary waves in the magnetotail, *J. Geophys. Res., 101*, 2685, 1996.

Petrukovich, A.A., V.A. Sergeev, L.M. Zelenyi, T. Mukai, T. Yamamoto, S. Kokubun, K. Shiokawa, C.S. Deehr, E.Y Budnick, J. Buchner, A.O. Fedorov, V.P Grigorieva, T. J. Hughes, N.F. Pissarenko, S.A. Romanov, and I Sandahl, Two spacecraft observations of a reconnection pulse during an auroral breakup, *J. Geophys. Res., 103*, 47, 1998.

Petschek, H.E., Magnetic field annihilation, in AAS-NASA Symposium on the physics of Solar Flares, *NASA Spec. Publ., SP-50*, 425, 1964.

Saito, Y., T. Mukai, T. Terasawa, A. Nishida, S. Machida, M. Hirahara, K. Maezawa, S. Kokubun, and T. Yamamoto, Slow-mode shocks in the magnetotail, *J. Geophys. Res., 100*, 23567, 1995.

Saito, Y., T. Mukai, T. Terasawa, A. Nishida, S. Machida, S. Kokubun, and T. Yamamoto, Foreshok structure of the slow-mode shocks in the Earth's magnetotail, *J. Geophys. Res., 101*, 13267, 1996.

Seki, K., M. Hirahara, T. Terasawa, T. Mukai, Y. Saito, S. Machida, T. Yamamoto, and S. Kokubun, Statistical properties and possible supply mechanims of tailward cold O^+ beams in the lobe/mantle regions, *J. Geophys. Res., 103,* 4477, 1998.

Shirai, H., K. Maezawa, T. Mukai, T. Yamamoto, Y. Saito, M. Fujimoto, and S. Kokubun, Entry process of low-energy electrons into the magnetosphere along open field lines: Polar rain electrons as field line tracers, *J. Geophys. Res., 103,* 4379, 1998.

Siscoe, G.L., and E. Sanchez, An MHD model for the complete open magnetotail boundary, *J. Geophys. Res., 92,* 7405, 1987.

Terasawa, T., M. Fujimoto, T. Mukai, I. Shinohara, Y. Saito, T. Yamamoto, S. Machida, S. Kokubun, A.J. Lazarus, J.T. Steinberg, and R.P. Lepping, Solar wind control of density and temperature in the near-earth plasma sheet: WIND-GEOTAIL collaboration, *Geophys. Res. Lett., 24*, 935, 1997.

A. Nishida, Institute of Space and Astronautical Science, 3-1-1 Yoshinodai, Sagamihara Kanagawa 229-8510, Japan

The Correspondence of EUV and White Light Observations of Coronal Mass Ejections with *SOHO* EIT and LASCO

B. J. Thompson[1], O. C. St. Cyr[2], S. P. Plunkett[3], J. B. Gurman[1], N. Gopalswamy[1,4], H. S. Hudson[5], R. A. Howard[6], D. J. Michels[6], and J.-P. Delaboudinière[7]

For over two years of operation, the LASCO coronagraph and the EIT imaging telescope have conducted observing campaigns intending to provide continuous coverage of global coronal dynamic phemonena. These compatible observations, which we call the "CME Watch," have been used to further our understanding of CME initiation and propagation, and to provide advance warning of potential Earth-impacting eruptions. This paper includes a discussion of the EUV signatures of CME's in the inner corona and attempts to optimally combine the EUV and white light observations. One result of this effort is the determination that it is highly likely that LASCO definitively observes the vast majority of coronal mass ejections. While previous coronagraph missions have not been sensitive enough to detect weak eruptions, LASCO is capable of imaging even the faint "halo" CME's, which are our most accurate indicator of geospace-impacting eruptions.

1. INTRODUCTION

As is always the case in defining a physical phenomenon, it is difficult to arrive at a definition that does not involve some assumptions about the physics in question. This has indeed been the case in the study of coronal mass ejections (CME's). The variety of manifestations obtained from white light, H-alpha, soft X-ray and EUV imaging have thus far precluded a purely empirically-based definition. For example, *Steinolfson* [1985] describes "coronal transients, or, for the subclass which involves expulsion of coronal material, coronal mass ejections." Unfortunately, this definition is not readily applied to EUV and soft X-ray data. As we will show, most of the EUV signatures of coronal mass ejections in the inner corona do not involve the direct observation of ejected material. These signatures include: "dimming" regions, which are presumably low-density regions depleted by the CME; "arcade" formation, which are closed loops consisting of material that is not ejected but instead is trapped in bright compact structures; and "EIT wave" events, which embody the reaction of the rest of the corona to the presence of a major transient eruption. These signatures, however, are based on the observational implications of massive ejections.

CME's are generally looked upon as a change in magnetic topology resulting from energetically favorable

[1]NASA Goddard Space Flight Center, Greenbelt, MD
[2]Computational Physics Inc., Naval Research Laboratory, Washington, DC
[3]Universities Space Research Association, Naval Research Laboratory, Washington, DC
[4]Department of Physics, Catholic University of America, Washington, DC
[5]Solar Physics Research Corp., Tucson, AZ
[6]E. O. Hulburt Center for Space Research, Naval Research Laboratory, Washington, DC
[7]Inst. D'Astrophysique Spatiale, Orsay, France

Sun-Earth Plasma Connections
Geophysical Monograph 109
Copyright 1999 by the American Geophysical Union

conditions. The continuous emergence of magnetic flux from the convection zone and through the photosphere necessitates the bulk expulsion of magnetic flux [*McComas, Gosling, and Phillips*, 1992; *Bieber and Rust*, 1995; *Low*, 1996], and CME's represent coronal reconfigurations which occur throughout the solar cycle. In this paper, the term "coronal mass ejection" will refer to the bulk removal of material from the corona in conjunction with the outflow of associated magnetic flux tubes. This definition implies that CME's are a phenomenon distinct from high-latitude solar wind flow where the material motion is along magnetic field lines, although there is still ambiguity regarding the characterization of transient massive phenomena such as jets and sprays and the characterization of possible sources of slow-speed solar wind flow.

There are several additional unresolved questions about the nature, structure and evolution of CME's. For example, *Hundhausen* [1998] points out that CME's have implied speeds which range from 10 to 2100 *km/sec*, defying any attempts to attribute the eruptions to a single simple physical mechanism. These "speeds span those of MHD modes in the corona; they range from a small fraction to several times the escape speed from the solar gravitational field. Thus, for example, any attempt to identify mass ejections with a particular wave mode seems doomed to failure. " The observations of the destabilization and early stages of CME eruption by the EUV Imaging Telescope (EIT) [*Delaboudinière et al*, 1995], on the Solar and Heliospheric Observatory (*SOHO*) spacecraft, can provide unique clues about their initiation and structure. This paper summarizes efforts to define EUV CME signatures and to determine their association with observed white light structures.

2. THE *SOHO* LASCO/EIT "CME WATCH"

The EIT data, when combined with observations of the Large Angle Spectroscopic Coronagraph (LASCO) [*Brueckner et al*, 1995; *Howard et al*, 1997], have served as a global coronal dynamics monitor, recording coronal transients and reconfigurations. We refer to this complementary observing program as the LASCO/EIT "CME Watch."

The LASCO/EIT "CME Watch" patrol has focussed on halo CME's as part of the Sun-Earth connections initiative [e.g. *Luhmann*, 1997]. A halo CME is an eruption which is assumed to have occurred along the Earth-Sun axis. The projection of the CME results in a brightening when extends partially or completely around the occulting disk. The "CME Watch" focusses on halo CME's for three important reasons:

1) Halo CME's have the greatest potential impact on near-Earth space weather, provided that the alerts are efficiently produced and disseminated. A typical CME can transit interplanetary space and reach Earth in 3 days; a fast CME may allow much less time to respond.

2) The "CME Watch" has provided the first opportunity to study halo CME's on a regular basis. Although it has been established that there is a strong correlation between halo CME's and the peak K-p (an index of geomagnetic activity) of magnetospheric storms and substorms, it is still not clear how to optimally combine EIT's CME observations with LASCO's imaging of the disturbed outer corona.

3) Halo CME's represent the remarkable opportunity to study a CME twice: a CME which is first detected through remote-sensing provided by coronal imagers has the potential to be recorded by interplanetary radio sensors such as the *WIND* spacecraft's WAVES instrument. The radio bursts can provide information on the CME's heliospheric location and propagation. As the CME completes its transit to Earth, a comprehensive suite of spacecraft and instrumentation await its arrival: the CME is detected *in situ*, revealing essential physical information such as density, temperature, charge states, magnetic signature and velocity profile. Finally, the ejection's interaction with the magnetosphere can result in severe agitation of geospace; from the solar photosphere, through the corona, across interplanetary space, to the outer magnetosphere, plasmasphere and finally Earth, the flow of energy and material across 93 million miles is traced in glorious detail.

Plate 1 contrasts a CME occurring at the solar limb (above) with a "halo" CME (below). Because the emission recorded by a white-light coronagraph is scattered with a strong angular preference for the plane of the sky, limb CME's tend to exhibit much more detail and structure, while halo CME's are more diffuse and faint. The viewing of halo CME's is also complicated by the fact that the corona is optically thin: a halo CME generally involves the integration over a variety of structures of varying density, while a limb CME is usually better resolved because the light is integrated along the line of sight along the denser outer portion of the CME. While an Earth-directed CME

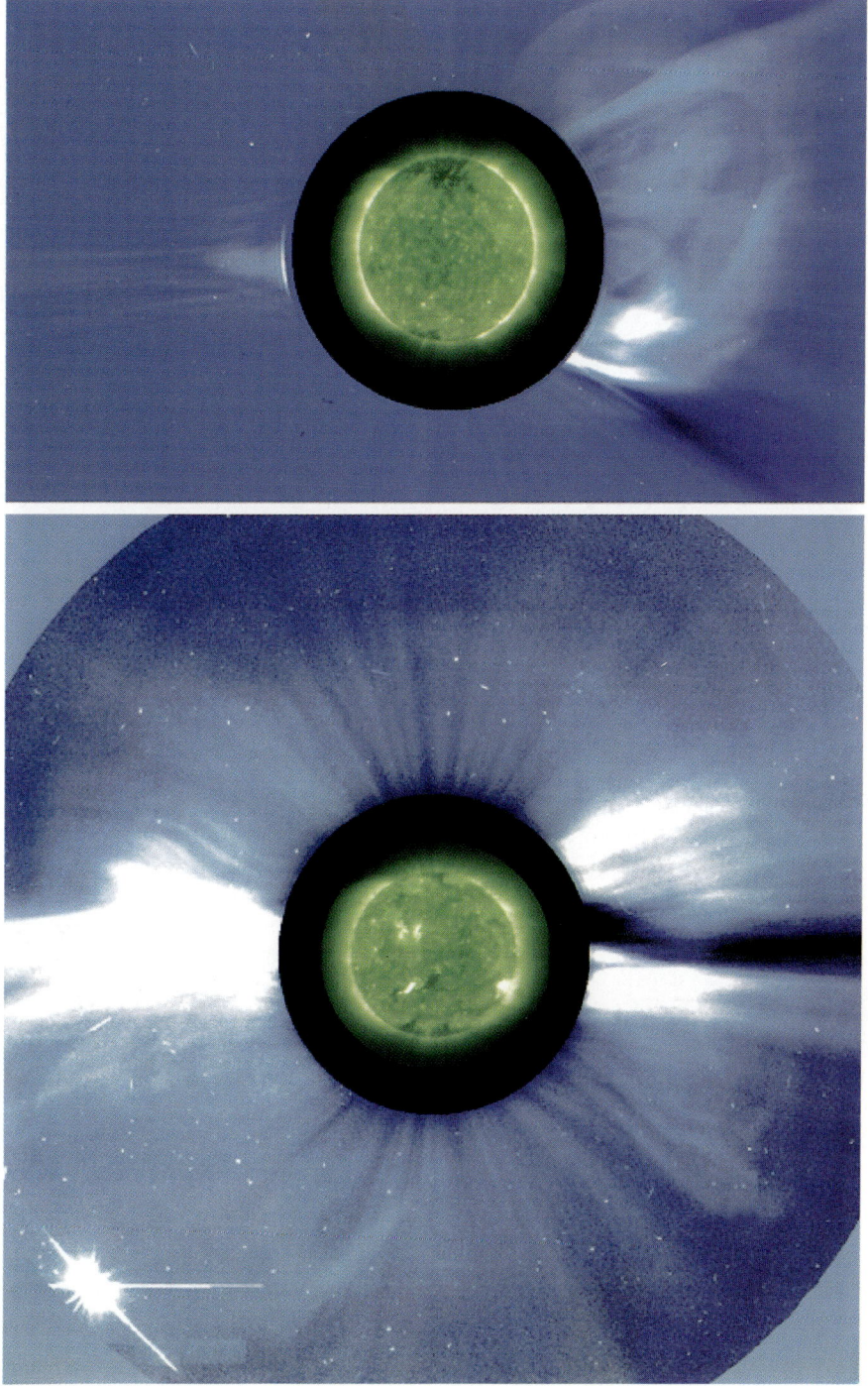

Plate 1. Composite images of two coronal mass ejections: a CME occurring at the limb of the Sun on 5 October, 1996 (top), and a head-on, "halo" CME on 7 April, 1997 (bottom). The inner disk observations are EIT 195 Å images. The outer coronal observations are LASCO C2 images. The EIT observations of the 5 October, 1996 CME (at 13:03 UT) show opening coronal features corresponding to the LASCO coronagraph data recorded 11:40 UT (see Plate 4). During the 7 April 1997 CME, EUV dimming regions (at 14:35 UT) near the active region Southeast of disk center accompanied the LASCO white light CME, which by 15:52 UT had a full 360-degree projection around the occulting disk.

represents the opportunity to directly view the eruption with EIT, staring down its axis, a limb CME can also be viewed with EIT, with the added benefit that LASCO can observe its propagation and evolution to 32 solar radii. This represents *15%* of the distance to Earth: the primary acceleration of a CME occurs within the first few solar radii, and outside that the propagation is very representative of the CME's interaction with the solar wind through its entire transit through interplanetary space. Ideally, a CME would be viewed from both angles, in profile and down its axis, to optimally combine the EUV and white light data, and to derive more information about structure and evolution. In the absence of such observations, attempts must be made to fully utilize the information provided by the available data.

3. WHITE LIGHT VS. EUV IMAGING OF CME'S

CME's are most readily viewed with a white light coronagraph for several reasons:

1) The source of the CME is the denser, lower altitude corona, and the average coronal density experiences a drastic decrease between typical imager and coronagraph fields of view. As the CME propagates to the field of view of the coronagraph, the material is seen in high contrast to the sparser outer corona.

2) The CME's typical "loop-like" topology, originating from closed field lines, is structurally different from the open quasi-radial field lines in the outer corona, providing an additional contrast.

3) White-light coronagraphs reflect integrated line-of-sight density, weighted by the Thomson scattering angle. Most CME's appear to be propagating radially by the time they approach the white light field of view, so the Thomson scattering angle does not drastically change. By contrast, spectral line imagers (H-alpha, soft X-rays, EUV) are sensitive to effects other than the density evolution: a change in temperature can cause the plasma to shift its emission in or out of a spectral bandpass, and sufficient velocity can cause the spectral line to doppler shift from observability. As a result, no single spectral disk imager sees the "whole CME" and even with several different data sets there is still much room for interpretation.

4) The CME's final destination is "away from the Sun," and a typical white-light coronagraph can survey the entire transit of an eruption. A disk imager usually only sees the base of an eruption, and it is difficult to distinguish whether the evolving field lines are actually part of an opening structure.

Between the *Skylab, Solwind*, and *Solar Maximum* missions, several thousand CME's have been observed by coronagraphs. Nonetheless, several basic questions regarding the generation, eruption, and propagation of CME's still exist. It has become clear that white light coronagraph imaging is not sufficient to address several aspects of CME research: the solution to some of these problems lies in the incorporation of disk imaging, particularly coronal disk imaging, and the improved field of view and sensitivity of the LASCO coronagraph.

One of the reasons disk imaging is necessary is because coronagraphs usually only see CME's when they are already in an advanced stage of eruption; often, the key signs of destabilization and the early stages of the instability are best captured with a coronal disk imager. *Fisher and Poland* [1981], for example, have monitored the evolution of structures in the coronagraph field of view, before, during and following eruption, but for most eruptions the clues to the timing and destabilization are hidden behind the occulting disk.

It is possible that a CME cannot be properly characterized without data from several types of instrumentation [e.g. *Hundhausen*, 1997]. White light coronagraphs, with sufficient sensitivity, should be able to resolve most coronal mass ejections. EUV imaging, particularly in the coronal wavelengths of *SOHO* EIT, records the evolution of the 1-2 million Kelvin corona, capturing the activity of "typical" coronal structures at moderate temperatures. Not only does EIT observe the erupting structures, but it also has sufficient signal in ambient coronal features to gauge the level of activity in regions not involved in the eruption. However, most coronal mass ejections result in localized heating of plasma, particularly the trapped plasma resulting from the magnetic reconfiguration. These regions are ideally viewed in soft X-ray wavelengths; soft X-rays not only show compact hot evolving loops but also are able to pinpoint large-scale structures which are participating in or are influenced by the coronal mass ejection. Finally, a large fraction of the CME mass can reside in the cool, dense prominence material; different spectral regimes than EUV or soft X-rays tend to optimally characterize prominence evolution.

This paper will attempt to summarize some of the attempts to combine EUV imaging of coronal mass ejections with white light observations, and will discuss some aspects of CME topology, destabilization, and dynamics which can be derived from these investigations. To date, 1997 is the only complete year of *SOHO* LASCO and EIT CME observations. The results discussed in this paper are based primarily on the 1997 CME statistics [*St. Cyr*, 1998a], with a few examples from 1996 and 1998.

4. EUV SIGNATURES OF CME'S

Because of the reasons described in the previous section, the occurrence of a CME is not as clear in EUV images as it is in white light coronagraph observations. A number of EUV CME signatures have been established, and although some phenomena (such as arcade formation and coronal dimmings) serve as strong indicators, there is no single diagnostic which is necessary and sufficient. Instead, we either look for a very strong example of one of these signatures, or particular combinations.

These EUV signatures were established, based on several principles (apart from the obvious "things we see during a white-light CME" criterion):

1) Several signatures of CME's have been established based on other forms of disk imaging, particularly with H-alpha and soft X-rays. H-alpha flares, filament eruptions, and their relationship to coronal mass ejections have been studied by several authors. *Hudson and Webb* [1997] give a description and several examples of soft X-ray signatures of coronal mass ejections.

2) Based on our understanding of CME's, we have expectations of how they would appear in the inner corona, with the EUV images being used as a representation of the "standard" million-degree corona and evolution. This includes the apparent opening of field lines, and signs of bulk motion of material away from the Sun.

3) Large-scale transient behavior (such as "EIT waves," described in this section) is often suspected of having a connection with CME's, primarily because most of the major transient activity in the corona can be linked to either a CME or a flare.

This section ennumerates and describes the diagnostics commonly applied to EUV images when determining the existence and extent of a CME.

4.1 Arcade Formation

EUV arcade formation provides a reliable diagnostic of magnetic structures involved in a coronal mass ejection. Bright arcade loops are observed to form across magnetic neutral lines which presumably was the location of the CME source region [*Webb*, 1976; *Kahler*, 1977]. Larger-scale loops can also appear, usually with one footpoint near a coronal hole boundary [*Zirker*, 1977; *Hudson, Acton, and Freeland*, 1996].

The CME represents a topological change, and these arcades are often attributed to the re-closing of magnetic field field lines and the energization of trapped plasma along the loops, though *Yokoyama and Shibata* [1998] discuss the ability of flare reconnection to influence adjacent, previously existing arcade loops. These correspond to the soft X-ray "long-decay enhancement" (LDE) flares, which *Hudson et al* [1996] and *Hudson and Webb* [1997] identify as a *Yohkoh* Soft X-ray Telescope (SXT) [*Tsuneta et al*, 1991] signature of a coronal mass ejection.

Several authors [*Sheeley et al*, 1975; *Webb*, 1976; *Rust and Webb*, 1977; *Kahler*, 1982, 1992; *Hudson et al*, 1996] have worked towards clarifying the relationship between LDE's, accompanying H-alpha "two-ribbon" flares, and CME's. The EIT observations confirm these results, and the flare/arcades fitting these descriptions, when observed, are a very strong indicator of a coronal mass ejection. The converse is not necessarily true, in that a coronal mass ejection can occur without accompanying observable arcade formation; however, this may correspond to a more gradual reconfiguration of magnetic field which does not result in significant thermal enhancement.

Some evidence exists [e.g. *Dryer*, 1996] that the statistics derived from and research based on LDE-associated CME's may only represent a particular class of CME's, and CME's associated with more impulsive flares may exhibit different structure and behavior. A discussion of the relationship between flares and CME's will not be included in this paper; their relationship, and the associated debate of "correlation vs. cause" have been approached by a number of authors [e.g. *Fisher and Poland*, 1981; *Kahler*, 1982; *Harrison*, 1986, 1991; *Gosling*, 1993; *Dryer*, 1996].

Plate 2 shows an example of post-CME local arcade formation, described more thoroughly in *Thompson et al* [1998a]. The first panel shows an EIT 195 Å image at 01:12 UT; the 195 Å bandpass is dominated by three Fe XII lines, and the peak temperature at which coronal emission occurs at this wavelength is estimated to be near 1.5 MK. The second panel shows the same region after the coronal mass ejection, at 07:12 UT. The third panel shows EIT 171 Å Fe IX/X at 07:00 UT, with a corresponding temperature of approximately 1 MK. The fourth panel shows EIT 284 Å Fe XV at 07:06 UT, with a temperature near 2 MK. The bright, horizontal loops in the later images show the arcade formation. The arcades first appear at the time of the eruption, continue to form and expand after the

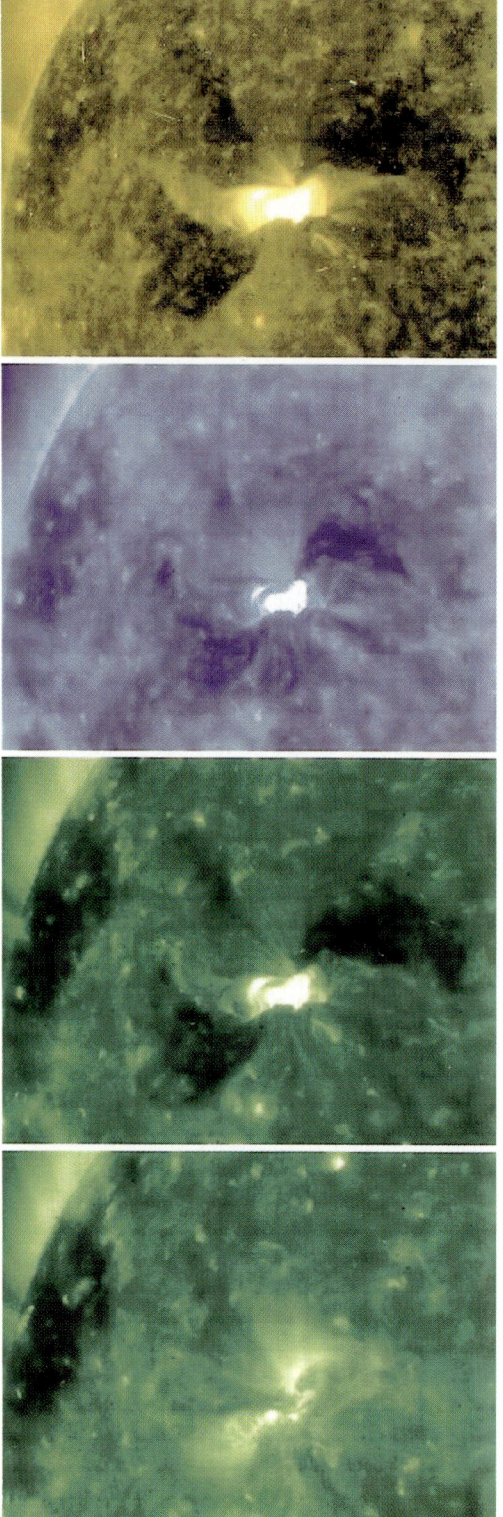

Plate 2. EIT EUV observations of "twin dimming" regions accompanying a CME on 12 May, 1997. The first two images are taken in the EIT 195 Å bandpass, at 01:12 and 07:12 UT, representing the pre- and post-eruption EUV corona. The third and fourth images are EIT 171 Å and 284 Å recorded at 07:00 and 07:06, respectively.

eruption. However, we must be cautious: the way in which the time of an eruption is defined can influence our conclusions about sequence and timing.

4.2 Coronal Dimmings

EUV dimmings of the corona, occurring on timescales ranging from minutes to hours, are another extremely reliable signature of a coronal mass ejection. Like the arcades described in the previous section, this association has been established in prior investigations [*Rust*, 1983, and references therein].

Rust [1983] examined the brightest *Skylab* LDE events, and determined that the majority of the events showed evidence of soft X-ray "voids" forming in association with the eruption. The voids were observed near the erupting region, either alone or in pairs, such as those shown in Plate 2. In one example, the darkened regions were the most distinct 2 1/2 hours after the eruption, and gradually disappeared over the following ten hours. In general, these coronal voids lasted less than 48 hours, and could decrease in emission until they were as dark as coronal holes.

Rust reasoned that since the interplanetary shocks associated with eruptions "are followed by a sustained period of high speed solar wind, and *recurrent* high speed solar wind streams originate in long-lived coronal holes [*Hundhausen*, 1972], then the transient solar wind speed increases might stem from transient coronal holes." This, combined with the coronal-hole-like decrease in emission observed in the void regions, provided his basis for labelling them "transient coronal holes."

Hudson et al [1996] determined that the dimming of the soft X-ray corona associated with a CME proceeded too rapidly to be explained by radiative or conductive cooling estimates. This led to the conclusion that the observed decrease in emission was due (at least in part) to the outflow of material, and not solely due to a shift in temperature. Although the results of their study were not in conflict with *Rust*'s speculation, the authors chose not to refer to the phenomenon as a transient coronal hole, preferring the more empirical label of "dimming."

The dimmings observed by EIT are by no means simple or standard. The "twin dimming" regions, shown in Plate 2, are distinctive in their symmetry and tendency to exhibit a spiraling, two-lobed shape. EIT's sensitivity and high signal-to-noise in *all* coronal regions allows a much better determination of shape, spectral variation, and evolution. However, these twin dimmings are only observable in approximately *25%* of all CME-associated dimming events in EIT, and dimmings are not observed in association with every CME.

The images shown in Plate 2 have been extended to include a portion of the North polar coronal hole and the

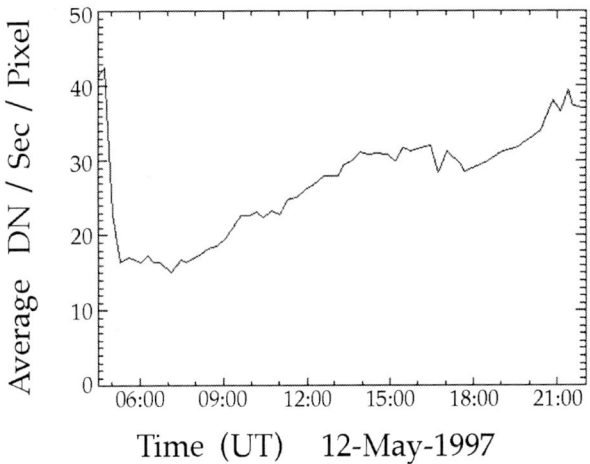

Figure 1. Temporal evolution of the Southern dimming region shown in Plate 2 (the Northern dimming region is not included). Plotted is the amount of data numbers (DN) per EIT pixel per second exhibited by the dimming region, over a period of more than 16 hours on 12 May, 1997.

solar limb; this allows a direct comparison of the "transient" coronal hole dimmings with a more static coronal hole. Photometrically, the emission recorded by EIT in all three of the coronal wavelengths (171, 195, and 284 Å), for at least a portion of the dimming region, has dropped to the same level as observed in the North polar coronal hole. When examining the transient dimming regions, a standard question involves the fractional decrease in emission over the area. However, to perform this estimate, one must define a "background level," which is to be subtracted beforehand. The standard level of background emission which might be chosen is the region of weakest emission on the solar disk, i.e. in coronal holes. Since the transient dimming has dropped to the polar coronal hole levels, the background of fractional decrease in emission approaches *100%* in all three of EIT's coronal wavelengths. This is a strong indication that the decrease in emission is primarily due to a decrease in density, and that the regions may have the physics of coronal holes.

Further evidence of the "transient coronal hole" nature of these dimmings may be derived from solar wind observations. Reports of trailing magnetic field lines which may map back to the newly opened field lines on the disk have been reported by *Neugebauer et al* [1997]. *Gosling et al* [1987] have long advocated counterstreaming *in situ* populations as a reliable indicator of CME field lines in the solar wind. However, there have yet been no explicit observations of high-speed solar wind originating at the transient dimming regions.

Figure 1 shows the evolution of the total emission in the transient dimming regions shown in the second panel of Plate 2, for EIT 195 Å Fe XII. The emission drops steadily for half an hour (beginning at 04:50 UT 12 May 1997), remains at this low level for a few hours, and then gradually returns to its original level. Transient brightenings and the expansion of the bright loops close to the CME neutral line may have influenced the measurement, *Thompson et al* [1998a] report that a comparison of the images appears to indicate that the dimming reached the minimum level of emission around 06:22. The images shown in Plate 2 are from the first complete four-wavelength set of observations after the eruption.

Sterling and Hudson [1997] demonstrated that the degree of soft X-ray dimming observed by *Yohkoh* SXT during a coronal mass ejection can only account for a fraction of the total mass of the white light CME. Furthermore, the assertion by *Thompson et al* [1998b] that the degree of EUV dimming does not scale proportionally with the mass of the observed white light CME is hardly surprising; both these results are indications that a significant fraction of the CME mass source can originate from regions which are not strongly emitting in EUV or soft X-rays, such as higher altitudes.

As an additional example of EUV dimmings in association with CME's, Plate 3 compares the EIT and *Yohkoh* SXT observations of an expanding loop on 12 February 1998. The *Yohkoh* SXT image, in the lower half of the figure, shows a loop which is pointing to the upper right. The loop continued to expand outward in subsequent images. The top figure shows the EIT 195 Å observation of this loop: this time visible as a dark loop feature. It is not clear whether the emission in EIT is decreased because the loop is too hot to be emitting in the EIT 195 Å bandpass, or whether another emission process such as absorption may be significant. Nonetheless, this EIT "dimming" observation cannot be interpreted in the same sense as the twin dimmings shown in Plate 2; the strong soft X-ray emission indicates that the dark regions in EIT are not due to a lack of material, but rather the presence of a feature which does not emit strongly in 195 Å.

Strong, rapidly decreasing (less than a few hours) regions of coronal dimming are regarded as a fairly definite indicator of the presence of a coronal mass ejection in EIT data. However, not all CME's (I am tempted to estimate that this number is less than *50%*) show unambiguous evidence of these structured dimming regions. In these cases, the presence or absence of other EUV CME signatures must also be considered. Furthermore, it is not clear whether the on-disk dimming regions correspond to the "dark cavity" in every CME observed in coronagraph data, or if they sometimes are the footpoints of the "bright front" of the CME [e.g. *Gopalswamy and Hanaoka*, 1998].

4.3 Expanding or Opening Features

Transient dimming regions, described in the previous section, are best observed on the solar disk. When viewed at the limb, the dimmings appear to correspond to dark opening regions often flanked by bright edges. These off-limb observations generally have a good correspondance with white light coronagraph data, as the CME cavity and the bright CME legs are visible in both.

The EIT data in Plate 4 correspond to the first CME in Plate 1, from 5 October 1996. The CME is an example of an eruption which occurred at the solar limb, and the 195 Å data in the left half of Plate 4 shows an EIT image at 13:04 UT, while the right half shows the same image with a previous image at 09:00 UT digitally subtracted from it. As in soft X-ray data, the clear presence of bright opening features and a dark evacuated region, occasionally accompanied by signs of outflowing material, provides strong evidence of a coronal mass ejection.

4.4 Filament Eruptions

All four wavelengths of EIT can detect the presence of a filament or prominence and any corresponding eruption. Usually, prominences are observed in the coronal images as a dark profile; the dark features are assumed to be due to cool dense material shadowing the bright emission behind it. In the cooler 304 Å He II wavelength (typically estimated as representing $5 - 8 \times 10^4$ K), the prominence can be seen either in emission or absorption. Plate 5 consists of *SOHO* EIT observations of a prominence, the left panel showing the 195 Å absorption features, and 304 Å He II in the right panel. Although the dark features in the 195 Å image are assumed to be due to absorption by cool, dense material, a careful comparison of the features *A-D* in both images indicates that they correspond to the absorption features also seen in the 304 Å image, indicating that the material imaged in 195 Å absorption is only a small fraction of the true prominence mass. Spectrometer observations of absorption features [e.g. *Kucera et al*, 1998] allow an analysis of the absorption profiles and comparisons.

The presence of the absorption features in the 195 Å image adds to the benefit of observing CME's with the coronal wavelengths. The dark features are observed cotemporal with bright evolving structures, allowing structural and timing comparisons. Frequently, an erupting prominence or filament can be recorded in 195 Å images, with the caveat the data may not reflect the true expanse of the prominence (as seen in Plate 5). Apart from coronagraph data there are few imagers which can "see" all the aspects of a CME, but for a few propitious EUV observations this is possible.

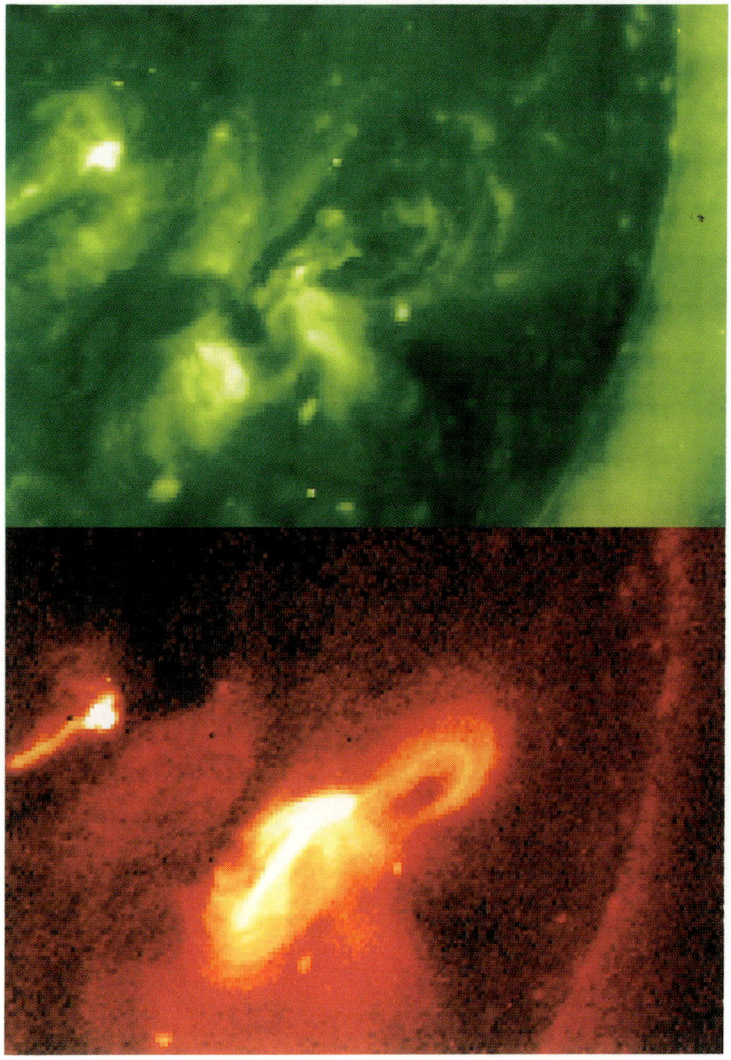

Plate 3. *SOHO* EIT 195 Å (above) and *Yohkoh* SXT (below) images of a CME in progress on 12 February, 1998. The bright soft X-ray loop in the SXT image corresponds to an expanding dark loop in the EUV image.

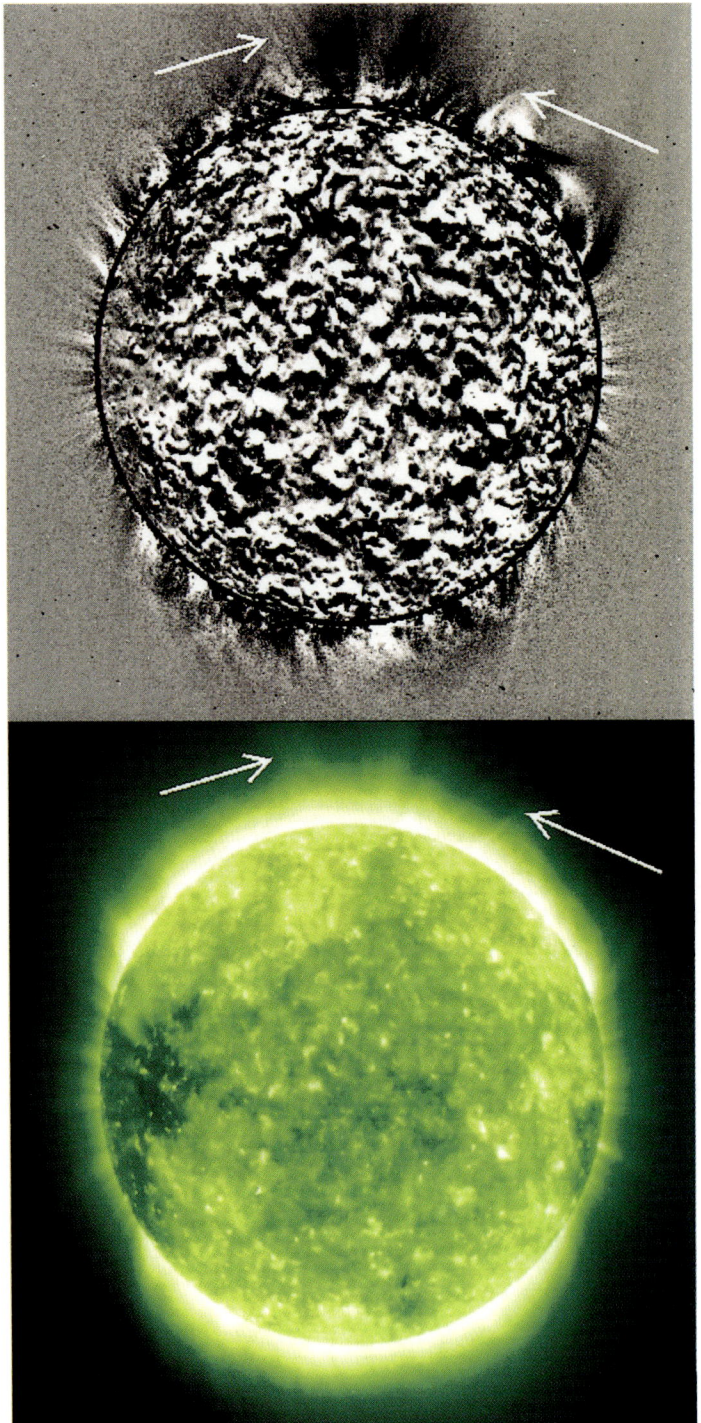

Plate 4. EIT 195 Å observations of the 5 October, 1996 CME (as shown in Plate 1). The left image was recorded at 13:03 UT, and the right image shows the difference between the first image and a previous image recorded at 09:00 UT (white represents an increase in emission, while black corresponds to a decrease). The arrows indicate opening features in the first image and an off-limb dimming in the second image.

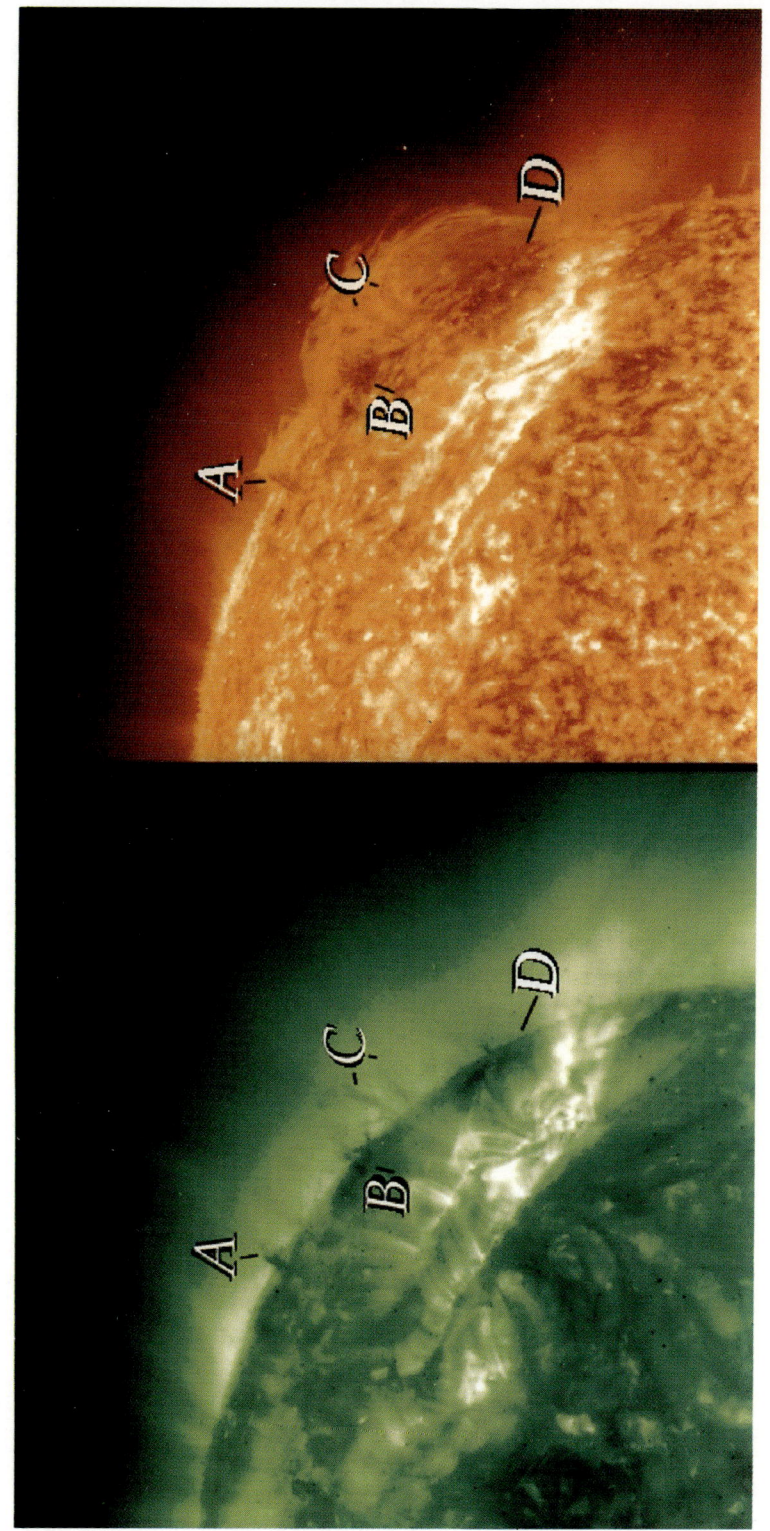

Plate 5. EIT EUV observations of a prominence in 195 Å (left) and 304 Å (right), recorded at 07:23 UT and 07:29 UT on 3 January, 1998. Dark features are labelled *A*, *B*, *C*, and *D* in both images.

4.5 "EIT Wave" Transients

When "EIT waves" were first observed, the assumption was that these phenomena were the coronal manifestation of Moreton waves [*Moreton*, 1961; *Smith and Harvey*, 1971], which were typically seen as moving wave fronts in H-alpha images. Among others, *Uchida* [1968] suggested that the Moreton wave phenomenon was not purely chromospheric in origin, but was instead the "ground track" of a three-dimensional wave front expanding in the corona. The EIT wave shown in Figure 2 shows a wave progression in a series of running-difference images (each successive image subtracted from the next image): in less than an hour, the wave has covered almost the entire visible solar disk, travelling at approximately 250 *km/sec*. The images in Figure 2 correspond to the arcade formation and dimmings shown in Plate 2; the running-difference technique is useful in showing transient changes of less than *25%*.

Over 100 of these wave transients have been recorded by EIT, and they are primarily interpreted as a signature of the impulse delivered to the ambient corona by the erupting field lines. Studies like that of *Sime et al* [1984] report that there is little evidence of lateral expansion of CME's, and that their expansion appears to be almost purely radial. However, this result is based on coronagraph observations, where the CME presumably has already undergone the destabilization process. In the low corona observed by EIT, there is evidence of lateral expansion during the very early stages of an eruption; this "kick" given to the corona is seen as a low-amplitude wave front propagating away from the erupting region.

EIT waves, using this interpretation, are thus not comprised of the CME material, and may have no magnetic connection to the CME. Most of the waves pass through the corona and do not appear to have any effect on the coronal features; at the limb, the wave passage can occasionally be resolved in the gentle deflection of features, like wheat blowing in a field.

There is some ambiguity as to whether these waves are truly the coronal manifestation of H-alpha Moreton waves, or whether a distinction has to be drawn between the sharp shock-like disturbance observed in the H-alpha and the more diffuse, low-amplitude wave fronts commonly observed by EIT, and exemplified in Figure 2. EIT has also observed extremely bright wave fronts in conjunction with the H-alpha observations, and it highly probable that the large density increase is indicative of a shock front, unlike the typical low-amplitude EIT waves. However, the bright fronts are more localized near the erupting regions, and low-amplitude wave fronts frequently appear after the bright ones have disappeared.

Sterling and Hudson [1997] report that *Yohkoh* SXT does not show evidence of these transients. However, for the observations thus far, EIT has had a better temporal cadence, and more importantly EIT has a very strong signal in the ambient corona, where the wave is propagating. Current estimates appear to indicate that EIT can observe a wave down to *5%* emission change, while SXT would require at least a *40%* perturbation and a 10-minute image cadence in the AlMg filter.

5. DISCUSSION: HALO CME'S AND GEOSPACE-IMPACTING ERUPTIONS

As general as this may sound, all CME studies are relevant to the studies of space weather and the potential geospace impact. The scattering geometry, upon which white light imaging relies, clearly favours eruptions at the solar limb, directed 90 degrees away from Earth (see Plate 1). These CME's are the best-resolved, while CME's directed along the Sun-Earth axis, termed "halo" CME's by *Howard et al* [1982], generally consist of a more diffuse structure encircling the occulting disk. While halo CME's are the most relevant eruptions for geospace studies, it is extremely difficult to determine their speed and volume, and aspects of their structure are difficult to identify because of the Thomson scattering angle, and because the viewing angle is along the axis of the CME, where many features overlap. However, *Paswaters et al* [1998] demonstrated that because the polarization brightness scattering profile varies differently from that of the total intensity, polarization brightness and white light images can be combined to derive some of this information. Unfortunately, this requires adequate coverage of the solar corona in both white light and polarization brightness, which are commonly unavailable.

Fortunately, halo CME's are ideal for disk imagers; *Yohkoh* SXT and *SOHO* EIT are able to observe the full extent of the CME in the low corona if the eruption occurs near disk center. Unfortunately, the clear preference of disk imagers for disk CME's hinders their comparison with coronagraph data, which is optimally viewed at the limb. Nonetheless, disk imaging data can provide a great deal of information about an Earth-directed CME. One major reason for the necessity of disk imaging data is that the Thomson scattering geometry of the corona is the same for Earthward and anti-Earthward propagating CME's: without disk observations, there is no distinction between the two directions.

Howard et al [1985] estimated that *2%* of all *Solwind* (1979-1985) CME's could be described as halo. For a CME to have a component along the solar meridian (and therefore along the Sun-Earth axis), the CME must originate at a longitude which is, on average, less half the CME's angular width away. Because halo CME's include both Earth-directed and anti-Earthward CME's, the implies

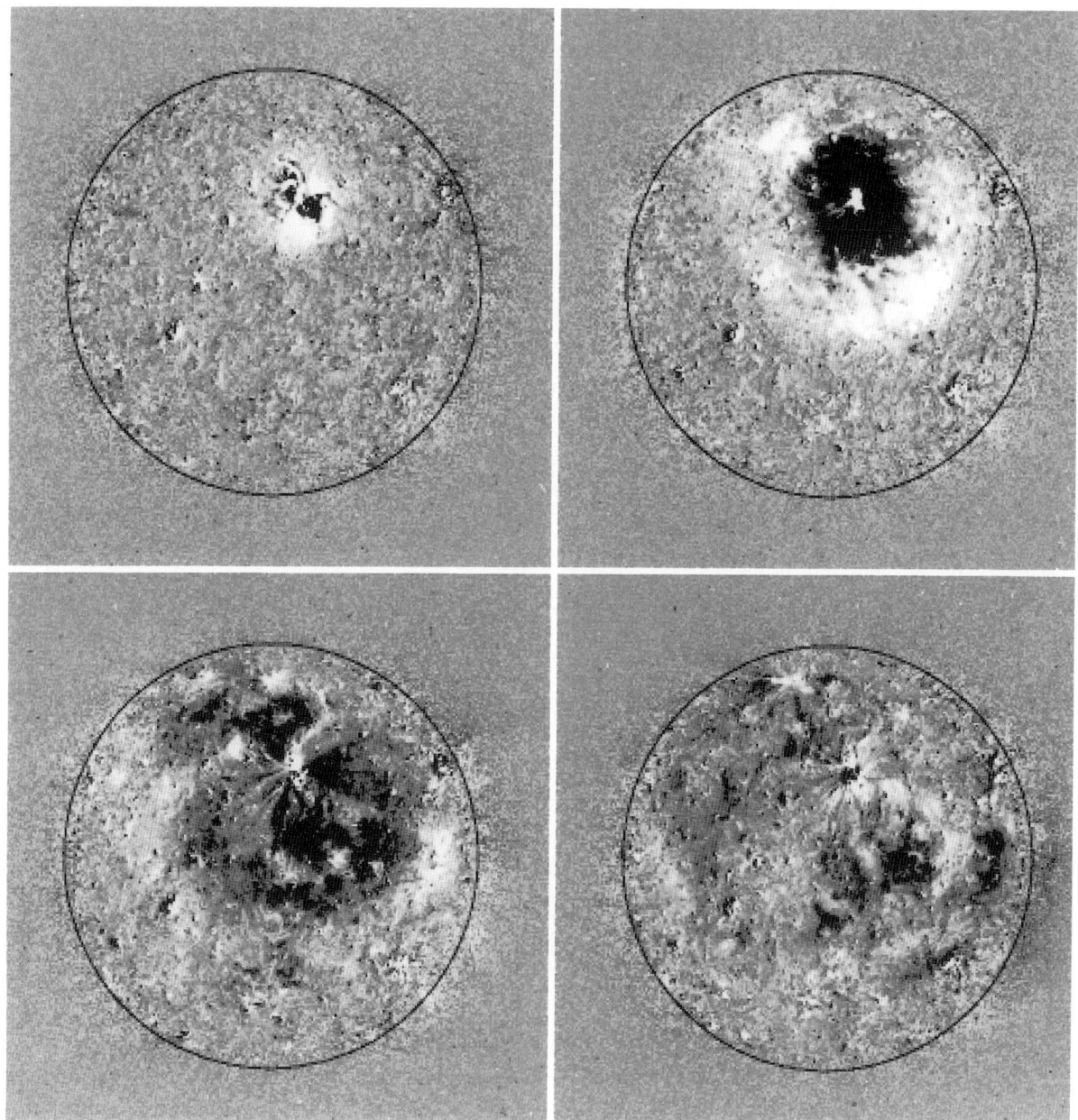

Figure 2. EIT 195 Å running-difference images of the 12 May, 1997 coronal "EIT wave," at 04:50, 05:07, 05:24 and 05:41 UT. Each image has had the previous image subtracted from it: bright regions represent an increase in emission since the last image, dark regions represent a decrease. The solar limb has been circled to enhance its visibility.

that the average longitudinal expanse of *Solwind* CME's is $360*.02*2/2$ (one "two" is for the half-angle, the other is for the two possible directions) = 7 degrees. Of course, the idea of a CME having such a thin longitudinal width is absurd; clearly the *2%* estimate is low because of the low scattering amplitude of halo CME's and the inability of the instrument to detect CME's below its sensitivity. *St. Cyr et al* [1998b] report that in 1997 *16%* of the CME's observed by *SOHO* LASCO, with its improved dynamic range, satisfied a "halo" criterion of having greater than 120 degrees projection around the solar disk. The average latitudinal width of the non-halo CME's was 52 degrees;

the longitudinal average (based on the same estimation which yielded *7%* for *Solwind* CME's) was 58 degrees. *Fisher and Munro* [1984] obtained a similar result using the relative brightness of the cavity and bright loop features of CME's: CME's, on a statistical average, are essentially cone-shaped, with comparable widths in both longitude and latitude. *Webb* [1988] confirmed this result by comparing the measured latitudinal widths of CME's with different neutral line orientations; he found little evidence of width variation with angle.

Another reason why limb studies of CME's are relevant to space weather research is because the propagation of the CME is more clearly resolved for limb events. *MacQueen et al* [1986] demonstrated that for 19 CME's observed by the *Solar Maximum Mission*, there was little or no evidence of non-radial propagation; an eruption propagated radially throughout the field of view. The *Skylab* (1973-1974) CME's exhibited only a minor equatorward deflection, which they attributed to the more dipolar structure of the corona at the time of the *Skylab* observations: the non-radial propagation was skewed toward the ecliptic plane, presumably because the large-scale coronal structure exerts pressure equatorward. *Plunkett et al* [1998] amended this result by not only examining the LASCO white light coronagraph data, but also the LASCO C1 Fabry-Perot coronagraph data and EIT EUV imaging data. The study of *Plunkett et al* yielded evidence for strongly non-radial propagation below 3 solar radii, and concluded that a coronal mass ejection's latitude of origin may not be the best indicator for its eventual direction of propagation. This interesting result appears to imply that a CME originating near the solar meridian, at an Earth-directed longitude, is more likely to intersect geospace because of this channeling effect; it is therefore the longitude, and not the latitude of the CME's origin which may be the best indicator of whether a CME may have space weather implications.

6. CONCLUSIONS

Unlike previous missions, there is evidence that the "CME Watch" campaign of LASCO and EIT captures nearly every coronal mass ejection. *16%* of all LASCO CME's have been identified as "halo" events, and the implication that the inferred average longitudinal width is even greater than the average latitudinal width indicates that LASCO is detecting the majority (if not all) of these faint halo eruptions. If LASCO were unable to detect some of these events, the inferred longitudinal width would be less than the range in latitudes, (as was the case with the *Solwind* halos), in conflict with the results of *Fisher and Munro* [1984] and *Webb* [1988]. The possibility that LASCO provides unambiguous coverage of coronal dynamics removes several important variables from the Sun-Earth connected system.

In general, one assumes that disk coverage is required to determine whether a CME may have geospace impact, as there is no distinction in coronagraph data between a CME propagating towards or away from Earth. Because some CME's have little or possibly no prominence material, eruptions can occur without producing a reliable signature in H-alpha images. Therefore, it is assumed that a coronal disk imager plays an integral role in a continuous CME campaign. Nearly every Earth-facing CME observed by LASCO appears to exhibit at least one of the EUV signatures in EIT; this is a somewhat circuitous assertion, because a CME which has no clear observability in EIT is often assumed to have occurred on the other side of the Sun. However, this argument is usually supported by the fact that the white light CME's without EUV signatures are typically lacking any other signature of an on-disk CME, such as a filament disappearance or an increase in soft or hard X-ray flux.

EIT does not serve solely as an affirmation of LASCO observations: major strides have been made towards the identification of coronal mass ejections in EUV data, and the combination of these observations with LASCO coronagraph data to produce a greater understanding of their relationship and correspondence. Most CME's only exhibit one or two of the signatures, and further studies will focus on the distinction between eruptions which exhibit different signatures. This work may result in a greater understanding of the similarities and distinctions between different CME's, and possibly a more definitive method of determining the properties of ejections which are most likely to influence geospace.

Acknowledgements. BJT would like to thank Bill Lewis for his assistance in preparing the paper, and Dominic Zarro and Joan Burkepile for insightful discussion.

REFERENCES

Bieber, J.W. and Rust, D.M. The escape of magnetic flux from the sun, *Ap. J.*, *453*, 911, 1995.

Brueckner, G.E., R.A. Howard, M.J. Koomen, C.M. Korendyke, D.J. Michels, J.D. Moses, D.G. Socker, K.P. Dere, P.L. Lamy, A. Llebaria, M.V. Bout, R. Schwenn, G.M. Simnett, D.K. Bedford, C.J. Eyles, The Large Angle Spectroscopic Coronagraph (LASCO), *Solar Phys.*, *163*, 357, 1995.

Burkepile, J.T. and O.C. St. Cyr, *A Revised and Expanded Catalogue of Mass Ejections Observed by the Solar Maximum Mission Coronagraph,* NCAR/TN-369+STR, NCAR, Boulder, 1993.

Delaboudinière, J.-P., and 27 other authors, EIT: Extreme-Ultraviolet Imaging Telescope for the *SOHO* Mission, *Solar Phys.*, *162*, 291, 1995.

Dryer, M., Comments on the origins of coronal mass ejections, *Solar Phys.*, 169, 421, 1996.

Fisher, R.R. and A.I. Poland, Coronal activity below 2 Rsun: 1980 February 15-17, *Ap. J.*, *246*, 1004, 1981.

Fisher, R.R. and R.H. Munro, Coronal transient geometry. I. The flare-associated event of 1981 March 25, *Ap. J., 280,* 428, 1984.

Gopalswamy, N. and Y. Hanaoka, Coronal dimming associated with a giant prominence eruption, *Ap. J., 498,* L179, 1998.

Gosling, J.T., E. Hildner, R.M. MacQueen, R.H. Munro, A.I. Poland, and C.L. Ross, The speeds of coronal mass ejection events, *Solar Phys., 48,* 389, 1976.

Gosling, J.T., D.N. Baker, S.J. Bame, W.C. Feldman, R.D. Zwickl and E.J. Smith, Bidirectional solar wind electron heat flux events, *J. Geophys. Res., 92,* 8519, 1987.

Gosling, J.T., The solar flare myth, *J. Geophys. Res.*, 98, 18937, 1993.

Harrison, R.A., Solar coronal mass ejections and flares, *Astron. Astrophys., 162,* 284, 1986.

Harrison, R.A., Coronal transients and their relation to solar flares, *Adv. Space Res.,* 11, 25, 1991.

Howard, R.A., D.J. Michels, N.R. Sheeley, Jr., and M.J. Koomen, The observation of a coronal transient directed at earth, *Ap. J., 264,* L101, 1982.

Howard, R.A., N.R. Sheeley, Jr., M.J. Koomen, and D.J. Michels, Coronal mass ejections, 1979-1981, *J. Geophys. Res., 90,* 8173, 1985.

Howard, R.A., G.E. Brueckner, O.C. St. Cyr, and 16 other authors, Observations of CMEs from SOHO/LASCO, in "Coronal Mass Ejections", edited by N.U. Crooker, J. Joselyn and J. Feynman, American Geophysical Union Monograph, Vol. 99, 17, 1997.

Hudson, H.S., L.W. Acton and S.L. Freeland, A long-duration solar flare with mass ejection and global consequences, *Ap. J., 470,* 629, 1996.

Hudson, H.S. and D. Webb, Soft X-ray signatures of coronal mass ejections, in "Coronal Mass Ejections", edited by N.U. Crooker, J. Joselyn and J. Feynman, American Geophysical Union Monograph, Vol. 99, 27, 1997.

Hundhausen, A.J., "Coronal Expansion and Solar Wind," Springer-Verlag, 1972.

Hundhausen, A.J., C.B. Sawyer, L. House, R.M.E. Illing, and W.J. Wagner, Coronal mass ejections observed during the Solar Maximum Mission: Latitude distribution and rate of occurrence, *J. Geophys. Res., 89,* 2639, 1984.

Hundhausen, A.J., J.T. Burkepile and O.C. St. Cyr, The speeds of coronal mass ejections: SMM observations from 1980 and 1984-1989, *J. Geophys. Res., 99,* 6543, 1994.

Hundhausen, A.J., The sizes and location of coronal mass ejections: SMM observations from 1980 and 1984-1989, *J. Geophys. Res,* 98, 13177, 1993.

Hundhausen, A.J., An Introduction, in "Coronal Mass Ejections", edited by N.U. Crooker, J. Joselyn and J. Feynman, American Geophysical Union Monograph, Vol. 99, 1, 1997.

Hundhausen, A.J., Coronal Mass Ejections: A Summary of SMM Observations from 1980 and 1984-1989, to appear in "The Many Faces of the Sun; Scientific Highlights of the Solar Maximum Mission," edited by K.T. Strong, J.L.R. Saba, and B.M. Haisch, Springer-Verlag, 1998.

Kahler, S., The morphological and statistical properties of solar X-ray events with long decay times, *Ap. J.,* 214, 891, 1977.

Kahler, S.W., The role of the big flare syndrome in correlations of solar energetic proton fluxes and associated microwave burst parameters, *J. Geophys. Res., 87,* 3439, 1982.

Kahler, S., Solar flares and coronal mass ejections, *Ann. Rev. Astron. Astrophys., 30,* 113, 1992.

Kucera, T. A., V. Andretta, A.I. Poland, Neutral hydrogen column depths in prominences using EUV absorption features, *Solar Phys.,* in press, 1998.

Low, B.C., Solar Activity and the Corona, *Solar Phys., 167,* 217, 1996.

Luhmann, J.G., CMEs and Space Weather, in "Coronal Mass Ejections", edited by N.U. Crooker, J. Joselyn and J. Feynman, American Geophysical Union Monograph, Vol. 99, *291,* 1997.

MacQueen, R.M., A. J. Hundhausen and C.W. Conover, The propagation of coronal mass ejection transients, *J. Geophys. Res., 91,* 31, 1986.

MacQueen, R.M., The three-dimensional structure of "loop-like" coronal mass ejections, *Solar Phys., 145,* 169, 1993.

McComas, D. J., Gosling, J.T., and Phillips, J.L., Interplanetary magnetic flux – Measurement and balance, *J. Geophys. Res., 97,* 171, 1992.

Moreton, G. F., Fast-moving disturbances on the Sun, *Sky and Telescope, 21,* 145, 1961.

Neugebauer, M., R. Goldstein and B.E. Goldstein, Features observed in the trailing regions of interplanetary clouds and coronal mass ejections, *J. Geophys. Res., 102,* 19743, 1997.

Paswaters, S.E., O.C. St. Cyr, R.A. Howard, G.E. Brueckner, D. Wang, Analysis of halo coronal mass ejections observed by *SOHO*-LASCO, abstract at the Spring 1998 Meeting of the American Geophysical Union, 1998.

Plunkett. S.P., D.J. Michels, R.A. Howard, R. Schwenn, G.M. Simnett, P.L. Lamy, On the non-radial propagation of coronal mass ejections, abstract at the 1998 COSPAR Science Assembly, 1998.

Rust, D.M. and D.F. Webb, Soft X-ray observations of large-scale coronal active region brightenings, *Solar Phys., 54,* 403, 1977.

Rust, D.M., Coronal disturbances and their terrestrial effects, *Space. Sci. Rev., 34,* 21, 1983.

Sheeley, N.R., Jr., J.D. Bohlin, G.E. Brueckner, J.D. Purcell, V.E. Scherrer, R. Tousey, J.B. Smith, D.M. Speich, E. Tandberg-Hanssen, R.M. Wilson, A.C. De Loach, R.B. Hoover and J.P. McGuire, Coronal changes associated with a disappearing filament, *Solar Phys. 45,* 377, 1975.

Sime, D.G., R.M. MacQueen and A.J. Hundhausen, Density distribution in loop-like coronal transients: A comparison of observations and a theoretical model, *J. Geophys. Res., 89,* 2113, 1984.

Smith, S.F. and K.L. Harvey, Observational effects of flare-associated waves, *Physics of the Solar Corona,* edited by C.J. Macris, D. Reidel Pub. Co, 156, 1971.

St. Cyr, O.C. and J.T. Burkepile, *A catalogue of mass ejections observed by the Solar Maximum Mission coronagraph,*

NCAR Technical Note NCAR/TN-352+STR, Boulder, 1990.

St. Cyr, O.C., *SOHO LASCO CME Catalog*, 1998a.

St. Cyr, O.C., *SOHO* observations of halo coronal mass ejections, abstract at the Spring 1998 Meeting of the American Geophysical Union, 1998b.

Steinolfson, R.S., Theories of shock formation in the solar atmosphere, in "Collisionless Shocks in the Heliosphere: Reviews of Current Research, " Geophys. Res. Monogr. Series Vol. 35, edited by B.T. Tsurutani and R.G. Stone, American Geophysical Union, p. 1, 1985.

Sterling, A.C. and H.S. Hudson, *Yohkoh* SXT observations of X-ray "dimming" associated with a halo coronal mass ejection, *Ap. J., 491*, L55, 1997.

Thompson, B.J., S.P. Plunkett, J.B. Gurman, J.S. Newmark, O.C. St. Cyr, D.J. Michels, *SOHO*/EIT observations of an Earth-directed coronal mass ejection on May 12, 1997, *Geophys. Res. Lett., 25*, 2465, 1998a.

Thompson, B.J., J.B. Gurman, J.S. Newmark, O.C. St. Cyr, H.S. Hudson, N. Gopalswamy, C.E. DeForest, M.A. Lyons, D.J. Michels, R.A. Howard, Dimming of the EUV corona associated with coronal mass ejections, abstract at the Spring 1998 Meeting of the American Geophysical Union, 1998b.

Tsuneta, S., L. Acton, M. Bruner, J. Lemen, W. Brown, R. Caravalho, R. Catura, S. Freeland, B. Jurcevich, J. Owens, The Soft X-ray Telescope for the Solar-A Mission, *Solar Phys., 136,* 37, 1991.

Uchida, Y., Fast-mode wavefronts and Moreton's wave phenomena, *Solar Phys., 4,* 30, 1968.

Webb, D.F., A.S. Krieger, and D.M. Rust, Coronal X-ray enhancements associated with H-alpha filament disappearances, *Solar Phys., 48*, 159, 1976.

Webb, D.F., Erupting prominences and the geometry of coronal mass ejections, *J. Geophys. Res., 93*, 1749, 1988.

Webb, D.F. and R.A. Howard, The solar cycle variations of the occurrence rate of coronal mass ejections and the solar wind mass flux, *J. Geophys. Res., 99*, 4201, 1994.

Yokoyama, T. and K. Shibata, Two-dimensional magnetohydrodynamic simulation of chromospheric evaporation in a solar flare based on a magnetic reconnection model, *Ap. J., 494*, L113, 1998.

Zirker, J., "Coronal Holes and High-Speed Wind Streams, " Colorado Ass. Univ. Press, Boulder, 1977.

B. J. Thompson, O. C. St. Cyr, S. P. Plunkett, J. B. Gurman, N. Gopalswamy , Code 682.3, NASA Goddard Space Flight Center, Greenbelt, MD 20771

H. S. Hudson, ISAS c/o Ogawara, 3-1-1 Yoshinodai Sagamihara-shi, Kanagawa 299, Japan

R. A. Howard and D. J. Michels, Code 7660, E. O. Hulburt Center for Space Research, Naval Research Laboratory, Washington, DC 20375

J.-P. Delaboudière, Inst. d'Astrophysique Spatiale, Université Paris XI, 91405 Orsay Cedex, France

GEOTAIL Substorm/Storm Studies

Rumi Nakamura[1]

Solar-Terrestrial Environment Laboratory, Nagoya University, Toyokawashi, Japan

The GEOTAIL spacecraft has surveyed a wide region in the tail between $X \sim -10\ R_E$ and $X \sim -220\ R_E$. The data collected allow one to study the evolution of the tail as related to substorms and storms both on a statistical basis as well as by means of detailed event analysis. During its distant tail mission GEOTAIL observed highly compressed field and enhanced occurrence of the magnetopause crossings associated with storms. Data from these disturbed condition have been used to estimate to what extent the distant tail is distorted from quiet condition due to enhanced solar-wind magnetosphere coupling processes. GEOTAIL monitored substorm-associated flow and field disturbances related to the near-Earth reconnection processes as well as plasmoid/flux rope. Particularly, the initial location and the evolution of the substorm disturbance in the midtail region have been extensively studied to evaluate different substorm models. Coordinated analysis with other ISTP satellites further enabled spatial and temporal evolution of the tail during substorms associated with different IMF conditions. This paper highlights some of the results dealing with tail dynamics during storms/substorms.

1. INTRODUCTION

Substorms and storms play a major role in the energy transfer process from the solar wind to the magnetosphere/ionosphere. GEOTAIL measurements, obtained from an extended region in the tail, provide an ideal dataset to study global tail dynamics. Three topics related to storms/substorms are discussed in the following sections by summarizing the relevant GEOTAIL observations: (1) Storm associated changes in the distant tail, (2) Timing and location of the near-Earth reconnection region, and (3) Response of the midtail to substorms and IMF disturbances.

2. STORM-ASSOCIATED CHANGES IN THE DISTANT TAIL

The average size and shape of the magnetosphere is in general controlled by the IMF orientation and the solar wind. During storm times solar wind and IMF conditions significantly change so that enhancement in the dayside reconnection rate and strong compression occur. Accordingly, the tail configuration is expected to be distorted from its average structure. Enhanced flux transferred from the dayside as well as the effect of the change in the solar wind pressure and solar wind flow direction can alter the orientation of the tail, tail cross-section, and energy content in the tail.

[1]now at: Max-Planck-Institut für extraterrestrische Physik, Garching, Germany

2.1. Magnetopause Crossings and the Structure of the Tail

During storm conditions, the occurrence of magnetopause crossings increases such that the magnetosheath is observed even in places close to the average location of the tail axis [e.g. *Nakamura et al.*, 1996]. The tail orientation is expected to fluctuate with respect to its nominal orientation. Thirteen storms occurred when GEOTAIL was in the distant tail region at X = –83 ~ –210 R_E and Y = 20 ~ –50 R_E between October 1993 and October 1994 [*Kokubun et al.*, 1996]. During all these storms, GEOTAIL observed both magnetosheath and magnetotail plasmas. Changes in the average magnetosheath parameters (density, speed, and direction) observed during these storms are found to be consistent with solar wind fluctuations expected from corotating interaction regions [*Nakamura et al.*, 1997].

Systematic signature in magnetosheath crossings was found by ISEE 3 observations in association with substorms [*Baker et al.*, 1984], i.e. magnetopause crossings from the magnetosheath(lobe) to the lobe(magnetosheath) during substorm growth(expansion) phase. The distant tail structure further changes in response to plasmoid encounters after near-Earth substorm onsets [*Baker et al.*, 1984]. These observations showed that the distant magnetotail changes diametrically in size also in a substorm time-scale. GEOTAIL observations also indicated that substorms cause variations of the boundary of the magnetotail predominantly at timescales of a few tens of minutes, while the flapping of the tail causes variations at timescales of hours [*Shodhan et al.*, 1996; *Kokubun et al.*, 1996]. Magnetopause crossings during storm times, therefore, occur as a consequence of the solar wind aberration or compression effects related to the large-scale solar wind structure causing the storm and also as a consequence of additional processes, such as substorms [*Williams et al.*, 1994] and local pressure enhancement in the solar wind [*Nakamura et al.*, 1996].

Using the magnetosheath crossing events, structures of the distant tail during geomagnetic storms can be obtained. Figures 1b and 1e show the location of the magnetopause crossings during the main phase and before the storm (quiet time) using GEOTAIL data for the thirteen storm events during its distant tail mission. The magnetopause locations are plotted in a solar wind aberrated coordinate system. Here, the flow direction in the magnetosheath is assumed to be parallel to the magnetotail axis. The distance from the tail axis is normalized by the average pressure indicated at the top of the plot. As a reference to the IMF field direction, the B_Y-B_Z direction of the magnetosheath field is also shown in Figures 1c and 1f. The average angle, $<\alpha>$, represents the average of the angle between $|B_Y|$ and $|B_Z|$, which is $\alpha = \arctan(|B_Z|/|B_Y|)$. During storm time, when the B_Z component of the magnetosheath field is larger compared with the B_Y component, the average tail cross section has a north-south elongated elliptical shape. The average tail for quiet time, on the other hand, is elongated toward the dawn-dusk direction. The B_Y component is larger than the B_Z component in the quiet time dataset. A flattened shape is expected from the significant contribution from B_Y, which has been reported in the ISEE 3 measurements [*Sibeck et al.*, 1986]. These observations imply that the anisotropy in the magnetosheath magnetic field pressure for different IMF orientation changes the shape of the tail so that north-south aligned tail was obtained during the main phase, when strong IMF southward is present. Such changes in the shape of tail was also obtained in global MHD models by *Ogino et al.*[1994] and *Nishida and Ogino* [1998] as shown in Figures 1a and 1d.

In addition to the storm-related magnetopause crossings, *Fairfield* [1993] has shown magnetosheath entry cases during periods of northward IMF and low Kp. By comparing global MHD model calculation and GEOTAIL observations *Frank et al.* [1995] have suggested that magnetosheath encounters were caused by a series of IMF rotations from northward to duskward. The difference in the distant tail structure obtained by GEOTAIL during northward and southward IMF is extensively discussed by *Nishida and Ogino* [1998].

2.2. Large Tail Lobe Magnetic Field Event and Energy Stored in the Tail

GEOTAIL observed a strong lobe field up to 53 nT in the distant tail during a period of time when the ring current was intensifying [*Kokubun et al.*, 1996; *Kokubun*, 1997]. Figure 2 shows Dst and the estimated parameters in the solar wind and magnetosphere during the November 3, 1993 storm using the average magnetosheath parameters associated with the magnetopause crossings. Pressure in the third panel is plotted in magnetic field units and represents the lobe field strength obtained by assuming pressure balance. The dimension of the tail shown in the fourth panel is obtained from the distance between GEOTAIL and the estimated tail axis direction using the instantaneous aberration. Before the main phase onset of the storms, when the Dst shows positive excursion, the enhanced solar wind pressure significantly reduces the radius of the distant tail compared with its quiet time value. During the storm main phase, the dimension of the tail is comparable to the quiet time value in spite of the high solar wind pressure. This is attributed to the enhancement of magnetic flux in the tail in association with southward IMF B_Z. An average energy of ~5 x 10^{15} J was calculated to be stored in the distant tail

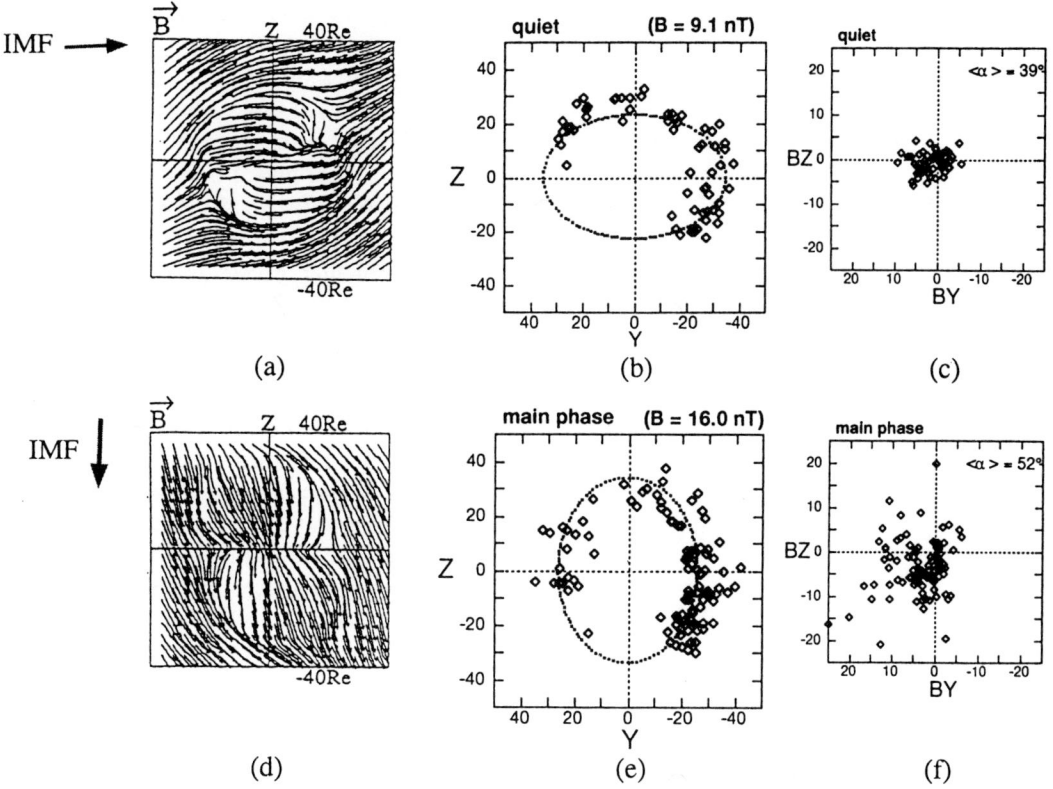

Figure 1. Magnetic field in the distant tail Y-Z plane obtained by *Ogino et al.* [1994] using an MHD model for (a) $B_Y > 0$ and (d) $B_Z < 0$ (Courtesy of T. Ogino). Location of the magnetopause crossings of GEOTAIL satellite in the aberrated Y-Z plane (b) before storms and (e) during the main phase of storms. Magnetosheath field in the aberrated Y-Z plane (c) before storms and (f) during the main phase of storms (from *Nakamura et al.*, 1997).

during the early main phase of storms, which is a comparable value to that stored in the midtail during an intense substorm growth phase [*Baker et al.*, 1981]. *Knipp et al.* [1998] showed that the ionospheric and ring current dissipation is ~4 x 10^{16} J smaller than the input energy from the dayside merging during the main phase. A part of the excess energy could be stored in the tail particularly during the early main phase of the storm. Although the dynamic pressure is high during this period and therefore the compression effect is important, the effect of magnetic fluxes that are transported to the magnetotail by the dayside reconnection prevent the tail radius from decreasing.

3. MAGNETIC RECONNECTION IN THE NEAR-EARTH MAGNETOTAIL

Although it is widely believed that magnetic reconnection takes place in the near-Earth magnetotail associated with substorm, the start timing of the reconnection commencement relative to a substorm onset on the ground is still a controversial problem [*Baker et al.*, 1996]. Location of the reconnection region can be obtained by examining relationships between B_Z and V_X polarity. Statistical studies using data from AMPTE/IRM satellite indicate that the near-Earth reconnection region is located tailward of 20 R_E [e.g. *Baumjohann et al.*, 1991] and those using data from IMP 6, 8 satellites obtained locations to be Earthward of 30 R_E [e.g. *Nakamura et al.*, 1994] during the substorm expansion phase. Since GEOTAIL has made an extensive survey of the plasma sheet at radial distances of 20 ~ 30 R_E, it is an ideal dataset to reexamine the location of flow reversals during substorms.

3.1. Fast Flows During the Expansion Phase

The direction of the fast flows in the plasma sheet during the substorm expansion phase was examined for 138 well-defined substorms when GEOTAIL was located in the magnetotail inside 50 R_E [*Nagai and Machida*, 1998]. Mid-latitude positive bays were used to identify the expansion

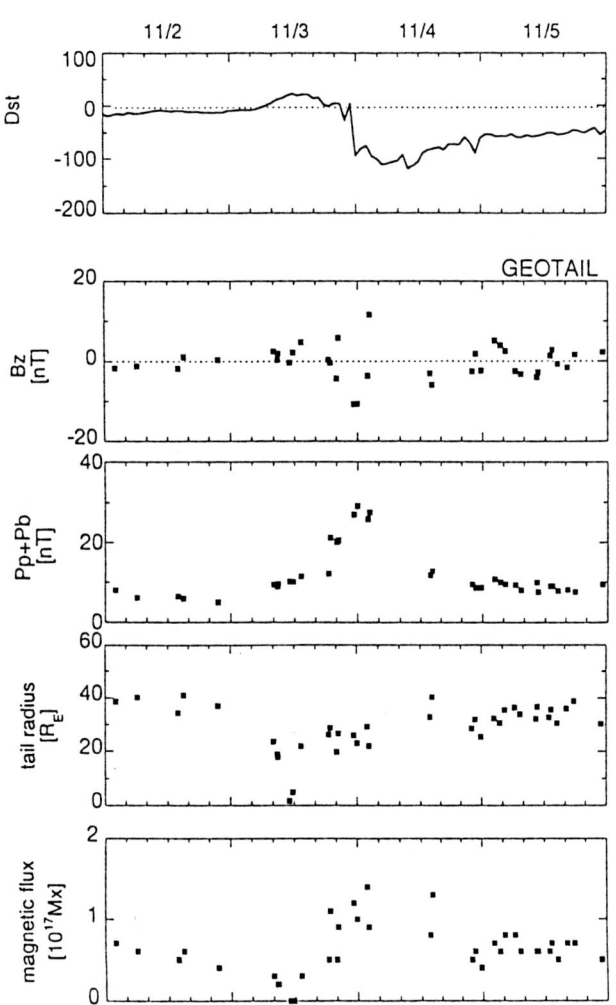

Figure 2. *Dst*, B_Z, total pressure in the magnetosheath as seen by GEOTAIL and estimated magnetospheric parameters during the November 3, 1993 storm using the average magnetosheath parameters for the magnetopause crossings. The third panel is plotted in magnetic field units and therefore represents the lobe field strength obtained by assuming pressure balance. The dimension of the tail shown in the fourth panel is obtained from the distance between GEOTAIL and the estimated tail axis direction using the instantaneous aberration. The bottom panel shows the estimated total magnetic flux from the magnetic field and dimension of the tail (from *Nakamura et al., 1997*).

phase, which was defined by the time interval between the initiation of a major positive change of the mid-latitude *H*-component and the time of the positive bay peak. Flow and magnetic field signatures observed by GEOTAIL during the expansion phase were divided into 6 categories: (a) tailward convection flows, (b) Earthward convection flows, (c) tailward non-convection flows, (d) Earthward non-convection flows, (e) stationary plasma sheet, and (f) tail lobe. Earthward (tailward) convection flow was defined as Earthward (tailward) flows when flow perpendicular to the ambient field is faster than 300 km/s and has positive (negative) B_Z. Earthward (tailward) non-convection flow corresponds to Earthward (tailward) flow with speed higher than 300 km/s and with positive (negative) B_Z and with flow perpendicular to the field slower than 300 km/s. Stationary plasma sheet events correspond to cases with no fast flows in the plasma sheet, although some of these events appeared when the results of moment calculation could be effected by energetic ions contributions (with energy exceeding the threshold of the LEP instrument). Lobe events correspond to cases when GEOTAIL remained in the tail lobe for at least the first 10-min interval of the expansion phase. Figure 3 shows the locations of these 6 types of GEOTAIL observations during the expansion phase. Tailward convection flows are observed mostly beyond 22 R_E, whereas Earthward flows are observed only inside 30 R_E. Thus, it appears that changes in the flow direction from Earthward to tailward occur somewhere between $X_{GSM} = -22\ R_E$ and $-30\ R_E$, which is consistent with the studies cited above. It should be noted that significant structure is present with flows also in the Y direction. The convection flows do not exist in the duskside and dawnside flanks of the plasma sheet throughout the expansion phase. On the other hand, fast convection flows are almost always observed during the expansion phase in the premidnight plasma sheet.

3.2. Fast Flows Close to the Time of Pi2 Onset

Nagai et al. [1998] studied flow characteristics during the period from −10 to +10 min relative to onset when GEOTAIL was located in the tail Earthward of $X_{GSM} = -50\ R_E$. Onset times of 321 substorm were identified based on Pi2 pulsation at Kakioka for 1000 ~ 1800 UT. Relative timing of the onset of the Earthward flows ($V_X \geq +300$ km/s) and tailward flows ($V_X \leq -300$ km/s) to the Pi2 onset was examined. It was found that in the premidnight region Earthward of 30 R_E, most of the tailward flows were observed to start simultaneous or earlier than the Pi2 onset. In the region between $X = -30\ R_E$ and $X = -50\ R_E$, the tailward flow started a few minutes after Pi2 onsets. Earthward flow, on the other hand, was observed only inside 30 R_E and some flows started in the premidnight sector before Pi2 onset. It was concluded that magnetic reconnection starts in the limited extent of the plasma sheet between $X_{GSM} = -22\ R_E$ and $X_{GSM} = -30\ R_E$ in the premidnight region and prior to an onset signature identified by Pi2 pulsation on the ground. These flow and field characteristics in

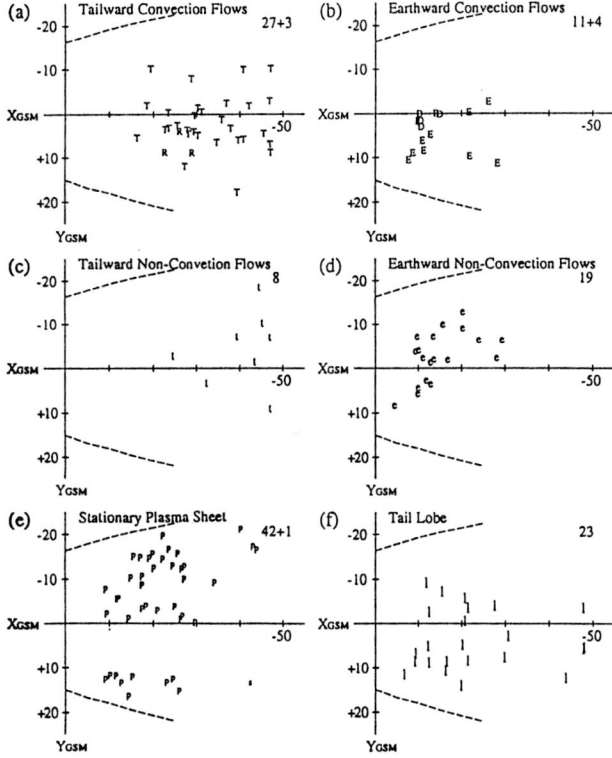

Figure 3. Positions of the GEOTAIL spacecraft in the aberrated GSM X-Y plane during the expansion phase of 138 substorms plotted separately for 6 types of observations: (a) 27 tailward convection flows, T, and 3 reversal events, R, (b) 11 earthward convection flows, E, and 4 dipolarization events, D, (c) 8 tailward non-convection flows, t, (d) 19 Earthward non-convection flows, e, (e) 42 stationary plasma sheet events, p, and one tailward convection with northward B_Z, s, and (f) 23 tail lobe events (from *Nagai and Machida*, 1998).

the midtail region are expected in the near-Earth neutral line model (see *Baker et al.* [1996]). The observed time delays of the flows relative to the onset were interpreted as the travel time from the reconnection site to the observed location [*Nagai et al.*, 1998] and/or longitudinal expansion of the magnetic reconnection site [*Nagai and Machida*, 1998].

3.3. Evolution of the Near-Earth Reconnection Region

Figure 4 summarizes the GEOTAIL observations associated with Pi2-onsets [*Nagai et al.*, 1998] and associated with the expansion phase deduced from positive bay activity [*Nagai and Machida*, 1998] as described in the previous sections. The occurrence rate of the plasma sheet events with tailward and Earthward fast flows, those without fast flows, and lobe events are shown. The occurrence rate of fast flows are significantly larger for the positive bay analysis (Figure 4b) versus the Pi2 onset analysis (Figure 4a). Further, tailward flow is observed more frequently than Earthward flow in association with Pi2 onsets (Figure 4a), but the occurrence rate of both flows are comparable during the positive bay enhancement (Figure 4b). Since the event selection criteria for the positive bay analysis could include flows for later time than the Pi2 onset analysis, these differences indicate some temporal evolution of the acceleration region. The difference in the ratio of tailward to Earthward flow in Figure 4 indicates that the probability of detecting flow events Earthward of the reconnection site is enhanced during the expansion phase more than the probability of events tailward of the reconnection site. Tailward transition of the acceleration region could therefore take place, in addition to longitudinal expansion suggested by *Nagai and Machida* [1998]. Such a difference in the tailward and Earthward flow occurrence between the early expansion phase and maximum of the AE substorm was also seen in the region $X = -25 \sim -35\ R_E$ region using the IMP-6, 8 dataset [*Nakamura et al.*, 1994].

Whether the X-line moves farther tailward to form the distant tail reconnection region as in the model originally proposed by *Hones* [1977], however, is still controversial. In fact, 80% of the tailward flows are followed by Earthward flows in the region Earthward of $-50\ R_E$ but only 40% of the tailward flows are followed by Earthward flows in the region at $X = -50 \sim -100\ R_E$ [*Nagai et al.*, 1997]. The result suggests that the X-line does not always reach the distant tail reconnection region, which is considered to be located beyond $-100\ R_E$ downtail [*Nishida et al.*, 1996 and

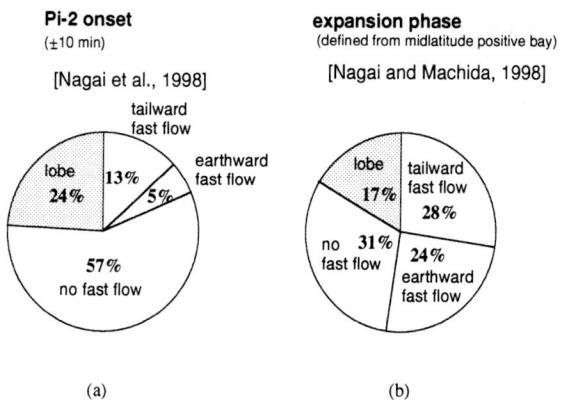

Figure 4. GEOTAIL observations associated with (a) Pi2-onsets [*Nagai et al.*, 1998] and (b) associated with the expansion phase deduced from positive bay activity [*Nagai and Machida*, 1998]. Occurrence rate of the plasma sheet events with tailward and Earthward fast flows, those without fast flows, and lobe events are shown.

reference therein]. The results obtained from energetic particle anisotropy analysis using both GEOTAIL and IMP 8 suggest that tailward retreat of the neutral line could in reality be the appearance of new acceleration regions at more distant sites in the tail and that multiple acceleration regions can exist at the same time [*Angelopoulos et al.*, 1996].

4. RELATIONSHIPS BETWEEN IMF AND MIDTAIL DISTURBANCES

On average, lobe magnetic field energy density is observed to enhance in association with a southward component of the IMF prior to substorm onset and rapidly recovers coincident with expansion onset [*Caan et al.*, 1975]. However, more detailed event analysis have obtained that the time variation of dissipation of the accumulated energy is not always the same. At some times the dissipation can occur more gradually during ongoing magnetic activity and the energy supplied by the solar wind can even exceed that being dissipated, thus causing the total tail energy to increase [*Fairfield et al.*, 1981]. It is therefore crucial to take the IMF condition into account to understand the evolution of the disturbance in the magnetotail and its relationship to the ionosphere.

4.1. Response of the Tail to the Enhanced Solar Wind Energy Input

A combination of ISTP satellites data and ground magnetic field data is an ideal dataset to study substorm evolution and its relationship to the IMF/solar wind. Figure 5 shows two examples of substorm intervals. IMF B_Z, solar wind electric field obtained by WIND, total pressure, dawn-to-dusk component of $V \times B$ field, and β obtained by GEOTAIL, and H-component from ground magnetograms in high-latitude region are plotted. At this time, GEOTAIL was located at $X = -22 \sim -26\ R_E$. Solar wind electric field is calculated as $E_{SW} = V\ (B_Z^2 + B_Y^2/2)^{1/2} \sin(\Theta/2)$, where $\Theta = \arctan(B_Y/B_Z)$. The pressure value estimated using solar wind parameters [*Fairfield and Jones*, 1996] is also shown as a reference (thin line in the middle panel). The onset time of the enhancement in the solar wind electric field is indicated by the solid lines, while the onset time of the enhancement in the plasma sheet electric field is indicated by the dotted lines. The dawn-dusk component of the magnetospheric electric field fluctuates much more than the corresponding solar wind electric field partly due to the changes in the plasma sheet thickness as well as flapping motion of the tail. Nonetheless, these magnetospheric electric field disturbances have a duration similar to that of the solar wind electric field enhancements with delay time of 45~100 min (see the shaded bars in the second and fourth panels showing the time interval of the solar wind electric field enhancements). It was shown that the plasma sheet is strongly turbulent (i.e. flows are dominated by fluctuations that are unpredictable, with rms velocities much larger than the mean velocities and with magnetic field fluctuations that are comparable to the mean field strength) [*Borovsky et al.*, 1997]. The observed characteristics shown in Figure 5, however, suggest that at least in this long-time interval, plasma sheet electric field disturbances are likely to be driven by the solar wind in a coherent fashion.

The pressure shown in Figure 5 starts to increase about 10 min after the solar wind electric field enhancement, except for event (v) on November 22, 1995, where the pressure stays constant. The enhancements in the pressure are most likely associated with magnetic flux enhancement in the midtail region [*Rybal'chenko and Sergeev*, 1985]. Onset of the dawn-to-dusk magnetospheric electric field enhancement is associated with a pressure decrease and westward electrojet enhancement for a-i, a-iii, b-i, and b-iii, so that energy dissipation in the tail occurs. On the other hand, magnetospheric electric field enhancements a-ii, a-iv, and b-ii, and b-iv happened during the recovery phase or before the ground magnetic activity. The pressure is increasing during these periods suggesting that the input rate from solar wind coupling exceeds that of the dissipation rate. The latter intervals do not correspond to strong solar wind input intervals. Therefore, it could be both the internal effect, such as magnetospheric field configuration, and the IMF input that determines the type of the dissipation process.

4.2. Effect of the Different Types of IMF Input

Response of the magnetosphere and ionosphere in the course of the westward electrojet evolution was studied for different types of IMF input. *Pulkkinen et al.* [1998] compared two substorms, in which the first substorm occurred during persistently negative IMF B_Z, whereas the second expansion followed a northward turning of the IMF. It was concluded that a continuous loading during the first substorm creates an inductive electric field, which stabilizes the tail. It was suggested that the IMF condition is important in explaining the evolution of the reconnection region after the onset.

Examination of the solar wind structure based on multipoint observations of the solar wind made it possible to examine the detailed interaction between the magnetosphere/ionosphere and solar wind. *Sergeev et al.* [1998] found some convection bay cases where an "incomplete"

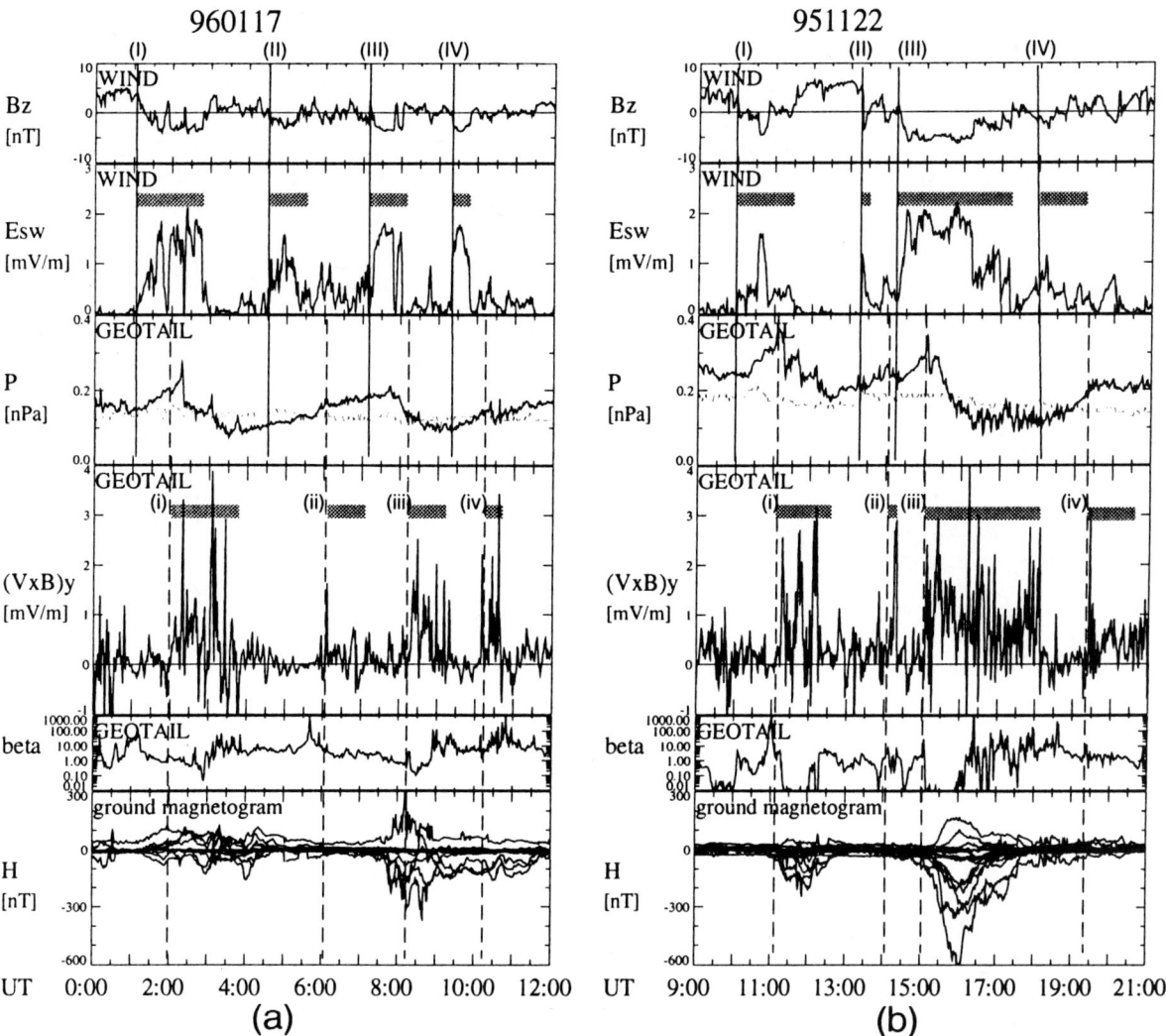

Figure 5. Two examples of substorm intervals on (a) January 17, 1996 and (b) November 22, 1995. IMF B_Z, solar wind electric field (Contribution from IMF B_Y is included) obtained by WIND, total pressure, dawn-to-dusk component of $V \times B$ field, and beta value obtained by GEOTAIL, and horizontal component of the ground magnetograms from selected high-latitude stations are plotted.

convection/current system was formed, i.e., a well-developed dawn vortex in the absence of well-defined dusk vortex. Such current system could be due to the dawn-dusk asymmetry in the southward IMF structure or to a highly slanted interplanetary discontinuity. Solar wind structures including localized ones, therefore, could produce different responses at the ionosphere and in the midtail than that found for normal substorm/convection disturbances.

5. CONCLUSIONS

Dynamic changes in the tail configuration during storms and substorms have been obtained using GEOTAIL data in the midtail as well as in the distant tail. It has been shown that IMF/solar wind information is essential to understand the state of the magnetosphere, which could be quite different according to the previous IMF/solar wind and substorm conditions. A combination of GEOTAIL data, other ISTP satellite data, and data from ground station networks —and comparison with modeling results— is an effective approach and should be further pursued to understand the evolution of the acceleration region as well as quantitative estimation of energy flow in the magnetotail during disturbed times.

Acknowledgments. The author gratefully acknowledges discussions with S. Kokubun, T. Mukai, T. Nagai, V. Sergeev, and

L. Bargatze. The author is especially grateful to the following individuals and institutions for providing data: R. P. Lepping, K. W. Ogilvie, J. T. Steinberg, and A. J. Lazarus for WIND data, T. Iyemori, K. Hayashi, O. A. Troshichev, S. I. Solovyev, K. Yumoto, K. Shiokawa, G. Rostoker, S. Nozawa, T. Hansen, Canadia Space Agency, Geological Survey of Canada, WDC-C2. The author wishes to thank A. Nishida and the entire GEOTAIL team.

REFERENCES

Angelopoulos, V., D. G. Mitchell, R. W. McEntire, D. J. Williams, A. T. Y. Lui, S. M. Krimigis, R. B. Decker, S. P. Christon, S. Kokubun, T. Yamamoto, Y. Saito, T. Mukai, F. S. Mozer, K. Tsuruda, G. D. Reeves, W. J. Hughes, E. Friis-Christensen, and O. Troshichev, Tailward progression of magnetotail acceleration centers: Relationship to substorm current wedge, *J. Geophys. Res., 101*, 24599, 1996.

Baker, D. N., E. W. Hones, Jr., P. R. Higbie, R. D. Belian, and P. Stauning, Global properties of the magnetosphere during a substorm growth phase: A case study, *J. Geophys. Res., 86*, 8941, 1981.

Baker, D. N., S. J. Bame, R. D. Belian, W. C. Feldman, J. T. Gosling, P. R. Higbie, E. W. Hones, Jr., D. J. McComas, and R. D. Zwickl, Correlated dynamical changes in the nea-Earth and distant magnetotail region: ISEE 3, *J. Geophys. Res., 89*, 3855, 1984.

Baker, D. N., T. I. Pulkkinen, V. Angelopoulos, W. Baumjohann, and R. L. McPherron, Neutral line model of substorms: Past results and present view, *J. Geophys. Res., 101*, 12975, 1996.

Baumjohann, W., G. Paschmann, T. Nagai, and H. Lühr, Superposed epoch analysis of the substorm plasma sheet, *J. Geophys. Res., 96*, 11605, 1991.

Borovsky, J. E., R. Elphic, H. O. Funsten, and M. F. Thomsen, The Earth's plasma sheet as a laboratory for flow turbulence in high-β MHD, *J. Plasma Physics, 57*, 1, 1997.

Caan, M. N., R. L. McPherron, and C. T. Russell, Substorm and interplanetary magnetic field effects on the geomagnetic tail lobes, *J. Geophys. Res., 80*, 191, 1975.

Fairfield, D. H., Solar wind control of the distant magnetotail: ISEE3, *J. Geophys. Res., 98*, 21265, 1993.

Fairfield, D. H., R. P. Lepping, E. W. Hones, Jr., S. J. Bame, and J. R. Asbridge, Simultaneous measurements of magnetotail dynamics by IMP spacecraft, *J. Geophys. Res., 86*, 1396, 1981.

Fairfield, D. H., and J. Jones, Variability of the tail lobe field strength, *J. Geophys. Res., 101*, 7785, 1996.

Frank, L. A., M. Ashour-Abdalla, J. Berchem, J. Raeder, W. R. Paterson, S. Kokubun, T. Yamamoto, R. P. Lepping, F. V. Coroniti, D. H. Fairfield, and K. L. Ackerson, Observations of plasmas and magnetic fields in Earth's distant magnetotail: Comparison with a global MHD model, *J. Geophys. Res., 100*, 19177, 1995.

Hones, E. W., Jr., Substorm processes in the magnetotail: Comments on "On hot tenuous plasmas, fireballs, and boundary layers in the Earth's magnetotail" by L. A. Frank, K. L. Ackerson, and R. P. Lepping, *J. Geophys. Res., 82*, 5633, 1977.

Knipp, D. J., B. A. Emery, M. Engebretson, X. Li, A. H. McAllister, T. Mukai, S. Kokubun, G. Reeves, D. Evans, T. Obara, X. Pi, T. Rosenberg, A. Weatherwax, M. G. McHarg, H. Chun, K. Mosely, M. Codreseu, L. Lanzerotti, F. Rich, P. Wilkinson, An overview of the early November 1993 geomagnetic storm, *J. Geophys. Res., in press*, 1998.

Kokubun, S., L. A. Frank, K. Hayashi, Y. Kamide, R P. Lepping, T. Mukai, R. Nakamura, W. R. Paterson, T. Yamamoto, and K. Yumoto, Large field events in the distant magnetotail during magnetic storms, *J. Geomag. Geoelectr., 48*, 561, 1996.

Kokubun, S., Dynamics of the magnetotail during magnetic storms: Review of ISEE 3 and GEOTAIL Observations, in *Magnetic Storms, Geophys. Monogr. Ser.*, vol. 98, edited by B. T. Tsurutani, W. D. Gonzalez, Y. Kamide, and J. K. Arballo, p.117, AGU, Washington, D. C., 1997.

Nagai, T., R. Nakamura, T. Mukai, T. Yamamoto, A. Nishida, and S. Kokubun, Substorms, tail flows and plasmoids, *Adv. Space Res., 20*, 961, 1997.

Nagai, T., M. Fujimoto, Y. Saito, S. Machida, T. Terasawa, R. Nakamura, T. Yamamoto, T. Mukai, A. Nishida, and S. Kokubun, Structure and dynamics of magnetic reconnection for substorm onsets with GEOTAIL observations, *J. Geophys. Res., 103*, 4419, 1998.

Nagai, T. and S. Machida, Magnetic Reconnection in the Near-Earth Magnetotail, in New Perspective on the Earth's Magnetotail, *Geophys. Monogr. Ser.*, edited by D. N. Baker, A. Nishida, and S. W. H. Cowley, AGU, Washington, D.C., in press, 1998.

Nakamura, R., D. N. Baker, D. H. Fairfield, D. G. Mitchell, R. L. McPherron, and E. W. Hones, Jr., Plasma flow and magnetic field characteristics near the midtail neutral sheet, *J. Geophys. Res., 99*, 23591, 1994.

Nakamura, R., S. Kokubun, Y. Kamide, T. Yamamoto, L. A. Frank, W. R. Paterson, E. Friis-Christensen, K. Hayashi, T. Iyemori, K. Yumoto, H. Lühr, and O. A. Troshichev, Observations of the magnetosheath near the nominal tail axis during the geomagnetic storm of January 25, 1993, *J. Geomag. Geoelectr., 48*, 557, 1996.

Nakamura, R., S. Kokubun, T. Mukai, and T. Yamamoto, Changes in the distant tail configuration during geomagnetic storms, *J. Geophys. Res., 102*, 9587, 1997.

Nishida, A., T. Mukai, T. Yamamoto, Y. Saito, and S. Kokubun, Magnetotail convection in geomagnetically active times. 1. Distance to the neutral lines, *J. Geomag. Geoelectr., 48*, 489, 1996.

Nishida, A., and T. Ogino, Convection in the Earth's magnetotail, in New Perspective on the Earth's Magnetotail, *Geophys. Monogr. Ser.*, edited by D. N. Baker, A. Nishida, and S. W. H. Cowley, AGU, Washington, D.C., in press, 1998.

Ogino, T., R. J. Walker, and M. Ashour-Abdalla, The effects of IMF orientation on the polar cap and magnetosphere, *Eos Trans. AGU 75*, Fall Meeting Suppl., 536, 1994.

Pulkkinen, T. I., D. N. Baker, L. A. Frank, J. B. Sigwarth, H. J. Opgenoorth, R. Greenwald, E. Friis-Christensen, T. Mukai, R. Nakamura, H. Singer, G. D. Reeves, and M. Lester, Two substorm intensifications compared: Onset, expansion, and global consequences, *J. Geophys. Res., 103*, 15, 1998.

Rybal'chenko, V. V., and V. A. Sergeev, Rate of magnetic flux buildup in the magnetospheric tail, *Geomag. Aeron., 25*, 378, 1985.

Sergeev, A., Y. Kamide, S. Kokubun, R. Nakamura, C. S. Deehr, T. J. Huges, R. P. Lepping, T. Mukai, A. A.

Petrukovich, J.-H. Shue, K. Shiokawa, O. A. Troshichev, and K. Yumoto, Short duration convection bays and localized IMF structures on November 28, 1995, *J. Geophys. Res.*, in press, 1998.

Shodhan, S., G. L. Siscoe, L. A. Frank, K. L. Ackerson, and W. R. Paterson, Boundary oscillations at GEOTAIL: Windsock, breathing, and wrenching, *J. Geophys. Res. 101,* 2577, 1996.

Sibeck, D. G., G. L. Siscoe, J. A. Slavin, and R. P. Lepping, Major flattening of the distant geomagnetic tail, *J. Geophys. Res., 91,* 4223, 1986.

Williams, D. J., A. T. Y. Lui, R. W. McEntire, V. Angelopoulos, C. Jacquey, S. P. Christon, L. A. Frank, K. L. Ackerson, S. Kokubun, and D. H. Fairfield, Magnetopause encounters in the magnetotail at distances of ~80 RE, *Geophys. Res. Lett., 21,* 3007, 1994.

Max-Planck-Institut für extraterrestrische Physik, Postf. 1603, 85740 Garching, Germany

New Results on the Polar Cap and PSBL Dynamics

G. Parks,[1] M. Brittnacher,[1] L. J. Chen,[1] M. McCarthy,[1] D. Larson,[2] R. P. Lin,[2] G. Germany,[3] J. Spann,[4] H. Reme,[5] and T. Sanderson[6]

A fundamental science goal of the International Solar Terrestrial Physics (ISTP) mission is to enhance understanding of how the solar wind couples its mass, momentum and energy into the magnetosphere. We present two new results from the UltraViolet Imager (UVI) and 3D Plasma experiments on spacecraft Polar and Wind that are relevant to this problem. First, we have found that the polar cap area which undergoes expansion and contraction can sometimes act independently of the interplanetary magnetic field (IMF). Second, we have discovered unidirectional ion beams in the plasma sheet boundary layer (PSBL) that propagate in the tailward direction immediately adjacent to the earthward propagating beam. Both of these observations are not predicted by current reconnection models. These new observations are indicating that other mechanisms may also be active.

INTRODUCTION

New observations are promised when scientists are given opportunities to conduct experiments with state of the art instrumentation. This is the case with the ISTP mission, which has given space physics an unprecedented tool to study the Sun-Earth connection [*Acuna et al.*, 1995]. A flotilla of coordinated spacecraft located in strategic regions of space is tackling fundamental problems of space physics: how the solar wind interacts with planetary magnetic fields, how planetary ionospheres couple to the plasma sheet and how mass, momentum and energy are transported and dissipated in magnetospheres.

Advanced techniques and highly innovative experiments on ISTP spacecraft are yielding exciting results and new discoveries. We focus on two important results that are relevant to furthering understanding of the polar cap [*Brittnacher et al.*, 1999] and plasma sheet dynamics [*Parks et al.*, 1998]. As discussed below, these results are fundamental for understanding the physical mechanisms, including the magnetic reconnection process. These new observations should be an inspiration and challenge to theorists and modelers.

POLAR CAP

The magnetic field in the polar cap, the region poleward of the auroral oval, is assumed to be open and connected to the interplanetary space. The reconnection models consider the polar cap area a fundamental quantity. Reconnection substorm models (see for example, *Russell and McPherron* [1973]) rely on the interconnection of the IMF with the magnetopause field

[1] Geophysics Program, University of Washington, Seattle, WA
[2] Space Sciences Laboratory, University of California, Berkeley, CA
[3] CSPAR, University of Alabama, Huntsville, AL
[4] Marshall Space Flight Center, Huntsville, AL
[5] C.E.S.R., Paul Sabatier University, Toulouse, France
[6] Space Science Department, European Space Agency, Noordwijk, The Netherlands

and predict that storage and dissipation of the magnetic energy will cause the polar cap area to increase and decrease [*Siscoe and Huang*, 1985]. These models also predict that open flux tubes in the polar cap are more effectively produced as the magnitude of the southward IMF component becomes larger. These predictions have been evaluated using the global UltraViolet Images (UVI) obtained on Polar.

At the outset we must recognize that the open flux region that defines the polar cap is not an experimental observable. In practice, different methods have been used to size the polar cap area. For example, a polar cap boundary was defined in terms of one kilorayleigh contours of auroral luminosity [*Frank and Craven*, 1988; *Kamide et al.*, 1998]. Others used ion and electron boundaries detected at polar altitudes [*Makita et al.*, 1985; *Troshichev et al.*, 1996] to determine the polar cap region.

Our definition of the polar cap is different. We define the polar cap area by a predetermined threshold auroral luminosity of photon flux at the camera of < 4 photons cm^{-2} s^{-1} (120 Rayleigh or ≈ 1 erg cm^{-2} s^{-1}). This is 8 times the minimum sensitivity of the camera, 0.5 photons cm^{-2} s^{-1}. All regions poleward of the auroral oval below this threshold value are included in the polar cap area. Oval-aligned and transpolar arcs in the polar cap that exceed the threshold are excluded. In sections where the oval disappears or is below the instrument threshold value, the polar cap is closed by curve fitting across the missing region using adjacent boundaries that are clearly resolved. Our method sums regions devoid of precipitation that can have very irregular shapes and are sometimes not simply connected.

We will use the UVI data from a substorm that occurred on January 9, 1997 to study the polar cap. The left side of Plate 1 shows a sequence of six images that cover the period of growth and expansion phases of a substorm. (The spacecraft wobble introduced by the despun platform has been deconvolved in these images.) These images, obtained at apogee (≈ 9 R$_E$), have been projected to 120 km altitudes and displayed in a magnetic frame of reference. The top three panels on the right side show the size of the polar cap area, amount of precipitated electron energy in the aurora and IMF (GSM coordinate) measured by WIND (the times are corrected for propagation time delay which is estimated to be ≈ 17 minutes). The dashed lines correspond to the times of the UVI images.

The "keogram" from UVI images provide additional information on the variations of the polar cap boundary with Magnetic Local Time (MLT). Keograms shown in Plate 1 come from four fixed MLT sectors of auroral activity plotted as a function of UT and magnetic latitude. The resolution here is 0.5 hour MLT and 0.5° Magnetic Latitude. The global images and keograms are used to quantify how the polar cap area changes relative to the IMF.

The aurora was initially fairly weak. The global precipitated energy flux in the time interval 05 UT to 0630 UT was steady at ≈ 10–15 GW (GW=gigawatt=10^{16} ergs s^{-1}) which is typical of quiet auroral conditions. Keograms show that the weak precipitation was diffuse (< 2 ergs cm^{-2} s^{-1}) and covered a broad latitudinal width (caution: the lower cutoff between 05 and 07 UT is due to the edge of the field of view of the instrument).

The polar cap area slowly increased from ≈ 6 to $\approx 8 \times 10^6$ km^2 between 05 and 0630 UT. The 33% increase is statistically significant (the standard deviation of measured values is $\approx 0.4 \times 10^6$ km^2, and including errors introduced by curve fitting, the estimated accuracy is $\approx 1 \times 10^6$ km^2). This increase is associated with the equatorward motion of a faint but discernible precipitation band seen in dusk and midnight MLT sectors. Note that the equatorward motion of the oval was not evident in the 06 MLT sector where both boundaries were fixed in latitude. The oval dynamics occurred when the IMF was fairly steady with B$_x \approx 6$ nT, B$_y \approx 1$ nT, and a small B$_z$, ≈ -0.5 nT.

The polar cap area began to increase more rapidly starting ≈ 0645 UT in association with a narrowing of the aurora, most prominent in the dusk sector. This time also coincided with the fading of the faint precipitation in the high latitude region (see dusk and midnight MLT keograms) to below the threshold level, ≈ 1 erg cm^{-2} s^{-1}. A maximum area of $\approx 10 \times 10^6$ km^2 is reached at ≈ 0715 UT and the area remained at this value until the onset of the substorm at ≈ 0747 UT when the polar cap area began to decrease rapidly. The expansion of the auroral activity into the polar cap region reduced the polar cap area to $\approx 4 \times 10^6$ km^2 by ≈ 0840 UT. Subsequently, the polar cap area remained constant while the precipitated energy continued to decrease. The polar cap area is usually the smallest at the time the expansion activity attains a maximum.

Note that there occurred two onsets of the poleward expansion, the first at ≈ 0747 UT and the second at ≈ 0804 UT. A poleward expansion was not observed in the dawn sector for the first onset. Also, the expansion was not observed for either onsets in dusk or noon sectors. The IMF during the interval of time ≈ 0645–0750 UT was varying, beginning with a large change in B$_x$ and B$_y$ while B$_z$ was $\approx 0 \pm 0.5$ nT.

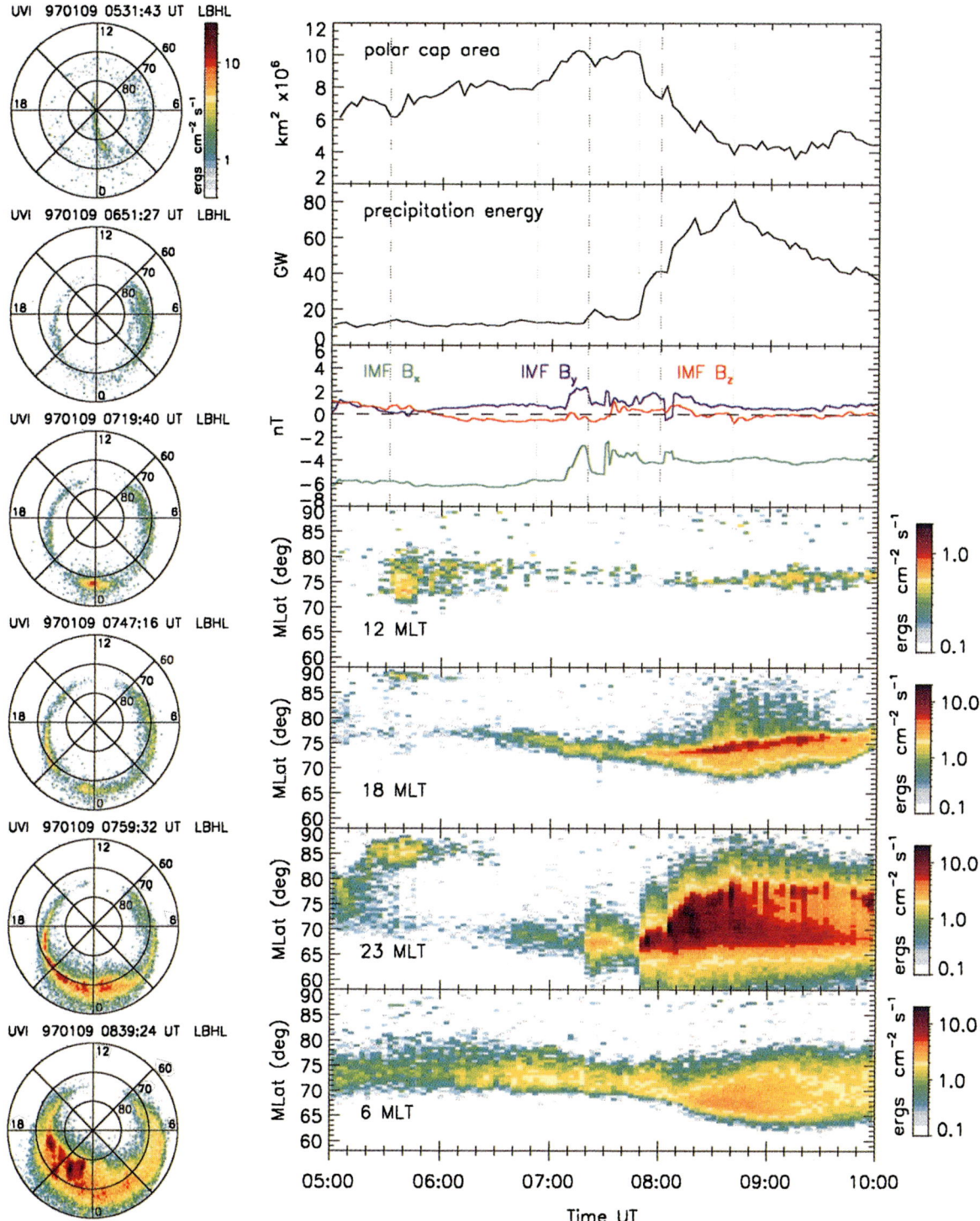

Plate 1. Selected UVI images from the substorm event of January 9, 1997. The times shown in the images correspond to the beginning time of an image integration period of 36.8 s. The top three panels on the right show calculated polar cap area, total precipitation energy integrated over the hemisphere, and the IMF. The bottom four panels are keograms at four different MLT sectors showing the behavior of the auroral energy flux in latitude and UT.

The January 9, 1997 substorm and other events with varying IMF conditions [*Brittnacher et al.*, 1999] show that the equatorward motion of the aurora and clearing of faint precipitation in the high latitude regions contribute toward increases in the polar cap area. The decrease in magnetic latitude of a few degrees on the dayside and about 3–4 degrees on the nightside is consistent with substorm models that predict that the expansion of the polar cap is due to the increased rate of merging of the IMF with the dayside magnetopause. However, there is a complicated MLT relationship between the motion of the poleward and equatorward boundaries and substorm phases. For example, both boundaries at midnight expand but the equatorward boundaries at dusk and dawn continue to move equatorward unimpeded, past the onset time at local midnight. This feature is not predicted by any reconnection models. Thus, reconnection models appear validated only in the noon and midnight sectors but not in the global sense.

The substorm on January 9, 1997 was associated with nondisturbed solar wind condition. The IMF B_z component was typically $\approx 0 \pm 0.5$ nT throughout the different phases of the substorm. However, the increase in the polar cap area was comparable to those observed when the IMF had much larger southward component [*Brittnacher et al.*, 1999]. Other UVI observations also show that the gain or loss in the polar cap area is not directly related to the magnitude of substorms. The independent behavior of the polar cap area and the southward IMF component is not consistent with substorm models that predict that large southward B_z component should be more effective in producing open flux tubes in the polar cap region. Note this substorm may be associated with the solar wind dynamic pressure (not shown; see also *Brittnacher et al.* [1998]). Also, merging substorm models predict a more direct relationship between the loss of polar cap area and the intensity of substorms since particle energization is a byproduct of the conversion of open to closed flux by reconnection.

Murphree et al. [1991] questioned whether the polar cap area derived from satellite images represents the magnitude of open flux. This question cannot be answered with certainty from observations at this time. Observations from UVI images caution against attributing all changes of the polar cap area to reconnection physics. Changes of the polar cap area during substorms are much more complicated than predicted and what all goes into polar cap dynamics is not fully known.

PLASMA SHEET BOUNDARY LAYER

A reconnection model of the geomagnetic tail was proposed more than thirty years ago [*Dungey*, 1961; *Axford et al.*, 1965]. More recent models have tied auroral processes to the geomagnetic tail and have postulated that ion beams and high-speed earthward flows would be produced as the magnetic stress energy is released from the reconnection of the two tail lobes [*Cowley*, 1984]. Both earthward-traveling ion beams and high-speed flows have been observed and they have been interpreted as products of the neutral line formation [*Forbes et al.*, 1981; *Takahashi and Hones*, 1988; *Baumjohann et al.*, 1990].

However, recent Wind observations show that unidirectional ion beams also propagate in the tailward direction immediately adjacent to the earthward propagating beams [*Parks et al.*, 1998]. They are different from the tailward propagating beams that are associated with counterstreaming beams [*Forbes et al.*, 1981], which has been interpreted as the reflected beam of the original earthward propagating beam. The existence of unidirectional ion beams propagating only in the tailward direction adjacent to the earthward propagating beam was not discussed by previous observations.

We use the observations of PSBL made by the Wind 3D plasma experiment on December 31, 1996 to demonstrate the existence of these new ion beams. The beams and the direction they travel can be clearly shown by use of energy-angle spectrograms. The bottom four panels of Plate 2 show these plots for four different energies. Here, the ordinate denotes the Φ angles (GSE) with 0° and 360° pointing in the earthward direction and 180° the tailward direction. The differential flux levels (see color bar) have been sorted according to the Φ-angles and plotted against the Universal Time.

As the spacecraft is engulfed by the expansion of the plasma sheet at 1230 UT, the spacecraft first encounters an earthward-traveling ion beam right as it crosses the outer edge of the PSBL. Immediately inside, a tailward-traveling ion beam is detected. Note that the tailward-traveling beams are as intense as the earthward-traveling beams. It is very unlikely that these are reflected beams of the original earthward-traveling beams which are normally found deeper inside the PSBL. There is no evidence of a large convective motion in the perpendicular direction of the magnetic field which would be required had the spacecraft penetrated deeper into the PSBL.

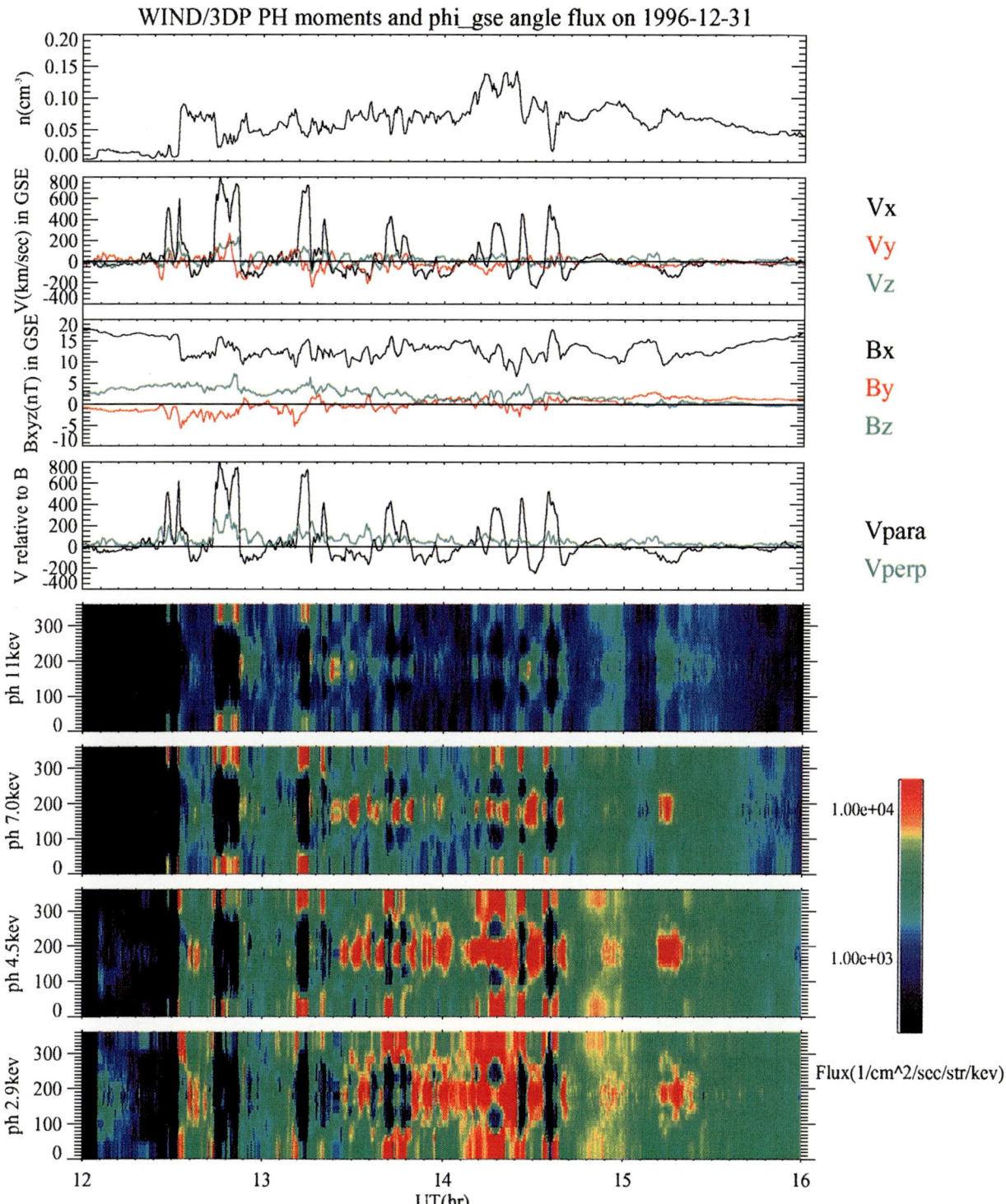

Plate 2. The top four panels show the ion density, mean velocity in GSE, magnetic field, and mean velocity relative to two directions of the magnetic field. The bottom four panels show energy-angle spectrograms of selected energy channels.

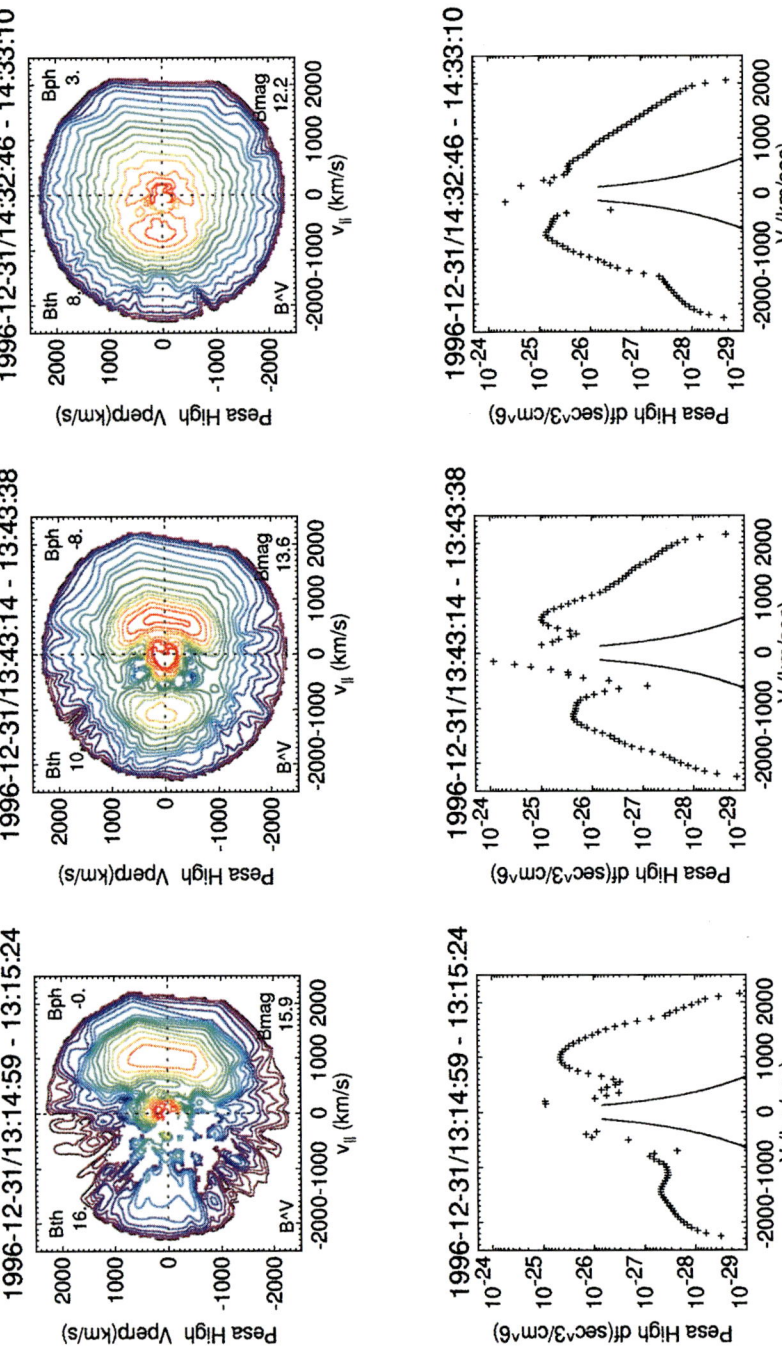

Plate 3. Isocontours of ion distribution functions in $(v_\parallel, v_\perp)$ space and cuts of the distribution function in the direction parallel to the magnetic field.

The spacecraft lingered near the outer boundary of PSBL as it oscillated back and forth which permitted multiple samplings of these ion beams. The boundary oscillation can be inferred from oscillations of the density and magnetic field, most clear in the B_x component. The alternating pattern of unidirectional beams traveling in the earthward and tailward directions is a consequence of this boundary oscillation. Counter-streaming beams at selected energies are seen at ≈ 1343 UT. These beams are associated with regions of higher density which we interpret as deeper inside PSBL.

The top two panels of Plate 2 show the density and mean velocities that have been computed as moments of the distribution function. Bursts of high mean velocities as large as 800 km s^{-1} with durations of minutes occur in the earthward direction ($+V_x$). To study the origin of these high mean velocities, examples of the phase space distributions from the three types of ion beams discussed above are shown in Plate 3. Constant phase space contours plotted in ($v_\parallel$, $v_\perp$) space and cuts of the corresponding distributions in the $v_\parallel$-direction are shown. The left column shows the beam propagating in the earthward direction, the center, counterstreaming beams, and the right, tailward propagating beams.

An important feature revealed by these plates is that the plasma distributions surrounding the earthward and tailward beams are different. The phase space plots together with the energy-angle plots reveal that plasma distributions in the region supporting the earthward beam consists predominantly of the beam distribution, whereas the plasma in the region of tailward-traveling beams consist of the beam plus an isotropic component. When these contrasting distributions are convolved in the calculation of mean velocities they yield > 400 km/s in the earthward direction and < 150 km/s in the tailward direction that are magnetic field aligned (Plate 2, fourth panel). These results indicate that studying the dynamics of PSBL using bulk moments alone can lead to an incorrect picture.

In conclusion, we have reported in this article new polar cap and PSBL features that are not currently explained by any models. These observations however are fundamentally important for understanding the dynamics of the polar cap and geomagnetic tail and must be included in models that predict how the solar wind interacts with the magnetosphere and the causal relationships between the geomagnetic tail and substorms.

Acknowledgments. We thank R. Lepping for use of the Wind magnetic field data. The research at the University of Washington is in part supported by NASA grants NAG5-3170 and NAG5-7114 and at UC Berkeley by NAG5-2815.

REFERENCES

Acuna, M., K. W. Ogilvie, D. N. Baker, S. A. Curtis, D. H. Fairfield, and W. H. Mish, The global Geospace Science Program and its investigations, *Space Sci. Rev.*, 71, 5, 1995.

Axford, W. I., H. E. Petschek, and G. L. Siscoe, Tail of the magnetosphere, *J. Geophys. Res.*, 70, 1231, 1965.

Baumjohann, W., G. Paschmann, and H. Luhr, Characteristics of high-speed ion flows in the plasma sheet., *J. Geophys. Res.*, 95, 3801, 1990.

Brittnacher, M., M. Wilber, M. Fillingim, D. Chua, G. Parks, J. Spann, and G. Germany, Global auroral response to a solar wind pressure pulse, *Adv. in Space Res.*, submitted, 1998.

Brittnacher, M., M. Fillingim, G. Parks, G. Germany, and J. Spann, Polar cap area and boundary motion during substorms, *J. Geophys. Res.*, submitted, 1999.

Cowley, S., The distant geomagnetic tail in theory and observation, in Magnetic Reconnection in Space and Laboratory Plasmas, Geophysical Monograph 30, American Geophysical Union, 1984.

Dungey, J., Interplanetary magnetic field and the auroral zone, *Phys. Rev. Lett.*, 6, 47, 1961.

Forbes, T. G., E. W. Hones, S. J. Bame, J. R. Asbridge, G. Paschmann, N. Sckopke, and C. T. Russell, Evidence for the tailward retreat of a magnetic neutral line in the magnetotail during substorm recovery, *Geophys. Res. Lett.*, 8, 261, 1981.

Frank, L. and J. Craven, Imaging results from Dynamics Explorer 1, *Rev. Geophys.*, 26, 249, 1988.

Kamide, Y., S. Kokubun, L. F. Bargatze, and L. A. Frank, The size of the polar cap as an indicator of substorm energy, *Phys. Chem. of Earth*, (in press), 1998.

Makita, K., C.-I. Meng, and S.-I. Akasofu, Temporal and spatial variations of the polar cap dimension inferred from the precipitation boundaries, *J. Geophys. Res.*, 90, 2744, 1985.

Murphree, J. S., R. D. Elphinstone, L. L. Cogger, and D. Hearn, Viking optical substorm signatures in Magnetospheric Substorms, AGU monograph 64, American Geophysical Union, 1991.

Parks, G., L. J. Chen, M. McCarthy, D. Larson, R. P. Lin, T. Phan, H. Reme, and T. Sanderson, New observations of ion beams in the plasma sheet boundary layer, *Geophys. Res. Lett.*, 25, 3285, 1998.

Russell, C. T. and R. L. McPherron, The magnetotail and substorms, *Space Sci. Rev.*, 11, 111, 1973.

Siscoe, G. and T. S. Huang, Polar cap inflation and deflation, *J. Geophys. Res.*, 90, 543, 1985.

Takahashi, K., and E. Hones, Jr., ISEE 1 and 2 observations of ion distributions at the plasma sheet-tail lobe boundary, *J. Geophys. Res.*, 93, 8558, 1988.

Troshichev, O. A., E. M. Shishkina, C.-I. Meng, and P. T. Newell, Identification of the poleward boundary of the auroral oval using characteristics of ion precipitation, *J. Geophys. Res.*, 101, 5035, 1996.

M. Brittnacher, Geophysics Program, AK-50, University of Washington, Seattle, Washington, 98195.

L. J. Chen, Geophysics Program, AK-50, University of Washington, Seattle, Washington, 98195.

G. Germany, CSPAR, University of Alabama, Huntsville, AL, 35899. D. Larson, Space Sciences Laboratory, University of California, Berkeley, CA, 94720-7450.

R. P. Lin, Space Sciences Laboratory, University of California, Berkeley, CA, 94720-7450.

M. McCarthy, Geophysics Program, AK-50, University of Washington, Seattle, Washington, 98195.

G. Parks, Geophysics Program, AK-50, University of Washington, Seattle, Washington, 98195; parks@geops.washington.edu.

H. Reme, C.E.S.R., Paul Sabatier University, 118 Route de Narbonne, 31062 Toulouse, France

T. Sanderson, Space Science Department, European Space Agency, Noordwijk, The Netherlands.

J. Spann, Marshall Space Flight Center, Huntsville, AL, 35899.

Global Energy-Resolved X-Ray Images of Northern Aurora and Their Mappings to the Equatorial Magnetosphere

David L. Chenette,[1] William L. Imhof,[1] Steven M. Petrinec,[1] Michael Schulz,[1] Joseph Mobilia,[1] John G. Pronko,[1] Michael A. Rinaldi,[1] John B. Cladis,[1] Frances Fenrich,[2] Nikolai Østgaard,[3] and Michael C. McNab[4]

The Polar Ionospheric X-ray Imaging Experiment (PIXIE) on the GGS/Polar satellite provides global energy-resolved (2–60 keV) X-ray images of Earth from an apogee altitude ~ 5×10^4 km. In the present study we analyze PIXIE data from the magnetic storm of 10–11 April 1997 (taken as prototypical). We show representative global X-ray images of the northern aurora, obtain representative spectra of the corresponding X-rays, measure (both globally and in hourly bins of magnetic local time) the radiated X-ray power corresponding to each image, extract optimal values for spectral parameters of the precipitating electrons that must have produced the X-rays, and calculate the effect of such precipitating electrons on height-integrated ionospheric Hall and Pedersen conductivities. Using representative global models of the magnetospheric **B** field, we map outstanding features of the auroral X-ray images along magnetic field lines to the equatorial magnetosphere in order to learn where the corresponding electrons must have resided immediately before precipitating. Finding that the most intense X-ray feature typically maps farthest out in the equatorial magnetosphere, we have used this feature (associated with an auroral arc or westward-traveling surge) to estimate the equatorial geocentric distance to the boundary between closed and open field lines in a model that treats this distance as an adjustable parameter.

[1]Space Physics Department, Advanced Technology Center, Lockheed Martin Missiles & Space, Palo Alto, California
[2]Space Sciences Laboratory, University of California, Berkeley, California
[3]Department of Physics, University of Bergen, Bergen, Norway
[4]Space and Environment Technology Center, The Aerospace Corporation, El Segundo, California

Sun-Earth Plasma Connections
Geophysical Monograph 109
Copyright 1999 by the American Geophysical Union

INTRODUCTION

A prime objective of the International Solar Terrestrial Physics (ISTP) program is to ascertain the flow of energy from the Sun to the Earth, as mediated by electrodynamic interactions between the solar wind and the Earth's magnetosphere. The Polar Ionospheric X-ray Imaging Experiment (PIXIE) aboard the GGS/Polar satellite supports this objective by providing (for the first time) energy-resolved global images of the bremsstrahlung X-rays produced by precipitating energetic (multi-keV) auroral electrons into the atmosphere. Such precipitating electrons provide an important path for energy to flow into the upper atmosphere, and they represent an important contribution to electrodynamic interactions of the magnetosphere with the ionosphere and upper atmosphere. The energy spectrum (and thus the total energy flux carried by such auroral electrons) can be determined from quantitative PIXIE measurements of the bremsstrahlung X-ray spectra.

The Polar satellite is a key component of the Global Geospace Science (GGS) mission [*Acuña et al.*, 1995]. The spacecraft was launched 24 February 1996 from Vandenberg AFB in California and follows a highly inclined elliptical orbit with apogee ($r = 8.87\ R_E$) over the northern polar cap and perigee ($r = 1.8\ R_E$) over the southern polar cap. The PIXIE instrument [*Imhof et al.*, 1995] is the first truly global imager at X-ray wavelengths. (Previous X-ray imagers [e.g., *Imhof et al.*, 1974; *Mizera et al.*, 1978] had operated from much lower apogee altitudes and had thus left much of the auroral oval outside viewing range on each pass.)

In this paper we show global images of auroral X-rays from the main phase of the geomagnetic storm that occurred 10–11 April 1997. We relate the time development and local-time distribution of electron precipitation during this event to other characteristics of these geomagnetic disturbances, as determined from ground-based and other correlative measurements. We also provide the first global-scale estimates of the electron energy flux and characteristic energy, as well as the ionospheric conductivity enhancements and their time variations during the main phase of this storm. Finally, using parametrically adjustable models for the magnetospheric **B** field, we map the inferred pattern of energetic electron precipitation from the ionosphere to the equatorial magnetosphere, so as to learn where the corresponding electrons had resided just before precipitating and where their magnetospheric source is likely to be. Taken together, the present results (all of which are subject to continuing analysis) serve as a starting point for more comprehensive global studies on the physics of the auroral magnetosphere.

EVENT STUDY: GEOMAGNETIC STORM
OF 10–11 APRIL 1997

The geomagnetic storm of 10–11 April 1997 was a moderate storm initiated by remnants of the well documented coronal mass ejection (CME) that had left the Sun at 1427 UT on 7 April 1997. The storm began its gradual commencement around 1300 UT on 10 April 1997 at most stations. Figure 1 shows provisional *Dst* and final *Kp* histograms for 10.5–11.5 April 1997. The most negative value of *Dst* during this event was −80 nT and occurred during hour 5 of 11 April. Several storms of similarly moderate intensity occurred during the first half of 1997. Storms in

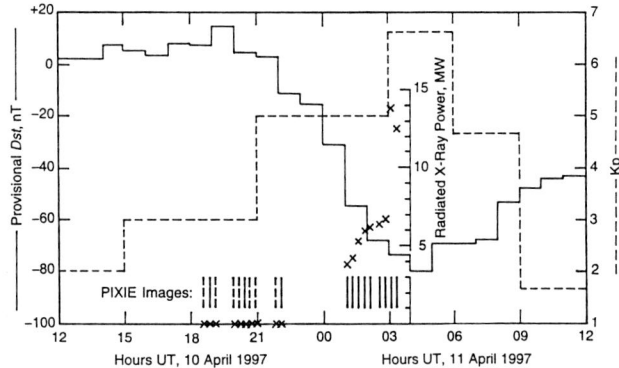

Figure 1. Geomagnetic indices *Dst* (solid histogram) and *Kp* (dashed histogram) for event of 10–11 April 1997. Dark vertical bands indicate time spans of available PIXIE images (to be shown in Plate 1 if band is solid, omitted if band is dashed). Data points (×) pertain to scale inserted at 0400 UT and indicate total (half-globally) radiated X-ray power inferred from corresponding images.

which the most negative value attained by *Dst* lies between −50 nT and −150 nT occur an average of almost once per month over the solar cycle [*McPherron*, 1995, p. 407].

PIXIE was operated during two separate intervals of several hours each (1832–2207 UT and 0102–0326 UT, respectively) during the apogee pass of 10–11 April 1997. (Our standard mode of instrument operation near apogee then was to leave PIXIE alternately "on" for about 5 minutes and then "off" for about 10 minutes in order to avoid unwanted electrical discharges that were occurring after about 20 minutes whenever PIXIE was left "on" continuously. Our standard mode of operation thus allowed us to obtain one good image about every 15–16 minutes during any north polar pass.) The time span of each 5-minute observation interval during the apogee pass of 10–11 April 1997 is indicated by a dark band on Figure 1. Data points (×) keyed to the scale in-serted at 0400 UT indicate the total (half-globally) radiated X-ray power corresponding to each image.

PIXIE can provide separate X-ray images for 128 different energy channels spanning 2–60 keV. Representative images shown in Plate 1 correspond to X-rays of 2–12 keV during the 10–11 April 1997 storm. These are actually superpositions of separate sub-images formed by auroral X-rays that must have passed through different openings in our aperture plates [cf. *Imhof et al.*, 1995, pp. 392–393]. The first row of Plate 1 shows three representative early storm-time X-ray images (centered about 97 minutes apart) from 10 April 1997. (This is just to show that our X-ray images from 1848 UT to 2207 UT on 10 April 1997 were at best

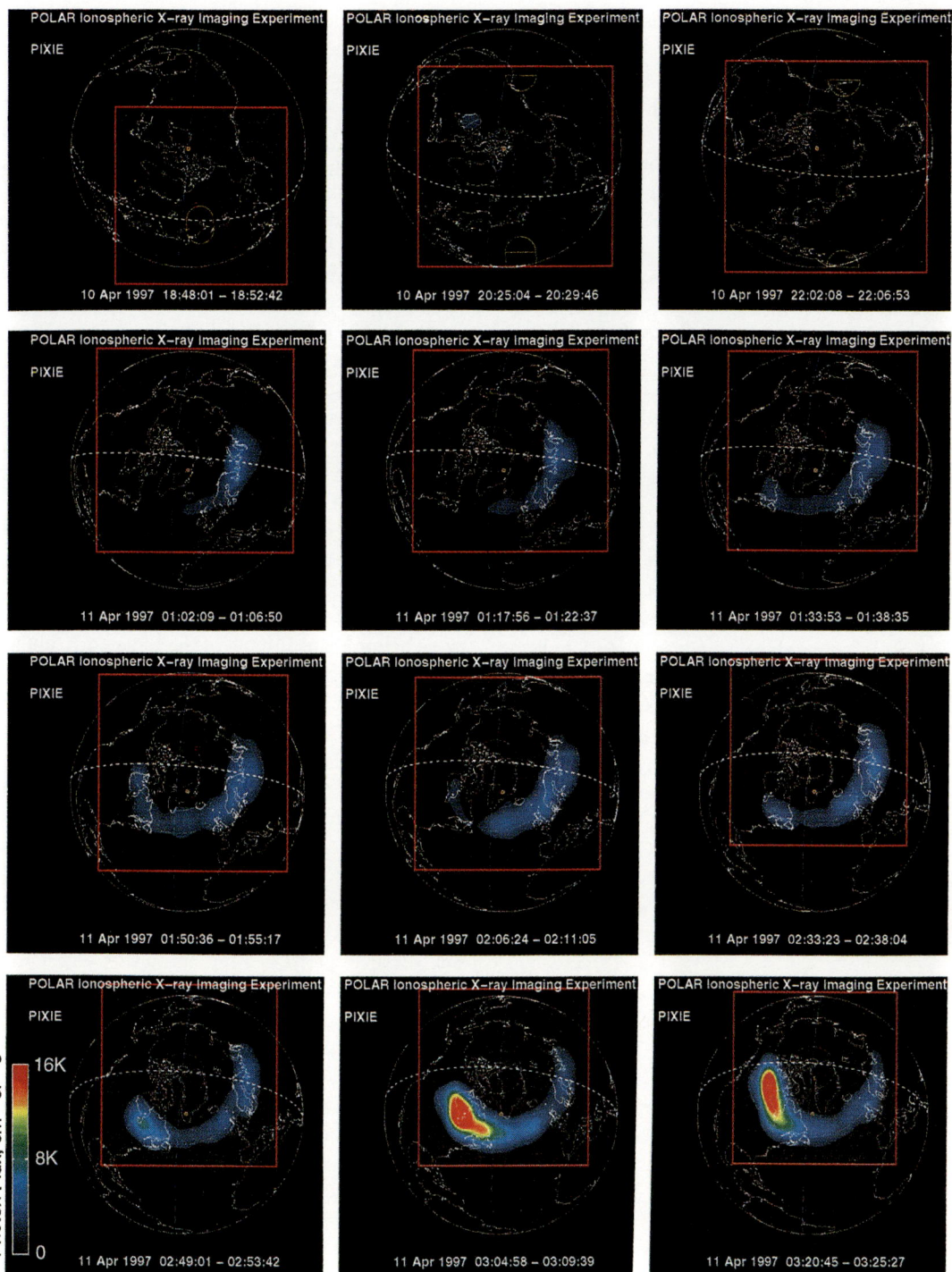

Plate 1. Representative X-ray images from anode 1 (2–12 keV) during event studied here. Coordinate contours indicate magnetic latitude at 10° intervals and magnetic local time (MLT) at 15° intervals. Dashed white curve marks day-night terminator (dayside above, nightside below). Red rectangle indicates PIXIE's field of view. Brownish circled dot (near Nuuk (= Godthåb) in image centered on 0307:18 UT) denotes subsatellite point.

faint and uninteresting, despite *Kp* and *Dst* evidence that the storm had already begun.) The auroral X-ray intensity was low (at background levels) throughout the day on 10 April 1997. The small compression (corresponding to the positive excursion in *Dst* around 2000 UT on 10 April 1997) had no discernible effect on auroral X-ray intensity. (The cartouche and demi-cartouches in row 1 outline image regions that might have been contaminated by X rays from the astrophysical source Circinus X-1, which we use for calibration. Such X rays would not really have passed through the Earth; our image reconstructions would just have assigned them to the wrong aperture.) Rows 2–4 of Plate 1 show nine consecutive stormtime X-ray images (mostly centered about 16 minutes apart) from 11 April 1997. These latter images (and especially the last three) are the ones interesting enough for further analysis.

The false-color scale in Plate 1 indicates X-ray intensity (photon flux) at anode 1 [cf. *Imhof et al.*, 1995, p. 395], which is sensitive to X rays in the 2–12 keV energy range. The same intensity scale is used in each image here. Images from time intervals on most other days are available at our continually updated Website (http://pixie.spasci.com). The terrestrial coordinate circles shown in Plate 1 indicate magnetic latitude (λ) and magnetic local time (MLT) at 100 km altitude, to which the land boundaries in Plate 1 are likewise projected. The noon-midnight magnetic meridian (MLT = 12 hr on the solid portion, 00 hr on the dashed portion) is shown as a blue curve. (Magnetic coordinate circles are more clearly legible in the images on our above-noted Website.) The dashed white curves in Plate 1 mark the day-night terminator, which passes very near the north magnetic pole in the last nine images shown (i.e., in rows 2–4). The day side corresponds to the upper part of each image in Plate 1.

The region of X-ray illumination in Plate 1 at first (just after 0100 UT, 11 April 1997) straddled the dawn meridian and spanned the range 01–08 hr MLT. The illuminated region then expanded its range westward into the evening sector over the next 1–2 hours (Plate 1, rows 2–3). The illuminated portion of the evening sector then continued to brighten in X rays, so that this became the region of maximum X-ray intensity by 0250 UT. The X-ray intensity there had greatly increased by 0307 UT, which corresponded to the time of maximum globally radiated X-ray power in Figure 1. Our last available image of this sequence (centered on 0323 UT) showed a westward displacement of the correspondingly bright X-ray feature in Plate 1, with a slightly reduced (but still high) globally radiated X-ray power in Figure 1.

The region of X-ray illumination in row 4 of Plate 1 extended from about 1900 MLT to about 0800 MLT at magnetic latitudes between 55° and 75°. The subregion of maximum X-ray intensity (coded red) appeared at magnetic latitudes between 60° and 70° in the last two panels. Its MLT range was about 2000–2200 in the 0307 UT image and about 1830–2130 in the 0323 UT image. This trend suggests a westward-traveling surge, which occurred over North America in the pre-midnight sector of MLT. The auroral intensification associated with this surge is evident already at 0251 UT in Plate 1, and perhaps even as early as 0235 UT. It occurred well before the most negative value attained by *Dst* during the storm (–80 nT during hour 5, cf. Figure 1) and thus well before recovery phase began. The region of least X-ray intensity around the auroral oval was centered as usual [e.g., *Chenette et al.*, 1993] on the afternoon sector in Plate 1. Auroral images at visible or UV wavelengths typically show a different pattern, in which the auroral intensity is more nearly uniformly distributed in MLT around the auroral oval.

Plate 2 shows how the nonzero values of half-globally radiated auroral X-ray power, indicated by crosses (×) in Figure 1, were distributed in MLT (divided into 15° bins centered on the indicated hours). As might have been expected from Plate 1, most of the radiated X-ray power came from the westward-traveling surge, whose centroid moved (according to Plate 2) from hour 21 MLT at 0307 UT to hour 20 MLT at 0323 UT. The surge developed over the course of 30 minutes (0250–0320 UT), but a precursor enhancement in the x-ray emission occurred as early as 0235 UT. The surge began at least an hour before *Dst* attained its most negative value (cf. Figure 1). Moreover, although the surge must have released considerable energy, the rate of change of *Dst* definitely moderated at the time of the surge (as if the surge were not associated with a major injection of hot plasma into the ring current). These observations support the suggestion of *McPherron* [1997, p. 137] that growth of the ring current is not dependent on hot plasma injections of the type characteristic of substorm expansion phase.

Plate 2 also shows a persistent maximum (primary until 0240 UT, secondary thereafter) in auroral X-ray intensity centered near dawn (hours 03–06 MLT). Perhaps surprisingly, the intensity associated with this "dawn" feature (integrated over 3–5 hours of MLT about its centroid) seems to have been least at 0307 UT, when the surge intensity was greatest. Plate 2 also confirms a gradual but persistent temporal increase in auroral X-ray intensity throughout the nightside range of MLT values from 0104 UT to 0307 UT.

The major local (nightside) and global increase in auroral X-ray intensity over just two hours of UT may be an important clue to the physical processes responsible for

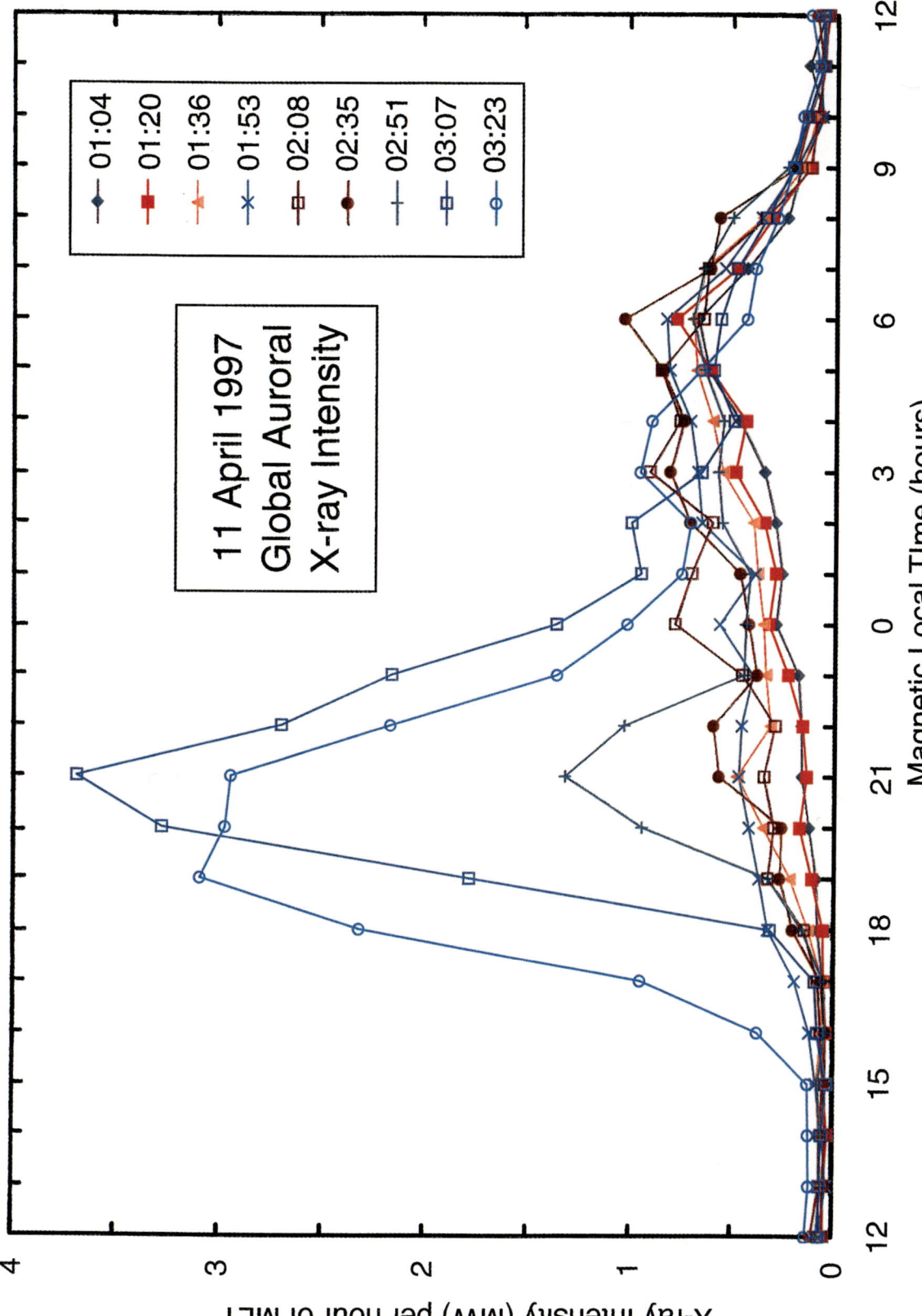

Plate 2. Contributions to total (half-globally) radiated X-ray power (cf. Figure 1) from 1-hr bins of MLT for selected UT values corresponding to images in Plate 1. Specified times (UT) refer to mid-points of accumulation intervals.

supplying and energizing the corresponding electrons. Such a sudden increase in auroral X-ray intensity (spanning many hours of MLT) could hardly be the direct consequence of a modulated electron source, for example, unless it occurred very near (e.g., within an hour's drift time of) the boundary between closed and open field lines (where the source might logically reside). Otherwise, the participating electrons must have belonged to an existing magnetospheric particle population suddenly subjected to a drastically changed dynamical environment.

INFERRED ELECTRON SPECTRAL PARAMETERS AND IONOSPHERIC EFFECTS

Results described above are based on measurements of total auroral X-ray intensity in the 2–12 keV range, which corresponds to 64 PIXIE energy channels [*Imhof et al.*, 1995, pp. 401–402]. To improve statistics we typically combine X-rays into energy bins of several channels each when constructing a spectrum. Since the X-ray spectrum produced by an incident electron of specified energy is known, we can in principle deduce the corresponding spectrum of precipitating electrons from the observed X-ray spectrum. If (as here) we treat the incident electron spectrum as exponential, for example, we can infer the best-fitting energy flux (or precipitating power) and characteristic (*e*-folding) energy of the precipitating electron spectrum that would have produced the X-rays. Our results constitute the first-ever determinations of electron energy flux and *e*-folding energy (E_0) as functions of MLT to be extracted from global auroral X-ray measurements that cover an entire hemisphere at the same time.

The X-ray energy spectrum shown histographically in Figure 2 corresponds to a 3-hr MLT bin centered on midnight in the image centered on 0307 UT in Plate 1. This is representative of the X-ray spectra we have analyzed so far. The solid line in Figure 2 represents the exponential electron spectrum that best replicates the numbers of counts recorded in the nine X-ray energy bins shown. We obtained this electron spectrum by applying the method of *Chenette et al.* [1993] and *Winningham et al.* [1993] to extract optimal estimates (minimizing χ^2) for the characteristic electron energy E_0 and area-integrated electron energy flux from the given X-ray spectrum. We thus extracted two electron spectral parameters from nine data bins consolidated from 42 X-ray channels in the 2–6.5 keV energy range. The dashed curve in Figure 2 represents (for comparison with histographic values) the smooth X-ray spectrum that would have resulted from the same exponential spectrum of precipitating electrons.

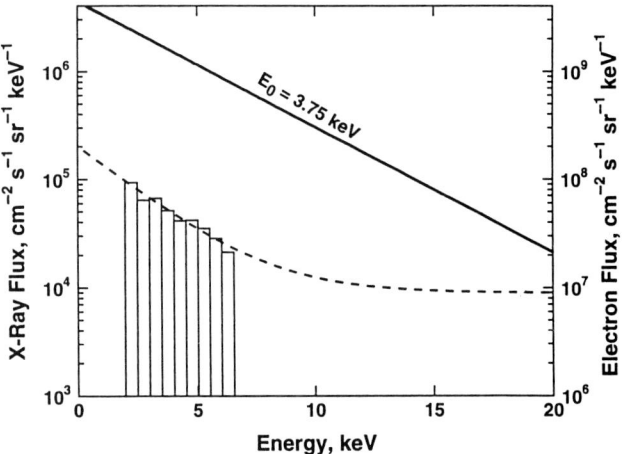

Figure 2. Histogram shows representative X-ray spectrum (2230–0130 MLT) from PIXIE image centered on 0307 UT, 11 April 1997. Solid line shows optimal (χ^2 minimizing) exponential electron spectrum. Dashed curve shows smooth X-ray spectrum that would result from optimal electron spectrum. Specified times (UT) refer to mid-points of accumulation intervals

We performed the same analysis on the 2–12 keV X-ray spectra corresponding to each of eight 3-hr MLT bins (with X-ray counts summed over magnetic latitude in the northern hemisphere) for the last nine images in Plate 1 (i.e., for all available images from 11 April 1997). The best-fitting values for electron power (area-integrated energy flux) and characteristic electron energy (E_0) that resulted from this analysis are shown in Figure 3. These results specify the exponential electron energy spectra that best fit the measured X-ray spectra. The same method can (of course) be used on other model shapes for the electron spectrum. Whether other postulated functional forms for the electron energy spectrum might have yielded still better fits to the measured X-ray spectra is a subject of continuing study, but we are reluctant (because of the trade-off between statistics and spatial/energy/time resolution) to extract too many adjustable parameters from our X-ray spectra. We assume for now that incident electron spectra are adequately described as exponentials (with spectral parameters dependent on MLT and UT) at least from a few keV to about 20 keV.

The upper panel of Figure 3 shows the inferred characteristic electron energy E_0 for each 3-hour MLT bin in each of the nine PIXIE images from 11 April 1997 in Plate 1. The lower panel shows the area-integrated precipitating electron energy flux (i.e., the precipitating electron power) for each 3-hour MLT bin in each image. The eight "data" points within each exposure interval in Figure 3 correspond

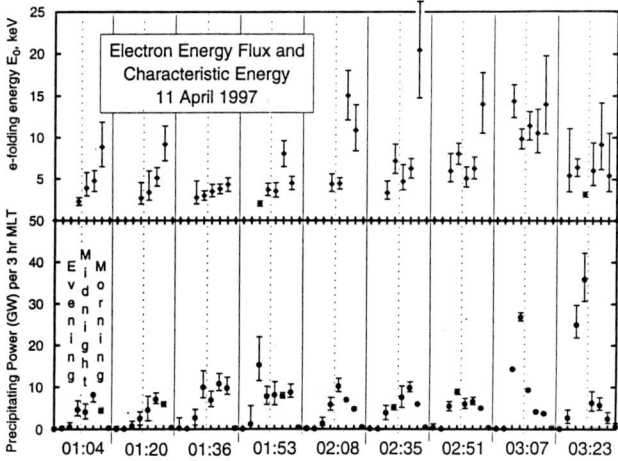

Figure 3. Electron spectral parameters determined as in Figure 2 for consecutive 3-hr bins of MLT from 11 April 1997 images (cf. Plate 1). Characteristic energy E_0 is indeterminate (plotted as zero) when precipitating energy flux is is too low (typically in afternoon quadrant). Specified times (UT) refer to mid-points of accumulation intervals

to bin centers that progress (left to right) eastward from noon (1200 MLT) through midnight (2400 MLT) to mid-morning (0900 MLT), as the vertically written words in the 0104 UT sub-panel indicate. Error bars reflect uncertainties in the parameters fitted through a least χ^2 analysis.

According to Figure 3, the precipitating power in keV electrons in the 1800–0600 MLT sector (dusk through midnight to dawn) increased steadily to about 3 GW per hour of MLT (~ 10 GW per 3-hr bin) during the two hours (~ 0100–0300 UT) when the ring-current intensity was clearly growing. There was, moreover, a significant and persistent hardening of the electron energy spectrum with increasing MLT in this sector. Total hemispheric power carried by precipitating electrons with energies above a few keV increased from about 21 GW to nearly 50 GW during the 0100–0300 UT interval. Precipitating power in the surge region exceeded total hemispheric power in the rest of the aurora, with peak power densities ~ 10 GW per hour of MLT. The surge spectrum was initially harder (E_0 ~10–15 keV) and then softened (E_0 ~ 5–10 keV) from 0100 UT to 0300 UT. (This last trend appears to be real, but e-folding energies determined for the surge region do have rather large uncertainties.)

The bremsstrahlung X-rays detected by PIXIE occur because the precipitating electrons characterized in Figure 3 decelerate as they penetrate the upper atmosphere and ionosphere. The main cause of electron deceleration here is ionization of the target medium (i.e., the neutral upper atmosphere). Precipitating electrons thus increase the number density of ionospheric electrons (and thereby increase the local Hall and Pedersen conductivities) by an amount that depends on their energy flux and also on the value of E_0 (since the altitude to which an electron will probably penetrate depends inversely on its incident energy). *Robinson et al.* [1987] have concisely summarized the dependence of resulting height-integrated Hall and Pedersen conductivities (Σ_H and Σ_P) on precipitating electron energy flux and characteristic energy by means of simple semi-empirical expressions. Conductance results corresponding to the spectral parameters in Figure 3 are shown in Figure 4. Although the major source of precipitating electron energy was associated with the surge, the weaker precipitation preceding the surge also must have had significant ionospheric effects. Using the relationships provided by *Robinson et al.* [1987], we estimate that the electron precipitation inferred from PIXIE data must have produced Pedersen and Hall conductances ~ 5–8 mhos over a wide range of MLT (from midnight to beyond dawn) during the 0100–0300 UT interval.

One reservation regarding our derivation of spectral parameters for precipitating auroral electrons in the present study from auroral X-ray spectra is that we have not yet taken account of X-rays produced other than by bremsstrahlung. In fact there should be a small but significant X-ray contribution from the K_α fluorescence of atmospheric

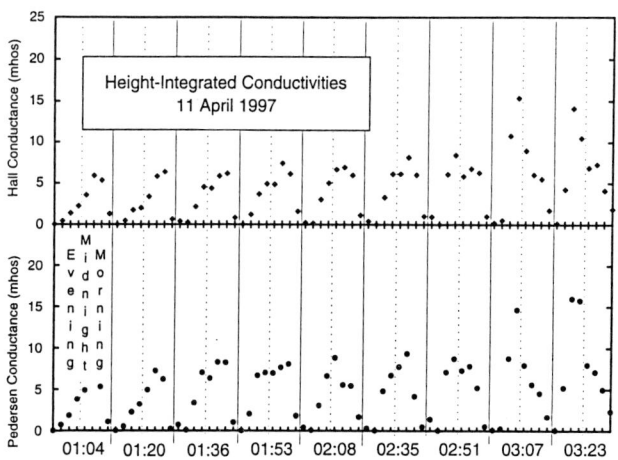

Figure 4. Height-integrated Hall and Pedersen conductivities, as given by semi-empirical formulas of *Robinson et al.* [1987], for electron spectral parameters shown in Figure 3. Conductances are not necessarily negligible even when E_0 is indeterminate (plotted as zero in Figure 3) because they scale as the square root of the precipitating energy flux.

argon excited by the same precipitating auroral electrons. This is problematic for our extraction of electron spectral parameters from the measured X-ray spectrum, since K_α fluorescence corresponds to a line at about 3 keV in the X-ray spectrum, whereas any thick-target bremsstrahlung X-ray spectrum is supposed to be monotonic. A peaked X-ray spectrum, in other words, could not be inverted by Green's-function methods to yield the corresponding electron spectrum unless cross-sections for line emission were made part of the analysis. Conversely, the spectral parameters extracted by minimizing χ^2 without accounting for line emission may be unreliable (contaminated by argon K_α fluorescence) even if the observed X-ray spectrum is nominally monotonic. We are (of course) working to include the effects of X-ray line emissions in future analyses of X-ray spectra measured by PIXIE.

MAPPING STUDY

Since the auroral X rays such as those detected by PIXIE are an important signature of the precipitation of plasmasheet electrons with energies ~ 2–50 keV, it would be interesting to know where in the magnetosphere these electrons had resided immediately before precipitating. Comparisons could possibly then be made with phase-space densities of directly observed plasmasheet electrons in such regions (e.g., from spacecraft in geosynchronous equatorial orbit). Such information regarding the dynamical history of participating particles would likely clarify the nature of transport processes that lead ultimately to auroral electron precipitation.

For this eventual purpose we have made field-aligned mappings from PIXIE images acquired during the representative magnetic storm of 11 April 1997 to the equatorial magnetosphere. Such mappings are necessarily model-dependent. Indeed, they can serve to test the field models used and to help evaluate adjustable modeling parameters, as well as to infer the origins of precipitating particles. For the present mapping work we have used version T96_01.FOR (available on the World Wide Web at http://www-spof.gsfc.nasa.gov/Modeling/T96.html) of the data-driven field model described by *Tsyganenko* [1995], as well as a parametrically adjustable version of the field model described by *Schulz and McNab* [1987, 1996]. The present study thus complements (for example) mappings of Viking ultraviolet (UV) auroral images [*Elphinstone et al.*, 1991] with the model of *Tsyganenko* [1987] as well as theoretical mapping studies [e.g., *Kaufmann et al.*, 1990; *Ding et al.*, 1994] based on various field models. (Such mappings need not be the same for X-ray features as for visible and UV features, since X-ray features involve higher-energy electrons.)

The images shown in Plate 1 have not been corrected for our detector's limited spatial resolution. Indeed, they have been further smoothed (relative to the pixel locations nominally recorded in the image plane of our instrument) in order to suppress some unwanted quasi-sinusoidal fine-structure (nonlinearity) in the relation between the voltage reported by our position sensor and the actual position at which a photon crossed the image plane [cf. *Imhof et al.*, 1995, pp. 396–398]. For these reasons the latitudinal extent of each X-ray intensity pattern in Plate 1 seems unrealistically wide, hence unsuited for direct mapping to the equatorial magnetosphere.

To test our hypothesis that the latitudinal extent of the X-ray images in Plate 1 is largely instrumental, we chose the second image from row 4 (centered on 0307 UT and reproduced in Plate 3a) as representative. We computed the mean value of sin λ among the photons recorded in each of 72 sectors (5° bins) of MLT for this image and then tentatively assigned all counts from a given MLT sector to the corresponding value of λ (geomagnetic latitude). The result is shown as a hypothetical source distribution in Plate 3b. Here the color scale ranges from red (up to 350 counts per 280-s accumulation interval) to violet (down to 20 counts) to black (< 20 counts). The bins in Plate 3b are 5° wide in MLT. They are everywhere equal in area and are approximately square at $\lambda \approx 67°$, where the X-ray intensity typically seemed greatest. (Our intention is to use such bins eventually for quantitative description of the X-ray intensity distribution in magnetic latitude and MLT, once we decide upon an optimal image-enhancement technique to infer "true" distributions of X-ray intensity from apparent distributions such as those seen in Plate 3a.)

Given the hypothetical source distribution in Plate 3b, we numerically projected the corresponding hypothetical intensities back through the central aperture of our PIXIE instrument to obtain the pseudo-image shown in Plate 3c. The resemblance between Plates 3a and 3c quite strikingly confirms that the apparent latitudinal spread of X-ray features in Plate 3a is largely an instrumental effect. If there were no difference at all in latitudinal width between the X-ray features in the Plate 3a and those in Plate 3c, then the "true" source distribution could not be distinguished (at our level of resolution) from the source distribution in Plate 3b. The X-ray feature in our selected PIXIE image (Plate 3a) actually is somewhat broader in latitude than the corresponding feature in our pseudo-image (Plate 3c), and so there is hope of eventually resolving the "true" source distribution with somewhat better resolution than we show in Plate 3b. For now, however, we feel justified in mapping the "hypothetical" source distribution from Plate 3b to the equatorial magnetosphere as a means of learning where the corresponding electrons came from.

The mean value of sin λ among photons recorded in any 5° sector of MLT does not necessarily correspond to the magnetic-latitude bin in which the most counts were recorded from that sector. We chose the former as being suitably "representative" of the photons in any MLT sector rather than decide what to do if the latitudinal distribution of X-ray intensity happened (for statistical or other reasons) to be bimodal in that sector. This is admittedly not the only way to infer an optimal source distribution from a PIXIE image, but it is somewhat reminiscent (sector by sector) of linear regression: The mean-square deviation of sin λ from the mean value of sin λ in any sector is less than the mean-square deviation of sin λ from any other value of sin λ. (Of course, the same could be said of λ and its mean value, but this would still lead to a source distribution which closely resembles Plate 3b.)

The multicolored contour projected onto the XY plane (GSM coordinates) in Plate 4a represents a mapping of the hypothetical source distribution from Plate 3b (with the same color scale) to the magnetic equatorial surface (locus of minima in field intensity B along field lines) in the T96_01 version of the field model described by *Tsyganenko* [1995]. (We have made plots similar to Plates 3b, 3c, and 4a also for several additional X-ray images from 11 April 1997, but there is enough room for just one representative example here.) Required values for the upstream solar-wind velocity, solar-wind pressure, and interplanetary magnetic field (IMF) were obtained from GGS/Wind plasma [*Ogilvie et al.*, 1995] and magnetometer [*Lepping et al.*, 1995] key parameter data (http://cdaweb.gsfc.nasa.gov/cdaweb/istp_public/). We selected 15-minute averaged plasma and **B**-field data from the time interval most nearly centered 55 minutes earlier than the X-ray image of interest in each case, since the solar wind velocity was about 450 km/sec at 0200–0400 UT that day. (The GGS/Wind spacecraft is stationed about 1.5×10^6 km upstream from Earth.) Required values for *Dst* were obtained from Kyoto University's provisional compilation (http://swdcdb.kugi.kyoto-u.ac.jp/dstdir/dst1/prov.html) for the hour corresponding to each X-ray image of interest. The magnetopause size, ellipsoidal shape, and orientation (as aberrated by Earth's orbital motion) in Plate 4a correspond to the measured upstream plasma parameters in the manner specified by *Tsyganenko* [1995]. Plate 4a shows that the most intense X-ray features (coded red in Plate 3) have mapped the farthest out (i.e., to $R \sim 15$) in the equatorial magnetosphere. This result suggests an association of the most intense X-ray emissions during this event with auroral arcs. That the less intense X-ray emissions (coded blue, violet, or black in Plate 3) have mapped to smaller values of equatorial R (~ 5–10) in Plate 4a suggests an association of such features during this event with the diffuse aurora. Dashed black portions of the multicolored mappings correspond to "source" intensities < 20 counts/sector. These are (as might have been expected from Plate 3) centered on the afternoon sector in Plate 4a.

The distinction between near-Earth and distant-tail mappings shows up more starkly in the magnetospheric source-surface model of *Schulz and McNab* [1987, 1996]. This is a less accurate representation of the magnetospheric **B** field than Tsyganenko's models, but it contains an adjustable parameter (called ξ_0^*) that literally specifies the geocentric distance to the nightside boundary between the model's closed and open (i.e., tail) magnetic field lines. The model contains two other adjustable parameters called b and ρ_∞^*, which measure (respectively) the geocentric distance to the "nose" of the magnetopause and the asymptotic radius of the quasi-cylindrical tail region, but these are held fixed (at 10 R_E and 17.7 R_E, respectively) in the present study. Another adjustable parameter is the angle ψ between dipole moment and solar-wind velocity, but this was very nearly 90° when the present X-ray data were taken and so is set equal to 90° (as in *Schulz and McNab* [1987]) for the present mapping study with the source-surface model. (Spherical-harmonic expansion coefficients $\{g_n^m\}$ for ξ_0^*/b = 0.6, 0.7, 0.8, . . ., 1.6 are available electronically by contacting either Schulz or McNab.)

The multicolored contours in Plates 4b and 4c represent mappings of the hypothetical source distributions from Plate 3b (with the same color scale) to the magnetic equatorial plane in the field model of *Schulz and McNab* [1996] for two distinct values of ξ_0^* (7 R_E in Plate 4b and 6 R_E in Plate 4c). The solid black curve in Plates 4b and 4c represents the magnetopause as specified by *Schulz and McNab* [1987]. The smooth dashed black curve perpendicular to the magnetopause in each panel marks the boundary between closed and open (i.e., tail) field lines in this model. Plate 4b shows that the hypothetical X-ray source distribution from Plate 3b resides entirely on closed field lines if $\xi_0^* \geq 7\ R_E$. (Since corresponding auroral features map to ever smaller values of equatorial R as ξ_0^* is increased, plots made for values of $\xi_0^* \geq 8\ R_E$ are uninteresting in the present context.) Plate 4c shows that the more intense X-ray emissions (coded red, orange, yellow, and green in Plate 3) map into the distant tail if $\xi_0^* \leq 6\ R_E$. (Only the lateral boundaries of mappings into the tail are shown explicitly in Plate 4c, since other field lines that map into the tail do not reach the equatorial plane.)

If indeed the most intense X-ray emissions (coded red in Plate 3) during this event are to be associated with auroral arcs, then Plates 4b and 4c suggest something between 6 R_E and 7 R_E as the most appropriate stormtime value (at least during this event) for the adjustable parameter ξ_0^* in the

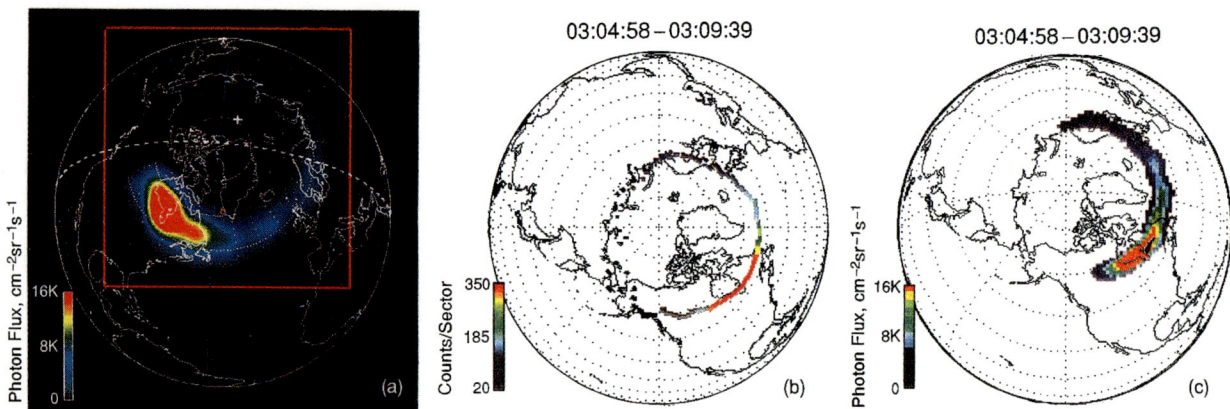

Plate 3. (a) Representative X-ray image from anode 1 (2–12 keV) during event studied here (03:04:58–03:09:39 UT, 11 April 1997). Coordinate contours indicate magnetic latitude at 10° intervals and magnetic local time (MLT) at 15° intervals. Blue curve marks noon-midnight magnetic meridian (MLT = 12 hr on solid portion, 00 hr on dashed portion). Dashed white curve marks day-night geographic terminator. White plus sign marks north geographic pole. Brown circle near Nuuk (Godthåb) marks subsatellite point. Red rectangle outlines PIXIE field of view. Provisional Dst values were −68 nT for hour 3 and −74 nT for hour 4. (b) Hypothetical X-ray source distribution, with all counts from each 5° MLT sector consolidated into a single magnetic latitude bin (corresponding to the mean value of $\sin \lambda$ among photons detected in that sector). Because of poor statistics, sectors with < 20 counts (coded black) show more scatter in their representative values of λ. Coordinate contours indicate geographic latitude at 10° intervals and geographic longitude at 45° intervals. (c) Pseudo-image obtained by projecting hypothetical source distribution from Plate 3b back through central aperture of PIXIE instrument. Color scale (after adjustment for the number of apertures) is the same as for Plate 3a. Coordinate contours indicate geographic latitude at 15° intervals and geographic longitude at 45° intervals.

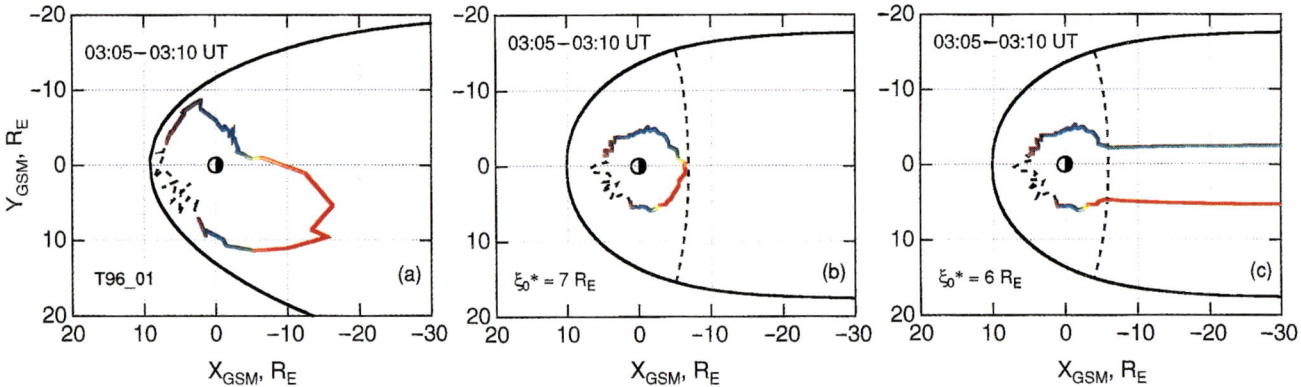

Plate 4. (a) Projected mapping (multicolored contour) of hypothetical source distribution from Plate 3b (with same color scale) to equatorial (minimum-B) surface in T96_01 version of *Tsyganenko* [1995] field model. Solid black curve represents the magnetopause in this model. (b, c) Mappings (multicolored contours) of hypothetical source distribution from Plate 3b (with same color scale) to equatorial plane in two different realizations of source-surface field model [*Schulz and McNab*, 1996], with aberration by Earth's orbital motion neglected. Solid black curve represents the magnetopause in this model, and smooth dashed black curve represents the boundary between closed and open field lines (inner edge of cross-tail current sheet) in each case. Jagged dashed black portions of the multicolored mapping contours correspond to "source" intensities < 20 counts/sector in all three panels.

source-surface model. The most intense X-ray features would then be associated with the boundary between closed and open (i.e., polar cap) field lines, whereas the less intense X-ray emissions (coded blue, violet, or black in Plate 3) that have mapped to smaller values of equatorial R (~ 5) in Plates 4b and 4c would be associated (as seems reasonable) with the diffuse aurora. Dashed black portions of the multicolored mappings correspond to "source" intensities < 20 counts/sector. These are (as might have been expected from Plate 3) centered on the afternoon sector in Plates 4b and 4c.

The above paragraphs summarize our interpretation of the present mapping results. We are inclined to identify the most intense features in our X-ray images with auroral arcs because these features map to the largest equatorial values of R (especially in the PM half of the nightside magnetosphere) in the optimally data-based field model of *Tsyganenko* [1995]. We can use this identification to estimate the best value for the critical parameter ξ_0^* ($\approx$ 6–7 R_E for the present event) in the mathematically idealized (source-surface) model of *Schulz and McNab* [1996], since ξ_0^* explicitly specifies the downstream distance to equatorial boundary between closed and open field lines in that model. (Features that map to this boundary in the source-surface model map in turn to the entire "downstream" length of the neutral sheet.) Features associated here with the diffuse aurora map to R ~ 5 in the equatorial region of the source-surface model.

SUMMARY

This has been a survey of early results from PIXIE, our X-ray imager aboard the GGS/Polar satellite. With PIXIE we have obtained the first-ever global images of the aurora at X-ray wavelengths. We have compiled a sequence of such images through the course of the magnetic storm that occurred 10–11 April 1997, and we have computed the total globally radiated X-ray power for each image in the sequence. The radiated power at first grew slowly (from near zero to 7 MW at 1–2 MW/hr) before suddenly doubling within 20 minutes to reach its maximum value ($\approx$ 14 MW) at least an hour before *Dst* attained its most negative value (–80 nT) for the storm.

By sorting the X-rays from each image into 1-hr bins of magnetic local time (MLT), we found a superposition of two overlapping but spatially dissimilar components: The sharp increase in global X-ray intensity corresponded to a spatially limited enhancement (~ 5 hr wide at half maximum) that traveled westward (as would a surge) through the evening sector. The other component, whose intensity grew slowly but steadily through the 2.5-hr image sequence, showed (as would the diffuse aurora) a somewhat broader maximum centered around dawn. Using the spectrum of X-rays from each hourly (15°) MLT bin, we have inferred optimal values for the energy flux and characteristic energy of the best-fitting exponential electron spectra. From these spectral parameters we have estimated the height-integrated Hall and Pedersen conductivities that would have resulted from the precipitation of such electrons.

Using representative global models [*Tsyganenko*, 1995; *Schulz and McNab*, 1996] of the magnetospheric **B** field, we have mapped outstanding X-ray features (as represented by the mean magnetic latitude among photons in each 1-hr MLT bin) along magnetic field lines to the equatorial magnetosphere for the most intense image of the present sequence. We have done this in order to learn where the corresponding electrons must have resided immediately before precipitating. Finding that the most intense X-ray feature (i.e., the one in the evening sector) typically maps (according to Tsyganenko's model) farthest out in the equatorial magnetosphere, we have used this feature (associated with an auroral arc or westward-traveling surge) to estimate the equatorial geocentric distance to the boundary between closed and open field lines in the source-surface model [*Schulz and McNab*, 1996], which treats this distance as an adjustable parameter.

We are still working to refine our X-ray images by adapting standard image-enhancement techniques to our instrument geometry. This step will (we hope) enable us to perform future mapping studies on fairly realistic iso-intensity contours rather than (as at present) merely on intensity maxima or other "representative" magnetic latitudes as functions of magnetic local time (MLT). We are also working to extract more reliable electron energy spectra from X-ray energy spectra in our still-growing PIXIE database. Characteristic energies of plasmasheet electron spectra are expected to vary with X and Y (GSM coordinates) in the plasma sheet, and such variation should express itself (via field-aligned mappings) in the spatial distributions of auroral X-ray intensities detected in different energy channels by our PIXIE instrument.

Acknowledgments. The authors thank N. A. Tsyganenko for having made his field models available to the magnetospheric community via the World Wide Web. We thank GGS/Wind principal investigators K. H. Ogilvie and R. P. Lepping, as well as Kyoto University's World Data Center C2 (WDC-C2), for making their data readily available on the World Wide Web. The coordinator (M. Schulz) of the mapping study especially thanks D. P. Stern, Y. I. Feldstein, and I. I. Alexeev for recent discussions on the philosophy of magnetospheric modeling. The present research was supported primarily by NASA contract NAS5-30372 for the analysis of PIXIE data from the GGS/Polar satellite. The incorporated work on field modeling has been supported also by NSF grant ATM-9119516 for the interpretation of Søndre Strømfjord radar data, the Independent Research and Development (IRAD) program of Lockheed Martin Missiles & Space, and the Aerospace Sponsored Research (ASR) program of The Aerospace Corporation.

REFERENCES

Acuña, M. H., K. W. Ogilvie, D. N. Baker, S. A. Curtis, D. H. Fairfield, and W. H. Mish, The Global Geospace Science program and its investigations, *Space Sci. Rev.*, *71*, 5–21, 1995; reprinted in *The Global Geospace Mission*, edited by C. T. Russell, pp. 5–21, Kluwer, Dordrecht, 1995.

Chenette, D. L., D. W. Datlowe, R. M. Robinson, T. L. Schumaker, R. R. Vondrak, and J. D. Winningham, Atmospheric energy input and ionization by energetic electrons during the geomagnetic storm of 8–9 November 1991, *Geophys. Res. Lett.*, *20*, 1323–1326, 1993.

Ding, C., T. W. Hill, and F. R. Toffoletto, Magnetic mapping and Birkeland currents in the Toffoletto-Hill and Tsyganenko magnetosphere models, *J. Geophys. Res.*, *99*, 17343–17350, 1994.

Elphinstone, R. D., D. Hearn, J. S. Murphree, and L. L. Cogger, Mapping using the Tsyganenko long magnetospheric model and its relationship to Viking auroral images, *J. Geophys. Res.*, *96*, 1467–1480, 1991.

Imhof, W. L., G. H. Nakano, R. G. Johnson, and J. B. Reagan, Satellite observations of Bremsstrahlung from widespread energetic electron precipitation events, *J. Geophys. Res.*, *79*, 565–574, 1974.

Imhof, W. L., et al., The Polar Ionospheric X-ray Imaging Experiment (PIXIE), *Space Sci. Rev.*, *71*, 385–408, 1995; reprinted in *The Global Geospace Mission*, edited by C. T. Russell, pp. 385–408, Kluwer, Dordrecht, 1995.

Kaufmann, R. L., D. J. Larson, and C. Lu, Mapping and distortion of auroral structures in the quiet magnetosphere, *J. Geophys. Res.*, *95*, 7973–7994, 1990.

Lepping, R. P., et al., The Wind magnetic field investigation, *Space Sci. Rev.*, *71*, 207–229, 1995; reprinted in *The Global Geospace Mission*, edited by C. T. Russell, pp. 207–229, Kluwer, Dordrecht, 1995.

McPherron, R. L., Magnetospheric dynamics, in *Introduction to Space Physics*, edited by M. G. Kivelson and C. T. Russell, ch. 13, pp. 400–458, Cambridge Univ. Press, New York, 1995.

McPherron, R. L., The role of substorms in the generation of magnetic storms, in *Magnetic Storms*, edited by B. T. Tsurutani, W. D. Gonzalez, Y. Kamide, and J. K. Arballo, pp. 131–147, Geophys. Monogr. 98, Am. Geophys. Union, Washington, D. C., 1997.

Mizera, P. F., J. G. Luhmann, W. A. Kolasinski, and J. B. Blake, Correlated observations of auroral arcs, electrons, and X rays from a DMSP satellite, *J. Geophys. Res.*, *83*, 5573–5578, 1978.

Ogilvie, K. H., et al., SWE, a comprehensive plasma instrument for the Wind spacecraft, *Space Sci. Rev.*, *71*, 55–77, 1995; reprinted in *The Global Geospace Mission*, edited by C. T. Russell, pp. 55–77, Kluwer, Dordrecht, 1995.

Robinson, R. M., R. R. Vondrak, K. Miller, T. Dabbs, and D. Hardy, On calculating ionospheric conductances from the flux and energy of precipitating electrons, *J. Geophys. Res.*, *92*, 2565–2569, 1987.

Schulz, M., and M. C. McNab, Source-surface model of the magnetosphere, *Geophys. Res. Lett.*, *14*, 182–185, 1987.

Schulz, M., and M. C. McNab, Source-surface modeling of planetary magnetospheres, *J. Geophys. Res.*, *101*, 5095–5118, 1996.

Tsyganenko, N. A., Global quantitative models of the geomagnetic field in the cislunar magnetosphere for different disturbance levels, *Planet Space Sci.*, *37*, 1347–1358, 1987.

Tsyganenko, N. A., Modeling the Earth's magnetospheric magnetic field confined within a realistic magnetopause, *J. Geophys. Res.*, *100*, 5599–5612, 1995.

Winningham, J. D., et al., The UARS particle environment monitor, *J. Geophys. Res.*, *98*, 10649–10666, 1993.

D. L. Chenette, W. L. Imhof, S. M. Petrinec, M. Schulz, J. Mobilia, J. G. Pronko, M. A. Rinaldi, and J. B. Cladis, Space Physics Department (O/H1-11, B/255), Advanced Technology Center, Lockheed Martin Missiles & Space, Palo Alto, CA 94304

F. Fenrich, Space Sciences Laboratory, University of California, Berkeley, CA 94720

N. Østgaard, Department of Physics, University of Bergen, N-5007 Bergen, Norway

M. C. McNab, Space & Environment Technology Center, The Aerospace Corporation, El Segundo, CA 90245

Numerical Cavity Mode Simulation and Polar Data From the January 1997 Magnetic Cloud Event

J. Goldstein, R. E. Denton, M. K. Hudson, W. Lotko,[1] and J. G. Lyon

Department of Physics and Astronomy, Dartmouth College, Hanover, NH 03755, U.S.A.

Data from the Plasma Wave Instrument (PWI) on the Polar satellite were used to construct an electron number density for the dawnside plasmasphere of January 11, 1997. This density profile was incorporated into a dynamic numerical simulation of a cold, ideal MHD plasmasphere on a dipole grid, with a broad-band excitation applied at the plasmapause. In the simulated plasmasphere, this impulsive perturbation produced a discrete spectrum of fast magnetosonic cavity modes, coupled to field line resonances (FLRs) at various harmonics. However, spectral analysis of electric and magnetic fields measured by the Polar satellite on January 11 does not provide convincing evidence for plasmaspheric cavity modes at the frequencies and local time predicted by the model.

1. INTRODUCTION

It has been proposed that the magnetosphere, bounded by the magnetopause and the ionosphere, and separated into the inner magnetosphere (i.e., the plasmasphere) and outer magnetosphere, can act as a resonant cavity for ultra-low-frequency (ULF) magnetosonic waves [*Kivelson et al.*, 1984; *Allan et al.*, 1986; *Zhu and Kivelson*, 1988; *Lee and Lysak*, 1989]. If this cavity mode hypothesis is correct, the inner and outer regions of the magnetosphere should each possess an intrinsic discrete spectrum of fast magnetosonic standing waves. The *plasmaspheric cavity mode* occupies the region between the ionosphere and plasmapause. The *outer cavity mode* resonates between the plasmapause and magnetopause. The third class of cavity mode oscillation,

---

[1] Thayer School of Engineering, Dartmouth College, Hanover, NH 03755, U.S.A.

Sun-Earth Plasma Connections
Geophysical Monograph 109
Copyright 1999 by the American Geophysical Union

the *"tunneling"* or *global mode*, extends from the ionosphere to the magnetopause, with comparable wave amplitudes in the plasmasphere and outer magnetosphere [*Zhu and Kivelson*, 1989]. This global mode arises as a result of tunneling through the Alfvén velocity barrier at the plasmapause. Furthermore, unless the wave fields are completely axisymmetric, individual cavity modes will be coupled to one or more standing toroidal Alfvén mode oscillations. This coupling is strongest at the field line resonance (FLR), where the frequency of the cavity mode matches the toroidal eigenfrequency of the local magnetic field line. In effect, the magnetosphere acts like a bell [*Kennel*, 1995], in which the fast-mode waves play the role of sound waves in air. When the bell is struck or tapped, it rings with a discrete spectrum of standing waves. To extend this analogy, the geomagnetic field lines can be envisioned as harp strings mounted inside the bell. Those strings whose natural frequencies match that of the ringing notes in the bell will vibrate in sympathy; individual field line resonances should act analogously to these selected resonating harp strings.

Despite the intuitive appeal of the theory, with few exceptions [*Kivelson et al.*, 1984, 1997] direct observational evidence for outer magnetospheric cavity modes is rare. This scarcity might be due to a number of causes; two are mentioned here. First, the existence of the magnetotail suggests that there is no stable enclosing boundary to confine outer cavity modes on the nightside; this reduces available locations for outer cavity resonances by half. Second, variability in the solar wind might mean that the magnetopause rarely sits still long enough to allow fast-mode standing waves to form. The plasmaspheric cavity suffers neither of these shortcomings; in principle, cavity modes might be a more common feature of this region. Recent observations suggest that Pc 3 and Pi 2 pulsations might be inner cavity modes [*Yumoto et al.*, 1992, 1994; *Takahashi et al.*, 1992, 1995; *Yumoto et al.*, 1990].

On a phenomenological level, cavity mode theory is fairly advanced; further work must be directed toward reconciling theoretical predictions with observations. One hypothetical mechanism for stimulating cavity modes is a strong perturbation to the magnetosphere such as occurs in sudden interplanetary impulses [*Yumoto et al.*, 1992], and storms. In this paper, we consider the possibility of plasmaspheric cavity mode excitation following the magnetic cloud event of January, 1997 [*Burlaga et al.*, 1998]. To study this event, we used a numerical MHD simulation that is qualitatively similar to previous models by *Lee and Lysak* [1989] and *Allan et al.* [1986], but which incorporates measurements made by the Polar Plasma Wave Instrument (PWI) [*Gurnett et al.*, 1995] to specify a zero-order plasma density. The results were compared with data from two other Polar instruments, the Electric Field Instrument (EFI) [*Harvey et al.*, 1995] and the Magnetic Field Experiment (MFE) [*Russell et al.*, 1995].

2. THE MODEL

The numerical model uses a finite-difference staggered leapfrog algorithm [*Taflove*, 1995] to advance the time-dependent linearized, cold ideal MHD equations. The solutions are obtained in dipole coordinates (as formulated by *Lee and Lysak* [1989]) on a two-dimensional mesh; azimuthal symmetry of the background density and dipole magnetic field is assumed. The linear wave fields **v** and **b** (velocity and magnetic field perturbations, respectively) are assumed to have a harmonic variation in the azimuthal direction, with wavenumber m. The simulation region is filled with a cold, perfectly conducting plasma with background density ρ_0. The boundaries of the cavity are perfect, rigid conductors at the ionosphere and the plasmapause. The equatorial ionosphere is located at $L = 1$, and the northern and southern ionospheres (located at a radius of $r \approx 1 R_E$)

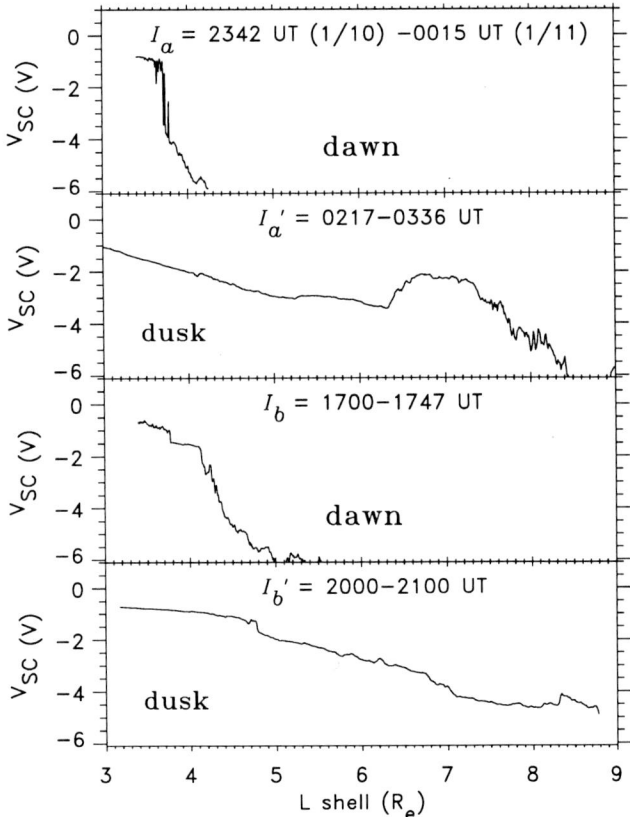

Figure 1. Polar spacecraft potential V_{SC} versus L-shell, measured by the Polar EFI on January 10-11, 1997, during four UT intervals. The intervals labeled "dawn" were selected for simulation in this study.

follow curves which are orthogonal to the dipole field lines. The determination of the plasmapause L value and plasmasphere mass density ρ_0 is described in the next section.

3. PLASMASPHERE CONFIGURATION, JANUARY 11, 1997

Shown in *Figure 1* are four plots versus L-shell of the Polar spacecraft potential V_{SC}, measured by the Polar Electric Field Instrument (EFI) on January 10-11, 1997. The spacecraft potential is correlated with ambient electron density [*Laakso and Pedersen*, 1998]; very approximately, values of potential V_{SC} $(V) = -2$, and -6 correspond to n_e $(cm^{-3})=100$, and 10, respectively. With this in mind, these plots can be used to roughly determine the shape and location of the plasmapause.

During the intervals $I_a = 2342 - 0015\ UT$ (first plot) and $I_b = 1700 - 1747\ UT$ (third plot), Polar passed through the dawn sector of the plasmasphere, approaching perigee from higher L-shells. The earlier dawnside plasmapause (I_a) is located at about $L = 3.7$, and is quite sharp and well-defined, dropping rapidly from $-1V$ to $-4V$ in under $0.1R_e$. This is consistent with the hypothesis that enhanced convection, due to the extended period of southward IMF on the previous day [*Burlaga et al.*, 1998], resulted in erosion of the outer plasmasphere flux tubes. The later dawnside profile (I_b) also shows a sharp edge at $L \approx 3.7$, but it levels off to a plateau of $-1.5V$ (L from 3.7 to 4.2), followed by a gradual decline to $-6V$. This suggests that some slow refilling of the eroded outer flux tubes occured in the 17 hours between I_a and I_b.

In the two duskside passes (second and fourth plots, primed intervals I'_a and I'_b, respectively), the profile at the plasmapause drops much more gradually. Since a weak density gradient is a poor reflector of fast magnetosonic waves, the two-dimensional numerical cavity mode simulations in this study were performed only for the intervals containing the two dawnside profiles, *i.e.* I_a and I_b.

To obtain a quantitative estimate of the electron density during intervals I_a and I_b, data from the Sweep Frequency Receiver (SFR-A) of the Polar Plasma Wave Instrument (PWI) were employed. *Panel (a)* of *Figure 2* shows the intensity ($[V/m]^2/Hz$) of the electric field (E_Z) parallel to the spin axis of Polar, versus frequency, during interval I_a. The dark trace that climbs upward in frequency from left to right is the signature of a noise band whose upper edge is f_{UH}, the upper hybrid resonance [*Carpenter et al.*, 1981]. This noise band was sampled at several points, and the electron number density n_e at these points was determined (see *panel (b)*) using direct measurements of the electron cyclotron frequency (f_{ce}) and the relations $f_{pe} = \sqrt{f_{UH}^2 - f_{ce}^2}$ and $n_e(cm^{-3}) \approx (f_{pe}(Hz)/9000)^2$. In *panel (c)*, the Polar spacecraft potential V_{SC} during interval I_a is shown for comparison. Using the L-values of Polar's trajectory in *panel (d)*, the electron density n_e is determined as a function of L, and plotted in *panel (e)*. For reference, the sampled data points are listed inside this panel, along with two nominal inner plasmaspheric number density values at $L = 1$ and $L = 2$. A similar determination of n_e versus L was made using Polar PWI data from the second dawnside interval I_b, but is not shown here.

As can be seen from *panel (e)* of *Figure 2*, the density profile from interval I_a contained a sharp density enhancement at $L \approx 3.5$, just inside the $L = 3.7$ plasmapause. The density enhancement is also visible in *panel*

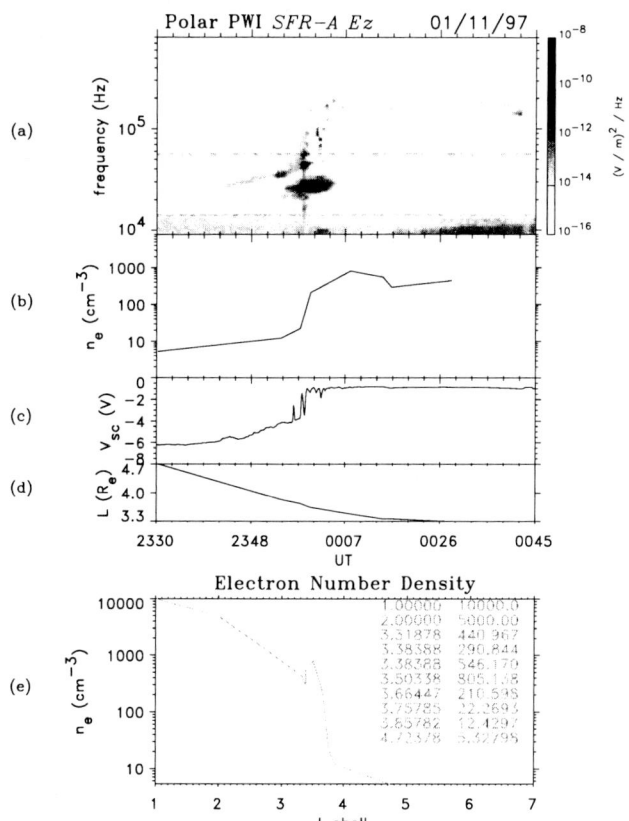

Figure 2. Determination of the electron number density in the plasmasphere on January 11, 1997, during the interval $I_a = 2330 - 0045$ UT.

(a) and *panel (b)*, between 0001 UT and 0016 UT. (It appears much broader in these top panels because Polar was moving slowly across L-shells during this part of its trajectory.) This density "spike" will be mentioned again in the discussion of the simulation results (next section).

In the simulation, the electron density was taken to be constant along the field lines. The mass density in units of m_P (the proton mass) was modeled by multiplying n_e by an empirical mass-weighting formula $W_0(r)$ [D. L. Gallagher, private communication; *Craven et al.*,1997]:

$$W_0(r) = (1.5918) - (0.1211)r + (0.0075)r^2 + (16.39)r^{-12} \quad (1)$$

This mass density function $\rho_0(L,r) = n_e(L) W_0(r)$ was used to specify the density inside the plasmaspheric cavity; a rigid conducting plasmapause at $L = L_p$

formed the outer boundary of the simulation. For the two dawnside UT intervals I_a and I_b, the outer boundaries were taken to be $L_{pa} = 3.7$ and $L_{pb} = 4.1$.

4. SIMULATION RESULTS

Simulations were performed using density profiles from intervals I_a and I_b; the results are displayed in *Plate 1a* and *Plate 1b*, respectively. Numerical solutions were obtained on a dipole grid of dimensions 90-by-90. In order to find the normal modes of the system, the plasmapause was perturbed with a broad-band compressional velocity impulse of ≈ 30 second duration. The pulse was localized near the equator. Azimuthal wavenumber $m = 3$ was assumed. FFTs were performed for all L values along the curve of sample points in the small plots of *Plate 1a* and *Plate 1b*. The three linearized magnetic field components are b_μ (the field-aligned or "compressional" wave), b_ϕ (the azimuthal or "toroidal" wave), and b_ν (the dipole-radial or "poloidal" wave). The results, after about 400 seconds, are the three magnetic field spectra of *Plates 1a* and *1b*, which are normalized to the maximum value of the weakest spectrum (in this case the poloidal component b_ν).

The spectra obtained are qualitatively very similar to those produced by the simulations of *Lee and Lysak* [1989]. The broad-band impulse produced fast-mode (compressional and poloidal) oscillations at discrete frequencies; the spectral lines evident in the top (b_ν) and bottom (b_μ) panels are identified as the cavity mode frequencies of the system. For both simulation intervals I_a and I_b (*Plates 1a* and *1b*, respectively), the $l = 1$ (fundamental) resonance is at about $35.0~mHz$, and in both simulations there are higher-harmonic resonances. The spectral lines in *1b* are at lower frequencies than those of *1a* because the plasmapause was further out during the later interval I_b. This effect is barely discernible for the $l = 1$ line, but gradually becomes more pronounced for higher l.

Where the fast-mode spectral lines intersect the toroidal eigenfrequencies, the cavity modes are coupled to toroidal field line resonances. At these locations, the otherwise weak toroidal response is greatly enhanced. The $l = 1$ cavity mode at about $35~mHz$ is coupled to the $n = 1$ FLR at $L \approx 2$, and there is an ($l = 1, n = 3$) resonance at $L \approx 3.1$. As in the results of *Lee and Lysak* [1989], the system prefers odd n harmonic FLRs because the initial velocity impulse was roughly symmetric with respect to the equator (which translates to antisymmetric magnetic fields).

The two sets of spectra *1a* and *1b* are strikingly similar, despite the fact that they are simulations of intervals which are 17 hours apart, with two different plasmapause locations and shapes. However, close examination of the fundamental resonance in *Plate 1a* (corresponding to the earlier interval $I_a \approx 2345 - 0015~UT$) reveals that it is actually composed of two spectral lines: one at 30 mHz, and the other at 35 mHz. The cause of this "double-fundamental" is the local density enhancement mentioned in the discussion of *Figure 2*. The presence of a density "spike," localized near the plasmapause, creates a sharp gradient in the Alfvén speed at $L \approx 3.3$; the 35 mHz fundamental is reflecting off of this extra gradient, and the 30 mHz resonance reflects off of the plasmapause boundary at $L = 3.7$. The double-fundamental, resulting from the inclusion into the model of a directly-measured number density, is a minor feature not found in previous similar models but which could exist in the real plasmasphere, where irregular structures are often found near the plasmapause.

To facilitate comparison with the Polar EFI (see next section), we show in *Plate 1c* simulation electric field spectra versus UT for the toroidal mode (e_ν) and fast mode (e_ϕ). These dynamic spectra were obtained by sampling the simulation data for interval I_a along a set of points that approximates Polar's trajectory during this time. The sample points are indicated in the small plot in *Plate 1c*; between 0000 and 0045 UT Polar skimmed the outermost L values of the dawnside plasmasphere, traveling towards higher (southern) latitude.

5. POLAR EFI AND MFE OBSERVATIONS

In *Plate 2*, data from the Electric Field Instrument (EFI) and Magnetic Field Experiment (MFE) on Polar are shown for interval I_a.

5.1. Fields Treatment

The top three panels of *Plate 2* contain the Polar spacecraft potential V_{SC}, and the fields E_X and E_Y, measured by the Polar EFI. The electric fields E_X and E_Y are in a coordinate system aligned with the background geomagnetic field: Z is along the background magnetic field, Y points in the azimuthal direction, and X is a radial coordinate perpendicular to both Y and Z and pointing away from the Earth. These fields were de-trended using a 6−minute running average, and the de-trended fields were Fourier-analyzed using FFTs, to produce the power spectra, also labeled E_X (*panel (e)*)

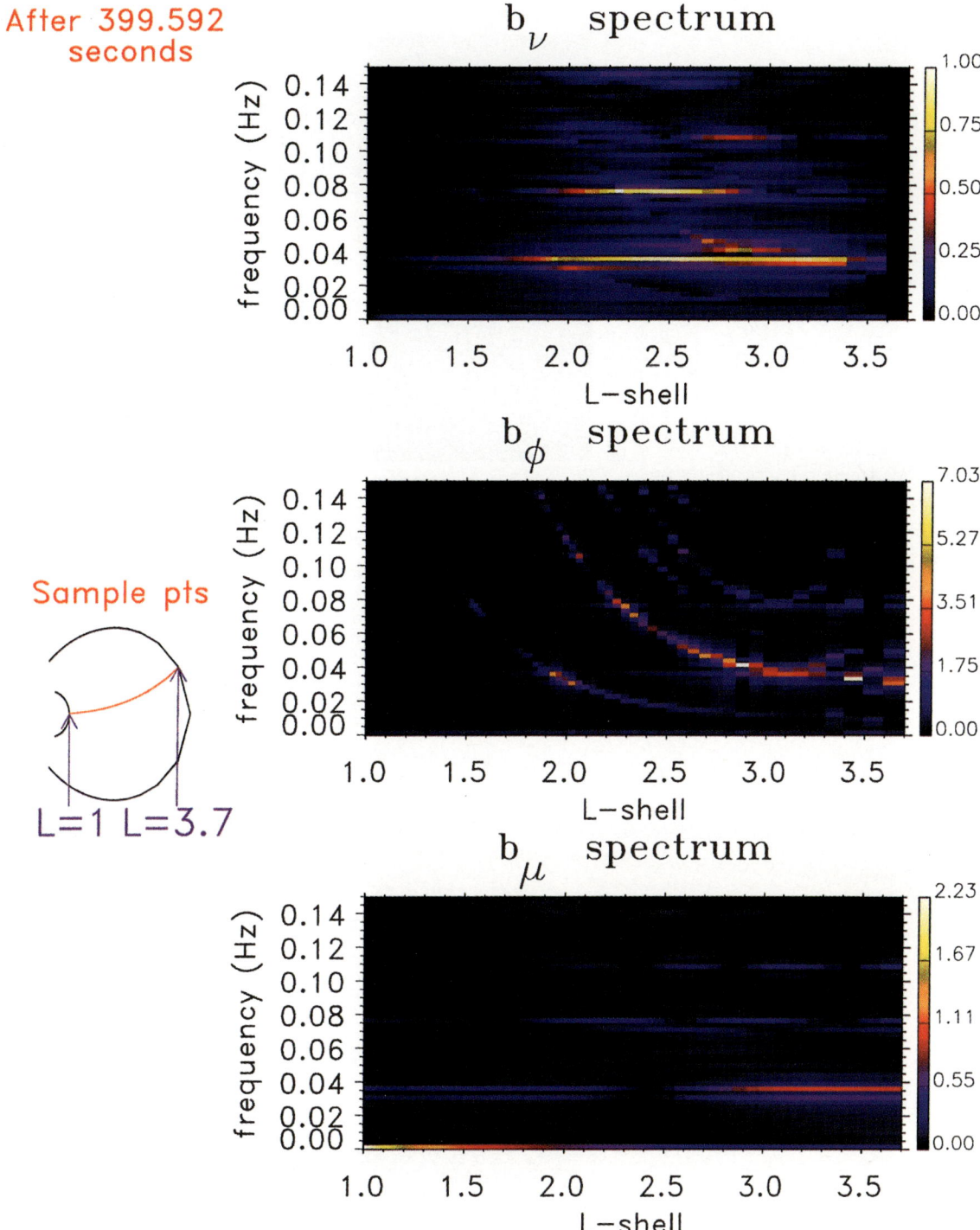

Plate 1a. Plasmaspheric response to a 30-second broad-band impulse, using the calculated number density from interval $I_a = 2330 - 0045\ UT$. After ≈ 400 seconds, a discrete spectrum is seen in the fast mode, coupled strongly to the $n = 1$ and $n = 3$ toroidal resonances. Only odd n toroidal resonances are visible.

82 JAN 97 CAVITY MODE SIMULATION AND POLAR DATA

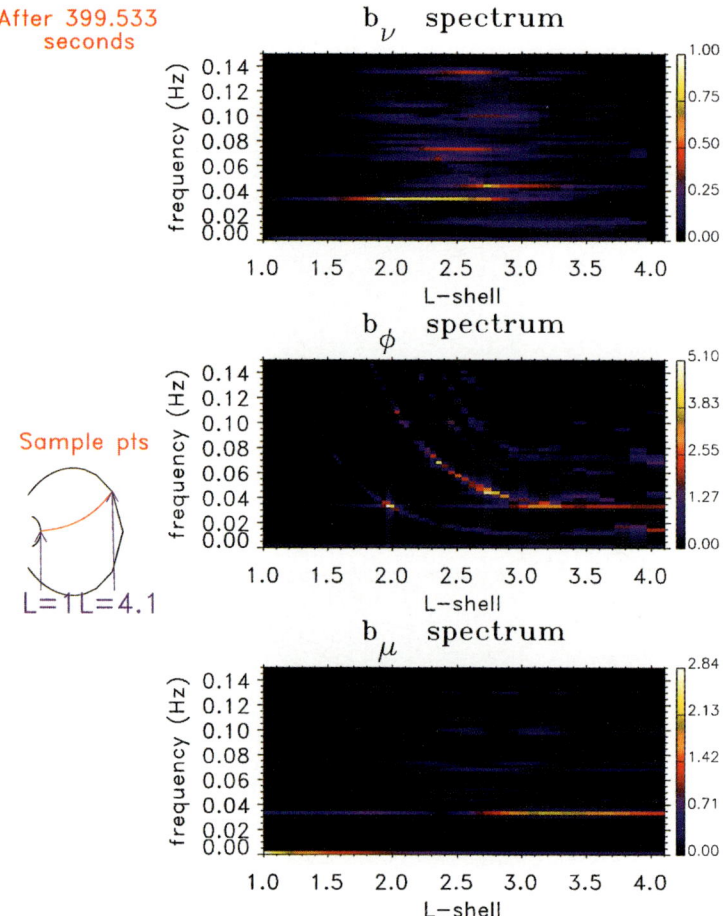

Plate 1b. Same as for *Plate 1a*, but using the number density from interval $I_b = 1700 - 1749\ UT$.

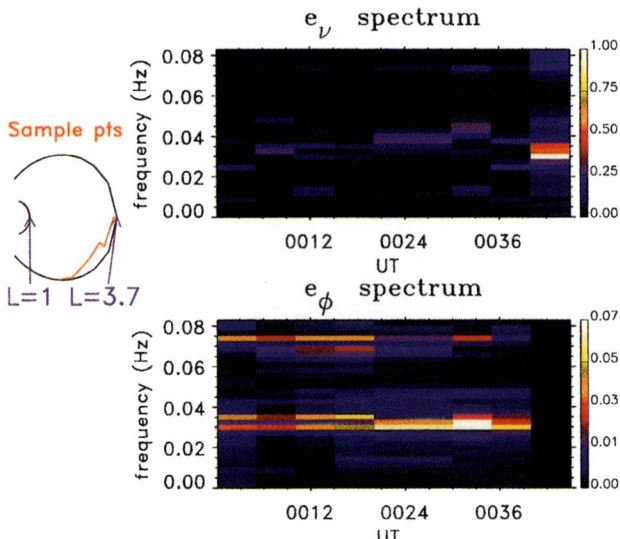

Plate 1c. Simulated electric field spectra for $0000 - 0045\ UT$, sampled along Polar's trajectory during this interval.

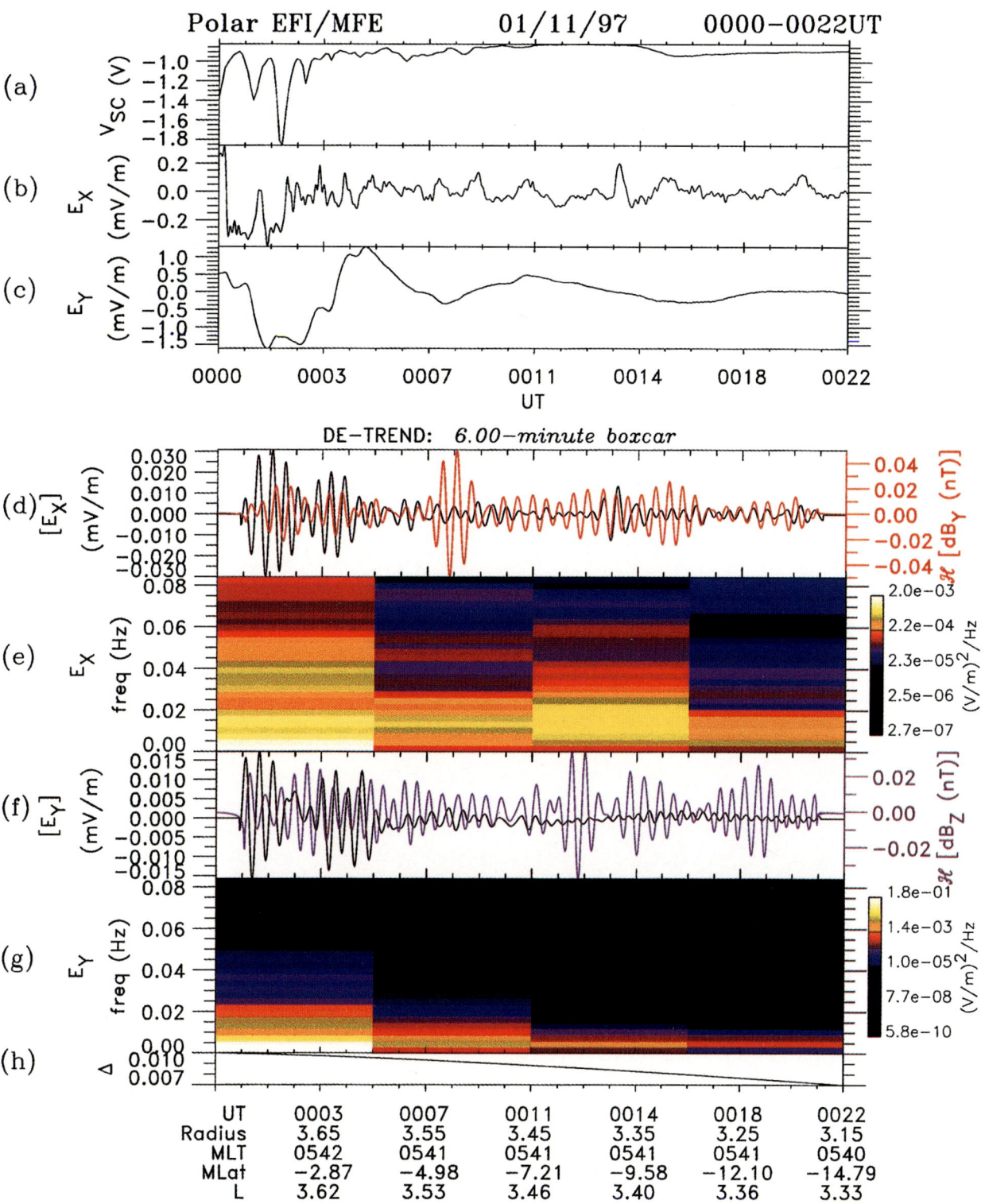

Plate 2. Data from the Polar EFI and MFE on January 11, 1997, 0000 − 0022 UT.

and E_Y (panel(g)). The de-trended fields were also bandpass-filtered to remove all frequencies outside the range $32 - 38\ mHz$; the filtered electric signals are labeled $[E_X]$ (panel (d)) and $[E_Y]$ (panel (f)), and are plotted in black.

The GSM magnetic fields measured by the Polar MFE were transformed to the same type of coordinate system as the electric fields; the Z coordinate was established by averaging the total measured field magnitude over 6 minutes. The signals were then de-trended, bandpass-filtered (with the same frequency window as the electric signals), and Hilbert-transformed (indicated by the script $\mathcal{H}$). The processed magnetic signals are plotted in color on the same axes as the filtered electric fields; the azimuthal component $\mathcal{H}[dB_Y]$ is in red (panel (d)), and the compressional $\mathcal{H}[dB_Z]$ is in purple (panel (f)).

Doppler shift due to the satellite's motion can be examined using the relation $f_P = f_S(1 \pm V_P/V_A)$. Here f_P is the frequency measured by the Polar satellite, f_S is the (true) frequency of the in situ signal, V_P is the instantaneous speed of the Polar satellite (relative to the background plasma, assumed to be corotating with the Earth), and V_A is a lower-limit estimate of the Alfvén speed, assuming a dipole magnetic field and a proton plasma with constant number density 1000 cm^{-3}). Thus the ratio $\Delta = V_P/V_A$ is an upper limit to the fractional uncertainty in f_P due to Doppler shift. As shown in panel(h), Δ is never more than 0.01, or 1%, for the interval I_a.

5.2. Comparison With Simulation Results

As shown in Plate 1c, cavity modes in the plasmasphere during the interval of Polar's passage would show up as spectral lines in the azimuthal electric field, coupled to toroidal field line resonances in the radial electric field. However, no resonances are evident in the E_Y spectrum (Plate 2, panel (g)) at any of the frequencies predicted by the simulation. In contrast to the numerical results, the wave power in the E_Y (fast mode) spectrum is mostly at $f \leq 5\ mHz$, and most of the power in the E_X (transverse/toroidal mode) spectrum is below $20\ mHz$.

To investigate the possibility that cavity resonances were excited, but were (relatively) too weak to show up clearly in the power spectra, bandpass-filtering was performed to select the fundamental frequency predicted by the model (panels (d) and (f)). The Hilbert transform shifts the phase of a signal by $\pi/2$. In the case of standing waves, the Hilbert transform of the magnetic wave should be in phase with the electric wave [Dubinin et al., 1990; Lotko and Streltsov, 1997], and for the fundamental cavity mode this condition should be satisfied during the entire UT interval. Since the filtered signals $[E_Y]$ and $\mathcal{H}[dB_Z]$ (panel (f)) do not seem to be in phase, fast-mode standing waves are not present at this frequency.

Although data are not shown here for interval I_b, the situation is similar to that of I_a. The data do not seem to indicate the presence of the cavity modes that are predicted by the model.

6. DISCUSSION

The quantitative similarity between the numerical solutions obtained for intervals I_a and I_b implies that the January 11, 1997 dawnside plasmasphere was stable over a 17-hour period, with the plasmapause boundary well defined by a sharp density gradient. Although in principle this configuration should provide a good resonant cavity, fast-mode resonances do not seem to be present in electric and magnetic field observations made by Polar while in the dawnside plasmasphere on this day.

Despite efforts to include an in-situ measured density profile in the simulation, it is possible that improvements to the model are necessary in order to achieve quantitative agreement with data. Three are listed here:

1. Our two-dimensional simulation surely cannot capture all of the dynamics of cavity mode formation and evolution. While 3D cavity modes have been studied [Lee and Lysak, 1991], the effects of a non-axisymmetric system need to be explored. In the real plasmasphere, it is likely that the fast-mode resonance condition is local-time dependent (as a consequence of the non-axisymmetry of the Alfvén profile [Kivelson et al., 1984]). If the plasmaspheric configuration used in the model is not valid over at least 8 hours of local time (i.e. one-third of the plasmasphere) then the assumption of azimuthal wavenumber $m = 3$ is flawed, and the normal mode frequencies from the simulation might be too low. Furthermore, the azimuthal "boundaries" of a cavity mode which occupies a finite span of local time might introduce energy loss, shortening its lifetime.

2. There are other limits to the lifetime of a plasmaspheric cavity mode. Ionospheric dissipation (due to Joule heating) is one. Leakage at the plasmapause boundary, due to the fact that an Alfvén gradient does not provide a perfect reflector, is another. Neither of these effects have been included in our simulation.

3. The plasmasphere is treated as a perfectly isolated cavity in our model. However, the results of Zhu

and Kivelson [1989] and Fujita and Glassmeier [1995] suggest that external stimulation of inner cavity modes might depend upon the intervention of the tunneling mode mentioned in section 1. Future simulations may need to incorporate the effects of coupling between the inner and outer cavities.

However, even if the limitations of the numerical model are ignored, there is another concern: without a suitable excitation mechanism, cavity modes will still not be found in the data. The January, 1997 magnetic cloud was characterized by a gradual rotation of the IMF from southward to northward; at the tail end of the cloud (around 0054 UT on January 11) the Wind spacecraft observed a dramatic density enhancement [Burlaga et al., 1998]. Yumoto et al. [1992] suggest that pressure pulses such as these ought to stimulate cavity modes. The extended period of southward IMF on January 10 served to strip off the outer flux tubes of the plasmasphere and sharpen the dawnside boundary, producing a configuration favorable for plasmaspheric cavity modes. However, by the time the pressure pulse seen by the Wind spacecraft reached the earth (about a half-hour later, or 0124 UT), Polar had already passed out of the dawnside plasmasphere. If the density enhancement seen by Wind at 0054 UT did stimulate dawnside cavity modes on January 11, they occured too late for Polar to observe them. When Polar returned to the dawnside plasmasphere 17 hours later, geomagnetic activity due to the pressure pulse had already tapered off.

In summary, it is possible that resonances were excited whose lifetimes and/or location(s) did not coincide with the passage of Polar through the dawnside plasmasphere on January 11. We note that although January 11, 1997 dawnside Polar EFI and MFE data does not seem to indicate cavity mode oscillations, examination of the dusk sector is underway. As mentioned in section 3, the gentle gradient of the duskside sector plasmapause is in general not a good reflector. However, the presence of a prominent density enhancement between $L = 6.3$ and $L = 8$ during the interval $0217 - 0336$ UT (see *Figure 1*) could act to confine waves, just as the density "spike" shown in *Figure 2* provided an extra reflecting boundary in the simulation.

Acknowledgments. We thank Professor Donald Gurnett for providing the Polar PWI data, Dr. Dennis Gallagher for providing the mass-weighting formula, Professor Forrest Mozer for providing the Polar EFI data and many helpful comments, and Dr. C. T. Russell for providing the Polar MFE observations. This work was supported by NASA grants NAG 5-2252, NAG 5-1098 (GGS), NAS 5-30371, NAG 5-3182, NGT 5-50153, the National Science Foundation under ATM-9622071, and the Dartmouth-UNH NASA Space Grant. The authors would like to thank the reviewer of this manuscript.

REFERENCES

Allan, W., E. M. Poulter, and S. P. White, Hydromagnetic wave coupling in the magnetosphere—Plasmaspause effects on impulse-excited resonances, *Planet. Space Sci.*, *34*, 1139, 1986.

Burlaga, L., et al., A magnetic cloud containing prominence material: January 1997, *J. Geophys. Res.*, *103*, 277, 1998.

Carpenter, D. L., R. R. Anderson, T. F. Bell, and T. R. Miller, A comparison of equatorial electron densities measured by whistlers and by a satellite radio technique, *Geophys. Res. Lett.*, *8*, 1107, 1981.

Craven, P. D., D. L. Gallagher, and R. H. Comfort, Relative concentration of He+ in the inner magnetosphere as observed by the DE 1 retarding ion mass spectrometer, *J. Geophys. Res.*, *102*, 2279, 1997.

Dubinin, E. M., P. L. Israelevich, and N. S. Nikolaeva, Auroral electromagnetic disturbances at an altitude of 900 km: the relationship between the electric and magnetic field variations, *Planet. Space Sci.*, *38*, 97, 1990.

Fujita, S., and K.-H. Glassmeier, Magnetospheric cavity resonance oscillations with energy flow across the magnetopause, *J. Geomagn. Geoelec.*, *47*, 1277, 1995.

Gurnett, D. A., et al., The Polar Plasma Wave Instrument, *Space Sci. Rev.*, *71*, 583, 1995.

Harvey, P., et al., The electric field experiment on the Polar satellite, *Space Sci. Rev.*, *71*, 583, 1995.

Kennel, C. F., *Convection and Substorms: Paradigms of Magnetospheric Phenomenology*, Oxford University Press, New York, NY, 1995.

Kivelson, M., M. Cao, and R. McPherron, A possible signature of magnetic cavity mode oscillations in ISEE spacecraft observations, *J. Geomagn. Geoelec.*, *49*, 1079, 1997.

Kivelson, M. G., J. Etcheto, and J. G. Trotignon, Global compressional oscillations of the terrestrial magnetosphere: The evidence and a model, *J. Geophys. Res.*, *89*, 9851, 1984.

Laakso, H., and A. Pedersen, Ambient electron density derived from differential potential measurements, in, *Measurement Techniques in Space Plasmas: Particles*, p. 49, 1998, ed. by J. E. Borovsky, R. F. Pfaff, and D. T. Young, Amer. Geophys. Union, Washington, D. C.

Lee, D.-H., and R. L. Lysak, Magnetospheric ULF wave coupling in the dipole model: the impulsive excitation, *J. Geophys. Res.*, *94*, 17097, 1989.

Lee, D.-H., and R. L. Lysak, Impulsive excitation of ULF waves in the three-dimensional dipole model: The initial results, *J. Geophys. Res.*, *96*, 3479, 1991.

Lotko, W., and A. V. Streltsov, Magnetospheric resonance, auroral structure and multipoint measurements, *Adv. Space Res.*, *20*, 1067, 1997.

Russell, C. T., R. C. Snare, J. D. Means, D. Pierce, D. Dearborn, M. Larson, G. Barr, and G. Le, The GGS/POLAR magnetic fields investigation, *Space Sci. Rev.*, *71*, 563, 1995.

Takahashi, K., S.-I. Ohtani, and K. Yumoto, AMPTE CCE observations of Pi 2 pulsations in the inner magnetosphere, *Geophys. Res. Lett.*, *19*, 1447, 1992.

Takahashi, K., S.-I. Ohtani, and B. J. Anderson, Statistical analysis of Pi 2 pulsations observed by the AMPTE CCE spacecraft in the inner magnetosphere, *J. Geophys. Res.*, *100*, 21929, 1995.

Yumoto, K., K. Takahashi, T. Sakurai, P. R. Sutcliffe, S. Kokubun, H. Luhr, T. Saito, M. Kuwashima, and N. Sato, Multiple ground-based and satellite observations of global Pi 2 magnetic pulsations, *J. Geophys. Res.*, *95*, 15175, 1990.

Yumoto, K., A. Isono, K. Shiokawa, H. Matsuoka, Y. Tanaka, F. W. Menk, and B. J. Fraser, Global cavity mode-like and localized field-line Pc 3-4 oscillations stimulated by interplanetary impulses (Si/Sc): Initial results from the 210° MM magnetic observations, in, *Solar Wind Sources of Magnetospheric Ultra-Low-Frequency Waves*, p. 335, 1994, ed. by M. J. Engebretson, K. Takahashi, and M. Scholer, Amer. Geophys. Union, Washington, D. C.

Yumoto, K., et al., Globally coordinated magnetic observations along 210° magnetic meridian during STEP period: 1. Preliminary results of low-latitude Pc 3's, *J. Geomagn. Geoelec.*, *44*, 261, 1992.

Zhu, X. M., and M. G. Kivelson, Analytic formulation and quantitative solutions of the coupled ULF wave problem, *J. Geophys. Res.*, *93*, 8602, 1988.

Zhu, X. M., and M. G. Kivelson, Global mode ULF pulsations in a magnetosphere with a nonmonotonic Alfvén velocity profile, *J. Geophys. Res.*, *94*, 1479, 1989.

R. E. Denton, J. Goldstein, M. K. Hudson, and J. G. Lyon, Department of Physics and Astronomy, Dartmouth College, Hanover, NH 03755, U.S.A. (email: denton@comet.dartmouth.edu; jerry@storm.dartmouth.edu; maryk@comet.dartmouth.edu; lyon@tinman.dartmouth.edu)

W. Lotko, Thayer School of Engineering, Dartmouth College, Hanover, NH 03755, U.S.A. (email: billl@cyberia.dartmouth.edu)

Polar/TIDE Results on Polar Ion Outflows

T. E. Moore[1], M. O. Chandler[2], C.R. Chappell[3], R. H. Comfort[4], P. D. Craven[2],
D. C. Delcourt[5], H. A. Elliott[4], B.L. Giles[1], J. L. Horwitz[4], C. J. Pollock[6], Y.-J. Su[4,7]

The ISTP Polar spacecraft is equipped with a unique plasma velocity analyzer system designed specifically for kinetic diagnostics of low-energy, low-density plasma ions. Such plasmas were previously unobservable in the polar cap region owing to their low velocities and the positive photoelectric charging of spacecraft in sunlight at low ambient plasma density. The thermal ion dynamics experiment (TIDE) incorporates seven large apertures, focusing electrostatic optics, and time-of-flight mass analysis, for enhanced sensitivity to low energy plasma ions. The plasma source instrument (PSI) limits and regulates the photoelectric charging of the Polar spacecraft at small potentials (~+2V). Together, TIDE and PSI have produced new observations of i) the mixing of solar and ionospheric plasmas in the cleft regions; ii) auroral heating and plasma transport; iii) solar illumination control of the polar cap ionosphere; iv) the downward motion of O^+ at lower altitudes throughout the polar cap region; v) the high altitude polar wind; vi) the high altitude convection of the polar outflows; vii) the unexpected dynamism of polar wind outflows; and viii) the supply of plasma to the plasma sheet. These observations indicate that most polar cap O^+ out flow originates in the dayside plasma upwelling region, creating a plasma fountain effect in the polar cap. The observations support the evaluation of consequences of the ionospheric source of plasma for magnetospheric dynamics and storm phenomena. Preliminary global modeling results indicate that ionospheric plasma is the dominant contributor to both the density and pressure of the plasma within a corresponding geopause that extends to the persistent neutral line in the central plasma sheet. TIDE and PSI have contributed fundamentally to our knowledge that the dissipation of solar wind energy is not limited to the ionosphere proper, but is distributed throughout a much larger geosphere of dominantly terrestrial origin.

[1]Laboratory for Extraterrestrial Physics, NASA Goddard Space Flight Center, Greenbelt, MD USA.
[2]Space Sciences Laboratory, NASA Marshall Space Flight Center, Code, Huntsville, AL USA.
[3]Deptartment of Media Relations, Vanderbilt University, Nashville, TN USA.
[4]Department of Physics, The University of Alabama in Huntsville, Huntsville, AL USA.
[5]Centre d'étude des Environnements Terrestre et Planétaires, Saint-Maur des Fossés, FRANCE.
[6]Department of Instrumentation and Space Research, Southwest Research Institute, San Antonio, TX USA.
[7]Now at: Los Alamos National Laboratories, Los Alamos, NM USA.

BACKGROUND

The earliest suggestions that terrestrial plasma escapes our planet supersonically were based upon theoretical considerations involving analogies with the solar wind [*Axford*, 1968; *Banks and Holzer*, 1968]. Based on the likely vacuum conditions in the wake of the Earth's magnetosphere, a polar wind of light ions was predicted to have supersonic outflow velocity when compared with very low thermal speeds characteristic of ionospheric plasma. Since then, polar wind outflows have been observed to begin in the topside ionosphere and to continue up to about 3.5 R_E altitude [*Ganguli*, 1996]. The significance of polar wind outflows was long thought to be limited to a high speed, low density phase of plasmaspheric refilling at high latitudes. Polar wind was not thought until recently to contribute significantly to the plasma sheet. Rather, accelerated outflows of ions from the auroral zones, including significant if not dominant components of O^+ ions, were thought more likely to contribute to plasma sheet and storm time plasmas, in part because they have already attained keV energies as they leave the auroral acceleration regions.

More recently, low energy observations led to the conclusion that the largest local fluxes of outflowing ions (mostly O^+ for solar maximum conditions) escape from the dayside cleft regions (including the cusps). The largest fluxes of O^+ were also found to be flowing at the lowest energies, less than 100 eV. In fact, the outflow was found to be a bulk phenomenon involving such strong heating that the core of the ion distribution exhibited temperatures of order 10 eV or higher at 1 R_E altitude. Test particle simulations of such outflows found that, owing to their starting positions on the dayside of the magnetosphere, such ions are able to travel long distances down the magnetotail. There, they are very strongly accelerated by their non-adiabatic interactions with the stretched and sharply curved magnetic field with associated convection electric field. Or, if they stray too far down tail, they may be lost entirely from the magnetospheric system. Those that are returned Earthward acquire plasma sheet-like energies, independent of how low their energies may have been upon arrival at the neutral sheet. H^+ ions also become plasma sheet-like upon interaction with the stretched neutral sheet.

Such considerations brought attention back to the polar wind outflows of light ions, extending over much of the high latitude ionosphere. These ions are flowing slowly, but fast enough to travel far enough down the tail to supply the plasma sheet with protons and helium ions in regions earthward of the most persistent neutral line, where the mantle plasmas are unlikely to be a strong source in the center of the plasma sheet ($Y_{GSM} \sim 0$). As these ions are accelerated to plasma sheet energies, they must be admitted and accounted for as a source of plasma in the plasma sheet.

TIDE and PSI [*Moore et al.*, 1995; *Moore et al.*, 1997] were conceived to provide competent plasma measurements as close as possible to zero energy, and consequently to reduce or eliminate spacecraft potential as a factor in observing low energy plasmas [*Comfort et al.*, 1998]. At the same time, the TIDE energy range was extended up to 450 eV, allowing for simultaneous direct observation of entering solar plasmas in the cusp/cleft regions, and of the auroral accelerated ion beams and conics at moderate energies. The results include a simultaneous look at mixed solar and ionospheric plasmas; a new perspective on auroral energized plasmas; a systematic sampling of the polar topside ionosphere at ~ 1 R_E altitude; and our first look at the high altitude behavior of the polar wind, which has been found to be much more dynamic and variable than expected. Altogether, a new data set has been collected that permits us to more comprehensively and quantitatively assess the transport of ionospheric plasma throughout the magnetosphere, and its relationship to solar plasma entry and transport processes.

CLEFT MIXING OF SOLAR AND IONOSPHERIC PLASMAS

Figure 1 shows an example of coexisting solar wind and plasmaspheric plasmas, observed in the dayside cusp region. Escaping plasmaspheric plasma was found to coexist with magnetosheath plasmas admitted inside the magnetosphere by reconnection in the high-latitude cusp region. Plasmaspheric material is accelerated significantly prior to observation at high latitude boundary layer locations, but is still distinct from the incoming magnetosheath plasma [*Dempsey et al.*, 1998; *Chandler et al.*, 1998; *Chandler et al.*, unpublished manuscript]. These observations of coexisting ionospheric and solar plasmas at the same location confirm what is qualitatively expected on plasma flux tubes connected to both plasma sources, given that reconnection creates such flux tubes, i.e. mixed components of solar and ionospheric plasmas [e.g. *Fuselier*, 1997]. The observations provide diagnostics on the connectivity of flux tubes and the location of reconnection sites along them. They also support comparison with a comprehensive quantitative theory of plasma transport in such plasma flux tubes (open to the magnetosheath), though such a theory has not yet been fully developed.

AURORAL PLASMA TRANSPORT

Plate 1 summarizes the observed characteristics of bulk transport of ionospheric plasma heated by auroral processes. The panels indicate the distribution of such events in location, flow velocity, temperature, and flux. Using the 0-450 eV energy range and 3D angular response of TIDE, ionospheric heating and outflow is well differentiated from magnetosheath plasma penetration into the dayside cusp ionosphere. Transverse ion heating and

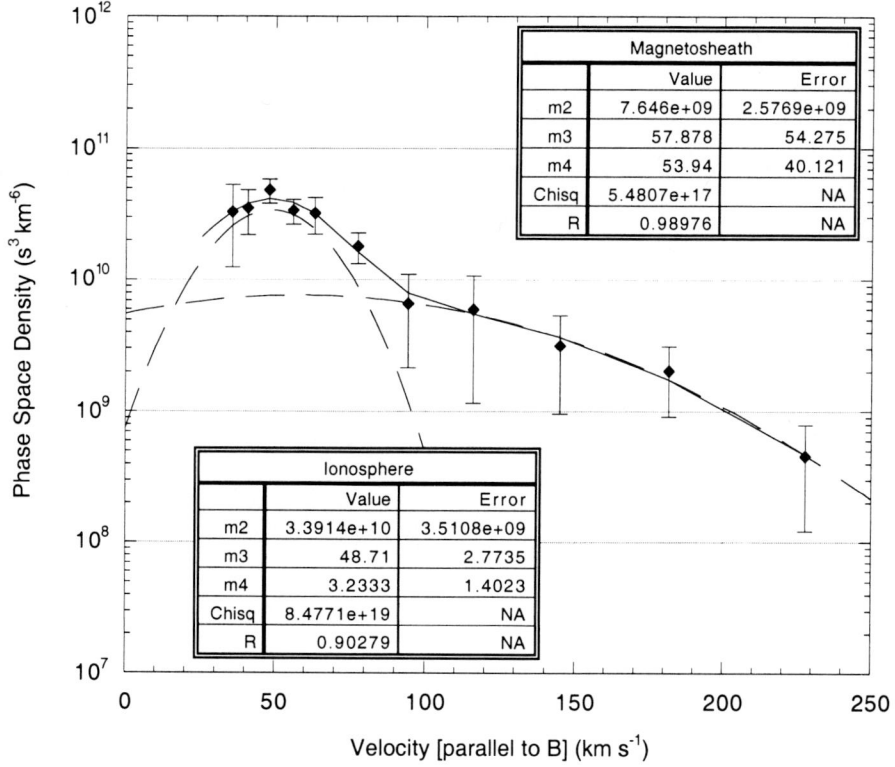

Figure 1. Velocity distribution in the TIDE energy range, showing combined low energy ionospheric and magnetosheath components, as indicated in the fit parameters (m2 = normalization; m3 = velocity in km/s; m4 = Temp. in eV). [after Chandler et al., 1998].

resultant outflow is found to be well-correlated in detail with UV auroral features. Ionospheric topside flows, conics and beams, are directly associated with auroral electric fields, plasma waves, and convection [*Hirahara et al.*, 1998a]. Event analysis of H$^+$ and O$^+$ conics in the polar cusp show angular and energy differentiation associated with a heating wall in the cusp [*Knudsen et al.* 1994; *Hirahara et al.*, 1998b]. The dayside upwelling ion region is active during these solar minimum conditions, producing strong upwelling fluxes of all species of ions, though not as enriched in heavy species as at solar maximum.

Auroral plasma density cavities observed at POLAR perigee (~1 R_E altitude) were also found to contain transversely heated and accelerated populations of ions. In cases where the spacecraft potential rose as high as ~+10V, a density drop as large as one to two orders of magnitude was found, relative to adjacent regions. In such events, the O$^+$ ions exhibit both conics and parallel flows [*Craven et al.*, 1998] having characteristic energies comparable to the spacecraft potential. This suggests that the density cavity results essentially from bulk energization of the ionospheric plasma. The plasma ions remain observable at auroral altitudes, with charging to levels not much greater than the typical ion energies.

CLEFT ORIGIN OF POLAR CAP O$^+$

Figure 2 illustrates the influence of solar illumination in control of polar wind H$^+$ fluxes and O$^+$ densities. Polar wind H$^+$ fluxes and O$^+$ densities at 5000km altitude are strongly correlated with increasing solar zenith angle (or day-night distance across the polar cap, in solar magnetic coordinates, X_{GSM}). This documents a strong solar illumination control of the plasma density, and thus the supply of both O$^+$ ions, and probably through charge exchange, H$^+$ ions as well [*Su et al.*, 1998a]. At this altitude in the polar cap, The O$^+$ plasma is a cold rammed distribution, while the H$^+$ and He$^+$ are in relatively well-developed outflow (though still transonic).

Figure 3 shows observations of topside O$^+$/H$^+$ plasma flow, showing that O$^+$ is on average downward moving in the polar cap. In contrast with the upwelling ion flows of all species in the dayside auroral zone, the polar cap O$^+$ plasma at perigee altitudes is on average a steady downward flow

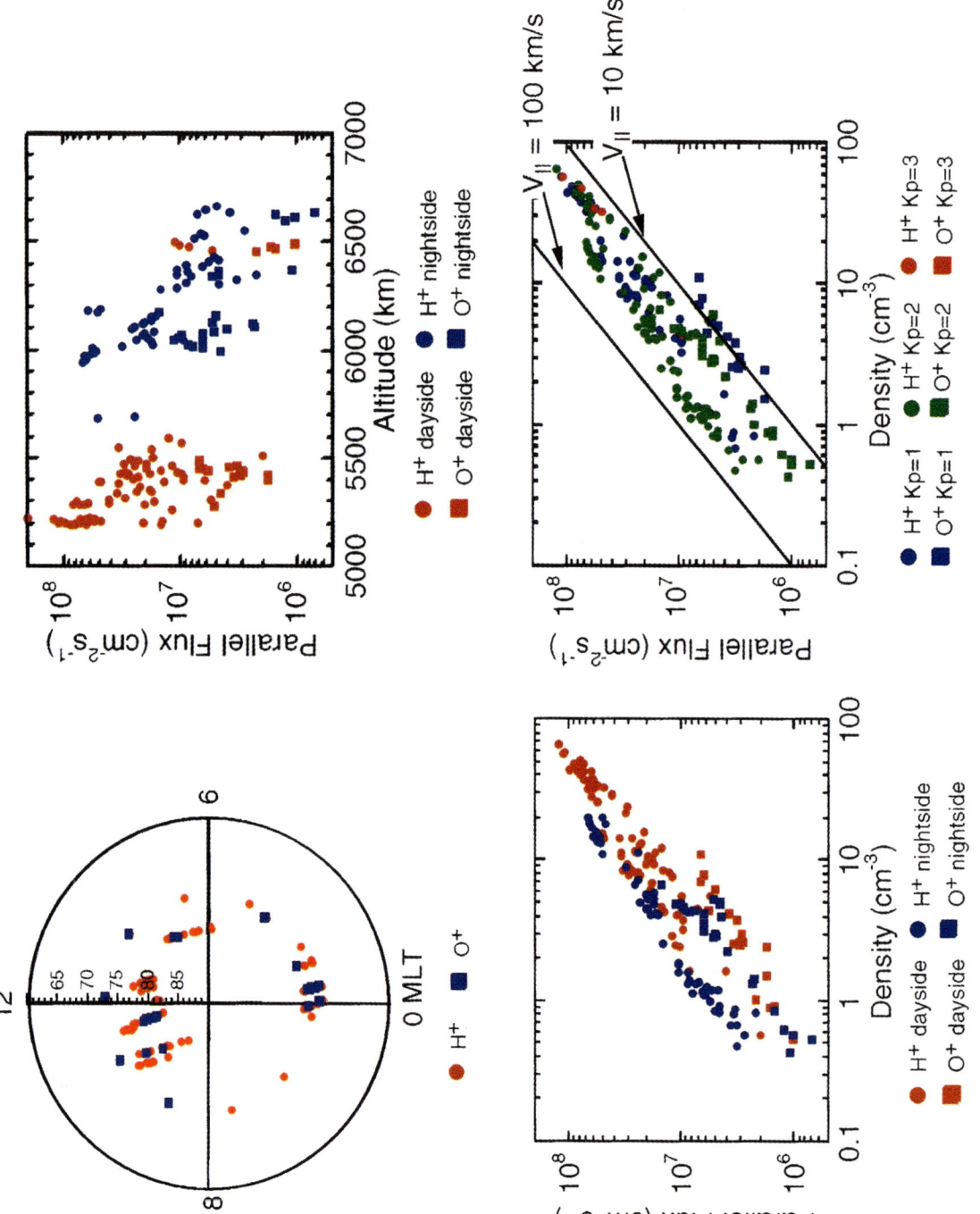

Plate 1. Characteristics of bulk transport of ionospheric plasma heated by auroral processes. The panels indicate the distribution of such events in location (near 1 R_E altitude), parallel flux, parallel velocity, and mean temperature. For these distributions, the mean $H^+ V_\parallel = 31.5 \pm 14.4$ km/s, mean $O^+ V_\parallel = 14.5 \pm 4.8$ km/s, mean $H^+ T_\perp = 11.7 \pm 4.6$ eV, and mean $O^+ T_\perp = 24.0 \pm 8.0$ eV [after Giles et al., 1997].

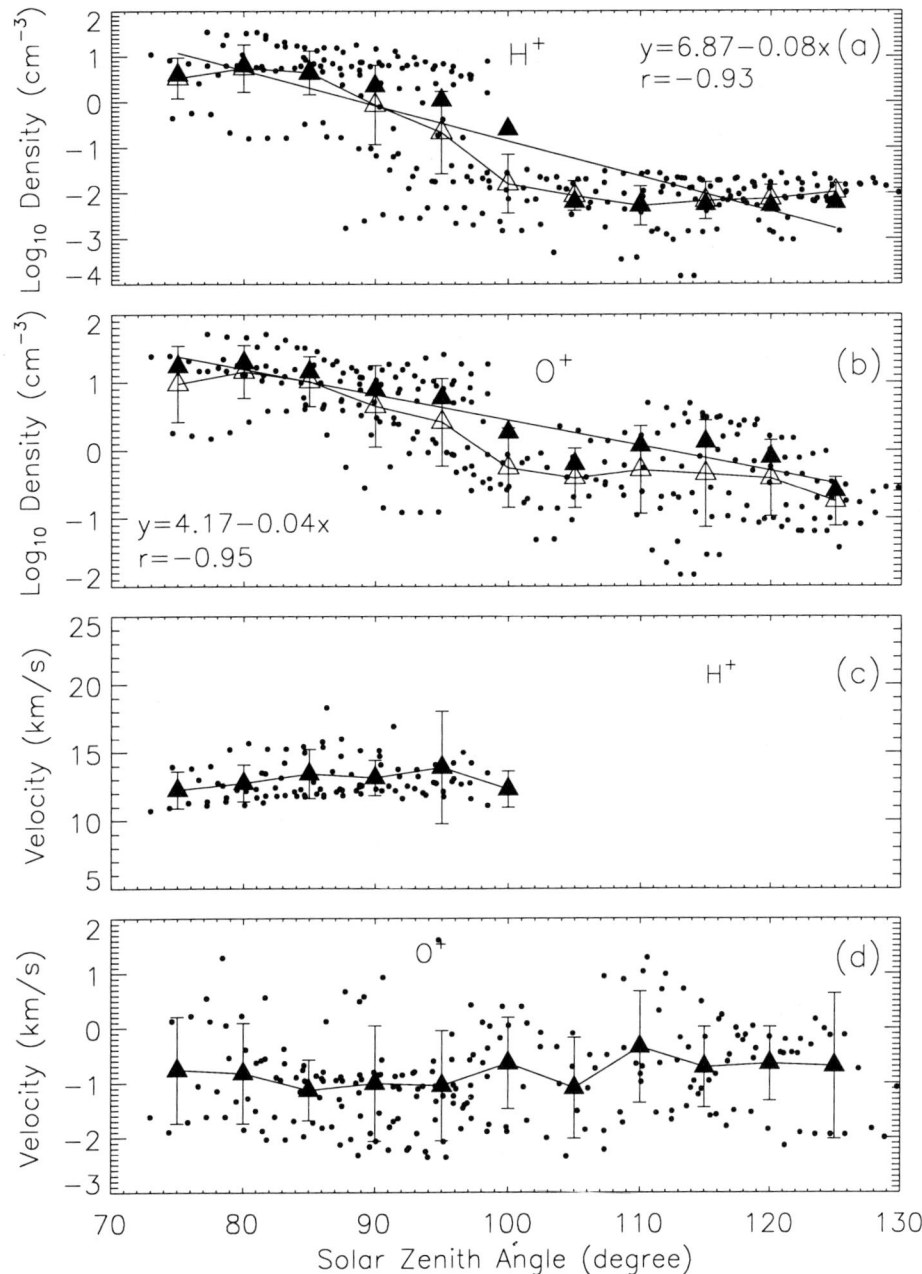

Figure 2. Observations of ~1 R_E polar wind density and velocity versus solar zenith angle, illustrating solar illumination control of polar wind H^+ and O^+ densities (and therefore H^+ flux at relatively constant velocity). [after Su et al., 1998a]

(with at most sporadic and localized upward flows). This shows unambiguously that the O^+ component of the polar cap plasma originates in the dayside upwelling region and does not flow up from the polar cap proper [*Su et al.*, 1998a]. It can be inferred from this that polar cap plasma tubes contain on average an oversupply of O^+ plasma relative to their low altitude boundary conditions, and yield O^+ back to the ionosphere even as they supply H^+ plasma outflows. Several recent theoretical studies have suggested that O^+ outflows can be expected from the polar cap if photoelectron effects are properly taken into account [*Ganguli*, 1996]. The present results suggest instead that

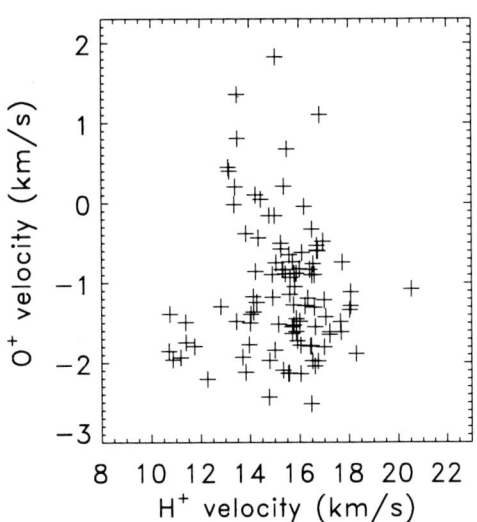 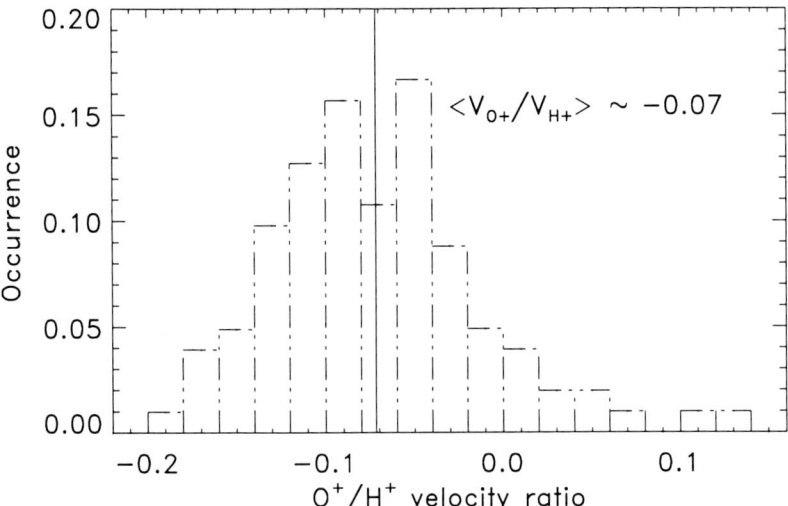

Figure 3. Observations of topside O^+/H^+ plasma flows, showing that O^+ is on average downward moving in the polar cap. [after Su et al., 1998a]

photoelectron effects have not been properly accounted for if they produce significant polar cap O^+ outflows. To date, few theoretical efforts have incorporated convecting flux tubes that are subjected to strong heating as they pass through the auroral zone. The downward motion that results when the heat source is removed in the polar cap may overcome photoelectron effects that would otherwise produce upflows.

HIGH ALTITUDE POLAR WIND

Figure 4 illustrates the observed characteristics of the high altitude polar wind. The polar wind has been confirmed to exist at very high altitudes for the first time by the TIDE-PSI system [*Moore et al.*, 1997]. The escaping polar wind is faster, much hotter, and richer in O^+, compared with the polar wind predicted by thermal outflow theories. It is every bit as pervasive in the polar cap as expected from these theories [*Su et al.*, 1998a]. The real polar wind likely differs from that which has been described by theory primarily as a result of significant energy inputs in the topside auroral ionosphere. As plasma flux tubes convect through and interact with auroral zone processes (that is, magnetospheric boundary layer or plasma sheet processes), these energy inputs increase the escaping mass flux primarily through low altitude heating, but with contributions from the production of ionization and topside electron heating by soft electron precipitation.

MAGNETOSPHERE-IONOSPHERE COUPLING

Figure 5 shows observations of plasma convection in the high altitude polar wind, showing anti-sunward convection during a period of generally southward interplanetary magnetic field (IMF) and vice versa for a period of northward IMF. In conjunction with the Plasma Source Instrument (PSI), TIDE has the unique ability to simultaneously measure both convection velocities and parallel flows in the polar cap at apogee (8-9 R_E). The convection of polar wind outflows in the polar cap was found to respond to interplanetary conditions in accordance with expectations of a four-cell convection pattern with sunward convection in the polar cap for northward IMF. Thus, the polar wind outflows are strongly influenced by the solar wind interaction at the boundaries of the magnetosphere. In particular, high latitude reconnection has the effect of reversing high latitude boundary layer flows. This effectively shuts down the internal supply of plasma to the plasma sheet (at least in the local hemisphere), and tends to allow the low latitude boundary layer to dominate the anti-sunward transport of boundary layer plasma, feeding the plasma sheet.

POLAR WIND DYNAMISM

Plate 2 illustrates the degree of dynamism that is often observed in the high altitude polar wind outflows. With the accumulation of additional polar cap passes with PSI operating, we have noticed that the polar wind is quite often very strongly dynamic in outflow velocity. Fluctuations of the outflow velocity by a factor of 3 or more are typical in some regions, while other regions are characterized by relatively steady polar wind. The spatial/temporal scale of these fluctuations is such that many of them are observed in the course of a single pass through the polar cap (few

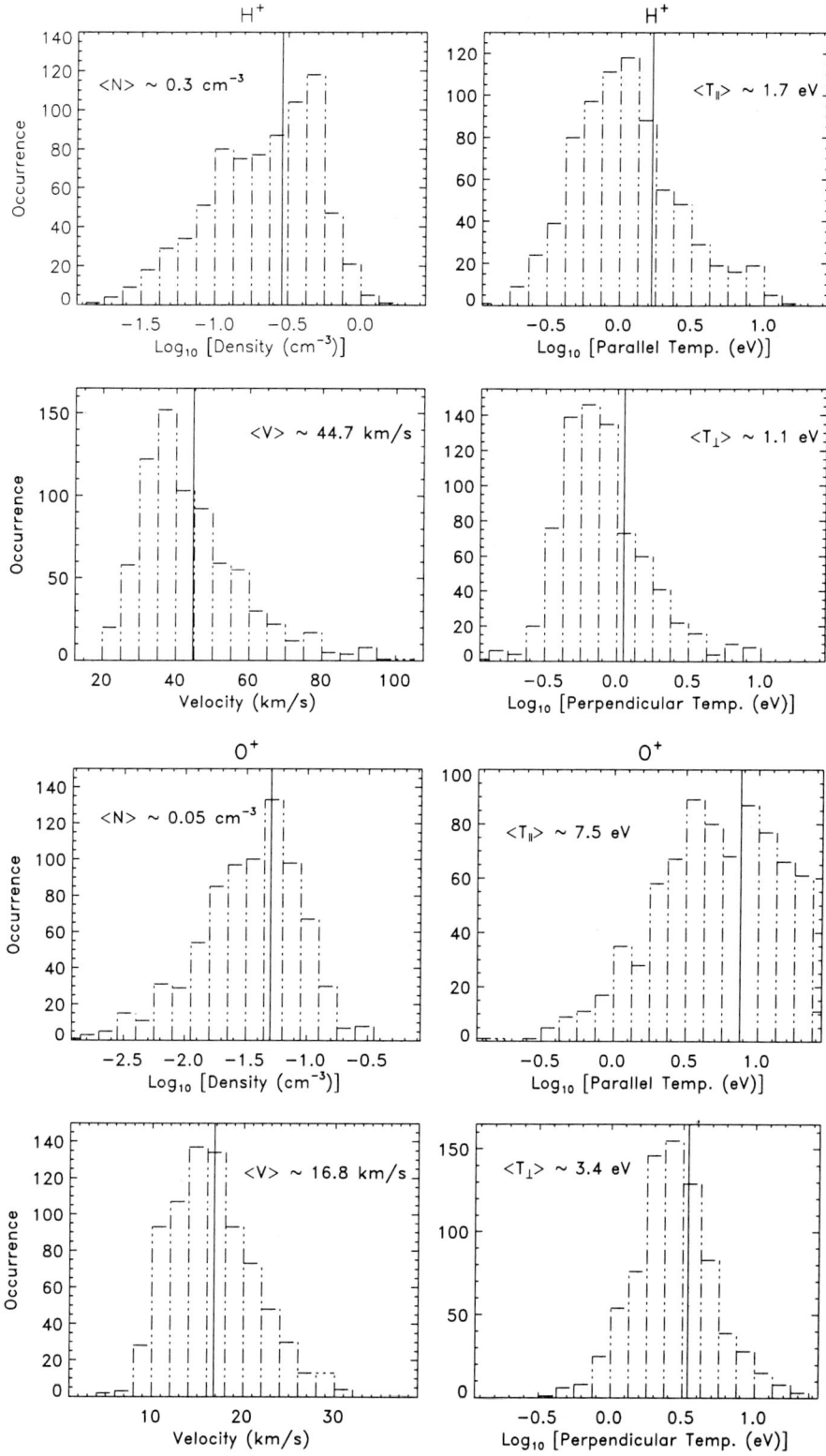

Figure 4. Observed characteristics of the high altitude polar wind, for H^+ (left panels) and O^+ (right panels), in density, velocity, parallel and perpendicular temperatures. [after Su et al., 1998a]

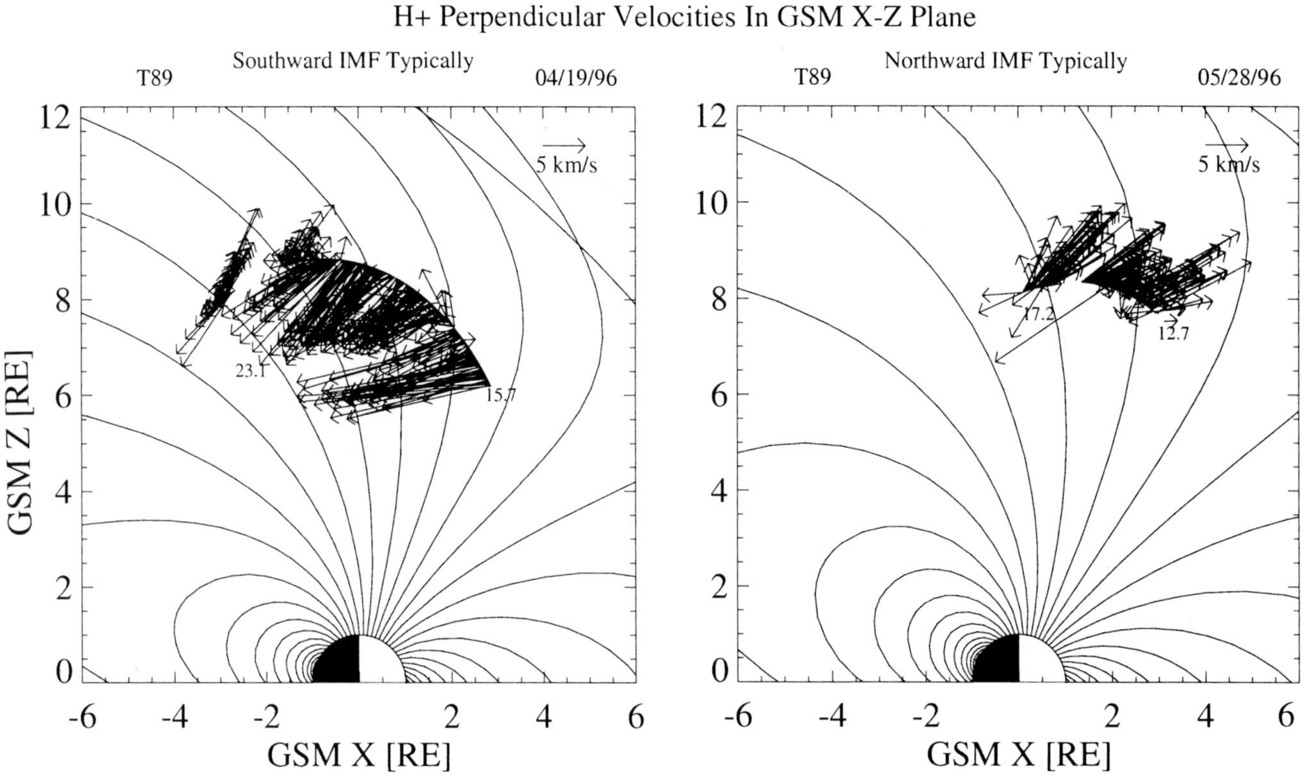

H+ Perpendicular Velocities In GSM X-Z Plane

Figure 5. Observations of plasma convection in the high altitude polar wind, showing antisunward convection during a period of generally southward IMF and vice versa during a period of northward IMF [after Elliott et al., 1997].

hours). Such fluctuations are significant in controlling the fate of the outflows insofar as they may be lost downstream beyond the persistent neutral line at high velocities, captured and strongly accelerated in the plasma sheet if they are somewhat slower, or recirculated directly into the plasmasphere if they are very slow.

Often the fluctuations do not occupy the entire polar cap but are somewhat stronger near dawn or dusk. Correlated (or anti-correlated) fluctuations often occur in the simultaneously observed polar rain electrons (E~100eV) [*Su et al.*, 1998b]. The fluctuations are not accompanied by obvious correlated changes in the differential flux or apparent composition of the plasma (ratio of H^+ and O^+ differential fluxes), and it is believed that these represent the result of high altitude energy inputs in to the flow. One model that has been proposed for such fluctuations involves the existence of a standing electrostatic shock in the flow that confines the atmospheric photoelectrons while accelerating ions outward as they fall through the shock [*Su et al.*, 1998a]. Under some conditions, it has been suggested by *Barakat et al.*[1997] that the shock may fluctuate in altitude and/or amplitude, leading to the observed fluctuations in velocity at a slowly moving observing point. Fluctuations of such large amplitude had not previously been anticipated in polar wind outflows, and have an influence on the fraction of the polar wind that is lost from the magnetosphere down the tail.

PLASMA SHEET SUPPLY

Figure 6 summarizes a categorization of the types of trajectories resulting from initial conditions within the range of typical polar wind velocities and locations of observation for the TIDE-PSI data set. The convection, field aligned velocity, and location of the polar wind are all important factors in the ultimate destiny of the flow and resultant role that it will play in global magnetospheric dynamics. Single particle simulations in realistic fields [*Giles et al.*, 1997] provide the means to evaluate polar wind significance and destiny. Flows with a very high ratio of parallel to tailward cross-field flow (or even sunward

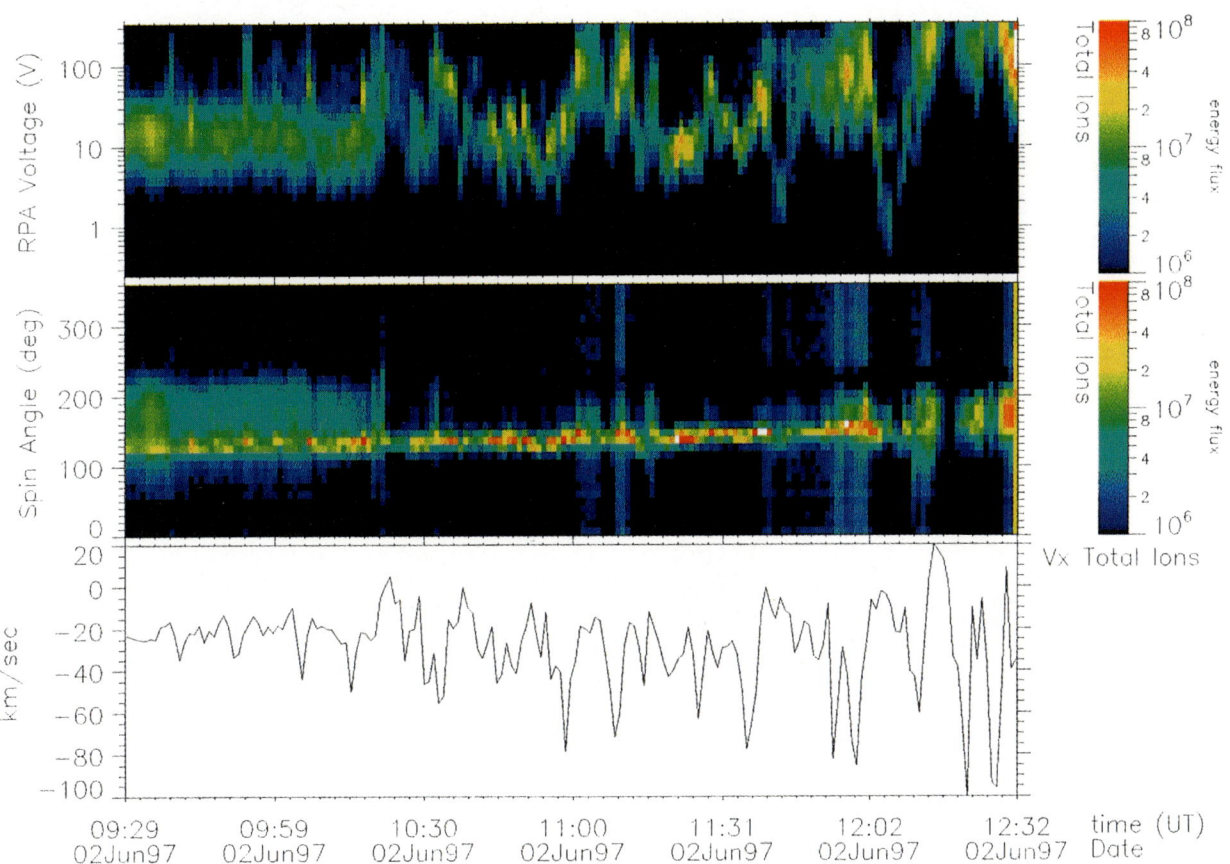

Plate 2. High altitude polar wind ion energy (upper panel) and spin angle distribution (middle panel) for an event exhibiting strong fluctuations of the parallel polar wind streaming velocity (lower panel).

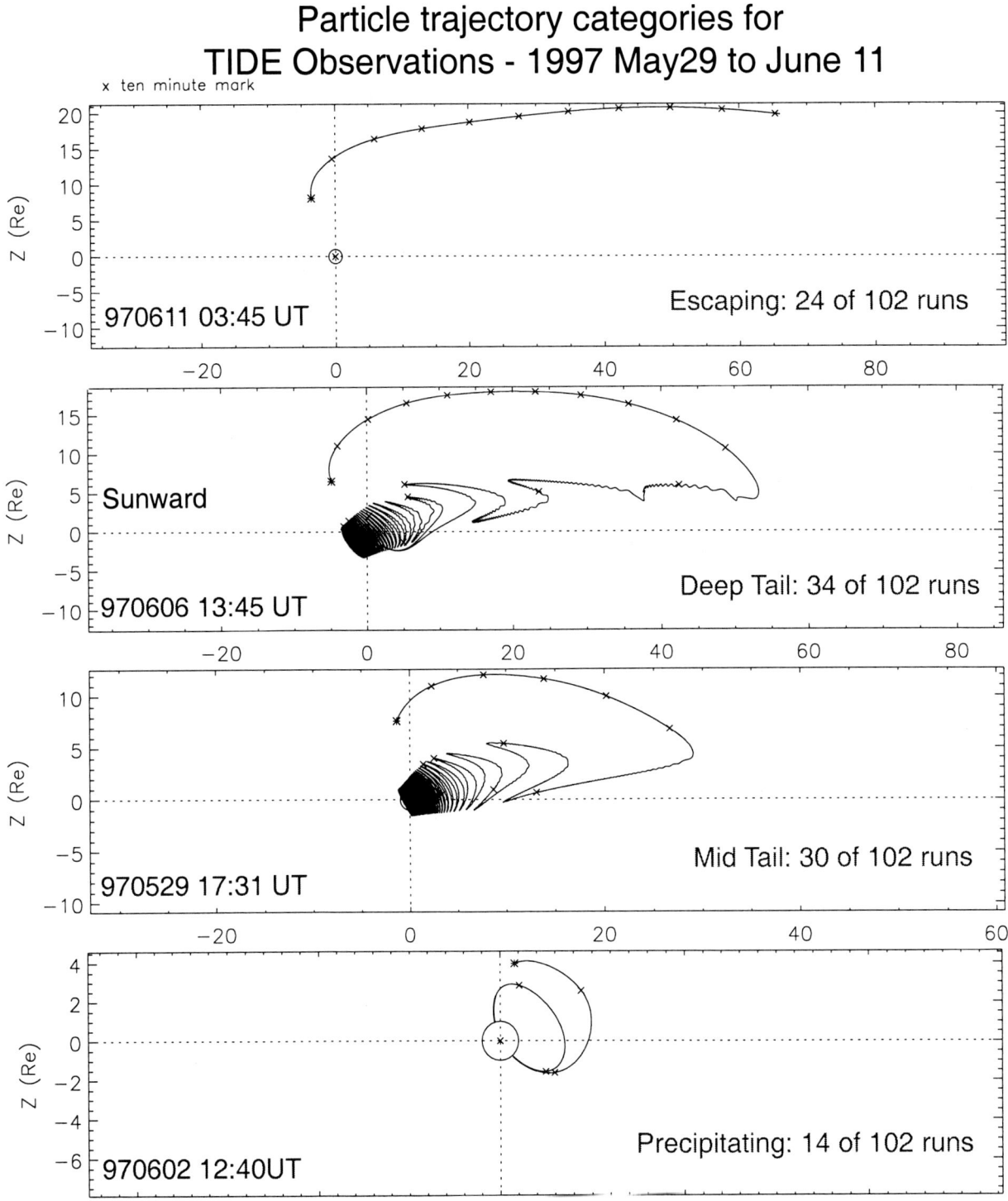

Figure 6. A categorization of the types of trajectories (projections in the GSM X-Z plane) resulting from initial conditions corresponding to typical polar wind velocities and locations of observation. [Giles et al., 1997].

flow, as described above) tend to be lost entirely from the magnetosphere through the tail lobes. Flows that are slow relative to their tailward convection tend to be recirculated within the inner magnetosphere supplying relatively cold plasma only to the plasmasphere. Flows like those observed from POLAR tend to enter the active parts of the plasma sheet, for typical convection conditions. Assuming their inertia does not load down the drivers of plasma sheet convection, H^+ ions in these outflows are energized instantly and unanimously on crossing the neutral sheet to plasma sheet, to energies of several keV, thus qualifying them immediately as a part of the plasma sheet ion population.

The omnipresence of polar wind outflow throughout the entire polar cap implies that some plasma will be provided to the active parts of the plasma sheet, almost independent of magnetospheric convection. For rapid polar cap convection, the denser, hotter and faster outflows of the dayside magnetosphere will be fed to the active plasma sheet regions inside of 30 R_E. For slower convection, the relatively low density outflows of the night side polar cap and the even lower density but energized outflows of the night side auroral zone will be fed to the active plasma sheet. For very slow or even reversed polar cap convection, polar outflows will be largely lost through the lobes or into the low latitude boundary layer and magnetosheath.

These results, combined with the observed cold, dense character of the plasma sheet for northward IMF [*Terasawa et al.*, 1997], suggest a paradoxical inference. It would seem that the hot low density plasma sheet during active convection conditions is actually supplied by the ionospheric polar outflows, while the cold dense plasma sheet found for northward IMF is supplied from the magnetospheric flanks by residual low latitude boundary layer circulation.

SUMMARY AND PLANS

The following noteworthy scientific accomplishments resulted from the main mission phase of TIDE-PSI data acquisition:
1. Plasmaspheric plasma outflows were found to coexist with magnetosheath plasmas in the cusp/cleft region, on reconnected field lines;
2. Transverse ion heating and resultant outflow was found to be correlated in detail with UV auroral features and associated electric fields and waves; Auroral plasma density cavities observed at POLAR perigee (~1 R_E altitude) were found to contain transversely heated and accelerated populations of ions with characteristic energy comparable to the s/c potential;
3. The polar wind H^+ flux and O^+ density near 1 R_E were found to be strongly controlled by solar illumination;
4. The O^+ component of the polar plasma was shown to originate mainly in the dayside cleft region rather than in the polar cap proper;
5. The 6-9 R_E altitude polar wind, observed for the first time, was found to be faster, hotter, deficient in He^+ and richer in O^+ content than in thermal outflow theories, consistent with signficant contributions from dayside plasma heating processes;
6. Convection of polar wind outflows in the polar cap was found to respond to interplanetary conditions in accordance with the creation of a sunward plasma flow in the polar cap for Northward IMF B_Z, with clear implications for transport of plasma to the plasma sheet.
7. The high altitude polar wind was found often to be dynamic in space and/or time, with fluctuations in velocity having amplitude of a factor of order three;
8. Using 3D test particle simulations, the polar wind outflow was shown to supply plasma to active plasma sheet regions where neutral sheet acceleration immediately accelerates the protons to plasma sheet energies of several keV.

Future work on the Polar/TIDE data set will include continued acquisition, processing, correlative analysis, and interpretation of low-energy plasma data. The objectives of these studies for the Solar Maximum Initiative focus on the variations of low-energy plasma transport within the magnetosphere. We have learned that ionospheric plasma outflows are a pervasive feature of the low density polar cap and lobe regions, as well as the boundary layers and auroral zones. We now seek to 1) document and understand the dependence of these outflows upon heliospheric conditions as solar activity increases through solar maximum, and 2) understand the influence of solar-dependent plasma flows upon the development of magnetospheric storms, through collaborations in various global modeling efforts.

DISCUSSION AND CONCLUSIONS

Plate 3 illustrates the need for a higher-dimensionality view of ionospheric plasma in convecting plasma flux tubes, that has become apparent from the TIDE-PSI results.

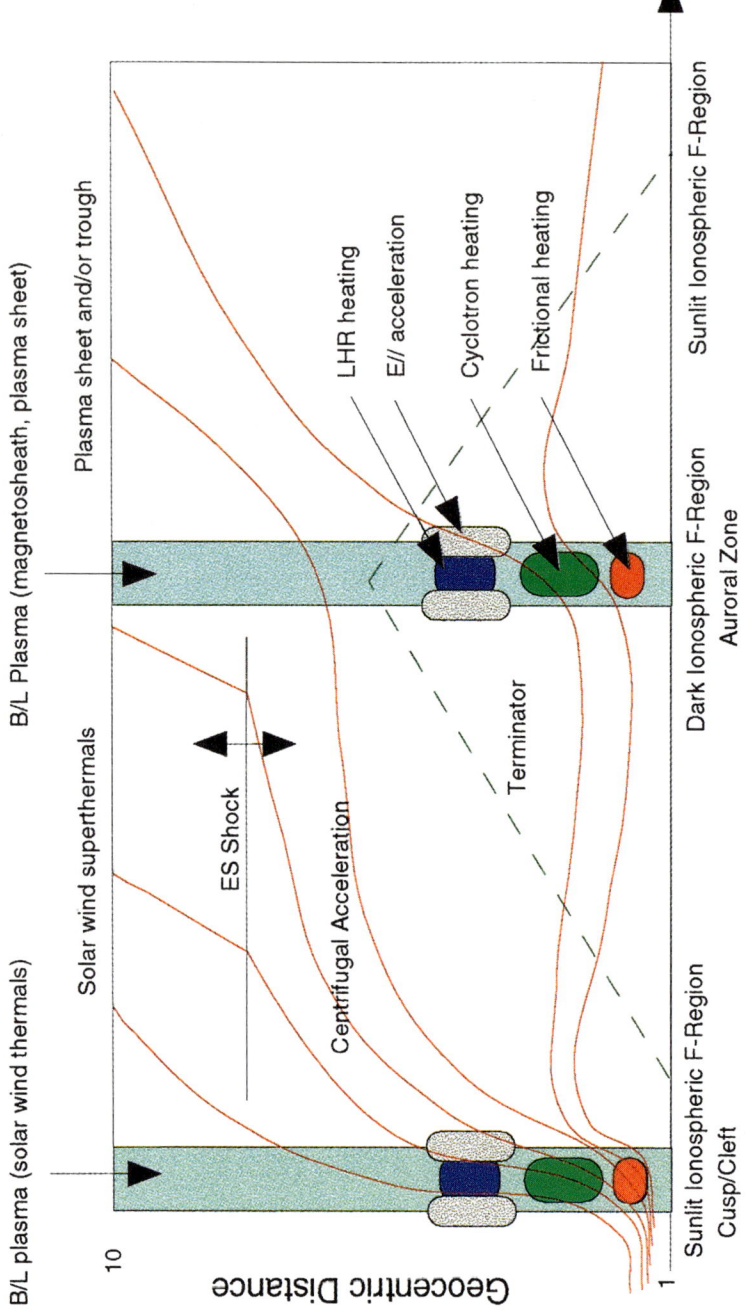

Plate 3. Schematic of plasma flux tube behavior during repeated high latitude convection cycles, including relevant low altitude and high altitude heating and acceleration processes. Flow streamlines are indicated as fine hairlines. The terminator is illustrated as a dashed line. The x-axis is periodic and wraps back to the origin.

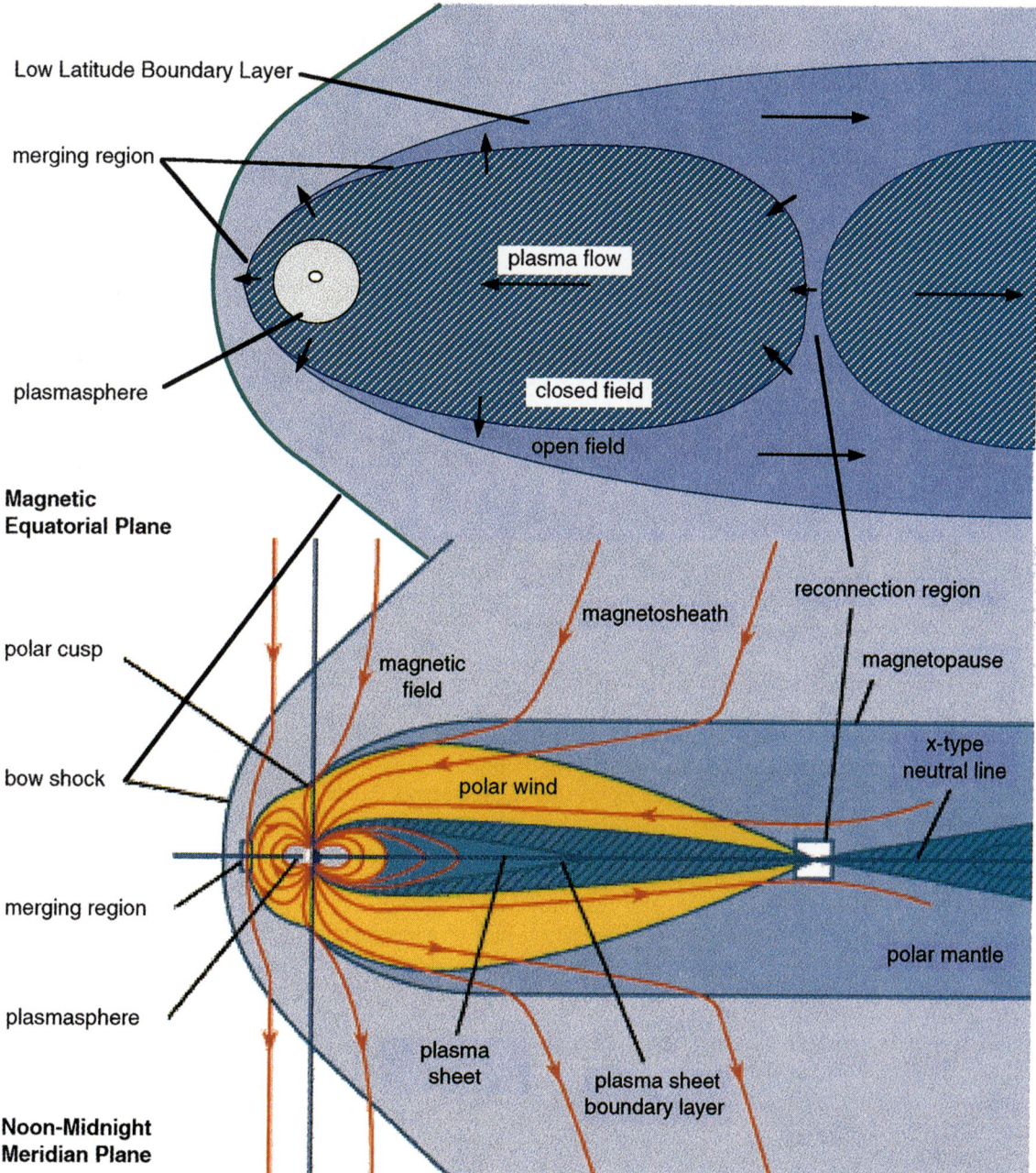

Plate 4. Schematic of ionospheric outflow destiny in the magnetotail, in relation to familiar structures in both the equatorial plane (upper panel) and the noon-midnight meridian (lower panel).

Here, the (periodic) convection cycle is represented along the x-axis, while the altitude profile along a flux tube is represented by the y-axis. The closed cycle convection is taken to begin immediately equatorward of the dayside auroral zone, or cusp/cleft region. Each plasma flux tube arrives at that position after a prior cycle around the prevailing circulation pattern, even if the precise circulation pattern has not been constant in magnitude or topology. Such an approach must take the place of the traditional one-dimensional description of a polar wind flux tube.

Plasma flux tubes circulate repeatedly around the high latitude convection flow pattern, and are repeatedly subjected to sunlight and and darkness, flow shears, associated current sheets, precipitating energetic particles, and turbulence that result from the continuing interaction with the flowing solar wind. Most relevant to plasma outflows is the input of significant electron and ion heating, as well as parallel electric field accelerations in regions of strong field-parallel electrical currents. These effects influence strongly the mass flux of the ionospheric outflows, principally through the addition of O^+ ions to the outflow. Flux tubes also undergo the cycle of elongation and relaxation that accompanies a trip through the magnetotail and plasma sheet.

At higher altitudes, the plasma flux tubes experience other effects including superthermal electron populations originating both in ionosphere (photoelectrons) and in the solar wind (polar rain electrons). These populations may conspire to create standing electrostatic potential structures across parts of the polar cap, that accelerate or decelerate the polar wind and heavy ion outflows. The plasma ions are also subject to centrifugal forcing owing to the interactions of the geoelectric and geomagnetic fields and the accelerations of the plasma convection frame. The result is a convection-driven "flinging" of especially the heavy ions down the tail. As noted earlier, such effects control the degree to which ionospheric plasma is lost to the downstream solar wind, accelerated by neutral sheet effects in the mid-tail, or circulated with little acceleration into the inner magnetosphere, becoming part of the refilling of the plasmasphere. All of these effects must ultimately be included in a complete theory of the terrestrial polar wind.

The increasing evidence that the ionosphere typically and continuously supplies plasma to the plasma sheet, has led to recent efforts to include an ionospheric fluid within global circulation models of the magnetosphere [*Winglee et al.*, 1997, 1998, unpublished manuscript]. This initial effort suggests that the pressure geopause encompasses much of the plasma sheet Earthward of the persistent neutral line, except during northward IMF conditions. That is, the ionosphere dominates the pressure in the active part of the plasma sheet for typical conditions. In view of this significant result, other efforts to include the ionospheric plasma as a dynamic component of the magnetosphere can be anticipated to follow, and are already beginning [*Song et al.*, 1998].

Plate 4 illustrates the destiny of the polar wind outflows as they, to varying degrees, escape down the tail lobes, become trapped in the earthward flow of the central plasma sheet, or are recirculated into the inner magnetosphere through the plasmaspheric trough region. The entire high latitude ionosphere can be thought of as differing from the low latitude ionosphere or plasmasphere mainly as a participant in the boundary layer flows of magnetospheric plasma. Because these flux tubes spend part of their convection cycle open to solar wind entry, the ionospheric outflows will be inhibited where solar wind pressure is signfiicant, for example in the low latitude boundary layer, where outer plasmaspheric or trough plasma mixes with solar plasma. In contrast, those flux tubes which reconnect through the cusp region pass into the high latitude boundary layer that constitutes the supersonic wake of the solar wind. Here there is a correspondingly vanishing solar wind plasma pressure that assures supersonic polar wind outflow.

This schematic view of ionospheric outflow morphology suggests that the ionospheric supply of plasma to the plasma sheet will be significantly augmented (if not displaced) by solar wind plasma as the interplanetary magnetic field rotates to northward, and the high latitude portion of magnetospheric convection shuts down. Conversely, when southward IMF causes the polar lobe flows to dominate magnetospheric convection, the central plasma sheet will be dominantly supplied with ionospheric outflows that have little solar wind content. Global models of the type that are now developing should be a signficant aid in understanding how the geopause extent varies over the more typical cases that are intermediate between these extremes.

Acknowledgements. This work was supported by the International Solar Terrestrial Physics Program at the Goddard Space Flight Center, under UPN 370-17-43. The authors are indebted to the entire ISTP Polar team and to the TIDE-PSI technical staffs of Marshall Space Flight Center, Southwest Research Institute, Los Alamos National Laboratories, Hughes Research Laboratories, and the Centre d'études des Environnments Terrestre et Planétaires.

REFERENCES

Axford, W.I., The polar wind and terrestrial helium budget, J. Geophys. Res. 73 (21), p.6855, 1968.

Banks, P. M., and T. E. Holzer, J. Geophys. Res. 73, 6846, 1968.

Barakat, A. R., H. G. Demars, and R. W. Schunk, Temporal behavior of the polar wind in the presence of hot electrons, Eos, 78(46), p.F502, 1997.

Chandler, M.O., P. D. Craven, B. L. Giles, T. E. Moore, C. J. Pollock, J.H. Waite, D.T. Young, J.L. Burch, J. E. Nordholt, J.R. Wygant, C.T. Russell, Plasma Transport in the Cleft, Entry Layer, and Lobes, Eos Trans. AGU, 77(46), F603, 1996.

Craven, P. D., V. N. Coffey, J. Wygant, T. E. Moore, "A search for low energy plasma in auroral density cavities", *Toward solar max 2000: The present achievements and future opportunities of ISTP and GEM*, Yosemite, CA, Febuary 10-13, [1998].

Comfort, R. H., T. E. Moore, P. D. Craven, C. J. Pollock, F. S. Mozer, and W. T. Williamson, Spacecraft potential control by PSI on the POLAR spacecraft, J. Spacecraft and Rockets, in press, 1998.

Dempsey, D. L., J. L. Burch, M. M. Huddleston, C. J. Pollock, J. H. Waite, Jr., M.Wüest, T. E. Moore, and E. G. Shelley, Reflected solar wind ions and downward accelerated ionospheric ions during the January 1997 magnetic cloud event, *Geophys. Res. Lett..*, 25(15), p.2979, 1998.

Elliott, H. A., et al., Ion outflow and convection in the polar cap and cleft as measured by TIDE, EFI, MFE, and TIMAS on POLAR, Eos 78(46), p.F623, 1997.

Fuselier, S. A., et al., Bifurcated cusp ion signatures: evidence for re-reconnection?, Geophys. Res. Lett., v.24(12), p.1471, 1997.

Ganguli, S.B, The Polar Wind, Revs. Geophys. 34(3), p.311, 1996.

Giles, B. L., T. E. Moore, M. O. Chandler, P.D. Craven, C. J. Pollock,The Upwelling Ion Source of Low Energy Ions: Initial Results from TIDE/PSI on POLAR, Eos Trans., AGU, 78(17), S293, 1997.

Giles, B L, C R Chappell, D C Delcourt, T E Moore, M O Chandler and P D Craven, Magnetospheric Plasmas - A Direct Measurement of the Ionospheric Source, Fall AGU, December 8-12, 1997.

Hirahara, M., J.L. Horwitz, T.E. Moore, G.A. Germany, J.F. Spann, W.K. Peterson, E.G. Shelley, M.O. Chandler, B.L. Giles, P.E. Craven, C.J. Pollock, D.A. Gurnett, J.S. Pickett, A.M. Persoon, J.D. Scudder, N.C. Maynard, F.S. Mozer, M.J. Brittnacher, and T. Nagai, Relationship of topside ionospheric ion outflows to auroral forms and precipitation, plasma waves, and convection observed by Polar, *J. Geophys. Res.*, 103(A8), p.17391, 1998.

Hirahara, M., J. L. Horwitz, T. E. Moore, M. O. Chandler, B. L. Giles, P. D. Craven, and C. J. Pollock, Polar observations of properties of H^+ and O^+ conics in the cusp near 5000 km altitude, in Geospace Mass and Energy Flow: Results from the International Solar-Terrestrial Program, AGU monograph, ed. by J. L. Horwitz, D. L. Gallagher, and W. K. Peterson, in press, 1998.

Knudsen, D. J., B. A. Whalen, T. Abe, and A. Yau, Temporal evolution and spatial dispersion of ion conics: Evidence for a polar cap heating wall, in Solar System Plasmas in Space and Time, ed. by J. L. Burch and J. H. Waite, Geophys. Monograph #84, AGU, Washington, DC., p. 163, 1994.

Moore, T. E., C. R. Chappell, M. O. Chandler, S. A. Fields, C. J. Pollock, D. L. Reasoner, D. T. Young, J. L. Burch, N. Eaker, J. H. Waite, Jr., D. J. McComas, J. E. Nordholt, M. F. Thomsen, J. J. Berthelier, and R. Robson, The Thermal Ion Dynamics Experiment and Plasma Source Instrument, Space Sci. Revs. 71, p.409, 1995.

Moore, T. E. C. R. Chappell, M. O. Chandler, P. D. Craven, B. L. Giles, C. J. Pollock, D. T. Young, J. H. Waite, Jr., J. E. Nordholt, M. F. Thomsen, D. J. McComas, J. J. Berthelier, W. S. Williamson, R. Robson, and F. S. Mozer, High-altitude observations of the polar wind, Science, 277, p.349, 1997.

Song, P., D. L. DeZeeuw, T. I. Gombosi, P. T. Groth, K.G. Powell, GGCM Phase 1 model runs with the BATS-R-US code, Eos 79(17), p.S325, 1998.

Su, Y.-J., J. L. Horwitz, T. E. Moore, M. O. Chandler, P. D. Craven, B. L. Giles, M. Hirahara, and C. J. Pollock, Polar wind survey with TIDE/PSI suite aboard POLAR, J. Geophys. Res., 103, in review, 1998a.

Terasawa, T., et al., Solar wind control of density and temperature in the near-Earth plasma sheet: WIND/Geotail collaboration, Geophys. Res. Lett., 24(8), p.935, 1997.

Su, Y-J., J. L. Horwitz, T. E. Moore, M. O. Chandler, P. D. Craven, B.L . Giles, C. J. Pollock, S. W. Chang, and J. Scudder "Polar wind and low-energy electron measurements at POLAR apogee", to be presented to Spring AGU meeting, Boston, Mass, May, 1998b.

Winglee, R. M., Mixing of solar wind and ionospheric plasmas in the magnetosphere as identified by a multi-fluid global model, Eos 78(46), pF613, 1997.

Winglee, R. M., M. Brittnacher, and G. Parks, Multi-component plasma flows during the May, 1996 substorm: a comparison between Wind observations and multi-fluid simulations, Eos 79(17), p.S321, 1998.

T. E. Moore and B. L. Giles, NASA Goddard Space Flight Center, Mail Code 692, Greenbelt, MD 20771 USA (e-mail: thomas.e.moore @gsfc.nasa.gov).

M. O. Chandler, P. D. Craven, NASA Marshall Space Flight Center, Mail Code ES83, Huntsville, AL 35812 USA (michael.chandler @msfc.nasa.gov).

C. R. Chappell, Department of Media Relations, Vanderbilt University, 110 21st St. S., Nashville, TN 37203 USA (e-mail charles.r.chappell@vanderbilt.edu).

R. H. Comfort, H. A. Elliott, J. L. Horwitz, *Y.-J. Su, Department of Physics, The University of Alabama in Huntsville, Huntsville, AL 35899 USA (e-mail: comfortr@cspar.uah.edu).

D. C. Delcourt, Centre d'étude des Environnments Terrestre et Planétaires, 4, Avenue de Neptune, Saint-Maur des Fossés, 94107 FRANCE (dominique.delcourt@cetp.ipsl.fr).

C. J. Pollock, Southwest Research Institute, 6220 Culebra Rd.., San Antonio, TX 78228-510 USA (cpollock@swri.org).

* now at: Space and Atmospheric Sciences, Los Alamos National Laboratories, Los Alamos, NM 87545 USA (e-mail: ysu@lanl.gov).

The Low-Latitude Boundary Layer: Application of ISTP Advances to Past Data

M. Lockwood and M. A. Hapgood

Rutherford Appleton Laboratory, Chilton, UK

The destruction of the four Cluster craft was a major loss to the planned ISTP effort, of which studies of the magnetopause and low-latitude boundary layer (LLBL) were an important part. While awaiting the re-flight mission, Cluster-II, we have been applying advances in our understanding made using other ISTP craft (like Polar and Wind) and using ground-based facilities (in particular the EISCAT incoherent scatter radars and the SuperDARN HF coherent radars) to measurements of the LLBL made in 1984 and 1985 by the AMPTE-UKS and -IRM spacecraft pair. In particular, one unexplained result of the AMPTE mission was that the electron characteristics could, in nearly all cases, order independent measurements near the magnetopause, such as the magnetic field, ion temperatures and the plasma flow. Studies of the cusp have shown that the precipitation is ordered by the time-elapsed since the field line was opened by reconnection. This insight has allowed us to re-analyse the AMPTE data and show that the ordering by the transition parameter is also due to the variation of time elapsed since reconnection, with the important implication that reconnection usually coats most of the dayside magnetopause with at least some newly-opened field lines. In addition, we can use the electron characteristics to isolate features like RDs, slow-mode shocks and slow-mode expansion fans. The ion characteristics can be used to compute the reconnection rate. We here retrospectively apply these new techniques, developed in the ISTP era, to a much-studied flux transfer event observed by the AMPTE satellites. As a result, we gain new understanding of its cause and structure.

THE MAGNETOPAUSE TRANSITION PARAMETER

The magnetopause transition parameter was based on the work of *Hall et al.* [1985] and *Bryant and Riggs* [1989] and exploits the observed anti-correlation of electron density and temperature, also noted by *Sckopke et al.* [1981] and *Phan et al.* [1997]. *Hapgood and Bryant* [1992] developed its definition and implementation. The electron density, N_e, is plotted as a function of an electron temperature, T_e on a log-log scale and the characteristic variation fitted with a polynomial. (In fact, slightly better results are usually obtained by using the perpendicular electron temperature, $T_{e\perp}$). The transition parameter τ is the percentage distance along that fitted curve, a value of 0 being ascribed to the magnetosheath end of the curve and 100 being at the magnetospheric end.

The anti-correlation of N_e and T_e over much (but not all) of the curve can be explained as the change in the moments with a changing ratio of the magnetosheath to the magnetospheric components of an electron gas. Thus, for example, an increased sheath component of the electron gas will decrease the temperature of the total distribution, whilst increasing the density. Almost any process which causes a mixing of the two electron populations (of which reconnection is just one example) could cause this. Thus the existence of a transition parameter is not surprising. What is

Sun-Earth Plasma Connections
Geophysical Monograph 109
Copyright 1999 by the American Geophysical Union

extraordinary is how well it orders independently-measures parameters like the magnetic field, the ion spectrum (and its moments) and the plasma flow [*Hapgood and Bryant*, 1992]. Data sequences in these parameters showing complex variations with observation time, t_s (such as, for example, would be obtained for a series of multiple full and/or partial boundary crossings) give simple variations, with very little scatter, when plotted as a function of transition parameter, τ. This ordering by the transition parameter was found to be effective in 41 out of 44 magnetopause crossings by AMPTE-UKS and neither surface waves nor flux transfer events (FTEs) disrupt it [*Bryant and Riggs*, 1988; *Hapgood and Lockwood*, 1995]. The success of the transition parameter, based only on the characteristics of the electron gas, in ordering the AMPTE data on the magnetic field and ion gas implied an underlying physical ordering of the particles and fields of the magnetopause boundary layer. However, the nature of that ordering and why it was present was not understood.

MODELLING THE INJECTED MAGNETOSHEATH ION POPULATION

Models of ion behaviour in the magnetosphere have recently been developed and successfully used to predict signatures of ion precipitation into the cusp ionosphere [*Onsager et al.*, 1993; *Onsager*, 1994; *Lockwood and Smith*, 1994; *Lockwood*, 1995; *Lockwood and Davis*, 1996b; *Lockwood et al.*, 1997]. These models allow for four main elements: (1) the spatial variations of the magnetosheath density and temperature (to date, gas dynamic predictions have been employed [*Sprieter et al.*, 1966]); (2) the evolution of reconnected field lines over the magnetopause, as predicted by *Cowley and Owen* [1989]; (3) the theory of the ion acceleration and distribution functions at the magnetopause current sheet [*Cowley*, 1982] and (4) the time-of-flight velocity filter effect of ion motion along convecting field lines [*Rosenbauer et al.*, 1975; *Reiff et al.*, 1977]. These models of magnetosheath ion injection and transport have been very successful in reproducing the distribution functions of the precipitating cusp ions at low and middle altitudes both during steady-state conditions [*Onsager et al.*, 1993 and *Lockwood*, 1997, respectively] and for periods when magnetopause reconnection is pulsed [*Lockwood and Davis*, 1996b; and *Lockwood et al.*, 1998, respectively].

A refinement of the model of *Cowley* [1982] has been introduced by *Lockwood et al.* [1996], who allowed for reflection of magnetospheric ions off the Alfvén wave (hereafter called a rotational discontinuity, RD) on the interior edge of the open LLBL, as well as at the main RD (i.e., the magnetopause itself) on the outer edge of the LLBL. The magnetopause is an RD emanating from the reconnection site and standing in the inflow into the magnetopause from the magnetosheath. Correspondingly, the interior RD stands in the inflow from the magnetospheric side of the boundary. The theory of Cowley allows for the reflection of ions that are incident on an RD by flowing along the reconnected field lines, as well as their transmission through the RD. Cowley assumed that 50% of the incident ions were reflected, and 50% transmitted, a ratio which was found to be roughly correct in the case studied by *Fuselier et al.* [1991]. Recently, we have been able to self-consistently evaluate these reflection coefficients by using kinetic theory of *Cowley* [1982], taking the moments of the predicted distribution functions and then iterating the reflection coefficients until the moments obey the fluid conservation equations for an RD (conservation of mass, normal momentum, tangential momentum and energy), as given by *Hudson* [1970]. With the addition of the interior RD, *Lockwood et al.* [1996] were able to model energetic ion precipitation at the equatorward edge of the cusp dispersion ramp, reproducing the observed spectra as well as the moments of the ion distribution. The model, with this extension, was also successfully employed by *Lockwood* [1997] and *Lockwood and Moen* [1996] to match observed ion precipitation distribution functions and fluxes, respectively.

The time-dependent version of the model computes the ion spectrum seen at a given location relative to the reconnection X-line, as a function to the time elapsed since reconnection, (t_s-t_o), where t_s is the time that a field line is observed and t_o is the time that it was reconnected. The importance of considering the precipitation as a function of (t_s-t_o) was revealed by studies of poleward-moving transients in the cusp, as seen by optical imagers [Sandholt et al., 1990], the EISCAT incoherent scatter radars [*Lockwood et al.*, 1993] and the Halley Bay and CUTLASS HF radars [*Pinnock et al.*, 1995]; *Neudegg et al.* [1998], as explained by *Lockwood and Davis* [1996b]. This concept has been tested using ISTP satellite data. *Lockwood et al.* [1998] applied the model to fit the energy-time spectrograms of injected cusp ions seen by the Hydra instrument on the Polar satellite at middle altitudes.. Specifically, they fitted the sawtooth form for ions which have been injected, mirrored below the satellite and were observed moving upward. This completely prescribed the predictions for downgoing, zero pitch-angle injected ions which reach the satellite directly. Thus comparison with the observed downgoing ions was a blind test which verified the model and that reconnection was taking place mainly in short pulses.

HOW THE TRANSITION PARAMETER WORKS

The ion model discussed above has been used by *Lockwood and Hapgood* [1997] to give an important insight into how the magnetopause transition parameter ordering works. This was achieved by returning to the AMPTE data, using the understanding and the model of the ion gas derived and tested using ISTP data. In particular, *Lockwood and Hapgood* [1998] have revisited a much-studied flux transfer event (FTE), observed by the AMPTE-UKS and -IRM satellites around 10:46 UT on 28 October 1984 during an outbound magnetopause crossing. The satellites were at a GSM latitude of 25.7° (northern hemisphere) and at a magnetic local time of 08:55 (i.e. in the mid-morning sector). They were separated by 180 km in a direction roughly aligned with the boundary-normal (as determined from the magnetopause crossing by UKS which took place considerably later, at 11:45-12:45 UT), with UKS closer to the boundary than IRM. This event was first reported by *Rijnbeek et al.* [1987] who noted its layered structure. Subsequently, it has been the subject of studies by *Farrugia et al.* [1988], *Lockwood et al.*, [1988], *Bryant and Riggs* [1989], *Sibeck* [1992], *Sibeck and Smith* [1992], and *Smith and Owen*, [1992]. *Rijnbeek et al.* [1987] and *Farrugia et al.* [1988] showed that there was a high-pressure core at the event centre predominantly due to particle pressure, but that outside this was a layer of high magnetic pressure and low particle pressure. The origin of this high pressure core of some FTEs has never been satisfactorily explained [*Paschmann et al.*, 1982].

The cusp ion model can be applied to the open low-latitude boundary layer at the magnetopause data, the only difference being that the satellite is relatively close to the reconnection site, compared to spacecraft at middle of low altitudes. Figure 1 illustrates the general principles by showing schematically an open LLBL (reconnection outflow layer) produced by (in this case steady) reconnection at X. The figure shows the separatrices s (which pass through X and for which the time elapsed since reconnection (t_s-t_o) is zero) and four other newly-opened field lines as they evolve away from X with increasing reconnection (t_s-t_o). Standing in the inflow to the magnetopause are the exterior and interior RDs (the dashed lines m and i). At a given distance from the X-line to the satellite, d, the (t_s-t_o) increases inside the LLBL as the satellite (UKS) approaches the exterior RD (i.e. the magnetopause): (t_s-t_o) is zero at the interior separatrix and reaches a maximum value (for that d) at the magnetopause. Outside the magnetopause, in the magnetosheath boundary layer (MSBL), (t_s-t_o) decreases again, reaching zero at the exterior separatrix. Outside the separatrices (t_s-t_o) is negative (i.e. the field lines have yet to be reconnected) but (t_s-t_o) has influence on neither the plasma nor the field. Ions reaching the satellite have a spread of field-aligned velocities and follow trajectories within the dark grey shaded wedge (the lowest energy ions having the longest flight time and having originated at the reconnection site, the highest energy ions (with fluxes that are detectable) crossing the magnetopause considerably closer to the

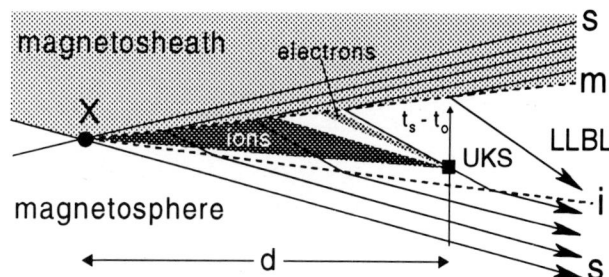

Figure 1. Schematic of AMPTE-UKS in the open LLBL, between the magnetopause (m, also referred to as the exterior RD, e) and interior (i) Alfvén waves emanating from the reconnection site (X-line), X. Field lines evolving away from X are shown at five elapsed times since they were opened (t_s-t_o), including zero for the magnetic separatrices (s). The spectrum of sheath ions reaching UKS has a spread of trajectories shown by the dark shaded wedge, the trajectories of the sheath electrons are much closer to field-aligned (lighter shaded wedge). The populations seen depend on UKS's depth into the LLBL, i.e. on the (t_s-t_o) at a given distance d from X.

satellite. The electrons reaching the satellite have a much higher field-aligned velocities and have trajectories which are closer to field-aligned and are within the lighter grey wedge. Because quasi-neutrality is maintained, the number of sheath electrons reaching the satellite is approximately the same as the number of sheath ions, the latter being a function of the distance d and the time elapsed since reconnection (t_s-t_o), i.e. it depends on how deep into the LLBL the satellite is situated.

Figure 2 shows the results of least-squares fitting the observed moments of the ion gas during this FTE, using the ion model discussed above. The distance between the satellite and the X line, d, is assumed to be $8R_E$ (see later). The plot shows the moments of the ion gas as a function of observation time t_s (given on the figure axis in seconds after 10:43UT). The histograms are the observed values and the lines are the fitted model values. The procedure adopted was to vary the value of the time elapsed since reconnection (t_s-t_o) at every observation time t_s, until the best fits to the ion number density N and temperature T were obtained. This prescribes the variation in the ion pressure, P, but the number density at energies above 1 keV, $N_{[E>1keV]}$, and field-parallel velocity V_{para} are independent tests of these fits. Figure 1 also shows the fitted (t_s-t_o). As a further test, the time-of-flight cut-off energy of the ions, E_{ic} $(=(m/2)\{d/(t_s-t_o)\}^2)$ is computed and compared with the observed value. The bottom panel of figure 1 shows the variation of the transition parameter τ during this event.

It can be seen that the model provides an explanation of the high ion pressure in the core of the event. Essentially, the field lines in the core have been opened for longer (large t_s-t_o) allowing more of the lower-energy ions to reach the satellite (lower E_{ic}), raising N.

and this explains why τ is able to order the ion and field data which also depend on (t_s-t_o). For a constant distance d, both τ and (t_s-t_o) are monotonic functions of the distance of the satellite from the magnetopause (the form of that function depending on the variation of the reconnection rate). It should be noted that in 41 out of 44 magnetopause crossings by AMPTE-UKS, the transition parameter was able to order independent magnetopause data. This implies that at least some newly-opened field lines coat most of the dayside low-latitude magnetopause most of the time, irrespective of the IMF orientation (i.e. an open LLBL is nearly always present).

IMPLICATIONS FOR UNDERSTANDING FTES

As well as producing good fits to the moments of the ion gas, as shown in figure 2, the ion model can reproduce the energy-time spectrogram for this FTE event (the observed and modelled spectrograms have been presented by *Farrugia et al.* [1988] and *Lockwood and Hapgood* [1998], respectively). This is true for the boundary layers of this structured event, as well as the event core. Of particular importance is the fact that the event boundaries show a continuous evolution of the ion gas from the magnetospheric population to that in the event core. This is explained using the ion model by the continuous variation in (t_s-t_o) with observation time t_s shown in figure 2. This eliminates the original "fossil flux tube" model of FTEs

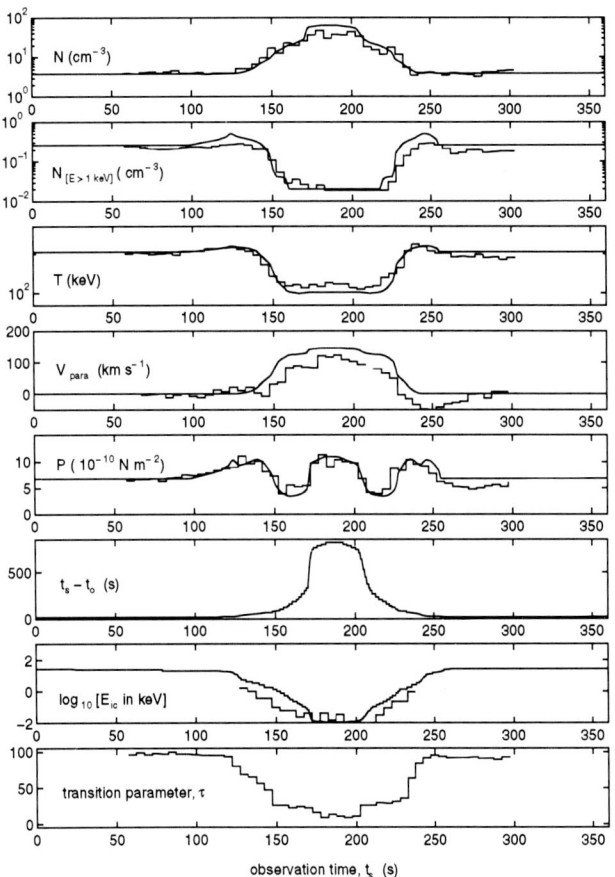

Figure 2. AMPTE UKS observations of an FTE on 28 October 1984, plotted as a function of observation time, t_s, which is zero at 10:43:00. Observed (histogram) and best-fit modelled (curves) moments of the ion gas are shown. From top to bottom: the ion density, N, observed in the instrument energy range of 100 eV-16 keV, the ion density in the energy range 1 - 16 keV, $N_{[E>1keV]}$, the ion temperature, T, the field parallel velocity, $V_{||}$, the ion pressure, P, the best-fit time elapsed since reconnection (t_s-t_o), the low-energy ion cut-off, E_{ic}, and the observed electron transition parameter τ.

Figure 3 shows the variation of the transition parameter τ with the best-fit (t_s-t_o) for the period shown in figure 2. It can be seen the plot follows the same locus for the entry of the FTE as it does for the exit. In addition, figure 3 shows the predictions of a simple model (dashed line) developed by *Lockwood and Hapgood* [1997]. In this model, the electron density at the satellite is controlled by a potential barrier between the magnetopause and the satellite, of magnitude such that the total electron density at the satellite is the same as that of the ion gas.

The key point is that the transition parameter τ has a simple variation with time elapsed since reconnection (t_s-t_o)

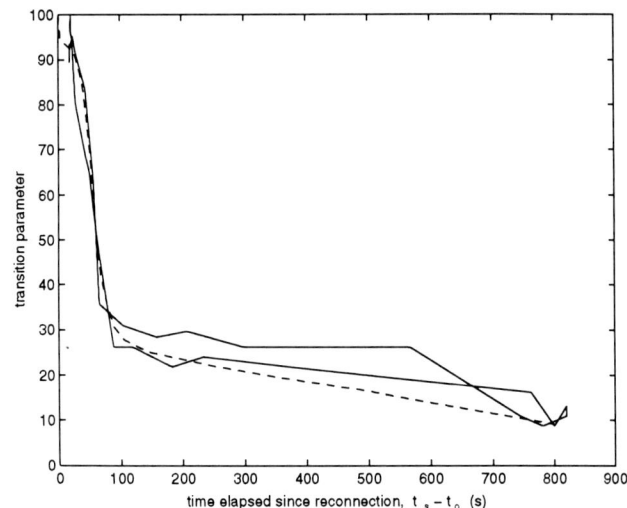

Figure 3. Solid line: hodogram showing the variation of the observed transition parameter τ with the time-elapsed since reconnection (t_s-t_o), from the fit to the ion data shown in figure 2. Dashed line: model prediction made by applying the transition parameter to simulted electron data using the ion model and a potential barrier to maintain quasi-neutrality.

[*Russell and Elphic*, 1978; 1979] as a cause of this event. This is because this model predicts a discontinuous jump in (t_s-t_o) from negative to positive values (with a corresponding jump in ion characteristics) as the satellite moves from the draped closed field lines to the open field lines of the fossil flux tube when the satellite enters the event. (The converse jump would be seen on leaving the event. The only way that this model could explain this event is if there were some additional mechanism to cause the continuous evolution of ion parameters across the event boundary layer. No such mechanism has yet been proposed and is, anyway, not necessary as the boundary layer is well explained as a variation of (t_s-t_o). In other words, figure 2 shows that the event is well explained as a brief entry of the satellite into the open LLBL.

The observed direction of field-aligned motion of the injected sheath ions and electrons shows that the field lines detected in the centre of this event were connected magnetically to the northern hemisphere. In other words, the outward boundary normal field component B_N is negative where these field lines thread the boundary. The nested nature of the signals seen by UKS and IRM enable us to quantify the size of the event and to determine the speed and direction of event motion [see *Lockwood and Hapgood*, 1998].

Figure 4 shows the two models of FTE formation that are consistent with this finding. Figure 4(a) shows the cylindrical 2-D reconnection pulse model discussed by *Southwood et al.* [1988] and demonstrated by *Scholer* [1988; 1989] using MHD simulations and by *Semenov et al.*, [1991; 1992a; b] using analytic theory. Figure 4(b) shows the pressure pulse model of *Sibeck et al.* [1990; 1992], with the important caveat that magnetopause reconnection must be ongoing throughout the event (possibility mentioned by Sibeck in his original paper). The satellite trajectory in the rest frame of the event is shown by the locus S. In both cases, the (t_s-t_o) of the field lines sampled by the satellite increase as the satellite is immersed deeper into the open LLBL. In the reconnection pulse model (figure 4a) this occurs because of a transient thickening of the open LLBL in response to a reconnection rate pulse. In the pressure pulse model (figure 4b) it occurs because of a transient compression of the magnetopause caused by a travelling enhancement of the magnetosheath pressure. Note that both cases show an indentation of the interior RD (i), but only in 4(b) is there a similar indentation of the exterior RD (e).

It is very difficult for a lone satellite in the magnetosphere to distinguish between these two possibilities on a case-by-case basis. However, we have been able to apply the method of *Lockwood and Smith* [1992], to determine the variation of the reconnection rate at which the open field lines seen in the event are produced.

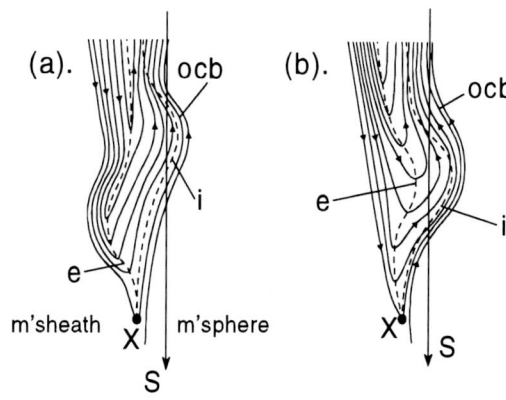

Figure 4. Explanations of the FTE event in terms of (a) the two-dimensional pulse model and (b) the pressure pulse model. X is the reconnection site; S is the satellite locus in the event rest frame; the dashed lines labelled e and i are the rotational discontinuities (RDs) standing on the inflow on the magnetosheath and magnetosphere sides, respectively; and ocb is the open-closed field-line boundary. (Note that the exterior RD was labelled m in figure 1).

This method was originally developed for ionospheric field lines in the cusp region, but has been modified by *Lockwood and Hapgood* [1998] to allow for the fact that the field at the magnetopause is compressible. The method has also been tested on simulated data by *Lockwood and Davis* [1996a]. The results are shown in figure 5, which shows the reconnection rate, computed from the variation of (t_s-t_o) derived in figure 2, as a function of the reconnection time, t_o. The plot shows data from both the satellite's entry into, and exit from, the event and these were found to agree when the distance d was iterated to 8 R_E. Using the inferred direction of event motion, this place the reconnection site within a few R_E of the subsolar point.

Figure 5 clearly shows that the reconnection rate was pulsed. The event core was reconnected in an earlier pulse (of which we see only the end as the satellite did not penetrate deep enough into the LLBL to see field lines opened any earlier), whereas the boundary layer (seen on both entry and exit) was reconnected in a pulse roughly 15 min. later, these field lines being draped over the bulge in the reconnection layer caused by the first pulse. However, this detection of a reconnection pulse is necessary, but not sufficient, for proof of the reconnection pulse FTE model (in the same way that the detection of a pressure pulse in the sheath would be necessary but not sufficient for proof of the pressure pulse model). However, it is an indication in favour of the reconnection pulse model, although the high magnetosheath densities required to model the event core do suggest that a pressure pulse may also have played some role.

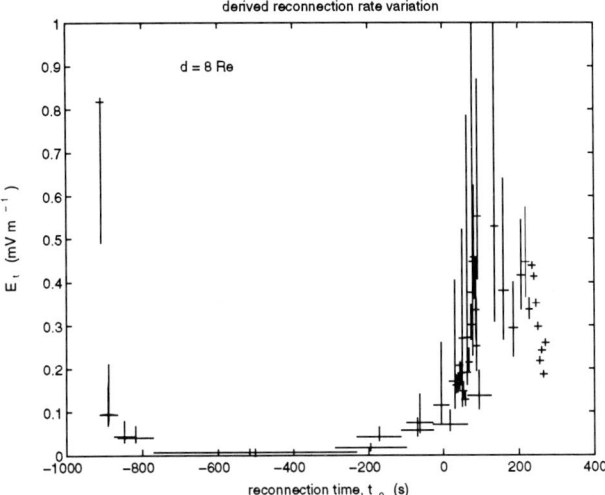

Figure 5. The reconnection rate E_t, shown as a function of reconnection time, t_o, calculated from t_s and $(t_s - t_o)$ for a distance between the X-line and the satellite of $d = 8\ R_E$, using the theory of *Lockwood and Smith* [1992], with modifications to allow for a compressible field. The magnetic flux transported over AMPTE-UKS in each 5-second integration period ($\Delta t_s = 5$s) is $BV_\perp \Delta t_s$, where $V_\perp$ is the magnitude of the field-perpendicular velocity and B the magnetic field strength. The reconnection rate $E_t = \Delta t_s BV_\perp / \Delta t_o$ [*Lockwood and Hapgood*, 1998].

IMPLICATIONS FOR UNDERSTANDING LLBL STRUCTURE

Because it is related to (t_s-t_o), the transition parameter can be used to identify structures in the LLBL, and so increase the number of observations of that structure in multiple intersections. This is particularly valuable when applying the tangential stress-balance test. *Lockwood and Hapgood* [1998] have generalised that test to allow for all Alfvénic disturbances, not just the RD which has been identified using the tangential stress-balance test to see if Whalén relation applies [*Paschmann*, 1979; 1986; *Sonnerup*, 1986].

Figure 1 shows a simple picture of an open LLBL with exterior and interior RDs, m and i, propagating into the inflow on the magnetosheath and magnetospheric sides, respectively. Similarly, there may be slow mode shocks and/or slow-mode expansion fans standing in these two inflow regions and some authors have suggested there may be a contact discontinuity where the two inflows meet [see review by *Lin and Lee*, 1993a; b]. Because these features have different field-aligned propagation speeds they can form a layered structure in the LLBL. In general, Alfvénic disturbances propagate at a speed V_A / f, where V_A is the local field-aligned Alfvén speed and f is a factor which depends on the type of disturbance [*Heyn et al.*, 1988] (for example, $f = 1$ for an RD; $f > 1$ for a slow mode shock or a slow mode expansion fan; and $f = \infty$ for a contact discontinuity). The vector subtraction of the field line velocity, $\underline{V}_f$, from the inflow velocity into the reconnection layer, $\underline{V}$, gives field-aligned flow speed of V_A / f. This is parallel to the field for the exterior discontinuity, but antiparallel for the interior one for this case with $B_N < 0$. This gives:

$$\underline{V} = \underline{V}_f \pm (\underline{V}_A / f) = \underline{V}_f \pm (\underline{B} / f) \cdot \{(1-\alpha)/\mu_o\}^{1/2}, \quad (1)$$

where α is the anisotropy factor ($\alpha = (P_\parallel - P_\perp)\mu_o/B^2$ where $P_\parallel$ and $P_\perp$ are the total field-perpendicular and field-parallel particle pressures) and ρ is the mass density ($\rho = Nm_i$, where m_i is the mean ion mass). The + and -, in this case with $B_N < 0$, relate, respectively, to exterior and interior disturbances, for which the field parallel flow in the field-line rest frame (the "de-Hoffman Teller frame" [*de Hoffman and Teller*, 1950], $V_\parallel'$, is positive and negative. Being a linear vector equation, equation (1) is valid for any component. Putting into a form equivalent to that used by *Paschmann et al.* [1979], but without actually applying the mass conservation condition for an RD (derived by *Hudson* [1970]):

$$\underline{V} = \underline{V}_f \pm (\rho_o/\rho)\underline{B} (1/f)\{ (1-\alpha)\rho/\mu_o\rho_o^2 \}^{1/2}. \quad (2)$$

Note that for $f = 1$, equation (2) reduces to the Whalén relation for an RD.

Lockwood and Hapgood [1998] used the transition parameter to isolate the field rotation on the edges of the event core and obtained negative slopes in the plots of the components of $\underline{V}$ against the corresponding component of $\underline{B}$, showing that this field rotation is an interior disturbance, and is not an exterior one because $B_N < 0$ (i.e. it is standing in the inflow from the magnetospheric side of the boundary). Equation (2) shows that the slope of the fits for the three components should be the same for an Alfvénic disturbance. *Lockwood and Hapgood* [1988] found slopes of -0.8±0.5, -0.7±0.5 and -0.7±0.5 km s^{-1}nT^{-1} for the L,M and N components.

If we consider an RD, $f = 1$ and the *Hudson* [1970] mass conservation condition for an RD applies, i.e. $(1-\alpha)\rho$ is constant. The theoretical slope from equation (2) is thus $\{(1-\alpha_o)/\mu_o\rho_o\}^{1/2}$ and the observed N_o and α_o yield a slope magnitude of 3.9±0.4 km s^{-1}nT^{-1} (the uncertainty arising from that in the mean ion mass, m_i). Therefore, this is not a successful application of the Whalén relation (for an RD) because the observed slope of magnitude 0.75±0.50 km s^{-1}nT^{-1} is inconsistent with $f = 1$, for any reasonable ion composition assumption. Nor is the structure a contact discontinuity, for which $f = \infty$: equation (2) predicts this would give a slope of zero (for any composition of the ion gas) which is also outside the observed range of 0.75 ± 0.50 km s^{-1}nT^{-1}.

Lockwood and Hapgood also investigated if the putative Alfvénic discontinuity could be a slow shock or a slow mode expansion fan. *Heyn et al.* [1988] show that $\eta<1$ for the former but $\eta>1$ for the latter, where:

$$\eta = (B_{t2}/B_{t1}) = \{1 + \beta(1 - P_2/P_1)\}^{1/2}, \quad (3)$$

where B_t is the discontinuity-tangential magnetic field and P is the particle pressure, and where the subscripts 1 and 2 refer to upstream and downstream of the discontinuity. From the sense of the slope, and because the field-parallel flow is negative in the de-Hoffman-Teller frame, we know that the upstream side is the magnetospheric side of this structure. For this event, $P_2>P_1$ and thus by (3) $\eta < 0$ and thus this structure is most likely to be a slow shock, rather than a slow-mode expansion fan. The plasma $\beta = 2P\mu_o/B^2 \approx 0.15$ and from the values of N, T_e and T_i upstream and downstream of the discontinuity, equation (3) yields $\eta \approx 0.92$. However a shock is rather surprising, considering the rather extended nature of the density change (estimated above to be of order 250 km) [see *Lin and Lee*, 1993].

The equations of *Heyn et al.* [1988] assume pressure isotropy ($\alpha = 0$), which is a good approximation on the edges of the event. From them, we can derive an expression for the factor f appropriate to a slow shock:

$$f = (\rho_1/\rho_2)^{1/2} \{1 + (1+\eta)/[\gamma\beta + (\gamma-1)(1-\eta)]\}^{1/2}, \quad (4)$$

where γ is the polytropic index. *Pudovkin et al.* [1997] use theory and past observations to estimate that γ is between 1.34 and 1.95 at the bow shock, but pressure anisotropy at the magnetopause means that the effective γ can be less than 1. We here use the relation:

$$\rho_1/\rho_2 = (P_2/P_1)^{1/\gamma} \quad (5)$$

for the ratios of the densities and pressures across the structure to estimate $\gamma = 1.2$. Using this and the mean $\beta = 0.15$, equation (4) yields $f = 2.7$ for a slow shock. Using the mean N and α of, respectively, 2×10^7 m^{-3} and -0.05, equation (2) yields theoretical slopes of 1.1 ± 0.1 km s^{-1}nT^{-1}. This value is still somewhat higher than the nominal observed value, but is consistent with it to within the uncertainties. Thus a slow-mode shock is a possibility.

For completeness, from the equations given by *Heyn et al.* [1988] we can also derive an expression for f for a slow-mode expansion fan:

$$f = (1 + V_{at}^2/C_s^2) = \{1 + B_t^2/(\gamma P\mu_o)\}^{1/2}, \quad (6)$$

where V_{at} is the boundary tangential Alfvén speed (corresponding to B_t) and C_s is the sound speed $= (\gamma P/\rho)^{1/2}$. This gives $f = 3.5$ for a slow-mode expansion fan and theoretical slopes for the stress-balance test of 0.9 ± 0.1 km s^{-1}nT^{-1}. This is within the range of possible experimental values of 0.75 ± 0.50 km s^{-1}nT^{-1} and similar to those for the slow shock.

From the above we find that the field rotation on the edges of the FTE core are a convecting structure and the consistency of the slope in the three components suggests it may be an Alfvénic discontinuity, but its speed of propagation is lower than an Alfvén wave (RD) and is most likely to be a slow shock. On the magnetospheric side of this structure $V_{//} < 0$ in the Earth's frame, as the flow is dominated by the escape of magnetospheric ions towards the magnetopause: nearer the magnetopause, within and on the other side of this structure, $V_{//} > 0$ as the flow is dominated by injected magnetosheath ions flowing Earthward.

CONCLUSIONS

The loss of Cluster was a devastating blow to ISTP studies of the LLBL. Because of it, we have turned our attention to a re-analysis of the AMPTE data, using new insights (such as of the importance of time elapsed since reconnection), techniques (such as that developed by *Lockwood and Smith* [1992] to compute the reconnection rate variation from cusp ion dispersion) and models (such as the injected ion model). There is a surprising wealth of new information to be gained in this way. To stress this point, we have shown how analysis of one magnetospheric FTE observed by AMPTE has:

- given an explanation of how the magnetopause transition parameter works
- explained the high particle pressure at the centre of such "core" FTEs
- explained the layer structure of the event in terms of reconnection rate variations
- identified the edge of the event core as an Alfvénic disturbance, most likely a slow shock, standing in the inflow to the reconnecting current sheet from the magnetospheric side
- shown that the field lines in the event core were reconnected in a pulse, providing support for the 2-dimensional reconnection pulse theory of FTEs.

Other results have been reported by *Lockwood and Hapgood* [1998].

Acknowledgements. This research was carried out under a grant from the UK Particle Physics and Astronomy Research Council. The authors are also grateful to the following Principle Investigators of the AMPTE mission: David Southwood (UKS magnetometer), Alan Johnstone (UKS ion detector), Götz Paschmann (IRM electron and ion detectors) and Hermann Lühr (IRM magnetometer). We also thank the staff of the CDHF and WDC at RAL for maintaining and archiving the AMPTE dataset and making it readily available to us.

REFERENCES

Bryant, D.A., and S. Riggs, At the edge of the Earth's magnetosphere: a survey by AMPTE-UKS, *Phil. Trans. R. Soc. Lond. A, 328,* 43 - 56, 1989.

Cowley, S. W. H., The causes of convection in the Earth's magnetosphere: a review of developments during IMS, *Rev. Geophys., 20,* 531-565, 1982.

Cowley, S.W.H., and C.J. Owen, A simple illustrative model of open flux tube motion over the dayside magnetopause, *Planet. Space Sci., 37,* 1461, 1989.

de Hoffmann, F., and E. Teller, Magneto-hydrodynamic shocks, *Phys. Rev., 80,* 692, 1950.

Farrugia, C. J., R. P. Rijnbeek, M. A. Saunders, D. J. Southwood, D. J. Rodgers, M. F. Smith, C. P. Chaloner, D. S. Hall, P. J. Christiansen, and L. J. C. Woolliscroft, A multi-instrument study of flux transfer event structure, *J. Geophys. Res., 93,* 14465-14477, 1988.

Fuselier, S.A., D.M. Klumpar, and E.G. Shelley, Ion reflection and transmission during reconnection at the Earth's subsolar magnetopause, *Geophys. Res. Lett., 18,* 139-142, 1991

Hall, D.S., D.A. Bryant, and C.P. Chaloner, Plasma variations at the dayside magnetopause, Proc. 7th ESA Symposium on Rockets and Balloons, ESA SP-229, 299-304, 1985.

Hapgood, M.A., and D.A. Bryant, Exploring the magnetospheric boundary layer, *Planet. Space Sci., 40,* 1431-1459, 1992.

Hapgood, M.A., and M. Lockwood, Rapid changes in LLBL thickness, *Geophys. Res. Lett., 22,* 77-80, 1995.

Heyn, M.F., H.K. Biernat, R.P. Rijnbeek and V.S. Semenov, The structure of reconnection layers, *J. Plasma Phys., 40(2),* 235-252, 1988.

Hudson, P.D., Discontinuities in an isotropic plasma and their identification in the solar wind, *Planet. Space Sci., 18,* 1611-1622, 1970.

Lin, Y., and L.C. Lee, The structure of reconnection layers in the magnetosphere, *Space Sci. Rev., 65,* 59-179, 1993.

Lockwood, M., The location and characteristics of the reconnection X-line deduced from low-altitude satellite and ground-based observations: 1. Theory, *J. Geophys. Res., 100,* 21791-21802, 1995.

Lockwood, M., Energy and pitch angle dispersions of LLBL/cusp ions seen at middle altitudes: predictions by the open magnetosphere model, *Ann. Geophysica, 15 ,* 1501-1514, 1997., 1997.

Lockwood, M., and C.J. Davis, An analysis of the accuracy of magnetopause reconnection rate variations deduced from cusp ion dispersion characteristics, *Ann. Geophysicae, 14,* 149-161, 1996a.

Lockwood, M, and C.J. Davis, On the longitudinal extent of magnetopause reconnection bursts, *Ann. Geophysicae, 14,* 865-878, 1996b.

Lockwood, M., and M.A. Hapgood, How the magnetopause transition parameter works, *Geophys. Res. Lett., 24,* 373-376, 1997

Lockwood, M., and M.A. Hapgood, On the cause of a magnetospheric flux transfer event, *J. Geophys. Res.,* in press, 1998.

Lockwood, M., and J. Moen, Ion populations on open field lines within the low-latitude boundary layer: theory and observations during a dayside transient event, *Geophys. Res Lett., 23,* 2895-2898, 1996.

Lockwood, M., and M.F. Smith, The variation of reconnection rate at the dayside magnetopause and cusp ion precipitation, *J. Geophys. Res., 97,* 14,841-14,847, 1992.

Lockwood, M., and M.F. Smith, Low- and mid-altitude cusp particle signatures for general magnetopause reconnection rate variations: I - Theory, *J. Geophys. Res., 99,* 8531-8555, 1994.

Lockwood, M., M. F. Smith, C. J. Farrugia, and G. L. Siscoe, Ionospheric ion upwelling in the wake of flux transfer events at the dayside magnetopause, *J. Geophys. Res., 93,* 5641-5654, 1988.

Lockwood, M, W.F. Denig, A.D. Farmer, V.N. Davda, S.W.H. Cowley, and H. Lühr, Ionospheric signatures of pulsed magnetic reconnection at the Earth's magnetopause, *Nature, 361 (6411),* 424-428, 1993b.

Lockwood, M., S.W.H. Cowley, and T.G. Onsager, Ion acceleration at both the interior and exterior Alfvén waves associated with the magnetopause reconnection site: signatures in cusp precipitation, *J. Geophys. Res., 101,* 21501 - 21515, 1996.

Lockwood, M., C.J. Davis, T.G. Onsager, and J.A. Scudder, Modelling signatures of pulsed magnetopause reconnection in cusp ion dispersion signatures seen at middle altitudes, *Geophys. Res. Lett.,* in press, 1997.

Onsager, T.G., A quantitative model of magnetosheath plasma in the low latitude boundary layer, cusp, and mantle, J.A. Holtet and A. Egeland (eds), *Physical Signatures of Magnetospheric Boundary Layer Processes,* pp. 385-400, 1994.

Onsager T.G., C.A. Kletzing, J.B. Austin, and H. MacKiernan, Model of magnetosheath plasma in the magnetosphere: cusp and mantle particles at low-altitudes, *Geophys. Res. Lett., 20,* 479-482, 1993

Paschmann, G., B.U.Ö. Sonnerup, I. Papamastorakis, N. Sckopke, G. Haerendel, S.J. Bame, J.R. Asbridge, J.T. Gosling, C.T. Russell, and R.C. Elphic, Plasma acceleration at the Earth's magnetopause: evidence for reconnection, *Nature, 282,* 243-246, 1979.

Paschmann, G., G. Haerendel, I. Papamastorakis, N. Sckopke, S. J. Bame, J. T. Gosling, and C.T. Russell, Plasma and magnetic field characteristics of magnetic flux transfer events, *J. Geophys. Res., 87,* 2159-2168, 1982.

Paschmann, G., I. Papamastorakis, W. Baumjohann, N. Sckopke, C.W. Carlson, B.U.Ö. Sonnerup, and H. Lühr, The magnetopause for large magnetic shear: AMPTE/IRM observations, *J. Geophys. Res., 91,* 11099-11115, 1986.

Phan, T.D., et al., Low-latitude flank magnetosheath, magnetopause and boundary layer for low magnetic shear: Wind observations, *J. Geophys. Res., 102,* 19,883-19,895, 1997.

Pinnock, M., A.S. Rodger, J.R. Dudeney, F. Rich, and K.B. Baker, High spatial and temporal resolution observations of the ionospheric cusp, *Ann. Geophys., 13,* 919-925, 1995.

Reiff, P. H., T. W. Hill, and J. L. Burch, Solar wind plasma injection at the dayside magnetospheric cusp, *J. Geophys. Res., 82,* 479-491, 1977.

Rijnbeek R. P., C. J. Farrugia, D. J. Southwood, M. W. Dunlop, W. A. C. Mier-Jedrejowicz, C. P. Chaloner, D. S. Hall, and M. F. Smith, A magnetic boundary signature within flux transfer events, *Planet. Space Sci., 35,* 871-878, 1987.

Rosenbauer, H., H. Grünwaldt, M.D. Montgomery, G. Paschmann, and N. Skopke, HEOS-2 plasma observations in the distant polar magnetosphere: the plasma mantle, *J. Geophys. Res.,80,* 2723-2737, 1975.

Russell, C. T., and R. C. Elphic, Initial ISEE magnetometer results: magnetopause observations, *Space Sci. Rev., 22,* 681-715, 1978.

Russell, C. T., and R. C. Elphic, ISEE observations of flux transfer events at the dayside magnetopause, *Geophys. Res. Lett., 6,* 33-36, 1979.

Sandholt, P.E., M. Lockwood, T. Oguti, S.W.H. Cowley, K.S.C. Freeman, A. Egeland, B. Lybekk and D.M. Willis, Midday auroral breakup events and related energy and momentum transfer from the magnetosheath, *J. geophys. Res., 95,* 1039-1061, 1990.

Scholer, M., Magnetic flux transfer at the magnetopause based on single X-line bursty reconnection, *Geophys. Res. Lett., 15,* 291-294, 1988a.

Scholer, M., Asymmetric time-dependent and stationary magnetic reconnection at the dayside magnetopause, *J. Geophys. Res., 94,* 15099-15111, 1989.

Sckopke, N., G. Paschmann, G. Haerendel, B.U.Ö. Sonnerup, S.J. Bame, T.G. Forbes, E.W. Hones, Jr., and C.T. Russell , Structure of the low latitude boundary layer, *J. Geophys. Res. 86,* 2099-2110. 1981

Semenov, V.S., I.V. Kubyshkin, H.K. Biernat, M.F. Heyn, R.P. Rijnbeek, B.P. Besser, and C.J. Farrugia, Flux transfer events interpreted in terms of a generalized model for Petschek-type reconnection, *Adv. Space, Res., 11,* (9)25-(9)28, 1991.

Semenov, V.S., I.V. Kubyshkin, V.V. Lebedeva, R.P. Rijnbeek, M.F. Heyn, H.K. Biernat, and C.J. Farrugia, A comparison and review of ready-state and time-varying reconnection, *Planet. Space Sci., 40,* 63-87, 1992a.

Semenov, V.S., I.V. Kubyshkin, V.V. Lebedeva, M.V. Sidneva, H.K. Biernat, M.F. Heyn, B.P. Besser, and R.P. Rijnbeek, Time-dependent localized reconnection of skewed magnetic fields, *J. Geophys. Res., 97,* 4251-4263, 1992b.

Sibeck, D. G., A model for the transient magnetospheric response to sudden solar wind dynamic pressure variations, *J. Geophys. Res., 95,* 3755-3771, 1990.

Sibeck, D.G., Transient events ion the outer magnetosphere, boundary waves or flux transfer events, *J. Geophys. Res., 97,* 4009-4026, 1992.

Sibeck D.G., and M.F. Smith, Magnetospheric plasma flows associated with boundary waves and flux transfer events, *Geophys. Res. Lett., 19,* 1903-1906, 1992.

Smith, M.F., and C.J. Owen, Temperature anisotropies in a magnetospheric FTE, *Geophys. Res. Lett., 19,* 1907-1910, 1992.

Sonnerup, B.U.Ö., I. Papamastorakis, G. Paschmann, and H. Lühr, The magnetopause for large magnetic shear: analysis of convection electric fields from AMPTE/IRM, *J. Geophys. Res., 95,* 10541-10557, 1986.

Spreiter, J.R., A.L. Summers, and A.Y. Alksne, Hydromagnetic flow around the magnetosphere, *Planet. Space Sci., 14,* 223-253, 1966.

Southwood, D. J., C. J. Farrugia, and M. A. Saunders, What are flux transfer events?, *Planet. Space Sci., 36,* 503-508, 1988

M.A.Hapgood and M. Lockwood, Rutherford Appleton Laboratory, Chilton, Didcot, Oxon., OX11 0QX, England, UK. (e-mail: m.lockwood@rl.ac.uk)

The Role of Magnetic Reconnection in Solar Activity

Spiro K. Antiochos and C. Richard DeVore

Naval Research Laboratory, Washington, D. C.

We argue that magnetic reconnection plays the determining role in many of the various manifestations of solar activity. In particular, it is the trigger mechanism for the most energetic of solar events, coronal mass ejections and eruptive flares. We propose that in order to obtain explosive eruptions, magnetic reconnection in the corona must have an "on-off" nature, and show that reconnection in a sheared multi-polar field configuration does have this property. Numerical simulation results which support this model are presented, and implications for coronal mass ejections/eruptive flare prediction are discussed.

1. INTRODUCTION

Magnetic reconnection is widely believed to be the most important process by which magnetic fields transfer energy to plasmas and, consequently, has been invoked as the explanation for a multitude of phenomena observed in space, astrophysical, and laboratory plasmas. Reconnection is the most likely mechanism by which the Sun's magnetic field powers the solar atmosphere, the chromosphere and corona. Reconnection is also the process by which the solar wind flow couples to planetary magnetospheres, and is the key process behind the eventual transfer of this energy to the Earth's ionosphere and lower atmosphere. Hence, it would not be unjustified to describe the ISTP as a program to study the effects of reconnection on the Sun-Earth Connection.

Reconnection is particularly attractive as an explanation for solar variability and activity because of the potential for both rapid dissipation of magnetic energy and the accompanying strong plasma dynamics. In general, one can categorize the energy release due to reconnection as taking four distinct forms, all of which are believed to occur commonly in the solar atmosphere. First, reconnection can result in direct plasma heating. This has been proposed as the process that heats both the non-flare [e.g., *Parker*, 1972, 1979, 1983; *Sturrock and Uchida*, 1981; *van Ballegooijen*, 1986; *Antiochos*, 1990] and flare corona [e.g., *Sweet*, 1958; *Parker*, 1963; *Petschek*, 1964; *Carmichael*, 1964; *Sturrock*, 1966] Second, reconnection can produce strong mass motions, the so-called reconnection jets. These have been proposed [*Karpen, Antiochos, and DeVore*, 1995, 1996, 1998] as the explanation for a large variety of transient dynamic phenomena ranging from spicules [e.g., *Blake and Sturrock*, 1985] to explosive events [*Dere et al.*, 1991] to large surges and sprays [e.g., *Rust*, 1968; *Herant et al.*, 1991; *Schmeider et al.*, 1994; *Yokohama and Shibata*, 1996] and X-ray jets [*Shibata et al.*, 1992]. Third, reconnection can lead to the acceleration of non-thermal particles. This process has been proposed as the explanation for the electron beams that give rise to the hard X-ray emission in flares [e.g., *Sturrock*, 1980 *and references therein*]. Finally, reconnection can produce MHD waves. This process has been proposed as the explanation for coronal heating [e.g., *Falconer et al.*, 1997] and for accelerating the solar wind [e.g., *Parker*, 1988; *Mullan*, 1990]. In fact, it is difficult to find a coronal phenomena that has not been attributed to magnetic reconnection!

Sun-Earth Plasma Connections
Geophysical Monograph 109
This paper not subject to U.S. copyright
Published in 1999 by the American Geophysical Union

Rather than considering one of these direct effects of reconnection, we discuss in this paper one of the most interesting and more subtle roles of reconnection — the initiation of major disruptions of the Sun's magnetic field. Large magnetic disruptions are generally observed as a coronal mass ejection (CME) and a flare associated prominence/filament eruption. We will refer to them as CME/eruptive flare (EF). These events are the most energetic (up to 10^{33} ergs) and destructive of solar disturbances. They are the main drivers of space weather and are a focus of study of the ISTP program. The May 2 1998 event, for example, was observed by SOHO/ISTP to consist of a filament eruption, fast CME (1,000 km/s) and an X-class flare that produced a strong proton storm at Earth and the strongest geomagnetic activity in recent years ($K_p = 9$). CME/EFs will be of particular interest in the upcoming maximum.

In addition to their key role in driving space weather, CME/EFs are important from the viewpoint of understanding basic space-plasma physics. They are the classic manifestation of solar activity, and the mechanism for their initiation and evolution has long been a central issue of solar physics research [e.g., *Sturrock*, 1980]. Space observations such as those from Yohkoh and SOHO have led to much progress in understanding magnetic disruptions, and a general framework for these events has emerged. A CME/EF is believed to consist of four steps: *(1.)* Stressed magnetic field slowly emerges from below the photosphere building up the free energy in the corona quasi-statically. The emerged field may be further stressed by the convective and rotational photospheric motions. *(2.)* Some triggering mechanism destroys the magnetic equilibrium of the corona, resulting in an explosive instability or loss of equilibrium. *(3.)* The magnetic field erupts outward, accelerating and ejecting plasma into the heliosphere. *(4.)* The field closes back down via reconnection to a less stressed state, heating the plasma and producing the intense X-ray burst.

In this four-step picture, a CME/EF can be thought of as simply the method by which the Sun rids itself of magnetic stress. Since the coronal conductivity is so high, the Sun has no choice but to eject any large-scale magnetic stress into the heliosphere. Note also that in this picture, a CME and an eruptive flare are part of a single phenomenon. The CME and prominence/filament eruption constitute step *(3.)*, the opening of the field. It should be emphasized that by opening we don't mean that the field actually disconnects from the Sun, only that the field becomes dynamic and expands outward without limit. During this step the free magnetic energy is used primarily to accelerate plasma and lift it against gravity. The X-ray or H_α flare constitutes step *(4.)*, the closing of the field. During this step the energy is used primarily to heat magnetically confined plasma and to accelerate nonthermal particles.

As with any astrophysical phenomenon, one can always find CMEs and flares that appear to be exceptions to the standard picture described above. Many CMEs, especially high-latitude slow ones, have no detectable X-ray emission. In this case the reconnection during step *(4.)* is presumably very slow, resulting in minimal heating. Much more difficult to reconcile, however, is the observation that intense X-ray flares sometimes occur with no apparent magnetic field opening i.e., no CME or prominence/filament eruption. Strong flare heating is widely believed to be due to fast reconnection, which generally involves the formation of a current sheet. It is the opening of the field during *(3.)* that forms a large current sheet, thereby, allowing rapid reconnection to occur during *(4.)*. In order to account for non-eruptive flares, it must be that either the field does erupt but produces undetectable plasma motions — this seems unlikely — or else large current sheets can form and fast reconnection can occur even in fully closed bipolar configurations. There are good arguments, however, against the formation of current sheets in such a magnetic topology [e.g., *van Ballejooigen*, 1985; *Antiochos*, 1987], furthermore reconnection is generally believed to be inhibited by line-tying effects for such topologies [*Dahlburg, Antiochos, and Zang*, 1991; *Velli, Einaudi, and Hood*, 1993]. More observational and theoretical work is clearly needed on the issue of apparently non-eruptive flares.

In this paper we adopt the standard 4-step picture above for a CME/EF and focus on Step *(2.)*, the triggering mechanism. Determining the flare trigger has long been one of the outstanding problems in solar physics [e.g., *Sturrock*, 1980]. In addition, it is clearly necessary to understand the triggering process if we are ever to develop physics-based CME and eruptive flare prediction schemes. We argue below that magnetic reconnection in a multi-flux magnetic topology is this long-sought-after triggering mechanism.

2. THEORETICAL RESULTS

Over the years many models have been proposed for the initiation of CMEs and EFs, the large majority of which are magnetically driven [e.g., *Sturrock*, 1989; *Moore and Roumeliotis*, 1992; *Mikic and Linker*, 1994; *Antiochos, DeVore, and Klimchuk*, 1998], although models in which gas buoyancy effects are important [e.g., *Low*, 1994; *Wolfson and Dlamini*, 1997] have also been discussed. Since the solar corona is generally measured to have a low plasma beta $< 10^{-2}$, it is almost certain that the energy for the eruption is stored in the magnetic field. This is especially true for the very energetic, fast eruptions which can have speeds exceeding 2000 km/s – of order the Alfven speed, far larger than coronal sound speeds.

In recent years, however, a major obstacle to the magnetic models has been pointed out by Aly and Sturrock. Their argument proceeds as follows. Aly (1984) in a landmark paper proved that the energy of any force-free magnetic field configuration in the corona is bounded above. This result implies that if the coronal field is stressed by photospheric motions, the energy of the field will not continue rising indefinitely even if the motions continue indefinitely. Aly showed that, in fact, the energy has an upper bound that is only of order the energy of the potential magnetic field with the same normal flux distribution at the photosphere. For example, a dipole field is bounded above by twice the potential field energy [*Aly*, 1984]. The physical reason for the energy bound is straightforward. The corona has an infinite volume, and so if even the footpoint motions become infinitely large, the field can simply expand outward toward infinity, thereby, keeping its energy finite. Note that this argument is equally valid if the magnetic stress is due to emergence from below rather than footpoint motions. As long as the evolution is quasi-static, the magnetic free energy and also the electric current, or magnetic stress, in the corona must have a strict upper bound [e.g., *Aly*, 1984; *Finn and Chen*, 1990]. Therefore, one must be careful in specifying boundary conditions for the corona at its photospheric base. For example, one is not free to specify arbitrarily the current entering the corona from the photospheric. In general, the safest and most physical boundary conditions are to specify footpoint motions, which is our approach in the simulations described below.

Building upon this work, Aly (1991) presented arguments that if the coronal magnetic field undergoes a quasi-static evolution, which is a good approximation for photospheric driving motions, then the least upper bound on its energy is the open field configuration in which every field line is stretched out to infinity. Sturrock (1991) independently presented similar arguments. Although their arguments do not constitute a rigorous proof, they are quite convincing and appear to be in good agreement with every numerical simulation performed to date [e.g., *Roumeliotis, Antiochos, and Sturrock,* 1994; *Mikic and Linker*, 1994; *Amari et al.*, 1996], (but see Wolfson and Low (1992) for a possible counter example). The Aly-Sturrock energy limit also seems physically intuitive. The basic assumption is that the coronal field is always free to lower its energy by an ideal outward expansion, if such an expansion is energetically favorable. Consequently, it should not be possible to build up quasi-statically significant energy above the fully open state, because as soon as the magnetic energy tries to exceed the open value, the field would simply expand to the open state. Once the field is open, further footpoint motions have no effect other than to propagate Alfven waves out into the heliosphere.

Although the energy limit seems perfectly reasonable, it also seems completely incompatible with CME/EF observations. If these events are magnetically driven, the energy in the pre-eruption magnetic field must be considerably larger than the open field state formed during the eruption, contrary to the energy limit. It is often claimed that one can circumvent the energy limit by simply opening part of the field. For example, one can leave the very low-lying field closed and open only above a certain height. We believe that this argument is spurious. While it is true that in their papers Aly and Sturrock concentrated on the fully open field case, as we discussed above, their result is based on quite general arguments that should hold even for a partial opening of the field. The basic point is that a slowly driven magnetic field is unlikely to erupt explosively to an open state that it can also reach by a process of ideal expansion. Indeed, EIT and LASCO/C1 observations from SOHO show that slow expansion is a general feature of the coronal magnetic field. Furthermore, we believe that most CMEs, which are typically slow and accelerate only up to slow-solar wind speeds, also represent merely a slow expansion of the field. On the other hand, there is no doubt that explosive eruptions do sometimes occur.

3. NUMERICAL RESULTS

An attractive solution to this apparent contradiction between theory and observation is to invoke reconnection. It seems likely that the extra freedom in the possible evolution of the field afforded by reconnection can somehow circumvent the energy limit. This is exactly what happens in 2.5D geometries. Mikic and Linker (1994) showed that footpoint shearing in a system with azimuthal symmetry and finite resistivity leads to the formation of disconnected plasmoids that are expelled rapidly from the corona with a significant energy drop. We emphasize again that, contrary to common misconception, the eruption in this case is not due the fact that the field opens only partially. The eruption occurs because the field opens to a state that it cannot access by an ideal expansion. The plasmoid carries off a large fraction of the shear of the closed field lines, so that the final state of the system is one in which the closed region contains much less shear than was originally imparted to this region by the footpoint motions. The field could not erupt to the state in which the shear in the closed field region stayed fixed.

This discussion leads to the main conclusion of this paper: the whole eruption process is controlled by the nature of magnetic reconnection in the corona. In order for plasmoid formation to result in an explosive eruption, the formation and hence the reconnection must proceed rapidly. The plasmoid must grow faster than the field expands outward. On the other hand, reconnection must not occur too easily, otherwise very little free energy could be built up. Therefore, to explain eruptions, magnetic reconnection in the solar corona must have a very particular form. It must have an "on-off" nature wherein it is basically off during step *(1.)* while the free energy is building up, but at a certain point switches on to a high value *and stays on*. It is this reconnection switch-on that is the flare trigger, step *(2.)*.

Since plasmoid formation is only a 2.5D phenomenon, it is not clear, however, whether reconnection can also lead to an eruption in a full 3D system. The 3D analogue to the 2.5D configuration is the so-called tether-cutting model [*Sturrock*, 1989; *Moore and Roumeliotis*, 1991]. The basic geometry of this model can be seen in Plate 1, which shows the results of a 3D MHD simulation of photospheric stressing of a coronal field. This simulation is a fully time-dependent calculation of our model for prominence formation [*Antiochos, Dahlburg, and Klimchuk*, 1994; *Antiochos*, 1995], and corresponds to a 3D generalization of the 2.5D CME calculations. The magnetic field in the figure is the result of applying a shear flow at the photosphere to an initially current-free dipole field. The plane of the photosphere in this simulation was taken to be a rectangle, part of which can be seen in the figure. The straight black line running lengthwise through the middle of the photospheric plane is the magnetic polarity reversal line. In agreement with observations [*Schmeider et al.*, 1996], the shear was chosen to be spatially localized near the polarity reversal line. The form of this shear can be seen in the contours of the normal component of B on the photospheric plane; for ease of viewing only one contour is shown on the far side of the bottom plane. Note that in the corona above this plane there are basically two types of field lines. Those originating in the shear zone (three thick lines) are stretched out and have concave up portions along their length, while those originating in the non-shear zone (three thin arrowed lines) appear to be dipolar. The mass of red lines in the center of the figure indicates where the magnetic field lines are concave up and represents the H_α mass of the prominence.

The idea behind tether-cutting is that reconnection leads to eruption by redistributing the shear so that it becomes concentrated on the furthermost field lines at the edge of the sheared region. The field lines around the center of the figure decrease their shear, while those near the edges increase. Such a transfer of shear to the outermost field lines is the physical analogue to plasmoid formation in 2.5D. If this shear transfer is sufficiently rapid, then eruption should occur. Hence, the 3D case is really not significantly different than the 2.5D system. In both cases the question of whether an eruption occurs comes down to the issue of the temporal behavior of reconnection in the solar corona

So far, our 3D simulations of tether-cutting show no evidence for an explosive eruption, which agrees with our previous work on the twisting of dipolar arcades [*Dahlburg, Antiochos, and Zang*, 1991]. However, this work is still in progress, and it may be that using different parameters, such as system size, form of shear, etc. will lead to eruption. Further studies of the tether-cutting model are needed before definitive conclusions can be reached.

Based on our work so far, however, we conjecture that the tether-cutting model does not lead to explosive eruptions. The basic reason for this result is that a bipolar field system as in Plate 1 is too simple topologically for fast reconnection to occur. We propose instead the "breakout" model [*Antiochos*, 1998; it *Antiochos, DeVore, and Klimchuk*, 1999] shown in Plate 2, which consists of a multi-polar field. The initial current-free field for this system is given by the vector potential:

$$\vec{A} = \frac{\sin\theta}{r^2} + \frac{(3 + 5\cos 2\theta)\sin\theta}{2r^4} \hat{\phi} \qquad (1)$$

This system has a more complex topology than that of Plate 1, it consists of four flux systems with separatrix surfaces and a null point in the corona, as shown in the first panel of Plate 2. There are now three neutral lines in the system, at solar latitudes of $\pm 45^\circ$ and at the equator. Note, however, that this field configuration arises from only four polarity regions on the photospheric surface, hence it is actually far simpler than what must be present in the real corona. Also it appears similar to what is often seen in the LASCO C1 images.

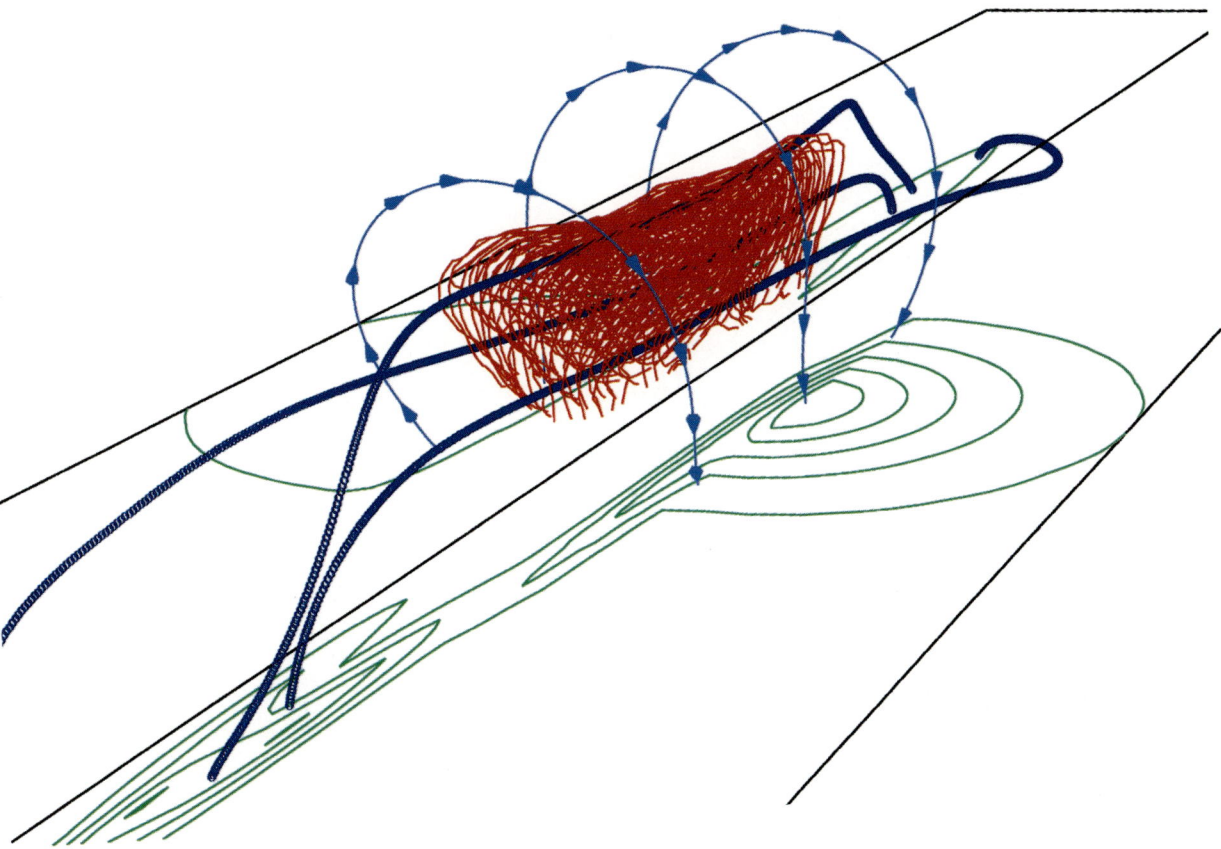

Plate 1. Results from a 3D MHD simulation of our model for prominence/filament formation and of the tether cutting model for coronal mass ejections/ eruptive flares. The configuration shown is that of an initial current-free dipole field that has been sheared by photospheric motions. The outline of the bottom boundary of our simulation box is shown along with the photospheric neutral line. Contours of normal-magnetic-field magnitude are plotted on the bottom boundary plane. Three field lines (thick dark-blue) with footpoints in the shear region and three field lines (arrowed, light-blue) with footpoints outside this region are shown. Also plotted are contours of positive-upward curvature of magnetic field (mass of red lines), which corresponds to the possible location of the prominence mass.

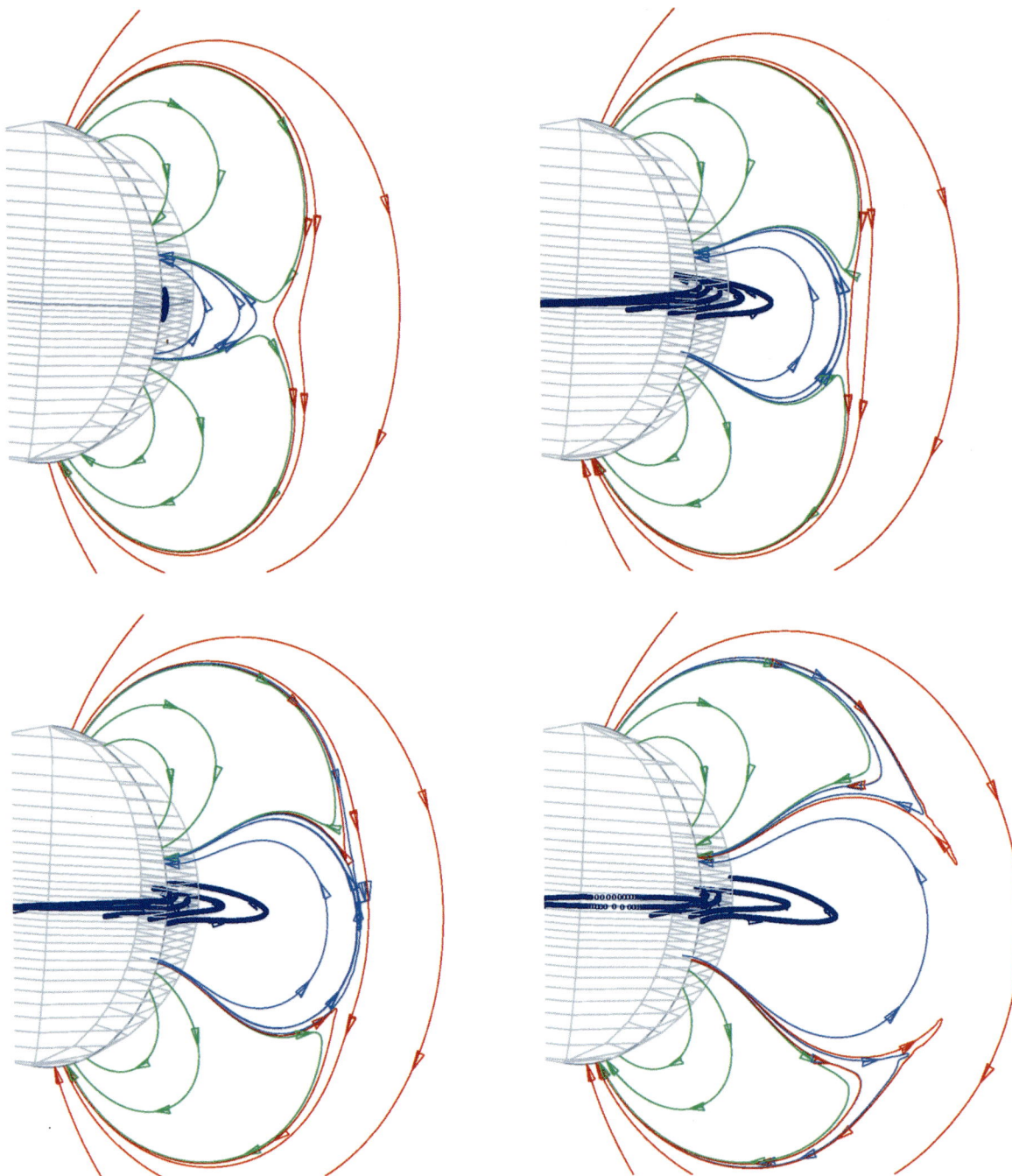

Plate 2. Results from a 2.5D simulation of the breakout model for coronal mass ejections and eruptive flares. The first panel (upper left) shows the initial current-free configuration which consists of a sum of a dipolar and an octopolar field. Photospheric shearing is applied only near the equatorial neutral line. The next three panels show the field after 17.5, 19.7, and 21.7 hours into the evolution. All field lines are traced from exactly the same footpoint positions in all panels. Note that in the last panel the overlying unsheared field is reconnecting and moving to the side, thereby allowing the sheared flux to erupt outward.

The effect of stressing this system by footpoint motions can be seen in Plate 2, which present the results from one of our 2.5D MHD simulations. For this case we selected the inner flux system (straddling the equator), as the one to shear. Again the shear is concentrated near the neutral line. The thick dark lines in Plate 2 correspond to the sheared prominence field lines of Plate 1. Basically, we have embedded the sheared arcade of Plate 1 into a more global field topology. As a result of being sheared, the inner flux system overlying the equator expands outward, just as the single arcade in Plate 1 expands outward. But now it begins to interact with the neighboring systems. As a result of this expansion the null point in the corona deforms into a neutral sheet, similar to the process originally proposed by Syrovatskii (1981). Eventually, this neutral sheet becomes sufficiently thin that reconnection begins. We see in the last panel of Plate 2 that some of the field lines that used to be in the inner and outer flux systems have transferred over to the sides. Once this reconnection starts it continues to accelerate because there are fewer unsheared field lines holding down the sheared ones. The rapidly accelerating transfer of flux from above the sheared field allows this field to "break out" and erupt open.

Note that our breakout model yields a very natural process for obtaining an "on-off" type of reconnection in the corona. In fact, we find that once the reconnection switches on, the rate becomes too large for our ideal code to simulate, but the essential result has been shown – in a multi-flux system reconnection will have the correct time-dependence to explain the initiation of CME/EFs, step (2.). Hence, we conclude that reconnection is responsible not only for releasing a large fraction of the flare energy during step (4.), it is also responsible for getting the whole event going.

Now that we have a likely candidate for the trigger mechanism, it is interesting to consider its use as a possible CME/EF predictor. There are two distinguishing features of the breakout model. One is that a multipolar configuration is needed for eruption. In fact, it is well known that complexity is needed to obtain eruptive flares, and magnetic complexity is already one of the main criteria being used for prediction. The second feature is that reconnection should occur high up in the corona before the eruption. We believe that it would be highly informative to use the resources of the ISTP to search for this reconnection during the upcoming solar maximum.

Acknowledgments. This work was supported in part by NASA and ONR.

REFERENCES

Aly, J. J., On some properties of force-free magnetic fields in infinite regions of space, *Astrophys. J., 283,* 349, 1984.

Aly, J. J., How much energy can be stored in a three-dimensional force-free magnetic field?, *Astrophys. J., 374,* L61, 1991.

Amari, T., Luciani, J. F., Aly, J. J., and Tagger, M., Very Fast Opening of a Three-dimensional Twisted Magnetic Flux Tube, *Astrophys. J., 466,* L39, 1996.

Antiochos, S. K., The topology of force-free magnetic fields and its implications for coronal activity, *Astrophys. J., 312,* 886, 1987.

Antiochos, S.K., Heating of the corona by magnetic singularities, *J. Italian Astron. Soc., 61,* 369, 1990.

Antiochos, S. K., Dahlburg, R. B., and Klimchuk, J. A., The magnetic field of solar prominences, *Astrophys. J., 420,* L41, 1994.

Antiochos, S. K., Solar drivers of space weather, *Astron. Soc. Pac. Conf. Series, 95,* 1, 1995.

Antiochos, S.K., The magnetic topology of solar eruptions, *Astrophys. J., 502,* L181, 1998.

Antiochos, S. K., DeVore, C. R., and Klimchuk, J. A., A model for solar coronal mass ejections, *Astrophys. J., 510,* in press, 1999.

Blake, M. L., and Sturrock, P. A., Spicules and surges, *Astrophys. J., 290,* 359, 1985.

Carmichael, H., A process for flares, in *AAS-NASA Symp. on Physics of Solar Flares,* edited by W. Ness, p. 451, NASA SP-50, Washington, DC, 1964.

Dahlburg, R. B., Antiochos, S. K., and Zang, T. A., Dynamics of solar coronal magnetic fields, *Astrophys. J., 383,* 420, 1991.

Dere, K. P., Bartoe, J.-D. F., Brueckner, G. E., Ewing, J., and Lund, P., Explosive events and magnetic reconnection in the solar atmosphere, *J. Geophys. Res., 96,* 9399, 1991.

Falconer, D. A. *et al.*, Neutral-line magnetic shear and enhanced coronal heating in solar active regions, *Astrophys. J., 482,* 519, 1997.

Finn, J. M., and Chen, J., Equilibrium of solar coronal arcades, *Astrophys. J., 349,* 345, 1990.

Herant, M., Pardo, F., Spiller, E, and Golub, L., Flares observed by the normal incidence X-ray telescope on 1989 September 11, *Astrophys. J., 376,* 797, 1991.

Karpen, J. T., Antiochos, S. K., and DeVore, C. R., The role of magnetic reconnection in chromospheric eruptions, *Astrophys. J., 450,* 422, 1995.

Karpen, J. T., Antiochos, S. K., and DeVore, C. R., Reconnection driven current filamentation in solar arcades, *Astrophys. J., 460,* L73, 1996.

Karpen, J. T., Antiochos, S. K., and DeVore, C. R., Dynamic responses to magnetic reconnection of solar arcades, *Astrophys. J., 495,* 491, 1998.

Low, B.C., Magnetohydrodynamic processes in the solar corona: flares, coronal mass ejetcions, and magnetic helicity, *Phys. Plasmas, 1,* 1684, 1994.

Mikić, Z. and Linker, J. A., Disruption of coronal magnetic field arcades, *Astrophys. J., 430,* 898, 1994.

Moore, R. L. and Roumeliotis, G., Triggering of eruptive flares: destabilization of the preflare magnetic field configuration, in *Eruptive Solar Flares,* edited by Z. Svestka, B. V. Jackson, and M. E. Machado, p. 69, Springer, Berlin, 1992.

Mullan, D. J., Sources of the solar wind - What are the smallest-scale structures? *Astron. Astrophys.*, *248*, 256, 1990.

Parker, E. N., The solar flare phenomenon and the theory of reconnection and annihilation of magnetic fields, *Astrophys. J. Supp.*, *8*, 177, 1963.

Parker, E. N., *Cosmical Magnetic Fields*, Clarendon Press, Oxford, U.K., 1979.

Parker, E. N., Magnetic neutral sheets in evolving fields. I - General theory. II - Formation of the solar corona, *Astrophys. J.*, *264*, 642, 1983.

Parker, E. N., Nanoflares and the solar X-ray corona, *Astrophys. J.*, *330*, 474, 1988.

Petschek, H. E., Magnetic field annihilation, in *AAS-NASA Symp. on Physics of Solar Flares*, edited by W. Ness, p. 425, NASA SP-50, Washington, DC, 1964.

Rust, D. M., Chromospheric explosions and satellite sunspots, in *Structure and Development of Solar Active Regions* (IAU Symposium 35), edited by K. O. Kiepenheuer, p. 77, Reidel, Dordrecht, 1968.

Schmieder, B., Golub, L., and Antiochos, S. K., Comparison between cool and hot plasma behaviors of surges, *Astrophys. J.*, *425*, 326, 1994.

Schmieder, B., Demoulin, P., Aulanier, G., and Golub, L. Differential Magnetic Field Shear in an Active Region, *Astrophys. J.*, *467*,, 881, 1996.

Shibata, K., *et al.*, Observations of X-ray jets with the Yohkoh Soft X-ray Telescope, *Publ. Astron. Soc. Japan*, *44*, L173, 1992.

Sturrock, P. A., A Model of the High Energy Phase of Solar Flares, *Nature*, *211*, 695, 1966.

Sturrock, P.A., *Solar Flares*, Colorado Assoc. Univ. Press, Boulder, CO., 1980.

Sturrock, P. A. and Uchida, Y., Coronal heating by stochastic magnetic pumping, *Astrophys. J.*, *246*, 331, 1981.

Sturrock, P. A., The role of eruption in solar flares, *Solar Phys.*, *121*, 387, 1989.

Sturrock, P. A., Maximum energy of semi-infinite magnetic field configurations, *Astrophys. J.*, *380*, 655, 1991.

Sweet, P. A., in *Electromagnetic Phenomena in Cosmic Physics*, edited by B. Lehnert, pp. 123-34, Cambridge Univ. Press, New York, 1958.

Syrovatskii, S. I., Pinch sheets and reconnection in astrophysics, *Ann. Rev. Astron. Aatrophys.*, *19*, 163, 1981.

van Ballegooijen, A. A., Electric currents in the solar corona and the existence of magnetostatic equilibrium, *Astrophys. J.*, *298*, 421, 1985.

van Ballegooijen, A. A., Cascade of magnetic energy as a mechanism of coronal heating, *Astrophys. J.*, *311*, 1001, 1986.

Velli, M., Einaudi, G., and Hood, A. W., Ideal kink instabilities in line-tied coronal loops - Growth rates and geometrical properties, *Astrophys. J.*, *350*, 428, 1990.

Wolfson, R. and Low, B. C., Energy buildup in sheared force-free magnetic fields *Astrophys. J.*, *391*, 353, 1992.

Wolfson, R and Dlamini, B, Cross-field currents: An energy source for coronal mass ejections? *Astrophys. J.*, *483*, 961, 1997.

Yokoyama, T., and Shibata, K., Numerical simulation of solar coronal X-Ray jets based on the magnetic reconnection model, *Publ. Astron. Soc. Japan*, *48*, 353, 1996.

S. K. Antiochos, Code 7675, Naval Research Lab, Washington, DC 20375-5352. (e-mail: antiochos@nrl.navy.mil

C. R. DeVore, Code 6440, Naval Research Lab, Washington, DC 20375. (e-mail: devore@lcp.nrl.navy.mil)

Unresolved Questions About the Structure and Dynamics of the Extended Solar Corona

William C. Feldman

Los Alamos National Laboratory, Los Alamos, NM

The most basic questions about the physical state of the solar corona remain unanswered after more than 50 years of study through remote-sensing observations. Although we know it to have temperatures that exceed 10^6 K, we do not know the linkage of mechanisms that both provide and dissipate the energy flux needed to sustain its existence. An abbreviated sample of basic questions is: What is the role of magnetic activity in the generation of plasma energy flux that may heat the corona? What are the size scales and temporal cadence of this activity? What is the role of suprathermal electrons generated by this activity in heating overlying layers? What is the role of waves in coronal heating processes? Do they provide the primary energy input from sources close to the photosphere or are they secondary to interactions in the corona between fine-scale plasma jets driven by, e.g. magnetic activity, that may lead to large-scale Alfvén and fast-mode waves, fast-mode shocks and plasma instabilities? If so, what drives the plasma jets and how do their properties depend on, and also shape, the macroscopic structure of the corona? How does the mix of semicoherent wave packets generated by the interaction between plasma jets evolve into a turbulent flow? What is the feedback of this turbulence on the suprathermal particle population and how does the energy flux that this population carries evolve with heliocentric distance? What mix of processes in the corona differentiate between low and high speed solar wind?

1. INTRODUCTION

Knowledge of the existence of an expanding solar atmosphere, the corona/solar wind, has been known for more than 50 years [*Biermann*, 1951]. Its extension to Earth and beyond exists as a supersonic expansion that continuously fills all of interplanetary space. Observations of all levels of the outer solar atmosphere and the solar wind from locations far from the Sun have been made from space for more than 35 years.

Yet, despite all of this activity, the sum total of information that these observations have provided has not been sufficient to understand the very basics of the nature of the corona. Specifically, we do not know the mechanisms that heat the corona to a million and more degrees K, and that accelerate its expansion to supersonic speeds to form the solar wind. Indeed, all we can presently specify is how much energy is needed to support all coronal energy losses and roughly where it must be deposited. The specific physical forms of this energy as a function of altitude and their physical links to one another and to processes in the lower solar atmosphere, are currently not known. In the following presentation, we will try to give an overview of what is known about the solar corona, and the limits of this information to delineate uniquely, the structure of the lower corona/solar wind portion of the solar atmosphere, and the physical processes that support its existence. We will then try to summarize the information that is needed in order to provide the experimental underpinnings for a theoretical understanding of the Sun's extended atmosphere. These experimental facts about the extended corona require an in situ investigation of the structure of the Sun's outer atmosphere, which constitutes the presently planned mission to the Sun, Solar Probe [*Feldman et al.*, 1989; *Axford et al.*, 1995; *Gloeckler et al.*, 1998].

2. GENERAL CLASSES OF CORONA AND SOLAR WIND

Attempts to classify all observations of the solar wind at heliocentric distances greater than 0.29 AU (the innermost distance that has been directly sampled by space probes) have identified three broad categories. The first such category is the transient solar wind, and the remaining two are apparently steady-state flows, the high speed solar wind, and the low speed solar *wind* [*Feldman et al.*, 1977; *Schwenn*, 1990]. However, these last two categories are structurally very different when studied in detail. Whereas the high speed solar wind has relatively constant characteristics, and therefore appears to be a single, structure-free entity, that of the slow solar wind is highly variable [*Bame et al.*, 1977]. Solar-generated transient disturbances, or transient solar wind flows, are generally embedded in the slow solar wind and thereby contribute in part to its variability (see, e.g., [*Gosling*, 1997], and references therein). The global connection of the two steady-state categories of flow to the Sun during solar minimum conditions is illustrated in Plate 1 using Ulysses data [*McComas et al.*, 1998]. Inspection shows that most of the high speed solar wind maps into the polar corona during this phase of the solar cycle, and the slow speed solar wind occupies a relatively thin band that straddles the Sun's magnetic equator. This band is seen to closely conform to the ecliptic plane near solar minimum.

The structure of the solar corona can likewise be classified into three categories; 1) the active corona, 2) the quiet corona, and, what looks at first glance to be no corona or holes in the corona, 3) the so-called coronal holes. Pictures that illustrate and delineate all of these regions were first published using data measured by Skylab experiments in the mid 1970s (see, e.g., [*Bohlin*, 1977; *Krieger*, 1977], and references therein). Although the active corona is often the solar origin of transient solar wind flows, coronal mass ejections, which are an important subclass of transient flows, are often generated by filament disruptions embedded within the nominally quiet corona. The slow speed solar wind maps into the quiet corona, which generally sits beneath the streamer belt (e.g., [*Hundhausen*, 1977]). This belt encircles the Sun at its magnetic equator near solar minimum but can wander considerably to higher latitudes at other times of the solar cycle. The high-speed solar wind maps into coronal holes [*Krieger et al.*, 1973].

3. TOTAL ENERGY AND MASS FLUX FAR FROM THE SUN

In situ observations of the solar wind have provided important constraints on models of coronal heating and solar wind acceleration by providing far-field boundary conditions. Characteristics of the high- and low-speed flows have been tabulated from a number of missions [*Feldman et al.*, 1977; *Schwenn*, 1983]. Meaningful connection with regard to physical constraints on the flow near the Sun is not possible for the slow-speed solar wind because of the fine scale, generally magnetically closed structure of the quiet corona, and the variability of the slow wind at large heliocentric distances. In addition, although the quiet corona is generally closed, its internal magnetic connectivity is continuously changing. As new flux emerges from below the photosphere, they break their internal connectivity and reconnect to surrounding magnetic fields [*Sheeley et al.*, 1975]. These new interconnections can sometimes couple photospheric fields that are concentrated at great distances. This situation does not appear to hold for the high-speed solar wind. We will therefore concentrate on the high speed solar wind because it affords our best chance for identifying its causitive factors through use of remote-sensing observations, which necessarily integrate small-scale plasma conditions over large distances.

Intercomparison of Yohkoh soft X-ray data, which delineate the extent of the polar coronal holes observed between about 1994 and 1997, and Ulysses observations of the polar solar wind at asymptotic distances, have provided our best figures for the particle flux (8.7×10^{13} cm^{-2} s^{-1}) and the total energy flux (11.5×10^5 erg cm^{-2} s^{-1}) just above the coronal base [*Feldman et al.*, 1997]. The uncertainties in determining both fluxes at the Sun from measurements made far from the Sun are dominated by uncertainties in the inferred solid-angle expansion factor of the high-speed flow. These uncertainties have been estimated from several independent coronal investigations [*Munro and Jackson*, 1977; *Wang and Sheeley*, 1995; *Gosling et al.*, 1997; *Cranmer et al.*, 1997] to be about $\pm 40\%$. This precision is not sufficiently good to allow a unique determination of the mix of heating mechanisms that support the solar corona.

4. THE CORONAL DENSITY PROFILE

Another possible source of information about coronal heating and acceleration mechanisms is the run of coronal density as a function of heliocentric distance. This profile has

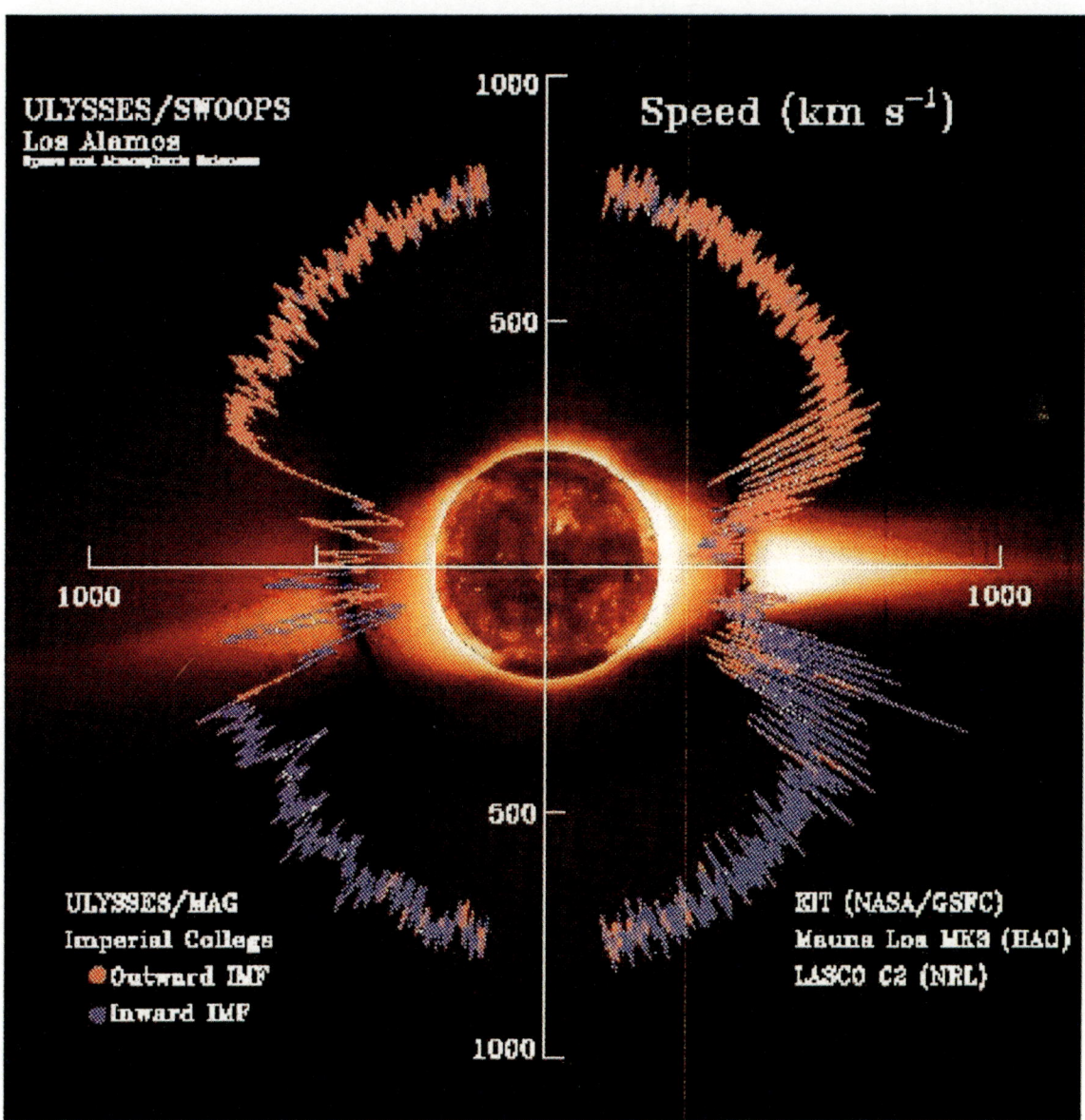

Plate 1. One-hour averages of the solar wind speed and magnetic polarity (red is outward pointing and blue is inward pointing) measured by Ulysses overlaid with three concentric images measured using the NASA/GSFC EIT instrument (center), the HAO Mauna Loa coronagraph (inner ring), and the NRL LASCO C2 coronagraph (outer ring).

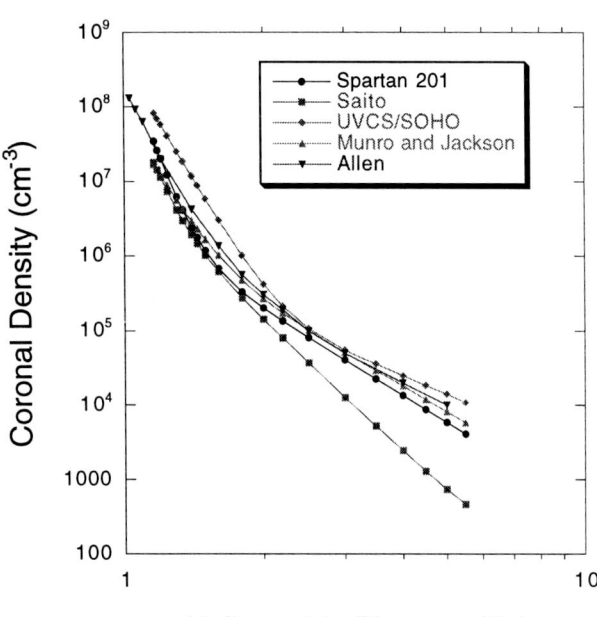

Figure 1. Coronal densities derived from the measured pB brightness of the polar corona over a time period spanning more than 30 years.

generally been inferred from the polarization brightness (pB) of the corona measured using coronagraphs. Here again, our best chance of a unique identification can be made by studying coronal holes because of their often-time large, time-stationary, and structure-free appearance. Five published determinations of inferred coronal densities above polar coronal holes are shown in Figure 1 [*Allen*, 1963; *Saito*, 1970; *Munro and Jackson*, 1977; *Fisher and Guhathakurta*, 1995; *Cranmer et al.*, 1998]. Inspection reveals significant differences. We conclude that either the density profile in coronal holes varies with time and solar conditions or that its determination uncertainty is significant. These variations contrast with those observed at large distances by Ulysses. Here conditions were sensibly constant over both solar poles over a time period spanning more than three years [*Feldman et al.*, 1996].

Uncertainties in the conversion of pB brightness to coronal density are estimated to range between 15% and 35% by Fisher and Guhathakurta [1995], which is seen to be considerably smaller than the differences between the profiles given in Figure 1. We can only conclude therefore, that time-variable conditions within coronal holes near the Sun are smoothed over before the flow exceeds heliocentric distances of about 0.3 AU, and/or that systematic errors in the inversion process used to convert measured pB brightness to density are larger than anticipated. Regardless, observed variations in coronal density profiles are sufficiently large to preclude identification from them of a unique coronal heating and acceleration mechanism. Specifically, these profiles cannot be used to discriminate between time-stationary, spatially-broad expansion models, and time-dependent pulsed models of a corona that consists of an ensemble of discrete, thin filaments of expanding plasma jets [*Feldman et al.*, 1997].

5. THE CORONAL SPEED PROFILE

Estimates of the radial speed profile of the high-speed solar wind also provide information that can help identify models of energy deposition and plasma acceleration in the extended corona. Several experimental techniques have been applied for this purpose; 1) interplanetary scintillation of radio waves that propagate through the corona [*Grall et al.*, 1996; *Coles et al.*, 1996; *Klinglesmith*, 1997], 2) Doppler shifts and widths of strong EUV lines (see e.g., [*Dere*, 1992, 1994; *Kohl et al.*, 1997; *Moses et al.*, 1997; *Innes et al.*, 1997; *Cranmer et al.*, 1998], and references therein), and 3) apparent motion of density irregularities that are assumed to act as markers of the bulk flow [*Shimojo et al.*, 1996; *Moses et al.*, 1997; *Sheeley et al.*, 1997]. The results of all three techniques seem to agree that acceleration of the apparently quasi-steady slow solar wind appears to be generally smaller, and extend over larger distances, than that of the high-speed solar wind. However, they have also shown that time and/or spatial variability occurs, occasionally, within all types of corona and solar wind flow types. This situation is expected in the slow wind because observations of the low-lying magnetic carpet that covers the quiet corona reveals constant activity. In addition, transient flows are known to be contained within the slow-speed wind. However, it does not seem to be consistent with our conception of the high-speed wind as a time-invariant and structure-free flow. Indeed, observations of transient flows embedded in the nominally open corona near the Sun changes our task from not having sufficiently precise information to uniquely identify the origin and dynamics of the extended solar atmosphere, to having too much information. The signatures of none of the transient events seen at low altitudes within coronal holes have been uniquely identified in the high speed solar wind far from the Sun (see, e.g., [*Feldman et al.*, 1997]). This disconnect cannot be dismissed easily because the sum total of these transient events supply a non-negligible fraction of the total mass and energy flux carried by the high-speed solar wind, yet is not sufficient to account for the high-speed expansion in its entirety (see, e.g. [*Dere*, 1992, 1994], and references therein). Nevertheless, the sum total input provided by these transients, and other presently unknown energy and momentum addition mechanisms, are evidently sufficient to accelerate the corona to its asymptotic speed of about 750 to 800 km s^{-1} by the time its constituent gas parcels reach 2.5 R$_s$ [*Grall et al.*, 1996; *Coles et al.*, 1996; *Klinglesmith*, 1997].

6. THE CORONAL TEMPERATURE PROFILE

The problem of coronal structure becomes more interesting when we turn to temperature. Spectroscopic observations of the low corona (at a heliocentric distance of 1.03 solar radii, R$_s$) reveal that the electron temperature within coronal holes is just under 10^6 K [*Feldman et al.*, 1998]. Models of the charge states of heavy ions measured by Ulysses are consistent with an electron temperature that maximizes at a value of about 1.2×10^6 K at about 1.3 R$_s$ in the south polar coronal hole [*Ko et al.*, 1997]. In contrast, measurements of hydrogen velocity distributions in coronal holes reveal temperatures that range as high as 4 to 6 $\times 10^6$ K at heliocentric distances between 1.8 and 3.5 R$_s$ [*Kohl et al.*, 1996, 1997; *Strachan et al.*, 1997]. Careful analysis of these distributions show further, that they are not isotropic, but have a temperature that is larger in the direction perpendicular to the magnetic field (B) than it is parallel to B.

These new, and unexpected observations have spawned a variety of one-dimensional models of the coronal expansion based on assumed steady-state, broadly-structured conditions (e.g., [*McKenzie et al.*, 1995; *Cranmer et al.*, 1997; *Hu et al.*, 1997]. We note that these models are not theories in that they have no predictive capability. They just summarize an important subset of the measured properties of the extended solar atmosphere within coronal holes. This task is accomplished by iterating parameters of assumed energy and momentum addition functions that are included in 1-D MHD codes until resultant density, speed, and temperature profiles agree with conditions inferred from measurements made at a variety of altitudes inside of about 4 R$_s$ and beyond about 1 AU. Satisfactory agreement between model and measured profiles are obtained. However, inversion of these profiles to infer heating mechanisms, is not unique. For example, the same conditions can also be consistent with a finely structured corona that consists of an ensemble of transient, filamentary flows that are stacked next to one another along the line of sight [*Feldman et al.*, 1997]. Both types of models provide very different views of coronal structure and its required heating and momentum addition mechanisms. We will return to this point later.

7. HEAVY ION ABUNDANCES

A diagnostic that offers the possibility of inferring conditions and their formative processes below the base of the corona is the run of heavy ion abundances. Observations in the solar wind far from the Sun reveal that abundance fractionation patterns in the low- and high-speed solar wind are different. The abundances of those elements having a first ionization potential that is higher than about 10.5 eV (with the exception of He) are close to their photospheric values in both high and low speed winds. However, elements having ionization potentials less than 10.5 eV are enriched by a factor of between 4 and 5 in the low-speed wind, but are roughly at their photospheric values in the high-speed wind [*Breneman and Stone*, 1985; *Bochsler et al.*, 1986; *Gloeckler and Geiss*, 1989; *von Steiger*, 1996]. A similar effect is inferred from spectroscopic measurements at the base of the corona (see, e.g., [*Feldman et al.*, 1998], and references therein). The abundance of He in the solar wind does not conform to this pattern. It is depleted by about a constant factor of 2 relative to the photosphere in the high speed solar wind, but is highly variable and, on average, depleted by a larger factor in the low-speed wind [*Feldman et al.*, 1977, 1996; *Schwenn*, 1990; *Barraclough et al.*, 1996].

The foregoing results imply that conditions below the coronal base must differ between the solar atmosphere that underlies the high- and low-speed solar wind. Models that attempt to explain this phenomenon consider various modes of ion-neutral separation mechanisms that operate just above the chromosphere (see, e.g., [*von Steiger*, 1996], and references therein). They all play on the longer ionization time implied by high first ionization potentials (FIP). Intercomparison between the measured FIP bias in magnetic loops as they emerge from the photosphere (no bias is observed), to that in unresolved fine-structure loops that exist below the base of the corona (which have a lifetime of between 10^4 and 10^5 s and results in FIP bias factors of between 2 and 2.5), to that on loops that comprise the quiet corona (which have lifetimes longer than a day and results in FIP bias factors of between 4 and 5), suggests that the FIP bias requires times longer than 10^5 s to set up [*Feldman*, 1998]. This suggestion provides a convenient explanation for the presence of the FIP bias in the slow-speed solar wind, which maps to the quiet, predominantly closed corona, and the absence of the FIP bias in high-speed wind, which maps to the magnetically-open coronal holes.

If the foregoing explanation of the FIP bias is correct, additional constraints need to be imposed on models of the high-speed solar wind expansion close to the Sun. Specifically, the expansion needs to be sufficiently rapid at altitudes below where all elements are ionized regardless of FIP, that a FIP bias does not have time to set up (about 10^5 s). This requirement translates to an acceleration sufficiently large that a rising gas volume reaches a height of 2×10^4 km (the base of the corona in coronal holes, [*Huber et al.*, 1974]) within 10^5 s. Most models of the high-speed solar wind satisfy this constraint.

However, this requirement is not sufficient to account for helium. Time-dependent models of the high-speed solar wind show that, regardless of ionization state, the helium abundance will build up at low altitudes to values of about 20% on timescales of 10 hours to a day. This build up provides a temperature buffer for the proton expansion [*Hansteen et al.,* 1994] and appears to be a necessary condition for producing solar wind flows that are not sharply dependent on coronal temperature, as observed.

Truly time-stationary conditions at low altitudes will then translate into the structure-free high-speed flow state observed routinely at large distances. However, these observations cannot account for sporadic injections of plasma volumes having large helium abundance (of order 20%) in response to small scale transients that are observed to occur routinely throughout coronal holes (see, e.g., [*Dere,* 1992, 1994], and references therein). The fact that large helium enhancements are not observed in the high-speed solar wind [*Barraclough et al.,* 1996; *Feldman et al,* 1996] implies that the cadence of transients must be considerably higher than one event in 10 hours (the time required to build helium enhancements in the low corona under steady-state conditions) on every field line that threads coronal holes. This condition appears to be satisfied. Specifically, the Sun is covered by about 3×10^4 explosive events at any one time [*Dere,* 1992]. Using the fact that each event has a characteristic size of about 1500 km and a time scale of about 60 s, implies that an explosive event occurs on every magnetic field line that threads the corona at average intervals of 7000 s, which is significantly less than 10 hours. This exercise implies that transient events must contribute importantly to the structure of the inner corona and the character of its expansion into interplanetary space. Otherwise, substantial He abundance enhancements should be observable in the high-speed solar wind, which are not observed.

8. BASIC INFORMATION NEEDED TO CONSTRUCT A THEORY

The foregoing discussion shows both that we know a lot about the structure of the solar corona and its solar wind extension to space, yet far from what is needed to construct a theory of coronal heating and acceleration that has predictive capability. Although we know about how much energy and momentum is needed to drive the solar wind and where it needs to be added to the corona, we do not know the specific forms of that energy and momentum as a function of heliocentric distance, and how they are linked to their formative processes at lower altitudes. For example, it has long been speculated that short-wavelength, high-frequency waves generated through numerous reconnection transients that occur within the chromospheric network, propagate to higher altitudes where they damp and heat overlying layers and thereby accelerate them upwards to form the wind (see, e.g., [*Axford and McKenzie,* 1992]). However, these waves have not been observed, and a model of their effects on overlying layers assume an unrealistic power spectrum at low altitudes [*Marsch and Tu,* 1997]. Another possible model of coronal heating postulates the corona is composed of an ensemble of finely filamented, strongly time-dependent plasma jets that are accelerated in magnetic reconnection events that occur at the margins of the chromospheric network (e.g., [*Feldman et al.,* 1997], and references therein). These jets then would interact with one another along $\underline{B}$, thereby generating ion cyclotron, Alfvén, and fast-mode magnetosonic waves (e.g., [*Montgomery et al.,* 1976] and perhaps shock waves, that then propagate through, and heat the corona over extended distances. Neither the remnants of the jets nor the waves that they no doubt generate, have been observed above the base of the corona.

Although waves provide the vehicle for coronal heating in both the foregoing viewpoints, these views differ fundamentally in the place of origin, and the spatial distribution of the waves and their dissipation products. Whereas the waves in the broadly-structured models originate from injection events at low altitudes above the photosphere, those in the finely-filamented models are injected throughout extended regions of the corona, at altitudes that likely exceed 15 R_s. Another fundamental difference in these viewpoints is the nature of the turbulent plasma state that develops in the extended corona. The primary effect of the dissipation of short wavelength, high-frequency waves that come directly from the photosphere should be to heat the ions perpendicular to $\underline{B}$ in the low corona as observed [*Kohl et al.,* 1997]. However, the evolution of an ensemble of plasma jets should generate intense bursts of suparathermal and energetic ions and electrons, and also heat the perpendicular to $\underline{B}$ component of the thermal ion population. These energized particle populations should then diffuse outwards, eventually transferring their energy to the thermal particle populations of the solar wind.

The purpose of the foregoing comments is not to denigrate any of the modeling efforts that have been pursued to date, which are laudable and useful, but to point out how much more information is needed to reach our goal of a theory of coronal heating and solar wind acceleration. All that we presently know indicates that waves must be an important ingredient for a definition of the structure of the corona. The real question is whether they are the primary vehicles of momentum and energy input, or whether they are secondary to whatever generates the waves in the first place. Another fundamental question is the basic spatial/temporal structure of the corona. Is the cadence of energy and momentum deposition sufficiently high relative to coronal relaxation rates, and sufficiently finely dispersed relative to coronal spatial scales, that the structure and dynamics of the corona can be well described as a broadly dispersed, quasi-time-stationary MHD gas. Or, does its basic nature require a description in terms of an ensemble of fine-scale filamentary jets whose relaxation times are longer than their creation times.

Regardless of the answers to the foregoing questions, the energy and momentum densities that are needed to support the corona are extremely large, and therefore should generate many, very interesting secondary effects. Based on our exploration of other plasma environments throughout the solar system, we expect a cascade that will lead to; 1) substantial fluxes of suprathermal electrons and ions, 2) a multitude of secondary waves, which may develop into a truly turbulent state, including perhaps, shock waves, and to 3) the considerable acceleration of energetic particles. These effects, in turn, should have a large, and direct affect on the plasma flow, which eventually develops into the relatively broadly-structured and docile wind that is observed far from the Sun.

All of these secondary effects should retain the signature of the original energy and momentum input mechanisms that initiated the cascade, to distances well beyond 4 R_s, the present design perihelion of the Solar Probe mission (see [*Gloeckler et al.* 1998] for a description of the Solar Probe mission and its measurement requirements). Of course, these distances depend on the cadence and spatial element size of the specific mechanisms that generate the corona. If waves directly from the chromospheric network are the basic vehicles of energy and momentum input, then much of this information will be lost by distances that may be as close as 4 R_s (see, e.g., [*Marsch and Tu,* 1997; *Kohl et al.,* 1997]). However, the parts of this input coming from explosive events (which we know must be there) should be resolvable out to distances beyond about 15 R_s [*Feldman et al.,* 1997]. Of course, the effects of suprathermal ions and electrons [*Scudder,* 1994, 1996; *Lie-Svendsen et al.,* 1997] should also be visible beyond 15 R_s.

Acknowledgments. This work was supported in part by NASA and conducted under the auspices of the U.S. Department of Energy.

REFERENCES

Allen, C.W., *Astrophysical Quantities,* p. 176, Athlone, London, 1963.

Axford, W.I., et al., *Close Encounter with the Sun*, JPL D-12850, 1995.

Barraclough, B.L., W.C. Feldman, J.T. Gosling, D.J. McComas, and J.L. Phillips, He abundance variations in the solar wind: observations from Ulysses, in *Solar Wind Eight*, edited by D. Winterhalter et al., AIP Conf. Proc., *382*, 277-280, 1996.

Biermann, L., Comet tails and solar corpuscular radiation, Z. Astrophys., *29*, 274, 1951.

Bohlin, J.D., An observational definition of coronal holes, in *Coronal Holes and High Speed Wind Streams*, J.B. Zirker, ed., pp. 27-70., Col. Assoc. Univ. Press., Boulder, 1977.

Breneman, H.H. and E.C. Stone, *Astrophys. J.*, *299*, L57, 1985.

Coles, W.A., M.T. Klinglesmith, and R.R. Grall, The solar wind velocity distribution near the Sun (abstract), EOS Trans. AGU, *77(46)*, Fall Meet. Suppl., F586, 1996.\

Cranmer, S.R., et al., An empirical model of a polar coronal hole at solar minimum, Astrophys. J., submitted, 1997.

Dere, K.P., Explosive events and magnetic reconnection in the solar atmosphere, in *Solar Wind Seven*, edited by E. Marsch, and R. Schwenn, pp. 11-20, Pergamon, Tarrytown, N.Y., 1992.

Dere, K.P., Explosive events, magnetic reconnection, and coronal heating, Adv. Space Res., *14(4)*, 13, 1994.

Feldman, W.C., J.R. Asbridge, S.J. Bame, and J.T. Gosling, Plasma and magnetic fields from the Sun, in *The Solar Output and its Variations,* edited by O.R. Whjite, pp. 351-382, Colo. Assoc. Univ. Press, Boulder, 1977.

Feldman, W.C., et al., Solar Probe, scientific rationale and mission concept, JPL D-6797, 1989.

Feldman, W.C., B.L. Barraclough, J.L. Phillips, and Y.-M. Wang, Constraints on high-speed solar wind structure near its coronal base: a Ulysses perspective, Astron. Astrophys. *316*, 355-367, 1996.

Feldman, W.C., S.R. Habbal, G. Hoogeveen, and Y.-M. Wang, Experimental constraints on pulsed and steady state models of the solar wind near the Sun, J. Geophys. Res., *102*, 26905-26918, 1997.

Feldman, U., FIP effect in the solar upper atmosphere: spectroscopic results, *Proceeding of Workshop on Solar Composition and its Evolution form Core to Corona, Bern,* Switzerland, Jan., 1998.

Feldman, U., U. Schühle, K.G. Widing, and J.M. Laming, The coronal composition above the solar equator and the north pole as determined from spectra acquired by the SUMER instrument on SOHO, Astrophys. J., submitted, 1998.

Fisher, R., and M. Guhathakurta, Physical properties of polar coronal rays and holes as observed with the Spartan 201-01 coronaograph, Astrophys. J., *447*, L139-L142, 1995.

Gloeckler, G., J. Geiss, in *AIP Conf. Proc.*, *183*, pp 49-71, C.J. Waddington, ed., (New York: AIP), 1989.

Gloeckler, G., S.R. Habbal, R.LO. McNutt, J.E. Randolph, A. Title, and B.T. Tsurutani, Solar Probe: A mission to the Sun and the inner core of the heliosphere, *The Physics of Sun-Earth Plasma and Field Processes,* J.L. Burch, ed., in press, 1998.

Grall, R.R., W.A. Colesk, M.T. Klinglesmith, A.R. Breen, P.J.S. Williams, J. Markkanen, and R. Esser, Rapid acceleration of the polar solar wind, Nature, *379*, 429-432, 1996.

Hansteen, V.H., E. Leer, and T.E. Holzer, Coupling of the coronal helium abundance to the solar wind, Astrophys. J., *428*, 843-853, 1994.

Hu, Y.Q., R. Esser, and S.R. Habbal, A fast solar wind model with anisotropic proton temperature, J. Geophys. Res., *102*, 14661-14676, 1997.

Huber, M.C.E., P.V. Foukal, R.W. Noyes, E.M. Reeves, E.J. Schmahl, J.G. Timothy, J.E. Vernazza, and G.L. Withbroe,

XUV observations of coronal holes, initial results from skylab, *Astrophys. J., 194*, L151, 1974.

Hundhausen, A.J., An interplanetary view of coronal holes, in *Coronal Holes and High Speed Wind Streams*, J.B. Zirker, ed., pp. 225-330, Col. Assoc. Univ. Press., Boulder, 1977.

Innes, D.E., P. Brekke, D. Germerott, and K. Wilhelm, Bursts of explosive events in the solar network, *Solar Phys., 175*, 341-348, 1997.

Klinglesmith, M.T., The polar solar wind from 2.5 to 40 solar radii: results of intensity scintillation measurements, Ph.D. thesis, Univ. of Calif., San Diego, 1997.

Ko, Y.-K, L.A. Fisk, J. Geiss, G. Gloeckler, and M. Guhathakurta, An empirical study of the electron temperature and heavy ion velocities in the south polar coronal hole, *Solar Phys., 171*, 345-361, 1997.

Kohl, J.L., et al., First results from the SOHO ultraviolet coronagraph spectrometer, *Solar Phys., 175*, , 1997.

Kohl, J.L., L. Strachan, and L.D. Gardner, Measurement of hydrogen velocity distributions in the extended solar corona, *Astrophys. J., 465*, L141-L144, 1996.

Krieger, A.S., A.F. Timothy, and E.C. Roelof, A coronal hole and its identification as the source of a high velocity solar wind stream, *Solar Phys. 29*, 505, 1973.

Krieger, A.S., Temporal behavior of coronal holes, in *Coronal Holes and High Speed Wind Streams*, J.B. Zirker, ed., pp. 71-102, Col. Assoc. Univ. Press., Boulder, 1977.

Lie-Svendsen, O., V.H. Hansteen, and E. Leer, Kinetic electrons in high-speed solar wind streams: formation of high-energy tails, *J. Geophys. Res., 102*, 4701-4718, 1997.

McComas, D.J., S.J. Bame, B.L. Barraclough, W.C. Feldman, J.O. Funsten, J.T. Gosling, P. Riley, R. Skoug, A. Balogh, R. Forsyth, B.E. Goldstein, and M. Neugebauer, Ulysses' return to the slow solar wind, *Geophys. Res. Lett., 25*, 1-4, 1998.

Marsch, E., and C.-Y. Tu, Solar wind and chromospheric network, *Solar Phys. 176*, 87-106, 1997.

McKenzie, J.J.F., M. Banaszkiewicz, and W.I. Axford, Acceleration of the high speed solar wind, *Astron. Astrophys., 303*, L45, 1995.

Montgomery, M.D., S.P. Gary, W.C. Feldman, and D.W. Forslund, Electromagnetic instabilities driven by unequal proton beams in the solar wind, *J. Geophys. Res., 81*, 2743-2749, 1976.

Moses, D., et al., EIT observations of the extreme ultraviolet Sun, Solar Phys., *175*, 571-599, 1997.

Munro, R.H., and B.V. Jackson, Physical properties of a polar coronal hole from 2 to 5 R_s, *Astrophys. J., 213*, 874-886, 1977.

Raymond, J.C., et al., Composition of coronal streamers from the SOHO ultraviolet coronagraph spectrometer, *Solar Phys., 175*, 645-665, 1997.

Saito, K., A non-spherical axisymmetric model of the solar K corona of the minimum type, Ann. *Tokyo Astron. Obs., 12*, 53, 1970.

Schwenn, R., The average solar wind in the inner heliosphere: structures and solaw variations, in *Solar Wind Five*, NASA Conf. Publ., CP-2280, 489-507, 1983.

Schwenn, R., Large-scale structure of the interplanetary medium, in *Physics of the Inner Heliosphere*, R. Schwenn and E. Marsch, eds., pp. 99-182, Springer-Verlag, Berlin, 1990.

Scudder, J.D., Ion and electron suprathermal tail strengths in the transition region: support for the velocity filtration model of the corona, *Astrophys. J., 427*, 446-452, 1994.

Scudder, J.D., Electron and ion temperature gradients and suprathermal tail strengths at Parker's solar wind sonic critical point, *J. Geophys. Res., 101*, 11039-11053, 1996.

Sheeley, N.R., Jr., J.D. Bohlin, G.E. Brueckner, J.D. Purcell, V.E. Scherrer, and R. Tousey, The reconnection of magnetic field lines in the solar corona, *Astrophys. J., 196*, L129-L131, 1975.

Sheeley, N.R., Jr., A volcanic origin for high-FIP material in the solar atmosphere, *Astrophys. J., 440*, 884-887, 1995.

Shimojo, M., S. Hashimoto, K. Shibata, T. Hirayama, H.S. Hudson, and L.W. Acton, Statistical study of solar X-ray jets observed with the Yohkoh soft X-ray telescope, *Publ. Aston. Soc. Jpn., 48*, 123-136, 1996.

Strachan, L., L.D. Gardner, P.L. Smith, and J.L. Koh., UV spectroscopy of the extended solar corona: results from UVCS/-Spartan, in *Scientific Basis for Robotic Exploration Close to the Sun*, edited by S.R. Habbal, AIP, CP 385, 1997.

von Steiger, R., Solar wind composition and chrtge states, in *Solar Wind Eight*, edited by D. Winterhalter et al., AIP Conf. Proc., *382*, 193-198, 1996.

Wang, Y.-M., and N.R. Sheeley, J., Solar implications of Ulysses interplanetary field measurements, *Astrophys. J., 447*, L143-L146, 1995.

Dr. William C. Feldman, Los Alamos National Laboratory, MS-D466, Los Alamos, NM 87545, wfeldman@lanl.gov

Models for Coronal and Interplanetary Magnetic Fields: A Critical Commentary

Kenneth H. Schatten[1]

Laboratory for Terrestrial Physics, NASA/Goddard Space Flight Center, Greenbelt, Maryland

The structure of the interplanetary medium is controlled by the coronal magnetic field from which the solar wind emanates. This field has been described with "Source Surface" (SS) and "Heliospheric Current Sheet" (HCS) models. The "Source Surface" model was the first to open the solar field into interplanetary space using volumetric coronal currents, which were a "source" for the IMF. The Heliospheric Current Sheet (HCS) model provided a more physically realistic solution. The field structure, without regard to sign, was dominated by a monopole-like pattern, which has been observed by Ulysses. Recently, Sheeley and Wang have utilized the HCS field model to calculate solar wind structures fairly accurately. Fisk, Schwadron, and Zurbuchen have investigated small differences from the SS model. These differences allow field line motions reminiscent of a "timeline" or moving "streakline" in a flow field, similar to the smoke pattern generated by a skywriting plane.

1. INTRODUCTION

Knowledge about coronal and interplanetary magnetic fields (CMFs and IMFs) began with the pioneering work of Parker, who deduced from his newly-developed solar wind theory that the IMF might be configured into an Archimedian spiral. This would occur owing to the dual constraints placed upon the IMF of maintaining a "frozen-in" field connection to the photosphere, rotating at roughly 27 days, and the solar wind, blowing radially outward at about 600 km/s, similar to the water pattern from a rotating garden hose. As we will see towards the end of this paper, frozen-in field constraints can still be a powerful tool in governing field geometry and topology,

---

[1] *Currently at: AI Solutions, Lanham, MD*

and lead to new insights into the time history of the three dimensional configuration of the IMF.

Advances in CMFs and IMFs in the 60's began on two different fronts. On the coronal front, Altschuler and Newkirk were making models of coronal magnetism, which allowed the photospheric field to be extended into the inner corona, using a purely potential "current free" model. These models gave a rough comparison with lower coronal features - loops and rays, but gradually diverged from observations higher in the solar atmosphere.

On the IMF front, Wilcox and Ness (1964) utilized more accurate observations from a new class of spacecraft to advance our understanding. The IMP1 (Interplanetary Monitoring Platform) spacecraft was the first to use a new, highly accurate fluxgate magnetometer to observe the IMF. What was found, was a new feature, which in some ways obeyed Parker's picture, but in another way, did not. A "sector pattern" was observed. In this pattern (see Figure 1), the field did lie along the predicted garden hose "spiral angle" of about 45 degrees. It was, however, persistently outward and inward for many days on end (about 7), forming a "sector", rather than varying as the photosphere does inward and outward, every day or two when a bipolar magnetic region (BMR) or unipolar magnetic region (UMR) crosses central meridian. These solar field designations are useful descriptions of large-scale field structures (Bumba and Howard, 1965).

This awkwardness or discrepancy was not really noted. Rather, the spiral angle was declared a success, and the new sector structure was hailed an important feature to classify solar wind phenomena (velocities, densities, etc.). In addition, geomagnetic effects fell into a new remarkable pattern when analyzed via the sector structure, and numerous other effects were examined as well. Surprisingly, it was not clear where the sector structure originated; the thinking (Wilcox and Ness, 1967) was that the sectors were formed on the Sun, since the Sun was the source of the solar wind. A model was needed to both understand the origins of sectors and to tie the potential field patterns in the inner corona to the Archimedian spiral sector structure. This was the goal of the "source surface" model.

2. SOURCE SURFACE MODEL FOR CORONAL AND INTERPLANETARY MAGNETIC FIELDS

Observationally, from the configuration of coronal steamers, and theoretically, from Parker's model, it was recognized that the coronal magnetic field was deviating from a potential totally current free model, obeying La Place's equation, with the field somehow being stretched towards the radial direction in the corona. This was clearly due to the effect of currents in the corona, extending the field outwards, and opening it up into interplanetary space. The question was how to describe this mathematically and physically. It seemed like a daunting task, as there were no direct field measurements within the corona. But what is an observer's nightmare, is a theoretician's dream. Without observational constraints, free reign was granted to employ any tactic, which might work, unconstrained by physical observations.

From this background, the source surface (SS) model was born (Schatten, Wilcox, and Ness (SWN), 1969). One placed a sphere (shown in Figure 1) called the "source surface" around the Sun, and insisted that from the photosphere to the source surface no currents flow, and that the field be governed by La Place's equation in this region. On the source surface, and outside it, currents restrict the field to being radial (and then stretch toward the spiral). This allowed the coronal field to be calculated from a "Green's function" solution. This clearly is very much a diode kind of solution, where the current is negligible inside the source surface, and totally dominates outside. Nevertheless, it is well posed (it can be solved from Maxwell's equations into a boundary value problem that has a mathematical solution) and has the correct number of boundary conditions (although they are mixed (see Schatten, 1968)): field conditions on one surface and constraints about two components on another - zero tangential field). In this model, the "source surface" was so named as being a surface, above which fields would extend outwards into interplanetary space, and so this surface, rather than the photosphere, could serve as a "source" for interplanetary field (as in the source of the Nile, but not nearly as dramatic).

Let us digress for a general discussion of flow patterns, as these will be important later. The IMF, because of the high conductivity of the solar wind plasma (the "frozen-in field condition" of Alfven's), is forced to move concurrently with the plasma. Let us consider what this implies. A field line in a fluid may be characterized as a general collection of points along this line in the fluid. The general description (see White, 1994) for the motion of paths are referred to as "ribbons." The specific example of a ribbon, which describes a line, which moves in space with the fluid, is referred to as a "timeline." A timeline may thus be regarded as the general description of a field line (and its motion) in a highly conducting fluid. A simplification of the timeline occurs when one point on the line is "rooted" at a specific location. This may occur, for example, when coronal field lines, although open at the source surface, are rooted more deeply in a region of relatively dense plasma, namely the photosphere. In such a case, the general case of the timeline reverts to a more specific form: a streakline.

A streakline is defined by meteorologists (see Hess, 1959; White, 1994) as being a line passing through the same point and parallel to the flow from that specific location. The equation for a streakline passing through a point, y, from time 0 to t, may be written as:

$$x = x[\varsigma(y,s),t],$$

with s as a time parameter, and $\varsigma(y,s)$ is defined as the initial position, ς, of a point passing through any other point, say y, at a later time, s. A familiar example of a streakline is the pattern of smoke given off by a chimney, which may differ markedly from the pathlines of the flow field.

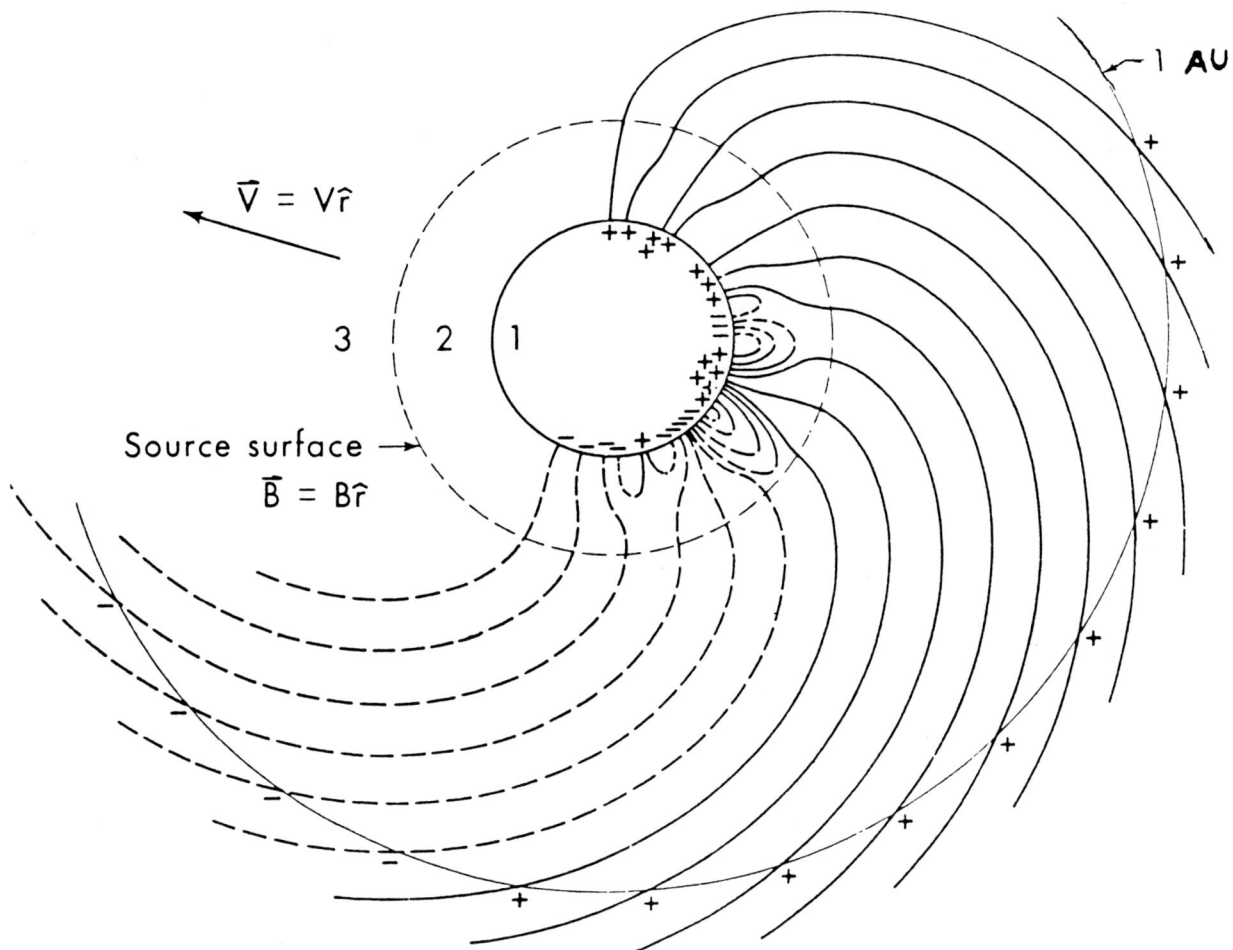

Figure 1. This figure (not to scale) shows the "Sector Structure" of the IMF and the "Source Surface" model. In the outer regions, the sector structure is seen, namely, the IMF lies along a garden hose angle for many days at a time in either an "outward" (+) or "inward" (-) field configuration. Shown in the center of the Figure is the physical basis of the Source Surface model. Boundary conditions are imposed in the photosphere by observations, but in the inner corona, the field obeys LaPlace's equation. Volumetric currents flowing on the source surface and in the outer corona constrain the field to be radial. A mathematical solution for this was obtained by S, W and N, 1969.

In addition, there are at least two other names given to describe flow patterns. The first is the familiar "streamline," a line is drawn parallel to the instantaneous velocity vector everywhere. The second is a "pathline" or "trajectory," which is the actual path taken by the fluid particles, a line along which any given fluid particle moves, e.g. the path traced by a given radioactive cloud released from a given location at a single moment of time. Simply because of its definition, a trajectory is the path of the fluid, which in the situation under discussion, is the path of the solar wind particles. The streakline differs from the pathline, because in the streakline the dye or smoke continues to issue forth from the same location, rather than move with the fluid.

One may consider differences between streaklines and pathlines in the following manner. A pathline is where a given particle will move in the future. A streakline (or timeline), on the other hand, shows where the fluid has emanated from, to arrive at any point along it. Hence, the streakline is more physical, since any point within the fluid can only "know" about its past, not its future which has yet to unfold, but does not yet exist. This explains why the field may follow a streakline or timeline, but not a streamline nor pathline. These latter two are unphysical

- the streamline deals only with the instantaneous present, and the pathline with the unknowable future.

Given, that the magnetic field configuration forms a timeline or streakline, to obtain the path of the particles, one needs to know how far along such a line the fluid has moved. We shall use the streakline term, since in the solar wind, the field lines are presumably rooted to the photosphere quite well. Streakline and pathlines can then provide a mutual set of constraints on the fluid and field. The "guiding" of the fluid by the field and vice-versa is governed by their relative strengths at different locations in the corona/solar wind. Where the magnetic field is strong, hence stiff, as in the inner corona, the trajectory of the particles will be guided along the field streakline, with the field serving as stiff wires upon which the solar wind moves. In this regime, the fluid behaves in a 1-D fashion like beads traveling outwards on wires. In the solar wind, beyond about 20 solar radii, where field becomes flaccid, the streakline of the field will twist to follow the twisted 3-D trajectory of the particles, as turbulence, shocks, etc. occurs. The streamline, being an instantaneous picture of the flow pattern is not very physical, except for non-existent steady state conditions, where these three flow lines become identical. The three flow lines will generally be different for time varying conditions. This will become important later. For now we shall simply assume that the solar wind carries the coronal field approximately radially into interplanetary space into an "open" configuration.

It is interesting that some conditions can greatly simplify the flow solutions. For example, the frozen-in field condition implies that the flow occurs only along the field line; hence is one-dimensional in this field coordinate system. Although too simple for the solar wind where shocks may occur, Landau and Lifshitz (1959) show that for arbitrary one-dimensional isentropic (no shock) gas flow, the Euler equations of motion for a fluid can be integrated to yield Bernoulli's equation. In these cases, each streamline would have a constant of motion along it as a boundary condition, the "velocity potential." Let us return, however, to the coronal models.

The source surface solution, surprisingly, yielded an understanding as to the origin of the sector structure. Despite the prevalent view that the sector structure had its origins within the Sun (Wilcox and Ness, 1967) and that it manifest itself in the photosphere, SS theory explains sector origins somewhat differently. The field of the Sun has many multipole moments. In the inner corona, the field is nearly potential. As one examines the field at larger radii, the field components decrease as $1/r^{n+2}$, where n is the multipole number. Hence, near the source surface, about 1 solar radius above the photosphere, the dipole and quadrupole moments dominate the higher order moments, and the effects of small-scale active region fields become less important. Thus the large scale IMF sectors appear to result from field line reconnection in the corona, as the field minimizes its energy density (minimal currents, etc.), leaving only the lowest order multipole moment fields to guide the solar wind flow into large scale, sector-like patterns.

Figure 1 shows the Source Surface model schematically, with some of the features discussed. We discuss some of the SS model's weaknesses, rather than its successes. The source surface size seemed to vary depending upon how one chose to fit conditions. For example, SWN used 1.6 solar radii, and Altschuler and Newkirk (AN) (1969), who adopted the source surface model with their own improved boundary fitting technique, used 2.5 solar radii. The former was better for IMF magnitudes; the latter for coronal shapes. Additionally, some coronal shapes disagreed with the calculated configurations. There were two kinds of obvious misfits: 1) after flares or CMEs were observed (during this era, known as magnetic bottles) field structures were sometimes affected, and 2) at high latitudes, non-radial streamers were clearly observed even in quasi-stationary conditions. Agreements were called "rather good," and Cowling (1969) even reviewed an eclipse structure prediction as "greatly daring" and "almost perfect" (had the streamers been drawn more nearly radial). Pragmatically, the streamers were artificially bent towards the ecliptic, as was usually observed outside several solar radii, but since the polar fields were weak, the coronal streamers actually agreed with SS theory, and were quite radial. These deficiencies led us to develop the heliospheric current sheet model. More recently, Mikic and Linker (1996) have advanced eclipse structure predictions by modeling the appearance quite realistically.

3. HELIOSPHERIC CURRENT SHEET (HCS) MODEL

Before considering the HCS model, let us discuss the Pneuman and Kopp (PK) (1971) model. PK offered a different approach to the solution of how the solar field would open up into interplanetary space. They offered an "exact MHD solution" under idealized MHD conditions, which for a single fluid MHD model, assumed that volumetric currents would open the solar field into interplanetary space. Highly regarded, because of its "exactness," nevertheless it did not seem to have the thin current sheets that are seen in interplanetary space. Current sheets rely upon the disparate properties of electrons and ions (a kinetic non-single fluid process, where the fluid elements have motions distinctly different from the bulk ideal fluid), and thus a modern two fluid MHD model is required to incorporate this aspect. Thus the model did not include this remarkably prominent feature seen in the solar wind. Namely, the solar wind can carry high currents on very thin sheets allowing the IMF to form "sectors" - a change of polarity with little change in field magnitude on a relatively small scale. The PK solution assumed a constant temperature plasma with only volume currents. Although this model was not used for 3-D calculations, volumetric currents must play some role and the PK method was a major advance.

Additionally, several people do consider volumetric currents primary (e.g. see the works of Stewart, G. A. and Bravo, S., 1996, Suess et al., 1996, and Wang, et al., 1998), and their significance relative to current sheets is highly debatable. It is interesting to note that the two-fluid MHD model of Mikic and Linker (1996) does allow current sheets and they are the primary source of coronal currents (Linker, 1998). Minimal differences between this model and the current sheet model appear to exist near the cusps of streamers.

As mentioned earlier, the solar wind carries the coronal field into an "open" configuration. To understand the field structure, one may ask: "What would the field configuration be of a magnetized highly-conducting flow in magnetostatic equilibrium, allowing for current sheets to form "open" configurations above some point?" For the solution of a "field" carried by a flow, one would want the field to be a timeline, and if tied to the photosphere, to be a "streakline" (which for steady state, is a streamline). Furthermore, in the inner corona, the field guides the flow, as magnetic forces dominate the inner corona, and ideally one wants to find a "force-free" configuration. Thus the field should not simply be radial, but rather only "open" in a force-free sense. This is a tall order since it is hard to describe mathematically. Nevertheless, the HCS solution emerged.

It was recognized that an algorithmic solution could be obtained by a multistep process, which managed to provide a force-free, idealized magnetostatic solution. This yielded a new feature - current sheets, which were ideal for describing sectors. The heliospheric "current sheet" model (HCS) (Schatten, 1971) is an attempt to: 1) overcome the pragmatic deficiencies of features which did not match the SS structures; and 2) solve the general theoretical problem of understanding the configuration of a field governed by magnetostatic forces and conducting fluid flow, which forced the field, not to be radial, but rather to be "open" above a certain point. The basis of the method uses two aspects of the field: 1) the magnetic stress tensor is "even" in B (it doesn't change sign as B

reverses), and 2) if the fields are all pointed outwards, they cannot reconnect or close upon themselves. Thus if we were to reorient the fields in a special manner, a force-free magnetostatic solution could be obtained. The solution would be current free, except for infinitely thin current sheets above some surface, still called the source surface, since it played the same role as the source surface in the SS and HCS models (all field lines above this SS were open). In the HCS model, however, no currents flowed on the source surface, and in fact, in the HCS model, no volumetric currents exist at all.

Let us outline the five step process (see Figure 2) for obtaining the solution: 1) calculate potential fields up to the source surface using photospheric boundary conditions ; 2) reverse inward pointing field lines, so all field lines point outwards on the source surface ; 3) fit the three components of the field there with a La Grange multiplier technique (used also in the photospheric fit) to obtain a Legendre expansion of the field on the source surface; 4) calculate potential fields above the source surface from the Legendre expansion, but now an artificial monopole term is introduced since all the fields were redirected outwards; and 5) reverse those field directions which were previously reversed so that now they have the correct outward polarity.

This last step removes the artificial monopole term used to create the solution, but introduces a thin current sheet at the reversal edge. This is the HCS, and it is now recognized that the IMF structure is dominated by this current sheet. This also leads to the now recognized property (from Ulysses) of the IMF at high latitudes being nearly a constant field magnitude (predicted from HCS theory, but not SS theory). The explanation is simple, with magnetic forces playing a guiding role in field and flow direction, the field assumes a constant magnitude with radial distance, as the monopole term dominates. Figure 3 shows an example of field lines calculated with the HCS model along with eclipse observations. Above about 1.6 solar radii, the fields remain "open," and one can see that in the outer corona the field lines move towards a constant field magnitude. Some structures (left and lower right) illustrate open non-radial behavior where the field pattern does better than the SS radial model.

4. RECENT ADVANCES IN CORONAL FIELD MODELING

Many improvements to the SS model have been made over the past three decades. Recently two studies have examined: A) field line constraints: motions, geometry, etc., consistent with the SS model, and B) the utilization of coronal field models to calculate solar wind properties:

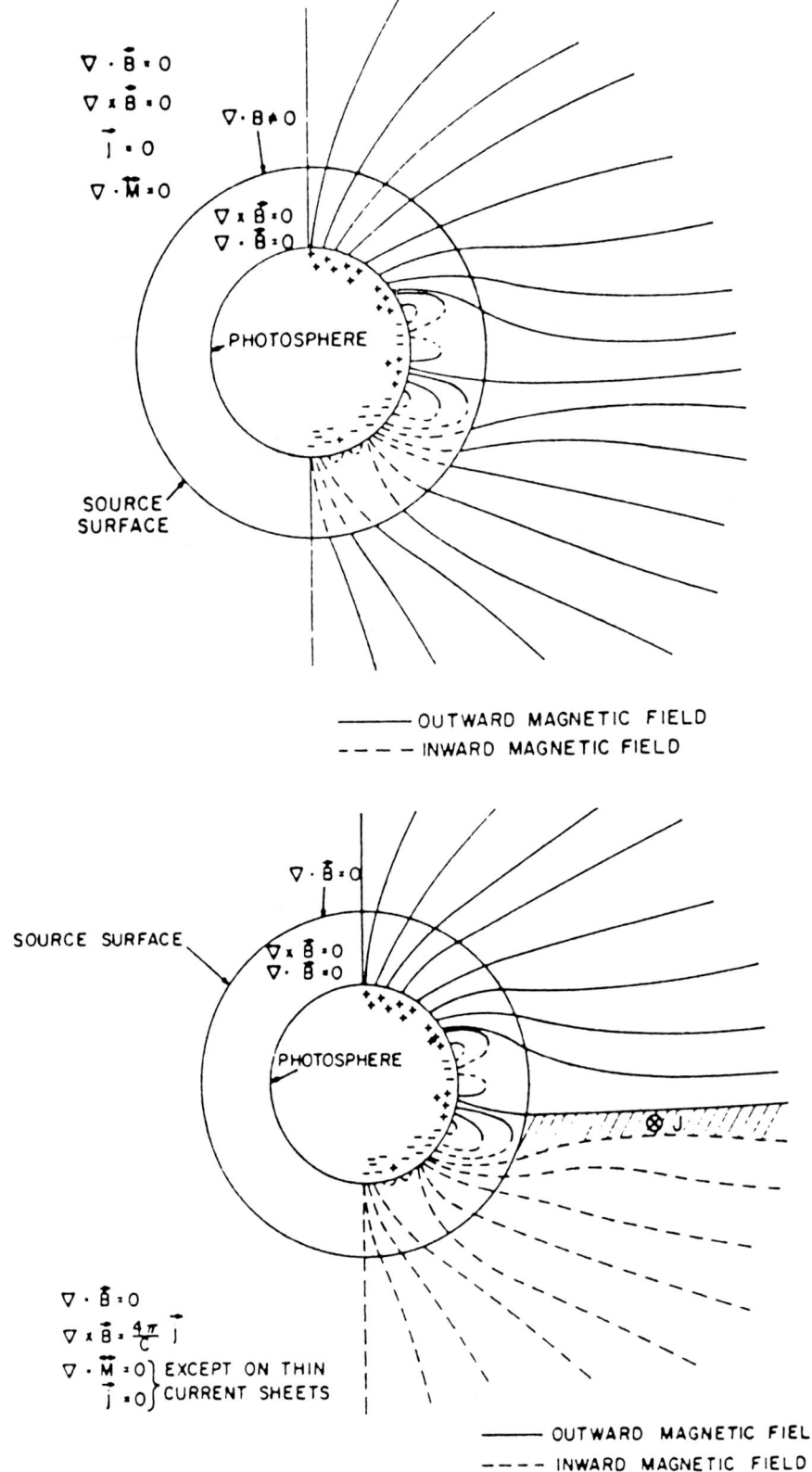

Figure 2. (top) Shown is step 2 in the current sheet model, where a potential field is computed up to the source surface, and then the field is reoriented so as to be always outward on this surface. The 3 field components are then refit so as to include a large monopole term. In step 5 (bottom), the field lines are reversed and this removes the monopole, but introduces a current sheet. With magnetic forces playing a guiding role in flow direction, the field assumes a constant field magnitude at large radial distance (as for a simple monopole).

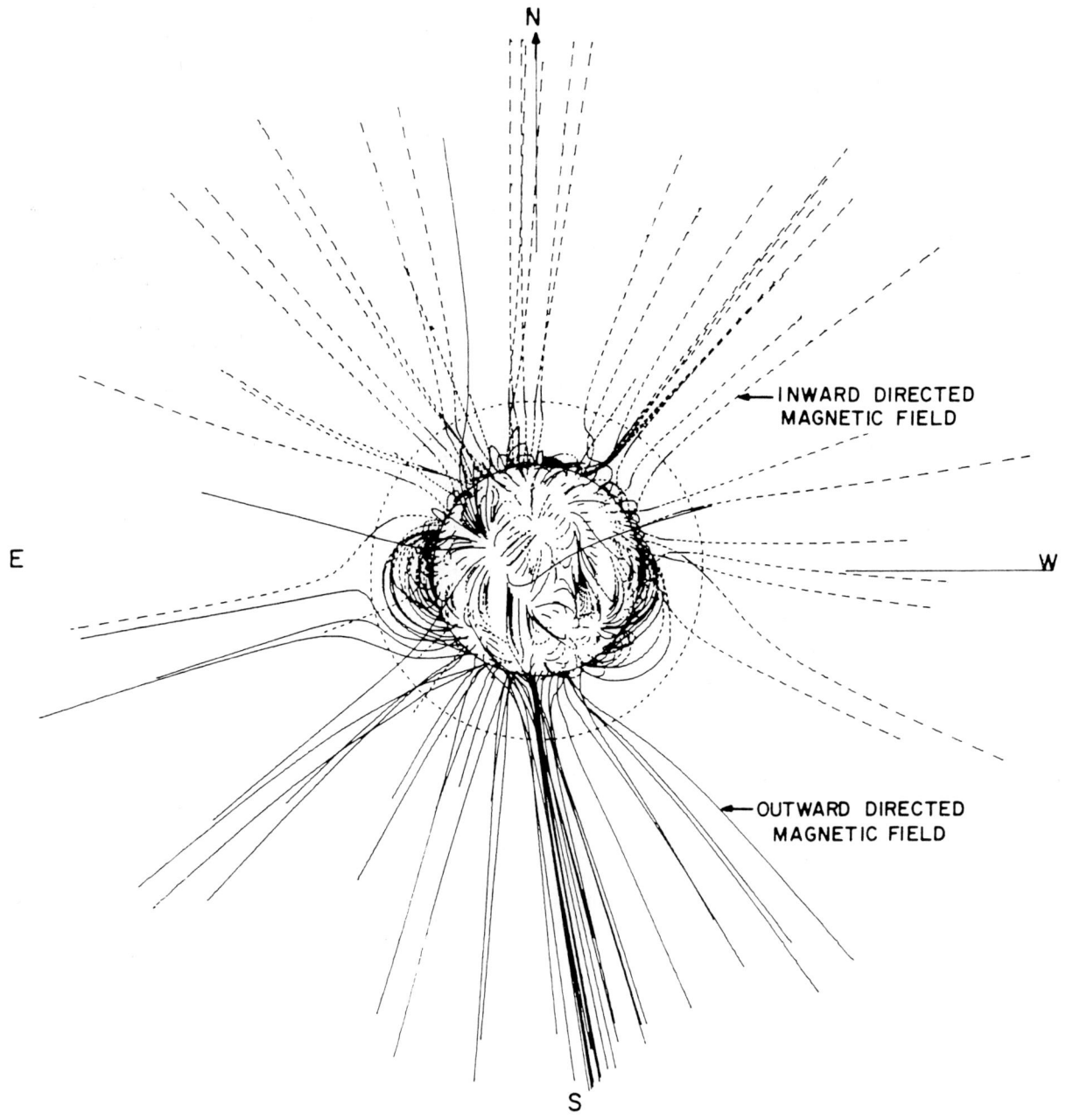

Figure 3. Shown is the computed field lines on top of structure of the eclipse of Nov. 12, 1966. Many features were said to line up well. Near the left side of the figure, non-radial streamers line up quite well with this model, and a poor fit would be obtained with SS modeling.

densities, velocities, etc. For improvements to the SS model, we refer the reader to the research and review information in Hoeksema, et al. (1982, 1994, 1996). Schulz developed a "non-spherical source surface," whereas a number of people, Hoeksema and colleagues, as well as Sheeley and Wang have improved our understanding of the coronal field, plus how to best fit photospheric boundary conditions. It seems that rather than a fit to the Line of Sight (LOS) field performed in the early days, a fit to an assumed radial field works better, as the photospheric plasma seems to bunch the field into tight radial bundles which then expand into the corona. Let us now discuss two recent advances in coronal fields above the photosphere.

4.1 Field Line Constraints –Fisk, Schwadron and Zurbuchen "Magneto-Motion" (FSZ-MM)

As discussed in the introduction, constraints associated with field lines can be used to ascertain field line behavior. Such aspects as their structure and motion are governed, in highly conducting fluids dominated by fluid dynamics, by the fluid motions. In this view, one uses the concept of "frozen field lines" developed by Alfven. Because of the high conductivity of the plasma, in steady state, the field tends to follow a "streamline" of the flow field. Parker utilized this concept to deduce the "garden hose spiral angle." Although this seems quite accurate, it does not now appear to be 100% correct under all conditions. The assumption of a steady-state photospheric boundary condition is a little simplified. Rather, we may want to consider general field line motions or a slightly more specific case of field lines "tied" to certain regions (e.g. photospheric regions and their "footpoint motions," the simplest being differential rotation).

As discussed earlier, timelines or streaklines seem better descriptors for field lines than streamlines. For the cases we are interested in, we also have a slight variation to the traditional streakline, where a line is tied to a fixed point, in that the field line may be frozen to a point on the photosphere; a location which itself is moving. One may ignore this for any particular field line by some change of coordinates, but not for all field lines. Thus a better description of an interplanetary field line might be a "moving streakline;" a pattern akin to that a skywriting plane might generate, where the wind velocity will affect the smoke pattern, but so too will the plane's gyrations (the source of the smoke). This, of course, enables the skywriting, since wind patterns aren't clever enough to spell without some help.

Let us examine how "footpoint motions" can affect a field pattern. Consider on the X-axis a single vertical fluxtube or field line emanating from a specific X location. It now moves in a left right pattern on the X-axis with a constant speed. Additionally, a purely vertical flow field extends the field upward in the Y-direction. The field geometry embedded on the conducting fluid would now form a zigzag pattern, despite no fluid motion of the outflowing gases in the X direction. This illustrates the differences between streaklines (the field paths) and trajectories (the fluid paths). In the same way, a non-stationary photospheric field can affect its associated streakline, which forms the IMF field pattern. Thus the normal Archimedian garden hose picture of Parker's requires some modification if the photospheric footpoint is considered non-stationary. In the Parker garden-hose picture, the fluid trajectory (in the stationary reference frame of the Sun) is an Archimedian spiral and the field streakline is identical.

For a changing photospheric field pattern, the motion of field lines needs to be considered in light of both footpoint motions in the photosphere and coronal field geometry. Time varying coronal field components can arise (in the absence of new flux originating below the photosphere) not from the average solar rotation, but from new field line motions associated with a distortion of pre-existing field line geometry due to solar differential rotation. This distortion is needed to simultaneously tie the fields to photospheric footpoints while undergoing continual (as well as sometimes discontinuous) reconfigurations necessary for the fields to maintain a "force free" geometry in the inner corona. Although the solar differential rotation may not seem much different from a rigid 27-day rotation, the differential rotation coupled with the frozen-field condition can cause an initial current-free potential field geometry to require currents and thus move into a higher energy state from which it must relax. The higher energy state then has the possibility for reconnection, and energy dissipation, either continuously or discontinuously. These geometries require different line tying geometries, and thus the differential rotation, however small, can be a primary driver (in the absence of new fields from below the photosphere) of reconnections, coronal activity, helicity, and CMEs (coronal mass ejections).

Following the work of Fisk et al. (1998), we can envision the field line motions by examining Figure 4. The coronal field has been observed to rotate rigidly (see Wang and Sheeley, 1994), hence FSZ consider, for discussion, the solar field to be approximated by a rigidly rotating dipole. We will discuss later, why the rigid coronal rotation is a reasonable approximation, even in the presence of solar differential rotation. Thus consider a reference frame rotating with the main solar dipole rotation rate, of roughly 27 days synodic. The main dipole field pattern (the sector structure) would be stationary in this frame, however, field lines could move within this pattern. Fisk, Schwadron and Zurbuchen (FSZ) (1998) and accompanying papers (Zurbuchen, et al., 1998, Schwadron, 1998) suggest the following kind of complex field rotations or motions, which we shall call FSZ "Magneto-Motion" (FSZMM).

In the absence of both solar differential rotation (SDR) and solar activity variations (SAV), all the field lines

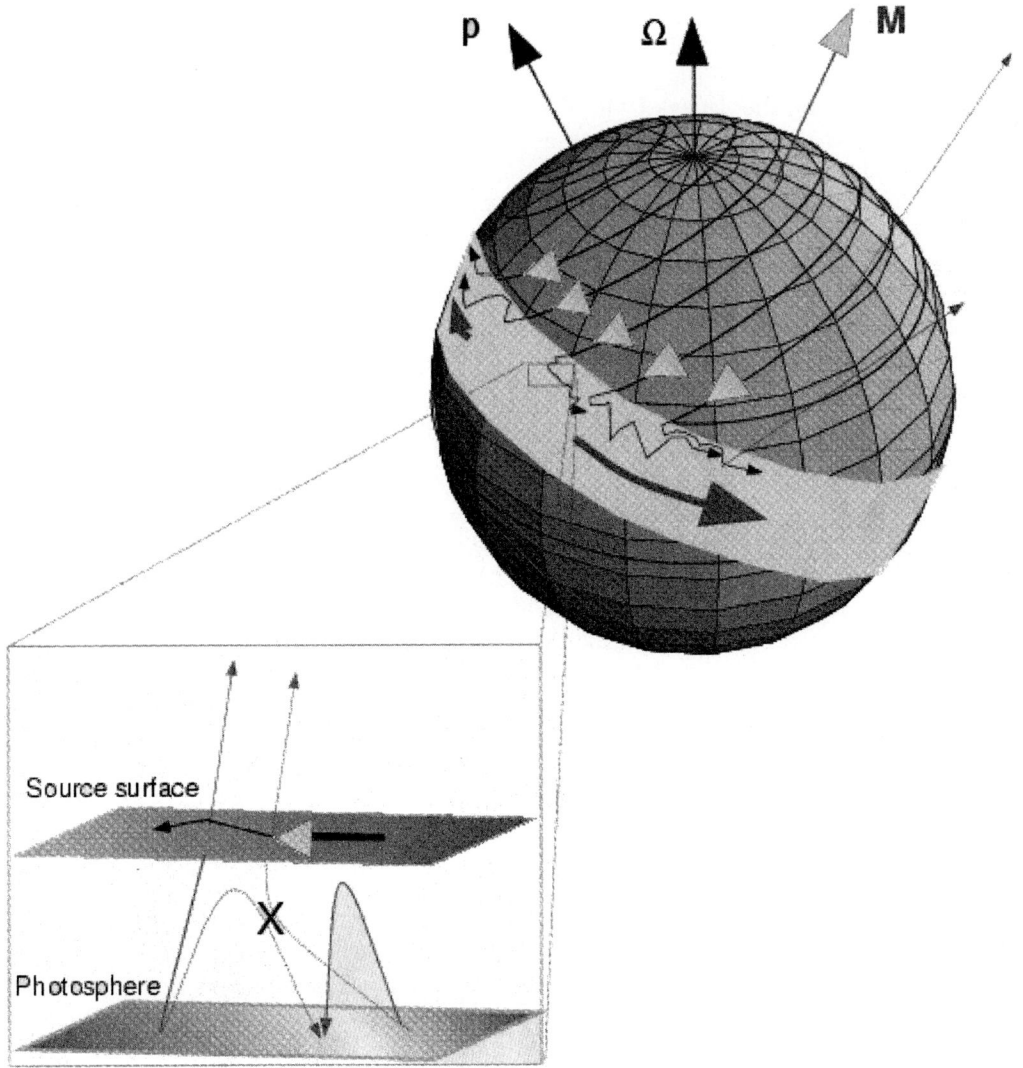

Figure 4. Illustrates the motion of magnetic field footpoints on the solar wind source surface. The Sun's dipole axis, M, and the rotation axis, Ω, are offset by 10 degrees here. Due to the non-radial expansion of the solar wind beneath the source surface, a field line rooted in the photosphere's pole will penetrate the source surface at the axis labeled p, offset from the rotation axis by 30 degrees, and hence the dipole axis by 40 degrees. There are three types of field lines: open, closed and those which undergo reconnection. The motion of the open field lines is shown, their footpoints move about the p axis in a retrograde sense indicated by the arrows. In the equatorial regions, the closed and reconnecting field lines move in a more complicated pattern.

would be fixed in this reference frame Since the SAV effects can cause large changes as new field lines may totally reconfigure the field geometry, we shall discuss only the SDR effects. For SDR effects, the particular field line passing through the axis of the Sun is roughly stationary. The high latitude photospheric fields would rotate backwards around this "skewed field line axis" in the reference frame. At lower latitudes, field lines could move between closed and open configurations, or just stay on closed structures. FSZ discuss both prograde and retrograde motions of the field lines needed to preserve a divergence-free field. The field lines undergo FSZMM, as they nutate up and down so as to preserve a divergence free field and the overall "sector" pattern of low multipole number on the source surface. Thus although the field lines rotate in the photosphere at a particular fixed heliographic latitude, due to differential rotation, the streaklines would move in heliomagnetic latitude; hence the lines would sometimes find themselves at "low" and other times at "high" heliomagnetic latitudes. In the streakline analogy presented earlier, we can envisage a skywriting plane in a storm, where a small motion of the plane results in a large effect on the smoke pattern.

Under the FSZ description, there are three classes of field lines (see Fig. 4): 1) open; 2) closed; and 3) "reconnecting field lines," which alternate between these two states as the Sun differentially rotates. The open field lines may be easiest to understand. They rotate about the skewed field line axis, wobbling up and down somewhat. The closed field lines are next. If the two footpoints rotate at the same rate, they just rotate about the Sun, with little change, however, more generally, they would be twisted in a force-free configuration, as the demands of their higher arcade components change. For case 3, the reconnecting field lines, let us consider an open member, which we may refer to as a "high" heliomagnetic member. As time passes, the field line would rotate as any other open field line, around the skewed field line axis, but progressing from a high to a low heliomagnetic state. Hence, as it progressed to lower magnetic latitudes, its fate would draw it close to a closed arcade, and then the field line would reconnect with an oppositely oriented field line. The two would then move as the closed field lines, deeper into the arcade, until its motion again drew it open.

In interplanetary space, the solar wind IMF field pattern would "rotate" more or less rigidly following the HCS or SS pattern, however, individual field line motions can move with respect to this pattern. This can affect the geometry of the field lines. They would undergo a separate motion within the sector pattern, the FSZMM. This is a kind of rotation or "arm-waving" pattern which has significant consequences in solar wind flow, as well as cosmic ray motions, since both inner coronal flow (below 20 solar radii) and the high energy particles too are guided by the field. In any case, the field line motion is a good area to investigate CME origins - even in the absence of new photospheric field activity.

Thus as we have seen recently, one observes some remarkable CME events even near solar minimum. These may not be explained by invoking new photospheric flux, but rather some events may be explicable from the demands of coronal field line reconnections. Investigations initiated by Fisk, Schwadron and Zurbuchen into the motion of field lines are a fruitful area for future research since this work has many unexpected implications in the areas of solar wind flow, CMEs, solar activity and cosmic rays.

4.2 Solar Wind Properties - Wang and Sheeley Flow Analyses

Wang and Sheeley (W & S) (1994, 1995, 1996, 1997) in a number of papers, have made significant advances in the understanding of solar wind fluid dynamics. They have been able to calculate fairly accurate solar wind properties (in the absence of large dynamic solar events) from the properties of the coronal magnetic field. Additionally, they have modeled the rotation of photospheric magnetic fields, providing some understanding for how coronal field rigid rotation can occur despite photospheric differential rotation.

There is not enough space to outline all W & S's excellent work; however, I will touch upon a couple of aspects. They have utilized the SS and HCS models described above as starting conditions. From this, they have calculated the "expansion factors" associated with the non-radial geometry of the field lines. Using this geometry and the relationship between the expansion factor and solar wind speed, they have successfully calculated numerous properties of the solar wind by including "wind-stream interactions." They have made numerous "stackplots" of solar wind properties showing surprising agreement with coronal fields obtained by Hoeksema from the Stanford Wilcox Solar Observatory (WSO). Figure 5 shows one such stackplot illustrating both the SS and HCS models. This simple stackplot shows only the sector structure: a) observed, b) CS - current sheet, and c) PFSS - potential field source surface. One can only say that there is general agreement. By examining high latitude plots from Ulysses, these authors, as well as Balogh, et al. (1992), have shown that the high latitude field obeys a pattern of nearly uniform field vs. latitude, which they refer to as "split monopole," consistent with the HCS model, but not the SS model.

They also discuss (Wang and Sheeley, 1991) the behavior of field structures on the Sun, and its coronal field implications. They discuss effects associated with differential rotation, annihilation of fields by diffusion, and meridional field transport. Their modeling is highly informative, yielding much insight into the origin of coronal field patterns from the underlying photospheric field configuration. Their insights contain much information on a variety of solar phenomena. For example, one may be able to understand the rigid coronal rotation (which is important to the Fisk et al. model) simply as follows. If the coronal field contains, as described earlier, primarily the lowest order solar multipole moments (e.g. the dipole moment), then its rotation is limited by these moments. Consider the field to contain only an offset dipole moment. In this case, it must rotate rigidly. The reason is that a dipole is a fixed shape or structure, hence must rotate as a rigid or solid object. Hence the pattern is forced to rotate rigidly (although its rotation rate could vary with time). In order to achieve significant differential rotation, as the photosphere does, the "pattern" of the object must contain many multipole moments (as is present in the photosphere, but not the corona).

5. CONCLUSIONS AND DISCUSSION

We now provide our impression of the physics of field lines in the heliosphere, focusing on the corona, and outline some ideas for future work. Let us first consider the magnetic field in the inner corona. A short distance above the chromosphere, the magnetic field from small-scale regions on the Sun (pores, spots, dark lanes, the "magnetic carpet", etc.) open up in the "canopy" layer to transform the photospheric field tangle into a more uniform corona field. This occurs because the density structure is sufficiently depleted to allow the field to spread out uninhibited by the coronal gases. Magnetic fields being great transporters of stress, which is nearly all they do, transmit the high convection zone stresses into the corona. With little plasma pressure to balance the stresses, the field stresses from different regions must themselves provide balance. Hence a "force-free" field configuration results. Taking the field to be force-free, we may simply think of this field in a potential sense, although field aligned currents do exist. The field-aligned currents are responsible for the magnetic motions (FSZMM) of individual field lines. Namely, despite the reasonable quality of the SS and HCS potential models, the field line motions go beyond these strict models, by moving about within some-such model. The individual field lines are forced from below by the demands of their photospheric footpoints, so that no single field line stays forever at the same place within the rigid coronal pattern, but field lines are continually opening and closing within the pattern.

Moving higher up in the corona, we now consider the solar wind - field interactions. As the plasma density decreases with altitude, various mechanisms (e.g. nanoflares, etc.) energize the plasma through accelerating and heating processes. When this occurs on a set of closed arcades, the inward magnetic stress is sufficient to contain the outward moving gases emanating from both sides of a coronal loop. If, however, the magnetic field has weakened to the point where the magnetic stresses are insufficient to contain the plasma (roughly where the transverse (non-radial) field energy density falls below the plasma energy density (see Schatten, 1972)), the plasma will "open" the field to form an escaping wind; this is the location of the Source Surface radius.

The Source Surface radius (near 1 solar radius) is significantly closer to the Sun than the traditional Alfven point (about 15 solar radii). The reason is as follows. The Alfven point is simply governed by where the Alfven velocity falls below the fluid speed. The SS radius is the location of the highest closed arches. Here one is only interested in the ability of the field to contain the escaping gases. Hence the radial component of the field plays no role, and only the transverse component of the field inhibits the gas from escaping. One can think of the pressure of the gas needing to exceed the field pressure at the top of a coronal arcade. The coronal arcade transverse field weakens as a high order multipole moment above the active region field, whereas the radial component of the field weakens only as $1/r^2$, hence the highest closed arches are considerably closer to the Sun than the Alfven point.

Continuing our saga of the coronal escaping gases, even though the fluid has most of its energy directed outwards, the magnetic field can still dominate the transverse (non-radial) pressure. Hence, the field still may retain a "force-free" field geometry and guide the flow into an open configuration. The remarkably different motions between ions and electrons allows the coronal plasma to form thin current sheets and allows the coronal field to "open" on a relatively small spatial scale.

Field lines then continue to guide the flow into a "split monopole" structure as the field pressure "equalizes" to retain a relatively uniform pressure independent of direction (latitude and longitude). The field geometry controls the manner in which the escaping gases accelerate. Plasma located near the central regions of strong-field , large coronal holes (e.g. the polar regions near solar minimum) find a high expansion ratio as the magnetic field opens widely, allowing a rapid expansion and high velocity wind to form. Other regions pressed from two sides, e.g. near coronal hole boundaries, find small divergence, or even converging field lines, and form slow solar wind.

Further from the Sun, in interplanetary space, the weakened field pressure (varying as B^2, hence as r^{-4}) is now dominated by the fluid motions. The outflowing gases interact in stream-stream interactions. The field, following a streakline, much as a die or marker in a stream, meanders to every turbulent whim of the flow patterns. The streakline flow occurs in the inner corona as well, where the field pressure dominates, however there, the stiffness of the field guides the flow velocity, whereas in interplanetary space the field stresses are unimportant, and the field serves only as a "marker."

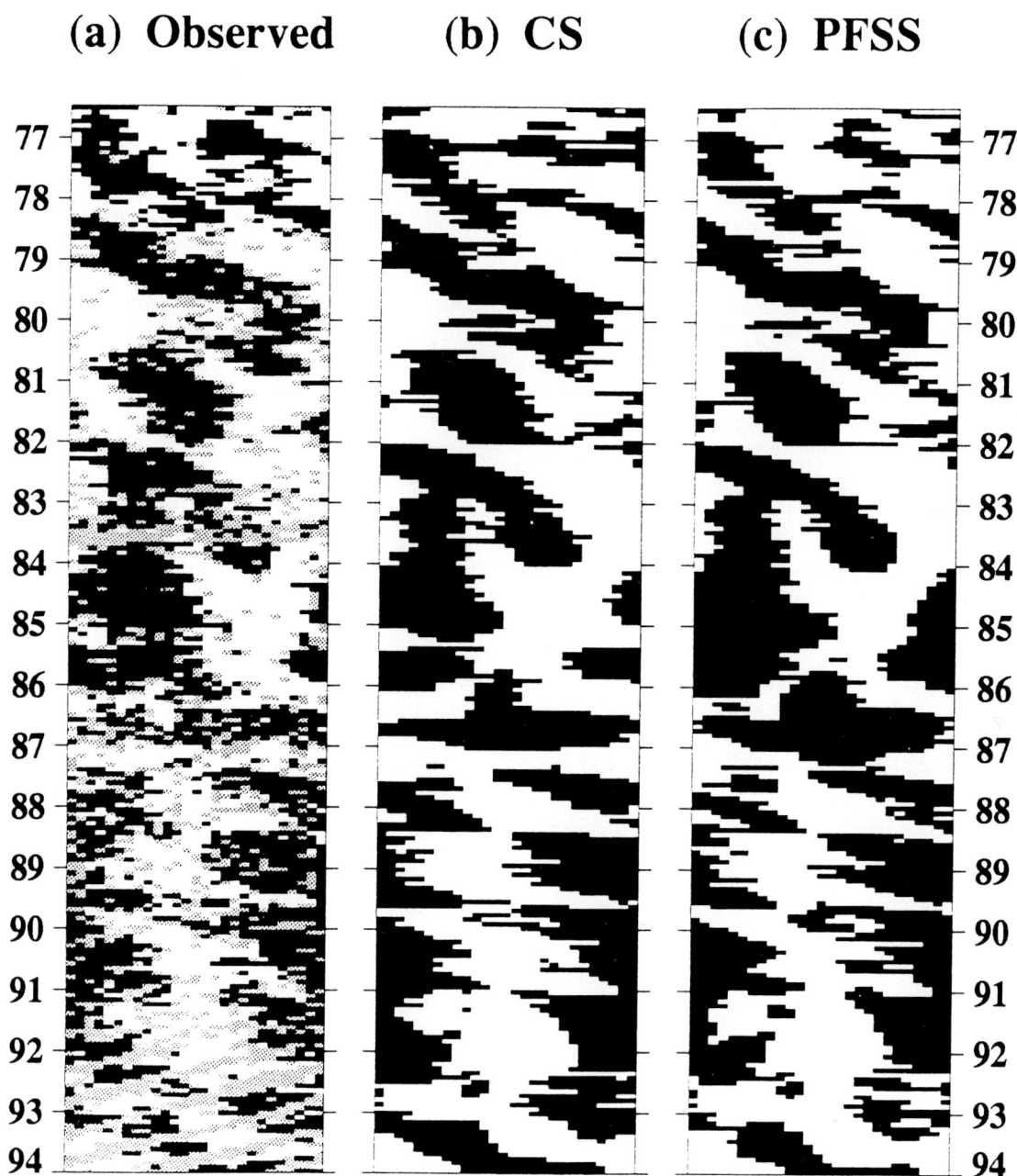

Figure 5. Variation of the interplanetary sector polarity at Earth from 1976-1993(after Wang and Sheeley, 1995). Daily averaged polarities are plotted in a 27 day row (Bartels format), with white (black) denoting outward (inward) pointing magnetic flux. (a) is observed polarities from a variety of spacecraft; (b) is calculated from the current sheet (CS) model; and (c) is using the potential-field source-surface (PFSS) model. The polarities between (b) and (c) are shifted by 4 days for the Sun-Earth transit time. As can be seen, at low heliographic latitude, both the CS and PFSS models do fairly well with the observed IMF sector structure.

Thus, we have seen the changing role of the magnetic field in the heliosphere. In the inner corona, the field totally dominates the structure, both containing the plasma and organizing its motions. Further out, the field no longer contains the fluid, but directs its motions. Still further out, in interplanetary space, the field lines only move with the fluid, much as passengers on a speeding train.

These views suggest some alternative methods to model coronal field lines, solar wind flows, and even look for CMEs. Basically, what we suggest is essentially a combined HCS modeling effort, together with the FSZMM and the W & S flow calculations. Nevertheless, it is a bit tricky (since with all the acronyms, it looks like alphabet soup), so let us outline the following methodology. It will be essentially to first calculate the field pattern, then enhance this with the field line motions, and then add the fluid motions along the field (out to the Alfven point), and then 3-D fluid dynamics beyond.

To understand this process, let us first ignore rotation. One starts with a photospheric field pattern; from this, ignoring rotation, one calculates the static coronal field from the HCS model. Near the Alfven point (about 10-20 solar radii), where the field lines lose their rigidity, one stops calculating the field pattern with this procedure. Now, one calculates the velocity vectors of the coronal plasma & solar wind flow from the diverging field structures just calculated, a la W & S. Within the Alfven point, one assumes 1-D flow motion along the field line. Beyond this, one allows the flows to interact in a 3-D fashion (stream-stream interactions, etc.). This provides a simple "snapshot" of the solar wind structure.

Next one adds the effects of rotation, both rigid and differential, a la FSZ (actually starting over again, as it all must be done self-consistently). To understand the process, first think about a totally rigid rotation, and then consider the reference frame for each field line at each location, separately. For the rigid case, it is required to calculate the flows and fields in the rigid rotating reference frame a la Parker's garden hose angle. Now, in this reference frame, for stationary conditions, the field lines would be governed by some coronal field model (e.g. SS or HCS), and in the prograde rigidly rotating reference frame, the field lines, owing to the stretching of the field by the outward motion of the plasma, in the local plasma frame, would be forced to flow retrograde (as in the spiral angle).

For the differentially rotating case, one needs to calculate the fields and flows in locally rotating reference frames (since the physics of particles, fields, etc. is always local). These are the frames in which the field lines are rotating. Note that close to the Sun, these prograde reference frames will be close to the photospheric differential rotation rate. Further out from the Sun, the rotation rate of the reference frames decreases, and in the solar wind, it eventually becomes retrograde, owing to the plasma motions! The field lines then need to be calculated in a self-consistent fashion (with both the global physics of their structure - from say HCS models, and the detailed pattern movements as the field lines move within the global field pattern). Field line rotations have been discussed by Fisk et al.; they approximate the rotation rates with a simple illustrative geometry. In a more exact model, rotation will vary from field line to field line associated with the Sun's differential rotation, and will decrease along the field line as well, etc.

Continuing with our model, having finished with a description for the field lines (pattern and motions), one then follows the fields within the Alfven point, and calculates fluid flows along them as beads on a wire (with a fluid energy or momentum equation). After the Alfven point, one simply lets the flows take over completely in 3-D, with their directions being affected by both the field pressure effects and their rotational motion effects.

Thus, the winds and fields are obtained consistent with their photospheric footpoint motions. For example, one allows the footpoints to differentially rotate, and follows the field line motions. These movements then affect the fluid motions. This will introduce added velocity fields. Additionally, field lines will become overwound and underwound, depending upon location. Overwound fields may need to "open up," yielding a CME, even in the absence of new photospheric flux. Hence there is much room for new understandings of interplanetary phenomena as a result of this paradigm, as opposed to the static pictures portrayed in the bare SS and HCS models. As an added note, it is interesting that one would not only expect in the FSZMM model to see field lines continuously "opening up," as seems evident in the dramatic pictures from SOHO, but additionally, one would expect a roughly equal number to close! These, however, would be more likely dark, having lost a lot of plasma, and hence might be a challenge to observe, but it is consistent with the model. With the field line motions envisaged, one would not expect field lines to continuously open, without a roughly equal number closing. Both processes could occur near the top of helmet streamers/coronal arcades, however, their longitudes would vary depend upon solar conditions – the relationship between large-scale and small-scale fields.

Acknowledgments. I would like to acknowledge the guidance of John Wilcox in his search for understanding large-scale solar structures, which led to the development of many of the ideas discussed in this paper. Additionally, Norman Ness, with his pioneering magnetometer observations of the interplanetary medium, along with his keen interest in scientific progress, played a pivotal role in the early development of these models. Further, I am appreciative of discussions with many solar/space scientists over the years, including Len Fisk, Jack Harvey, Karen Harvey, Todd Hoeksema, Jon Linker, Zoran Mikic, Gene Parker, Phil Scherrer, Nathan Schwadron, Mike Schulz, Neil Sheeley, Yi-Ming Wang, Xuepu Zhao, Thomas Zurbuchen, and others who have worked on large scale solar magnetism. This work was supported by NASA RTOP 344-12-53-78.

REFERENCES

Altschuler, M. D. and Newkirk, G. Jr., Magnetic fields and the structure of the solar corona I: Methods of calculating coronal magnetic fields, *Solar Phys., 9, 131-149, 1969.*

Balogh, A. et al, *Astron.& Astrophys. S., 92, 221, 1992*

Bumba, V. and Howard, R. F., *Astrophys. J., 141, 1502, 1965.*

Cowling T. G., Halley Lecture, "The Solar Wind", *The Observatory Vol. 89, No. 973, Dec. 1969, pp217-224.*

Fisk, L. A., N. A. Schwadron, and T. H. Zurbuchen, On the slow solar wind, submitted, *Astrophys. J., 1998.*

Hess, S. L., Introduction to theoretical meteorology, Holt, Rinehart and Winston, *p 201, 1959.*

Hoeksema, J. T., Wilcox, J. M., and Scherrer, P. H., Structure of the heliospheric current sheet in the early portion of sunspot cycle 21, *J.G. Res., 87, 10331, 1982.*

Hoeksema, J. T., Zhao, X. P., and Scherrer, P. H., Prediction of coronal and heliospheric magnetic fields: The promise of SOI-MDI on SOHO, p 76, *Solar Wind 8 Conference, AIP Conference Proc. 382, 1996.*

Hoeksema, J. T., and Zhao, X. P., Prediction of coronal and interplanetary magnetic fields, *Proc. 8th Int. Sym. Solar-Terr. Phys., 19, 1994.*

Landau, L. D. and Lifshitz, E. M., Fluid Mechanics, *Pergamon Press, pp. 10, 386, 1959.*

Linker, J. A., private comm. *1998.*

Mikic, Z. and Linker, J. A., The large scale structure of the solar corona and inner heliosphere, in *Solar Wind 8, AIP Press, p 105, 1996.*

Pneuman, G. W. and Kopp, R. A., Gas-magnetic field interactions in the solar corona, *Solar Phys. 18, pp. 258-270, 1971.*

Schatten, K. H., A model of interplanetary and coronal magnetic fields, *Ph. D. Thesis, UC Berkeley, 1968.*

Schatten, K. H., Prediction of coronal and interplanetary magnetic fields, *Solar Activity Observations and Predictions,* ed. McIntosh and Dryer, *MIT press, p 179, 1972.*

Schatten, K.H., Current sheet magnetic model for the solar corona, *Cosmic Electrodynamics, 11, 4, 1971,* also in *Asilomar Conference on the Solar Wind, 1971.*

Schatten, K. H., Wilcox, J. M. and N. F. Ness, A model of interplanetary and coronal magnetic fields, *Solar Physics, 6, 442, 1969.*

Schwadron, N. A., A model for pickup ion transport in the heliosphere in the limit of uniform hemispheric distributions, *J. Geophys. Res.,* in press, *1998.*

Stewart, G. A. and Bravo, S., Self consistent MHD modeling of the solar wind from polar coronal holes, in *Solar Wind 8, AIP Press, p 145, 1996.*

Suess, S. T. et al., Volumetric heating in coronal streamers, *J. Geophys. Res., 101, 19957, 1996.*

Wang, A. H., et al., Global model of the corona with heat and momentum addition, *J. Geophys. Res. 103, 1913-1922, 1998.*

Wang, Y.- M., and N. R. Sheeley, Jr., The rotation of photospheric magnetic fields, *Ap. J, 447, 399-412, 1994.*

Wang, Y.- M., and N. R. Sheeley, Jr., Solar implications of Ulysses interplanetary field measurements, *Ap. J, 447, L143-6, 1995.*

Wang, Y.- M., S. H. Hawley, and N. R. Sheeley, Jr., The magnetic nature of coronal holes, *Science, V 271, 464-9, 1996.*

Wang, Y.- M., and N. R. Sheeley, Jr., The high-latitude solar wind near sunspot maximum, *Geophys. Res. Lett., V 24, 3141-4, 1997.*

White, F. M., *Fluid Mechanics,* Third Edition, *McGraw-Hill, 1994.*

Wilcox, J. M. and Ness, N. F., Quasi-stationary corotation structure in the interplanetary medium, *J.Geophys.Res., 70, 5793-5805, 1965.*

Wilcox, J. M. and Ness, N. F., Solar source of the interplanetary sector structure, *Solar Phys., 1, 437-445 1967.*

Zurbuchen, T. H., N. A. Schwadron, and L. A. Fisk, Direct experimental evidence for a magnetic field with large excursions in latitude, *J. Geophys. Res.,* in press, *1998.*

Code 926, Laboratory for Terrestrial Physics, Goddard Space Flight Center/NASA, Greenbelt, MD 20771

Currently at: AI Solutions, Suite 215, 10001 Derekwood Lane, Lanham, MD 20706

A Multi-Spacecraft Study of Solar Wind Structure at 1 AU

K. I. Paularena, J. D. Richardson, and F. Dashevskiy

Center for Space Research, Massachusetts Institute of Technology, Cambridge

G. N. Zastenker and P. A. Dalin

Space Research Institute, Russian Academy of Sciences, Moscow

Studies of the correlations between solar wind parameters observed at different spacecraft have two complementary foci. The recent interest in space weather prediction and the launch of ACE as a real-time solar wind monitor have precipitated studies on how well such monitors actually predict solar wind conditions at Earth. Studies of magnetospheric response to solar wind activity generally assume 100% correlation between the solar wind observed at the monitor, wherever it may be, and the solar wind affecting the Earth environment. The second focus is on the intrinsic scale lengths of plasma variation in the solar wind. These lengths may provide information on the scale sizes of solar wind source regions and/or interaction lengths for the propagating solar wind. Also, an examination of the lags which give the best correlations shows that average front normals are neither radial nor perpendicular to the magnetic field, but are roughly halfway between.

1. INTRODUCTION

In the ISEE-3 era near the 1981 solar maximum, studies of interplanetary magnetic field (IMF) and plasma correlations found that correlations were often good but also often not good. *Crooker et al.* [1982] split ISEE-1 and ISEE-3 64-s averaged magnetic field data into 800 2-hour periods and found, for those periods where lags could be defined, good correlations (correlation coefficient, r, > 0.8) 25% of the time and poor correlations (r < 0.5) 25% of the time. Average correlations were higher when the magnetic field variance was high and when the spacecraft separation was less than 90 R_E in the plane perpendicular to the Earth-Sun line. *King* [1986] compared ISEE 3 and IMP 8 magnetic field and plasma data from L1 and near Earth. He found 1 σ differences between speeds and densities at these two spacecraft of 18 km/s and 30%, respectively (i.e., the probability that the ISEE 3 and IMP 8 speeds were within 18 km/s was 70%) and found that the plasma correlations did not vary with separation perpendicular to the Earth-Sun line. *Paularena et al.* [1998] correlated solar wind flux data from IMP 8, WIND, and INTERBALL-1 for four months in 1995 and 1996. They found average correlations of 0.71 which were independent of spacecraft separation; the best organizer of correlations was the value of flux and the standard deviation of flux. *Richardson et al.* [1998] compared plasma data from ISEE-3 and IMP 8 and found a strong dependence on radial separation of the spacecraft for separations larger than 150 R_E and that high density standard deviation was the best predictor of good correlations. *Collier et al.* [1998] compare WIND and IMP 8 IMF measurements from January to July of 1995 and find that correlation are worse than in the Crooker et al. [1982] study, which they attribute to solar cycle effects. *Richardson and Paularena* [1998] found that plasma fronts were oriented approximately halfway between an alignment perpendicular to the radial direction and one along the IMF direction. In this paper we use recent data from WIND, IMP 8, and INTERBALL near solar minimum and compare the results with ISEE-3 and IMP 8 data obtained near solar maximum (1978-1980).

2. DATA ANALYSIS

All the data sets were despiked and time regressions and bad data periods were removed. As in previous work [*Paularena et al.*, 1998 and *Richardson et al.*, 1998], we use the Pearson linear correlation coefficient for comparison of the data sets. To calculate the correlation coefficients, data from the upstream spacecraft are time-shifted to the downstream spacecraft X distance using the observed speed and the X separation (the advection shift) $t = (X_1 - X_2)/\bar{V}_1$, where X_1 and X_2 are the locations of the spacecraft in X and $\bar{V}_1$ is the average solar wind speed observed by one of the spacecraft of the pair during the chosen period. Because INTERBALL-1 does not routinely produce speed information, the IMP 8 or WIND speeds were used to calculate the advection shift for correlations with INTERBALL-1. For the IMP/WIND and IMP/ISEE correlations, the WIND or ISEE-3 speed was used since both WIND and ISEE-3 were generally upstream (sunward) of IMP 8. Examination of ~2.5 years of IMP 8 and WIND data show that, for 6-hour averages of solar wind speed, differences were greater than 10 km/s only 10% of the time, and greater than 20 km/s only 2% of the time. Similar results are expected for ISEE-3 speed measurements since speeds are the most accurately measured plasma parameter for all instruments.

The data are divided into 6-hour time periods and shifted as a block. This prevents time regressions from occurring if the speed changes abruptly during a time period. Data from the spacecraft with the faster sampling time are then interpolated to the spacecraft with the longer sampling time. Next we performed correlations between the data sets as a function of lag and thus determined the lag which gave the best correlation between data sets. Correlations between all pairs of spacecraft are calculated in the same manner.

3. RESULTS

Figure 1 shows an example where data from IMP 8, INTERBALL, and WIND are available. The positions (X,Y,Z) in GSE are (19,-29,12) for IMP 8, (19,20,11) for INTERBALL, and (77,1,0) for WIND. The top 3

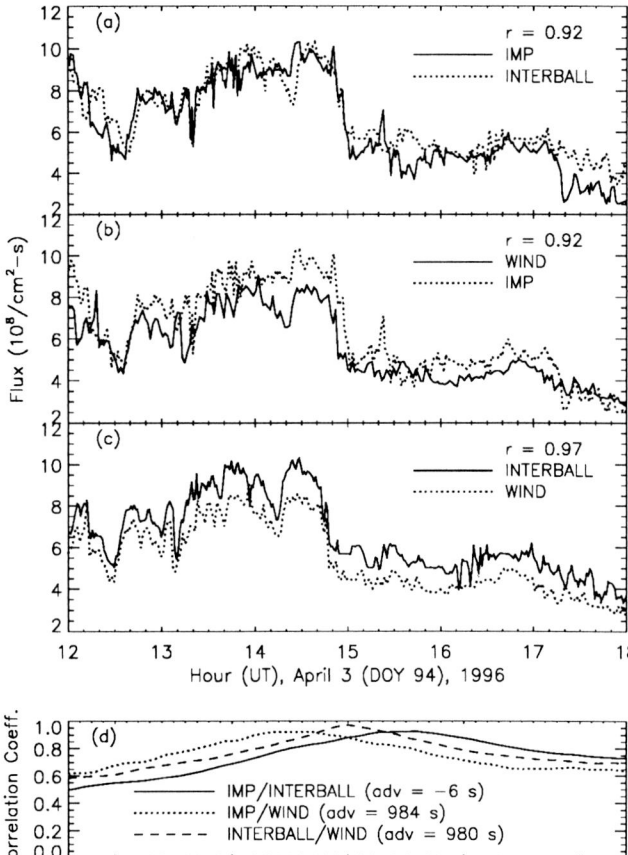

Figure 1. A period where the flux data show a great deal of agreement but the structures appear to be complex in geometry. Data shown are from April 3 (DOY 94) 1996. (a) IMP 8 and INTERBALL-1 data. (b) IMP 8 and WIND data. (c) INTERBALL-1 and WIND data. (d) Correlations as a function of lag for all three pairs.

panels show the data for each spacecraft pair time shifted to give the best correlation (after interpolation) and the bottom panel shows the correlation coefficient as a function of lag. The correlation is generally very good, although differences between plasma observed by different spacecraft are also very apparent, especially after 1500 UT. Although there are obvious flux differences between the WIND and IMP 8 or INTERBALL measurements, these differences are due only to density discrepancies. Because density magnitude does not affect Faraday cup measurements of speed, the speeds are essentially the same (to within less than 1%) when similar solar wind is being observed.

The correlation coefficients in this example are larger than usually found. Figure 2 shows a plot of the speed correlation versus time. The average correlation is

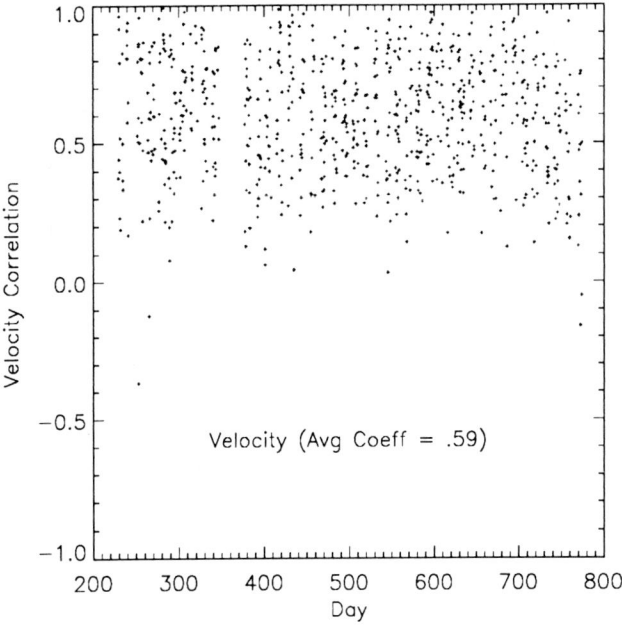

Figure 2. Time series of the speed correlation coefficients showing that the scatter is large and not systematic.

in the correlations for the IMP/INTERBALL/WIND pairs may be due to selection criteria differences; see *Paularena et al.* [1998] for details. The addition of the IMP/WIND comparison gives a much better perspective since a wider range of separations is covered. Correlation coefficients are roughly constant for separations of up to 200 R_E, then start decreasing fairly precipitously. Thus the scale length for solar wind variations in the radial direction seems to be about 200 R_E and spacecraft monitors beyond this distance are less effective (over 6-hour intervals) at predicting solar wind conditions at Earth. Assuming radial solar wind propagation at 400 km/s, 200 R_E of X separation corresponds to a scale length on the Sun of about 6000 km. Variations in the correlation coefficients with Y and Z separations of up to 100 and 30 R_E, respectively (the limitations of the data sets used), are small.

Figure 5 shows the percentage of 6-hour periods having flux correlation coefficients falling within bins 0.05 in width. The IMP/INTERBALL/WIND data set shows many more good (> 0.8) and many fewer poor (< 0.5) correlations than the IMP/ISEE and IMP/WIND data sets. This may be due to the overlapping nature of the three pair correlation results – when two good correlations exist, a third must also exist. It may also be due to the stringent selection criteria used, as mentioned above. Although the percentage of good correlations from the IMP/ISEE and IMP/WIND results are similar to those seen in the IMF correlation results

about 0.6. The correlation coefficients show no systematic variation with time and little consistency from 6-hour period to 6-hour period. These low correlations are contrary to a general community sense that solar wind is very similar from spacecraft to spacecraft. We believe this perception arose because most comparisons have been done over longer time scales. Figure 3 shows the correlation coefficient as a function of the length of time period over which the data are binned. For time periods of more than 30 days, both the speed and density correlation coefficients are greater than 0.9. As the lengths of the time periods decrease, the correlation coefficients decrease to about 0.5 for 2-hour time periods. Thus, although features on the scale of a solar rotation are well-correlated, on shorter scales the solar wind varies significantly over distances comparable to the spacecraft separations studied here.

Figure 4 shows the flux correlation coefficients as a function of spacecraft X separation. The data are divided into 10 R_E bins depending on the X separation of the spacecraft. Average correlation coefficients and 1 σ errors of the mean are shown for the IMP/INTERBALL/WIND data (open squares), the IMP/WIND data (grey squares) and the IMP/ISEE data (solid squares). Within the errors, the data from both spacecraft pairs are consistent, indicating that the solar wind scale length has not changed significantly from solar maximum to solar minimum. Differences

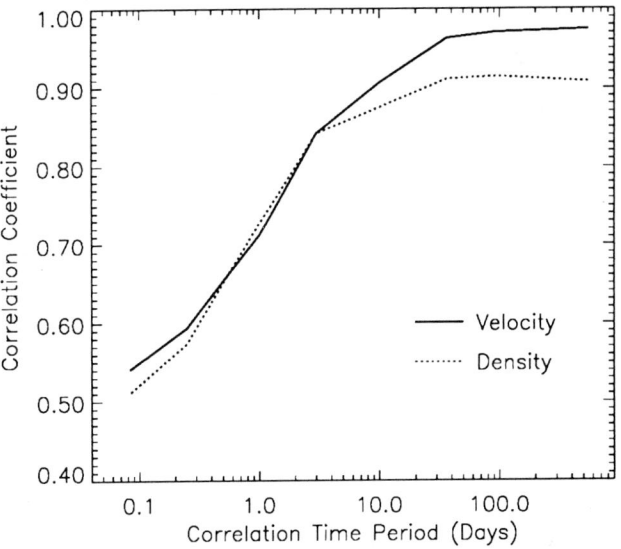

Figure 3. The correlation coefficients between IMP 8 and ISEE 3 speeds and densities as a function of the time period of the correlation.

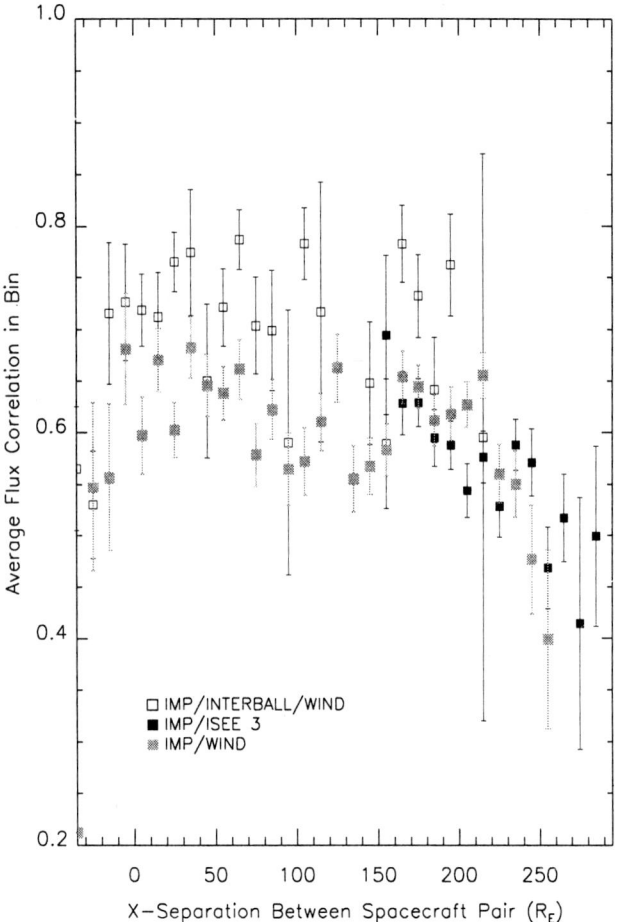

Figure 4. The correlation coefficients as a function of spacecraft separation along the X direction and 1σ standard errors.

of *Crooker et al.* [1982] (approximately 25% of the 2-hour periods had correlations less than 0.5 and approximately 25% had correlations greater than 0.8), there are more periods with poor correlations.

We investigated the variation of the correlation coefficients with solar wind parameters. The correlation coefficients did not vary with solar wind speed and varied at best weakly with velocity standard deviation. The correlations did vary with both density and flux and the standard deviations of density and flux. (Since the correlation coefficients are not a function of velocity, the variation with flux must be due to the density). The left column of Figure 6 shows the sharp increase, from ~ 0.4 to ~ 0.9, of the correlation coefficients as the flux standard deviation increases; the data are split into high and low density subsets for comparison. The right column of Figure 6 shows the results of splitting the density data into high and low standard deviation subsets and examining the correlations as a function of density. The results show clearly that the correlation coefficients depend on the standard deviation of density, not the density itself. However, for densities greater than $10/cm^3$, correlation coefficients are $\sim 50\%$ higher than they are for densities less than $4/cm^3$.

The additional lags, after advection shifting, between arrival of the same plasma (as determined by the time shift which gives the best correlation) provide information on the orientation of the plasma front [*Richardson and Paularena*, 1998]. We define an equivalent X separation (X_{sep}) as the lag times the solar wind speed. Figure 7 shows X_{sep} as a function of the Y separation of the spacecraft for time intervals with correlation coefficients > 0.8 (to select the most reliable lags). The

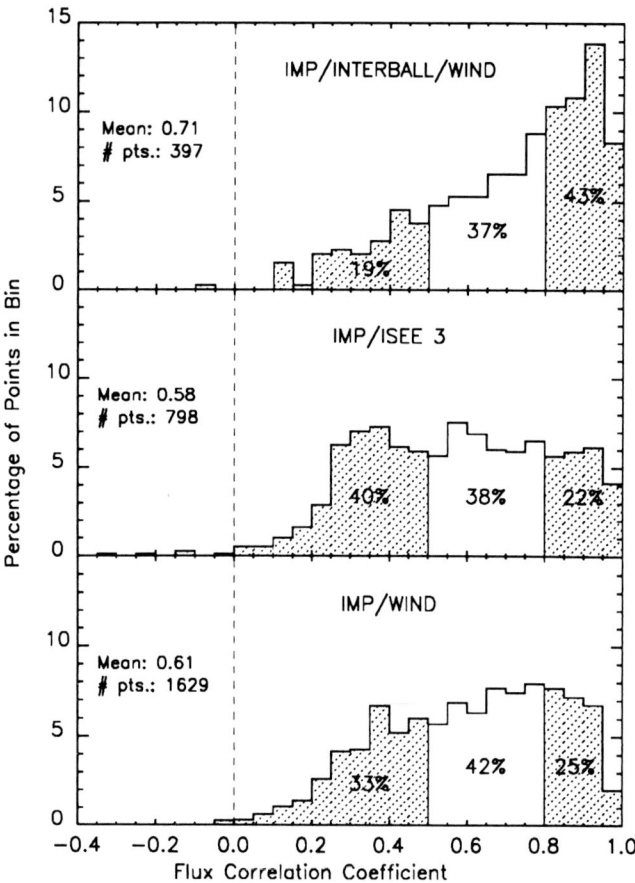

Figure 5. A histogram showing the percentage of segments having the flux correlation within bins 0.05 wide for the various spacecraft pairs. Shaded areas show the percentage of segments having correlations below 0.5 or above 0.8, for reference to the *Crooker et al.* [1982] results. Rounding results in the total in the top panel being slightly less than 100%.

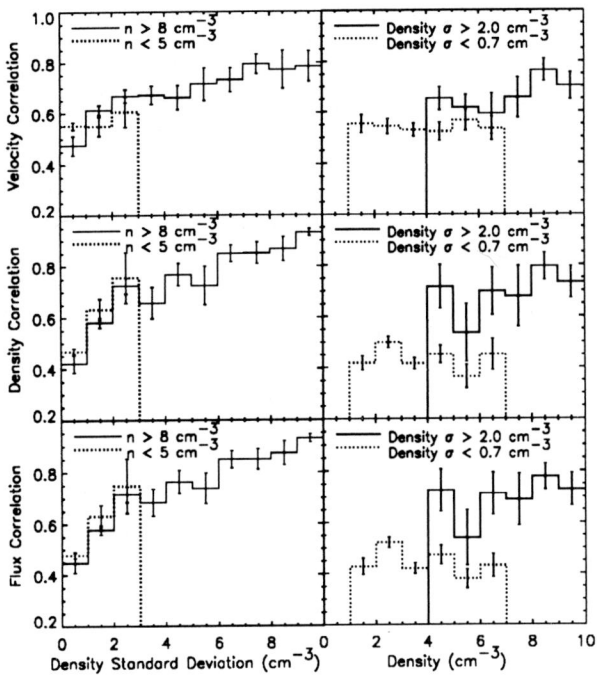

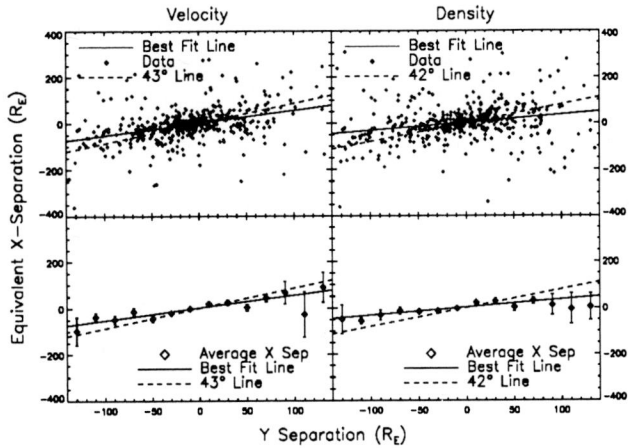

Figure 7. A scatter plot of the equivalent X separation for velocity and density (top panels) versus the spacecraft separation showing intervals where the correlation coefficient for both density and speed is greater than 0.8. The bottom panels show averages of the equivalent X separation over 20 R_E bins of Y separation and standard errors. In all panels the best linear fits are shown by the solid line, IMF angle fronts by the dashed lines.

Figure 6. The correlation coefficients as a function of the solar wind density standard deviations for high and low density cases and as a function of solar wind density for high and low standard deviation cases and 1σ standard errors.

best linear fit (solid line) gives an average front angle of $21° \pm 3°$. The average IMF orientation for the periods plotted is shown by the dashed line and is a much worse fit to the data. The bottom panels show X_{sep} averaged over 20 R_E bins and associated errors and again show that the best fit angle is much less than the average IMF angle. If solar wind variations were solely due to time dependence of the source, plasma fronts would have 0° orientations and arrival time would not depend on spacecraft Y separation. If solar wind variations were due solely to longitudinal variation of the source, the plasma fronts would be parallel to the IMF. The result that the average front angle is between these two extremes suggests that time and longitude dependence at the source are of comparable importance in driving the solar wind features observed near Earth.

4. DISCUSSION AND SUMMARY

We find that at times solar wind plasma is nearly identical at different spacecraft and at other times it is very different. The average correlation coefficient between 6-hour periods of data from various spacecraft pairs is about 0.65. Longer periods show higher average correlations, but there is no pattern of periods with consistently high or low correlation. The scale length for variations of the solar wind in the radial direction is about 200 R_E; correlation coefficients decrease rapidly beyond that distance. Variations of the correlation coefficients with Y and Z are not significant over the distances studied. These results imply that the L1 point (230 R_E) is near the limit for useful monitoring - if one could position a satellite further in front of Earth the increase in warning time would be offset by a decrease in accuracy. The best predictor of whether an upstream monitor is a reliable forecaster of solar wind changes downstream in the density standard deviation. For purposes of space weather forecasting the density could be used as a (less accurate) proxy for this parameter. We find that the plasma fronts are oriented at roughly half the average IMF angle and suggest this implies that time-dependent and longitudinally-dependent changes on the Sun are equally important for producing solar wind changes.

Acknowledgments. The authors thank A. Lazarus and J. Steinberg for providing WIND plasma data via the MIT plasma group web page, R. Lepping and A. Szabo for use of the WIND magnetic field data, and NSSDC for providing the ISEE 3 data. This work was partially supported by NASA contract NAG5-5360 (INTERBALL Guest Investigator) to MIT, by the NSF space weather program under grant ATM-9613935, and was made possible in part by Award RPI-246 of the U.S. Civilian Research & Development Foundation for the Independent States of the Former

Soviet Union (CRDF) to MIT and IKI, as well as by RFBR 95-02-03998 to IKI. The IMP 8 and WIND data analyses were supported by NASA contracts NAG5-584 (IMP) and NAG5-2839 (WIND) to MIT.

REFERENCES

Collier, M. R., J. A. Slavin, R. P. Lepping, A. Szabo, and K. W. Ogilvie, Timing accuracy for the simple planar propagation of magnetic field structures in the solar wind, *Geophys. Res. Lett., 25,* 2509-2512, 1998.

Crooker, N. U., G. L. Siscoe, C. T. Russell, and E. J. Smith, Factors controlling degree of correlation between ISEE 1 and ISEE 3 interplanetary magnetic field measurements, *J. Geophys. Res., 87,* 2224-2230, 1982.

King, J. H., Solar wind parameters and magnetospheric coupling studies, in *Solar Wind-Magnetosphere Coupling,* edited by Y. Kamide and J. A. Slavin, pp. 163–177, Terra Sci., Tokyo, 1986.

Paularena, K. I., G. N. Zastenker, A. J. Lazarus, and P. A. Dalin, Solar wind plasma correlations between IMP 8, INTERBALL-1, and WIND, *J. Geophys. Res., 103,* 14,601-14,617, 1998.

Richardson, J. D., and K. I. Paularena, The orientation of plasma structure in the solar wind, *Geophys. Res. Lett., 25,* 2097-2100, 1998.

Richardson, J. D., F. Dashevskiy, and K. I. Paularena, Solar wind plasma correlations between L1 and Earth, *J. Geophys. Res., 103,* 14,619-14,629, 1998.

___

F. Dashevskiy, K. I. Paularena, and J. D. Richardson, Center for Space Research, Massachusetts Institute of Technology, Room 37-651, Cambridge, MA, 02139. (email: kip@space.mit.edu)

G. Zastenker and P. A. Dalin, Space Research Institute, Russian Academy of Sciences, 84/32 Profsoyuznaya, 117810 Moscow, Russia. (email: gzastenk@arc.iki.rssi.ru)

Plasma Entry, Transport, and Loss in the Magnetosphere and Ionosphere

Patricia H. Reiff

Department of Space Physics & Astronomy, Rice University

The plasma in the magnetosphere has three main sources: the solar wind, the ionosphere, and secondaries from the extended atmosphere. Although the solar wind has long been considered the most important source, the contribution from the ionosphere is very significant and at certain times and certain locations may dominate. Transport of plasmas through the magnetosphere is predominantly given by **E**×**B** convection, but flows along the magnetic field direction are significant, and gradient and curvature drifts are large for energetic particles. The global pattern of convection, as observed in the ionosphere, is to first order controlled by the Interplanetary Magnetic Field (IMF), but time variations are important. In the magnetospheric tail, convection is nonuniform in time and in space; "bursty bulk flows" may be the dominant method of flux transfer back towards the Earth. Loss of magnetospheric plasmas occurs via several routes: convection paths intersecting the magnetopause, exhaust down the magnetotail, either by parallel flows in the lobes or reconnection-driven antisunward flows and plasmoid ejection; precipitation into the ionosphere; and charge exchange with the extended atmosphere. The spacecraft that will fly during the upcoming solar maximum will provide new eyes to view the plasmas in the magnetosphere, complementing the many exciting discoveries of the ISTP era.

INTRODUCTION

As early as the turn of the century it was recognized that, because of Earth's magnetic field, charged particles from deep space can only access the Earth's vicinity near the poles. Particles deep within the magnetic field are trapped there, exhibiting three kinds of periodic motion, listed in increasing order of period: cyclotron (making tight circles around the field); bounce (along the field from pole to pole) and drift (longitudinal motion around the Earth) [e.g. Parks, 1991]. Alfvén put forth the useful concept of "frozen-in flux", showing that if there are no electric fields along the magnetic field direction, then low-energy plasmas which start together on a given field line, stay together, as the particles all drift at the same speed in a direction perpendicular both to the Electric (E) and Magnetic (B) fields. This "E×B drift" is the dominant large-scale motion of low-energy particles. A consequence of frozen-in flux is that plasmas coming from the sun are connected to interplanetary field lines, and thus cannot be found on terrestrial field lines without a violation of one or more of the assumptions of ideal magnetohydrodynamics (MHD).

SOLAR WIND PLASMA ENTRY

To a first approximation, the plasmas from the sun do not enter the Earth's near vicinity: of the flux of solar wind particles ($1 - 30 \times 10^8$ / cm^2-s) incident on a circle the size of the magnetospheric cross-section (approximately 20 R_E

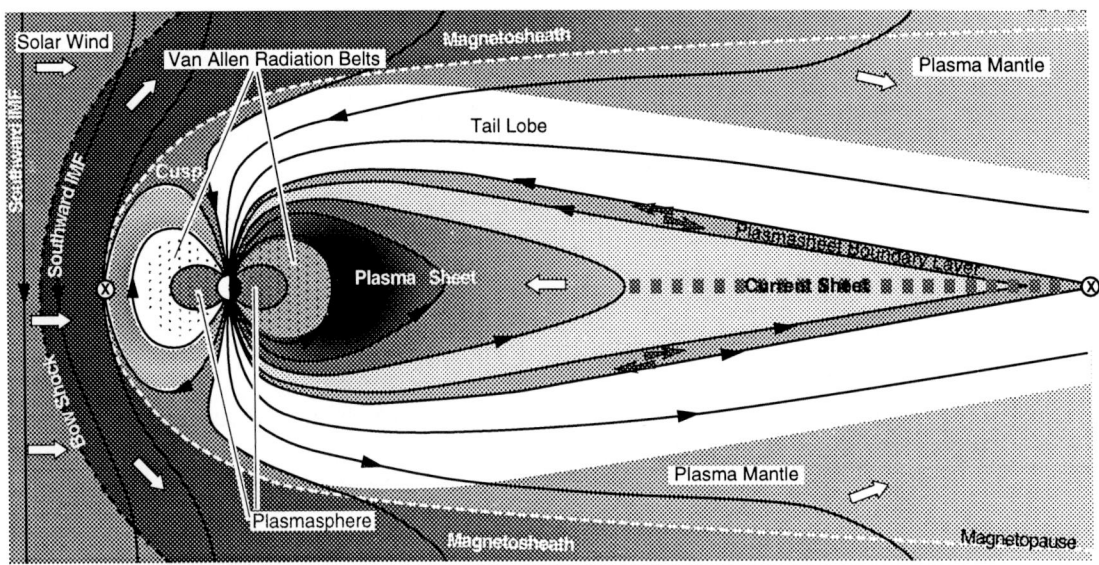

Figure 1. Schematic noon-midnight cut of the Earth's magnetosphere, showing principal regions and interconnected magnetic field lines. Solid arrows show field lines; open arrows show ion flow vectors; arrows in plasmasheet boundary layer show ion beams. (Adapted from Weiss et al., 1992; original from T. W. Hill].

radius), less than 1% of the available particles ($\sim 10^{29}$ /s) actually convect through the inner magnetosphere, whereas a larger fraction of the incident flux is observed in the boundary layers adjacent to the magnetopause. Boundary layers are frequently observed inside the magnetopause with a thickness of ~ 1 R_E with an average density and velocity about half that of the nearby magnetosheath. If that layer is uniform at all latitudes and all times, then the flux through the boundary layers is about 4% of the solar wind flux.

Two questions naturally arise: how does the solar wind plasma cross the magnetopause and why is a larger fraction observed flowing downtail in the boundary layers than is observed returning sunward? The first question we will address here; the second will be discussed in the "loss" section. The boundary layers at high latitudes in the tail and on the dayside are apparently on open field lines, i.e. field lines with one end intersecting the earth's surface and one end linking into interplanetary space. The process of transferring flux between fields of two different topologies (in this case between interplanetary field lines and open field lines) is called *reconnection* [*Vasyliunas*, 1975]. A schematic of the process is shown in Figure 1, which is a noon-midnight cross-section of the magnetosphere. The location where the closed field lines of Earth touch (and become connected to) the interplanetary field lines is called the dayside x-line. For southward IMF, steady-state conditions (as shown here) the dayside x-line is continuous across the dayside magnetopause, travels down the flanks of the magnetopause, and crosses the magnetotail at the nightside x-line. Its intersection with the noon-midnight plane is shown as the two circled "x's". The plasma on the newly-opened field lines proceeds tailwards from the dayside to the nightside. The magnetic field in the lobes then recloses, reconnecting at the nightside portion of the x-line. The plasma is then accelerated earthwards and returns to the dayside out of the noon-midnight plane, completing the convection cycle.

Plasmas can gain entry to low altitudes by flowing down along the newly-opened field lines, creating the open portion of the boundary layer and the magnetospheric cusp [*Shelley et al.*, 1976; *Reiff et al.*, 1977]. However, the entire polar cap is not filled with high-density solar wind plasma; tailward of about x=+1 R_E, the magnetosheath becomes supersonic. Although the ions continue to cross the high latitude magnetopause, they cannot "buck the flow" sufficiently to precipitate to low altitudes [*Reiff et al.*, 1977]. Since the ions cannot easily reach low altitudes, the electrons (by quasi-neutrality) may not either - a small ambipolar electric field is set up near the magnetopause to retard the electron entry; only the high-energy tail of the electron distribution drizzles into the central polar cap.

The velocity filter effect, in conjunction with the restricted entry site, means that for a given observation site, only ions with a particular range of pitch angles for a given energy can be observed [*Burch et al.*, 1982]. Numerical calculations of both ion entry [*Onsager et al.*, 1995; *Xue et al.*, 1997a,b] and electron entry [*Wing et al.*, 1996] have been quite successful in reproducing the observed particle distributions. This entry process is not steady state [e.g., *Lockwood*, 1995] - both variability in magnetosheath parameters and variability in the reconnection efficiency can cause multiple injection signatures [*Xue et al.*, 1997a,b]. Lockwood and Smith argue that changes in the low-energy cutoff in the ion dispersions indicate a non-steady reconnection rate, with one minute "on" and up to 7 minutes "off" being typical pulses of reconnection. In the *Xue et al.* [1997] paper, the best fit was for 1 minute "on" and two minutes "off".

Another study which favors pulsed (but large-scale in local time) reconnection was a near conjunction of Polar (moving slowly equatorward) and FAST (moving mostly in local time) [*Peterson et al.*, 1998]. Dispersions of several time scales were observed in that study. Since the Polar spacecraft was moving equatorward, a simple increase of ion energy with time should have been observed if the injection had been steady state. On top of that large-scale increase, however, several subinjections are observed, each decreasing in energy with time as if the injection were temporal [Figure 2]. Another study by *Fuselier et al.* [1997] argued from high-time resolution Polar TIMAS data that apparently multiple reconnections can occur on the same field lines at different locations. This is apparently the same effect noted by Xue at al.: since there is no net flux crossing the magnetopause between pulses of reconnection, a new burst of reconnection starts on the *same* field line as the end of the previous burst of reconnection, thus yielding what appear to be two sequential injections on the same field line.

At higher altitudes the cusp becomes much more turbulent, in an "entry layer" [*Haerendel et al.*, 1978]. Recently, the Polar spacecraft has been exploring this entry layer and has found a number of crossings that exhibit turbulent flow. One fascinating discovery has been the existence of very energetic (MeV) particles at the high latitude cusp [*Chen et al.*, 1998]. These may be trapped in the field minimum there [*Lyons et al.*, 1998] or trapped between the low-altitude field maximum and the kink at the magnetopause [*Curran and Goertz*, 1989]. These electrons may be responsible for the low-intensity but energetic electrons with a "trapped" pitch angle distribution observed in the low-altitude cusp [*McDiarmid et al.*, 1976; *Reiff et al.*, 1977]. Although Chen *et al.* argued that these ions are accelerated locally, an alternative explanation with a bow shock source is given by Chang *et al.* [1998.]

For most orientations of the IMF, the site of reconnection is near the subsolar point, although the angle of the x-line is tilted away from the equatorial plane if the IMF has a significant Y-component. If the IMF is strongly northward, however, the site of reconnection may move to near the dayside cusp. In several cases, the Polar spacecraft (which has an apogee of 9 R_E geocentric) has crossed the dayside magnetopause. The flows and magnetic fields observed in one well-documented case indicate "reverse reconnection" [*Russell et al.*, 1972]: connection of the cusp field lines with a northward-directed IMF [*Urquhart et al.*, 1998; *Savin et al.*, 1998], confirming earlier observations by *Gosling et al.* [1991]. Reverse reconnection is important because it facilitates injection of magnetosheath particles onto closed dayside field lines. This may be the dominant route of populating the closed dayside boundary layer [*Nishida*, 1991].

The other proposed manner of populating the closed dayside boundary layer is cross-field diffusion, either as single-particle scattering or turbulent mesoscale diffusion. Apparently an adequate intensity of the appropriate wave has been identified in the boundary layers [e.g., *Tsurutani et al.*, 1998; *Drakou et al.*, 1994]. Even at the Bohm diffusion rate, however, diffusion across the closed boundary layer should result in a maximum of only about 10 kV of convection potential [*Hill*, 1979].

The fact that the plasma sheet density is well correlated to the solar wind density when the IMF is southward and not particularly well with the solar wind velocity [*Borovsky et al.*, 1998] leads one to conclude that magnetic merging is the principal route of solar wind access to the magnetosphere.

IONOSPHERIC PLASMA SOURCES

The ionosphere is the other significant source of magnetospheric plasma. Some, in fact, have argued that it is the dominant source [*Chappell et al.*, 1987]. The ionosphere supplies plasma to the magnetosphere via several routes. The first is the cold "polar wind", <10 eV H^+ evaporating from the ionosphere. These ions have so little energy that they are typically repelled by spacecraft electric fields in the low density magnetospheric lobes. The Polar spacecraft has solved this problem (and for the first time measured these ions) by the use of a plasma emitter to control the

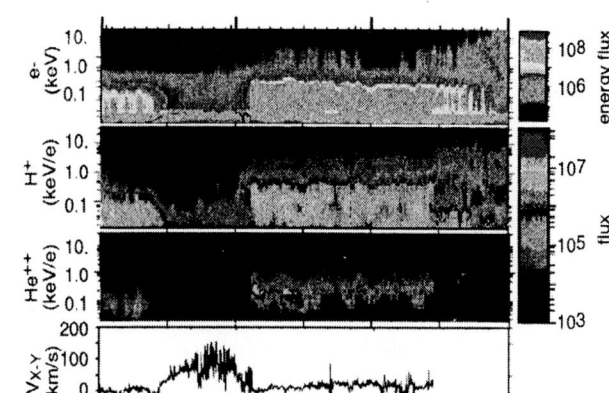

Figure 2. Cusp ion dispersion from Polar spacecraft showing sequential ion injections in an equatorward pass [from *Peterson et al.*, 1998].

spacecraft potential [*Moore et al.*, 1995]. These ions, confined to a narrow "source cone", can populate the plasmasphere only after pitch-angle scattering increases their pitch angles.

Warmer ions typically leave the ionosphere in the auroral zone (see the recent paper by *Hirahara et al.* [1998] and references therein). On the dayside, Joule heating from strong cusp flows creates a "fountain" of ~10 eV heavy ions (~50 tons per day) leaving the ionosphere [*Horwitz and Lockwood*, 1985; see *Arnoldy et al.*, 1996, for recent work]. Most of these ions convect tailward and reenter the ionosphere in the polar cap or nightside auroral zone [Figure 3]; but some are additionally accelerated by resonant waves [e.g. *Shelley et al.*, 1997] and become more energetic (100 - 1000 eV) H^+ and O^+ ion "conics", so named because their distribution function peaks at a pitch angle between 90° and 120°. These ions have sufficient energy to travel significantly along the field line during a magnetospheric convection cycle, and can reach the current sheet in the near-earth tail [*Delcourt et al.*, 1990].

Even more energetic (1 - 10 keV) upgoing ionospheric ions are observed above auroral arcs, in "ion beams" [*Möbius et al.*, 1998 and references therein]. These ion beams are the high-altitude evidence of a parallel electric field below the spacecraft [see, e.g. *Mizera and Fennel*, 1977; *Lu et al.*, 1992]; but small-scale (km-size) structures can be seen which are not obviously related to the large-scale potential drops [*McFadden et al.*, 1998]. To first order, the energy of the upgoing Oxygen equals that of the upgoing Hydrogen and Helium, indicating the importance of the parallel electric field. A persistent excess of the O^+ energy over H^+ is observed, however, which has been suggested to be caused by a two-stream instability attempting to equalize their velocities; however, the persistence of this effect over several orders of magnitude of relative concentration may point to another cause, for example, the pre-acceleration of O^+ in the ionosphere [*Möbius et al.*, 1998]. These ions, which reach the equatorial plane in the near-earth tail, can be dynamically important (see, e.g. *Daglis et al.* [1996]). In disturbed times, the plasma sheet and ring current can be up to 50% O^+ [*Young*, 1983], and this mechanism can directly populate these regions. Compression and time-varying electric fields from storms and major substorms can then inject these plasma sheet ions deep into the inner magnetosphere [*Harel et al.*, 1981].

SECONDARY PLASMA SOURCES

Already-trapped or precipitating plasmas can also ionize the Earth's atmosphere, creating secondary ions and electrons. The largest such source of ionization occurs in the

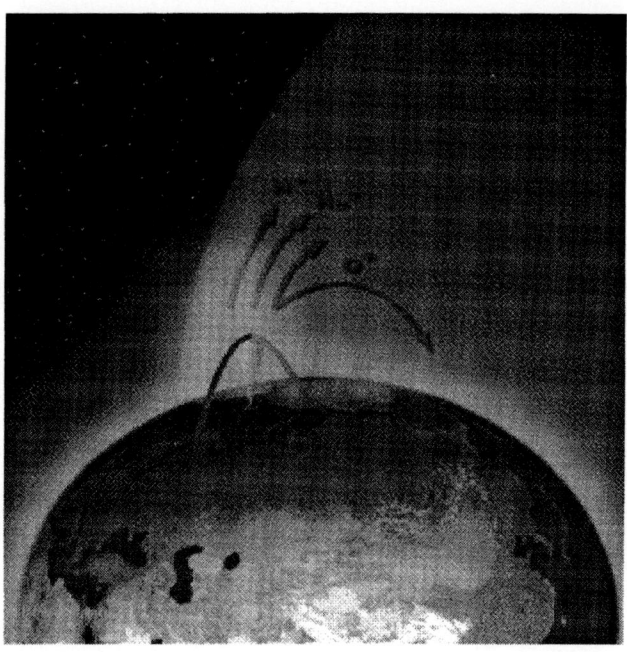

Figure 3. Cleft ion "fountain" as observed by the TIDE instrument on Polar [courtesy T. Moore].

auroral zone. As a rule of thumb, each precipitating auroral electron can create one electron-ion pair for each 35 eV of incident energy at the top of the atmosphere. Most of those ions, of course, remain in the ionosphere, but a few populate the magnetosphere directly.

Another source of cold ions which has not been heretofore considered important is the creation of cold ions from charge exchange with the ring current. The ring current's principal loss mechanism is charge exchange with the neutral atmosphere. The highest neutral density along an ion bounce path, and therefore the most likely location for such loss, is where the particles mirror. For nearly-isotropic distributions, such as the central plasma sheet, the loss rate is largest near the top of the ionosphere. For more highly trapped distributions, such as the inner plasma sheet and ring current, the ions (and therefore the losses) are more concentrated at the equator. The fast neutrals that result from such charge exchange are not gravitationally bound and escape on straight line trajectories [Figure 4]. Detection of these energetic neutrals, NAI for neutral atom imaging, is therefore a new and exciting way to study the Earth's magnetosphere remotely [*Williams et al.*, 1992]. The "IMAGE" mission, set to launch in January 2000, is the first mission designed entirely for remotely sensing the magnetosphere. Neutral atom imaging is the basis for three of the sensors.

Some low-resolution neutral atom images [Figure 5] have already been derived from Polar data [*Lui et al.*, 1996;

Henderson et al., 1997], and a number of missions using NAI techniques are being developed or proposed (e.g. TWINS; Astrid).

For each such neutral, however, there is also a corresponding cold (< 10 eV) ion. These ions can populate the plasmasphere directly (since the ones from ring current loss are born in the equatorial plane and are not confined to the atmospheric source cone) and thus may be a significant source of the plasmasphere. Recent work by Borovsky [personal communication, 1998] confirms this idea. He notes that the plasmasphere typically shows a two-stage refilling rate, refilling slowly at the outset and more rapidly later on. He suggests that the early slow refilling is from the charge exchange source (which can provide a few tons per day of H^+). The later faster refilling, then, is caused by the buildup of plasmaspheric density enhancing the coulomb scattering of the polar wind, allowing it to become trapped as well. This is an interesting model and deserves some additional work.

ION TRANSPORT

The motion of magnetospheric plasma is dominated by large-scale $E \times B$ convection. Plasma drifts across the magnetic field caused by magnetospheric electric fields are independent of mass or charge; thus plasmas which start on the same flux tube stay on the same drift path with one another. This is the idea behind frozen-in flux, a simplification first brought forward by Alfvén. This is a very powerful tool to characterize convection by the motion of field lines. To the extent that magnetic field lines are equipotentials, frozen-in-flux allows one to infer convection over the

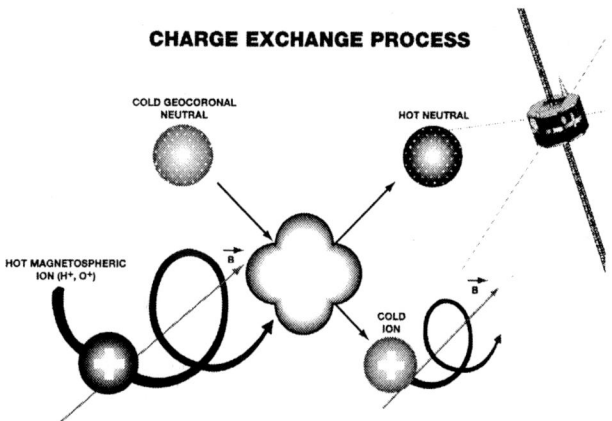

Figure 4. In the charge exchange process, a hot ion (black circle with "+") strips an electron from a cold neutral (gray circle), resulting in a fast neutral (black circle) and a cold ion (gray circle with "+"). The fast neutral travels straight paths, and can be detected remotely, as by the IMAGE spacecraft (top right). Courtesy J. L. Green (http://image.gsfc.nasa.gov).

154 MAGNETOSPHERIC PLASMA ENTRY, TRANSPORT AND LOSS

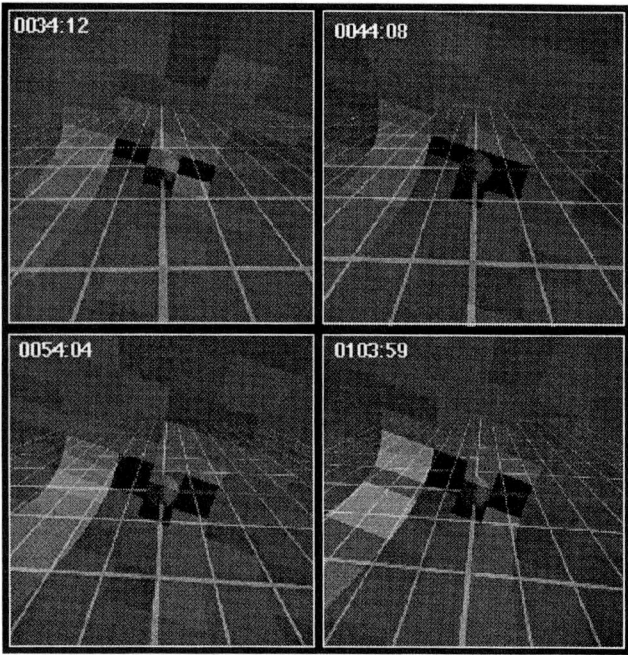

Figure 5. First Polar ENA image of the ring current [from *Lui et al.*, 1996].

These techniques are especially useful for understanding multiple-cell convection patterns and "theta arcs", which can change rapidly with time [*Maynard et al.*, 1998] and in fact may be the signature of polar cap flux opening after a period of Northward IMF [*Newell and Meng*, 1995].

These more recent studies which allow determination of the large-scale convection pattern on very short time scales have elucidated the role of "lobe cells", convection cells apparently completely contained within the polar cap, presumably driven by magnetic merging of already open tail lobe fields with a Northward IMF [*Reiff and Burch*, 1985]. Initially quite controversial, the presence of such lobe cells (which can have a quite different potential in the northern from the southern polar cap, depending on IMF B_y and season), has now been confirmed by magnetometer and superDARN observations. At first thought impossible by seemingly requiring the solar wind to flow backwards [*Hill*, 1994], they are now generally understood by means of "overdraping" the open flux across the dayside magnetopause, and lobe cells are now often observed in MHD models of the magnetosphere as well [*Crooker et al.*, 1998]. For IMF with large B_y and northward IMF, one observes

large volume of the magnetosphere by measuring the convective flow of the ionosphere, which is considerably easier and faster to traverse. However, the presence of an electric field $E_\parallel$ parallel to the magnetic field can disconnect the motions of plasmas above from those plasmas below the field. Such plasma slippage is common in the auroral zone, and reduces the convection electric field gradient below the arc and thus the need for field-aligned current [*Lyons*, 1981]. A recent study of the effects of parallel and inductive electric fields on mapping on the convection from the magnetosphere to the ionosphere has been done by Hesse et al., [1997].

A great deal of work has been done in measuring ionospheric convection, first in the "pattern recognition" mode whereby typical convection patterns are deduced by contrasting flow measurements made under various IMF conditions [e.g. *Heppner and Maynard*, 1987]. Later work was more quantitative, averaging ionospheric flows or electrostatic potentials for various conditions and then calculating from them the typical convection patterns [e.g. *Rich and Hairston*, 1994; Figure 6, from *Weimer*, 1996]. More recent work creates instantaneous (or near-instantaneous) 2-D convection patterns by using radar velocity measurements over a large area [e.g., *Ruohoniemi and Greenwald*, 1996] or by merging spacecraft overflight data with flows inferred from magnetometer perturbations [e.g., *Lu et al.*, 1997].

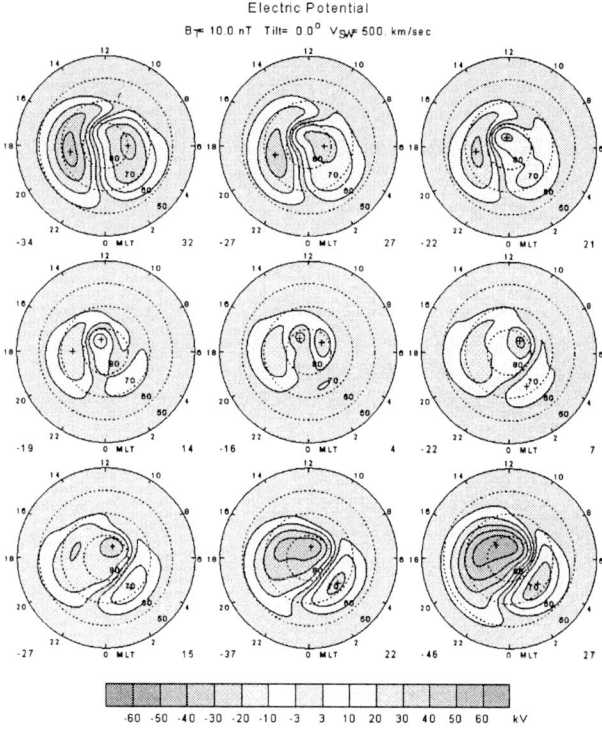

Figure 6. Convection potential patterns as a function of the direction of the IMF inferred from ionospheric electric field measurements [*Weimer*, 1996].

in the models a convection cell at very high latitudes, the inner portion of which is completely on open field lines.

The role of the neutral atmosphere in regulating ionospheric flow (and therefore influencing magnetospheric convection) has also seen great strides in recent years. Often ignored, for example, when determining ionospheric electric fields from magnetic perturbations, the neutral wind can make a large difference in inferred Joule heating rates. For example, the neutral wind typically reaches a speed of about 30% of the ion velocity after a few hours [*Killeen et al.*, 1984]. The neutral winds have even been observed to flow sunward (against the solar forcing, but in the same direction as the ion flow) when the IMF was strongly northward for several hours [*Killeen et al.*, 1985]. That neutral speedup, if it occurs at the Pedersen conductivity height, reduces the ionospheric current by 30% (which depends on the electric field in the rest frame of the neutrals) and the Joule heating rate by nearly a half over the value calculated by ignoring neutral winds.

Consider a sequence when the polar cap convection reversal moves equatorward (as it will when the IMF turns southward). The ion flow in the newly-reversed region is antisunward, whereas the neutrals are still moving sunward. This will lead to a strongly enhanced ionospheric current (again, since it is the electric field in the rest frame of the neutrals that counts), with a considerably larger than expected Joule heating rate and need for field-aligned currents. A similar effect occurs when the convection boundary moves poleward. Thus time-variable polar cap convection generally requires more field-aligned current and more energy input than steady state convection requires.

The neutral wind "flywheel" has been suggested as a reason why the polar cap flow does not stop instantaneously when the IMF turns northward (the other, more likely, explanation is that reconnection at the tail neutral line continues and drives convection by closing open polar cap field lines and reducing the size of the polar cap). The neutral wind may also explain the differing conclusions reached by various experiments in determining the "saturation" electric field. In our early work, using hourly-averaged interplanetary data, we found that the cross-polar cap potential drop responded nearly linearly to increases in the solar wind electric field, up to a value of about 160 kV, where the measured potential appeared to saturate [e.g. *Reiff and Luhmann*, 1986]. In our more recent work, when we had many thousands of DMSP orbits from which to select, we required interplanetary conditions to be steady for 4 hours. In that study, we found no apparent saturation effect in the observed potential [*Boyle et al.*, 1997].

When we tested the predicted (heavy line) versus observed (solid dots) polar cap potential for a case with a strong (20

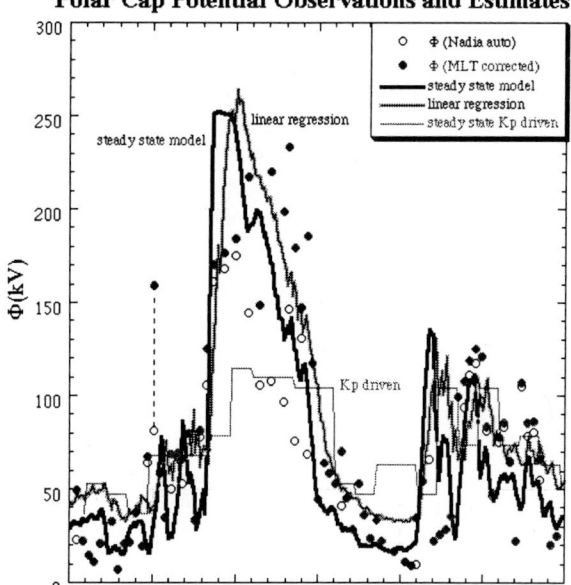

Figure 7. Measured and predicted polar cap potentials for the cloud event of October 1995 [from *Boyle*, 1997]. DMSP-derived measured potential drops are shown as open circles. A correction is applied to account for the spacecraft not crossing the potential maxima. The corrected data are shown as solid dots (in most cases the corrected data overlay the raw data). The potential drop expected had the IMF been steady for 4 hours is shown as a heavy line (formula from *Boyle et al.* 1997). Note that the agreement of the measurements with the model is very good except for times near the peak. The measured potential increases rapidly to approximately 160 kV at 291.9, but then rises more slowly with a ~2 hour time constant. When the corrections are large, (between 292.2 and 292.5), the solid dots should be considered upper limits and the open dots lower limits.

nT) southward IMF step, we found that the measured potential rose quickly to 160 kV, and then rose more slowly after that, with a ~2 hour time constant [Figure 7, from *Boyle*, 1997]. These results can be reconciled by invoking a field-aligned current limitation rather than a potential limitation. In the case of the early work, when we used one-hour averages of interplanetary data as the independent variable, the fields were frequently unsteady when the IMF was large and southward and a saturation effect was observed. In the Boyle *et al.* study, when we imposed a 4-hour steadiness criterion to the interplanetary measurements, the neutral winds had an opportunity to equalize with the ions and the maximum current limit was not reached; thus that study yielded the asymptotic potential (the potential which is reached in the case of steady IMF for an extended period). In the case of the stepwise southward IMF shift, the measured potential drop followed the asymptotic prediction only up to a potential of about 160 kV, where it may have reached the maximum available current. The potential then rose more slowly over the next few hours. This second, slower rise may be the signature of the neutrals speeding up; for a given maximum current, the maximum potential drop increases as the neutral wind increases.

Hill et al. [1976] argued that a maximum field-aligned current can be supported, because that current must close in a layer near the magnetopause. If that current becomes too large (approximately 10% of the local Chapman-Ferraro current), then the internal field becomes seriously distorted (and therefore can inhibit merging). Their simple arguments yielded a maximum potential of about 160 kV for reasonable ionospheric conductances. Now consider the effect of the neutral wind. Over the course of 2-3 hours, the neutral wind can reach, say, 30% of the ion speed, reducing the need for current by 30%. Therefore, the maximum supported potential will rise to ~200 kV, approximately the level we observed in Figure 7. Furthermore, the maximum potential will depend on the magnetic field just inside the magnetopause. Therefore, for times of high solar wind pressure, the maximum current (and thus the maximum potential) will also rise. This may explain the interesting new results of Greenwald [personal communication, 1998] who observes a dependence of polar cap potential on density, but only in times of large southward B_z, when the potential nears it maximum value.

ENERGY-DEPENDENT DRIFTS

More energetic particles do not obey simple $\mathbf{E} \times \mathbf{B}$ convective drift; for them, gradient and curvature drifts are very important, and lead to energy and species-dependent access to the inner magnetosphere. These differences are critical for the generation of region 2 Birkeland currents [e.g. *Jaggi and Wolf*, 1973]. Kinetic modeling of inner magnetosphere dynamics, and its merging with more traditional MHD models is a major thrust of the GEM (Geospace Environment Modeling) program.

PARTICLE LOSSES

The most important loss process for inner-magnetospheric particles is charge exchange loss. For nearly isotropic populations, this loss occurs at the foot of the field lines, making an atmospheric loss cone. For ions the loss altitude is typically 400 km; for electrons, 110 km, but both are energy-dependent (the more energetic particles are lost at lower altitudes). For highly trapped populations such as the ring current, the loss is strongest near the equatorial plane.

For more distant populations, transport losses become important. The plasmasphere can be lost by convection in tongues out the dayside if the polar cap potential increases [e.g. *Spiro et al.*, 1981]. Plasma sheet ions in the mid- to far-tail can be lost downtail via large plasmoids ejected by substorm processes [*Slavin et al.*, 1998]. Stretching of the tail field lines in the substorm growth phase can result in rapid curvature drifts across the magnetotail, ejecting ions from the dusk side and electrons from the dawn side. In fact, adiabatic processes increase plasma pressure on tail field lines, restricting their access to the inner magnetosphere until some loss process occurs [*Erickson and Wolf*, 1980]. This results in sporadic sunward convection in "bursty bulk flows" [*Angelopolous et al.*, 1997].

Finally, not all plasmas that are observed inside the magnetopause remain inside. The "velocity filter effect", which is observed in the magnetospheric cusp, was first observed in the plasma mantle [*Rosenbauer et al.*, 1975]. The highest-energy cusp ions bounce and either are lost into the magnetosheath or are found in a layer just inside or outside the magnetopause. Only the lowest-energy ions can drift from the high latitude magnetopause to the tail neutral sheet inside of the tail x-line within a convection cycle [*Delcourt et al.*, 1990]. Thus the cislunar near-equatorial plasma mantle is typically observed only on the side of the magnetotail which has the largest convection velocity (dependent on IMF B_y) [*Hardy et al.*, 1979]. The fact that the high-speed mantle plasma may reach the equatorial plane in the distant tail may influence the thinning of the plasma sheet [*Hill and Reiff*, 1980]. The mantle plasma may blow the x-line downtail and restrict the return flow.

CONCLUSIONS

Our knowledge of each of these processes has taken rapid steps in the past few years as a result of ISTP-era observations. It is clear that multipoint measurement of plasma transport by the ISTP suite will yield fertile ground for data analysis for quite a few years. In the upcoming era, however, we will have new eyes to remotely observe many of these processes. The charge exchange process (discussed earlier) will allow monitoring of the energetic plasmas in the inner magnetosphere by the ENA instruments on IMAGE, TWINS, Astrid, and other missions. By combining the view from several directions (and by using clever analysis techniques), we will be able to reconstruct the 3-D density distribution in the magnetosphere.

Another powerful way in which we will be able to monitor ion transport is by using EUV imaging of the plasmasphere, using resonant scattering of sunlight by He^+. This will allow for the first time a global view of the plasmaspheric deformation, expansion, and refilling. The IMAGE spacecraft will be able to probe the plasmasphere in two independent and complementary ways – by using the 304Å emission, and also by radio sounding [Reiff et al., 1994]. In the radio sounding process, a succession of encoded radio bursts is transmitted. When they reach a surface where the local plasma frequency is equal to the transmitted frequency, the radio waves are reflected. If that surface of constant plasma frequency (which is also a surface of constant plasma density) is normal to the line of sight from the spacecraft, the radio signal is returned to the spacecraft and the remote density structure can then be determined. This procedure has been used for years in sounding the ionosphere from below, and a few spacecraft have performed "topside" sounding of the ionosphere, but IMAGE will be the first spacecraft to feature sounding in the magnetospheric cavity.

In summary, the magnetosphere still has secrets to reveal and the missions of the upcoming century, combined with ground-based observations and a dynamic theory program, will be the key to our continued increase of our knowledge of magnetospheric processes.

Acknowledgments. This work was funded in part by NASA under grant NAG5-3216 and under the IMAGE mission by SWRI subcontract AO-95-OOS-2. The author thanks W. K. Peterson for helpful comments.

REFERENCES

Angelopoulos, V., et al., Magnetotail flow bursts: association to global magnetospheric circulation, relationship to ionospheric activity and direct evidence for localization, *Geophys. Res. Lett., 18,* 2271, 1997.

Arnoldy, R. L., K. A. Lynch, P. M. Kintner, J. Bonnell, T. E. Moore and C. J. Pollock, SCIFER: Structure of the cleft ion fountain at 1400-km altitude, *Geophys. Res. Lett., 23,* 1869-1872, 1996.

Boyle, C. B., Polar cap response to the magnetic cloud event of 18-21 October 1995, Ph.D. Thesis, Rice University, 1997.

Boyle, C. B., P. H. Reiff, and M. R. Hairston, Empirical polar cap potentials", *J. Geophys. Res., 102,* 111 - 125, 1997.

Burch, J. L., P. H. Reiff, R. A. Heelis, J. D. Winningham, W. B. Hanson, C. Gurgiola, J. D. Menietti, R. A. Hoffman, and J. N. Barfield, Plasma injection and transport in the mid-altitude polar cusp," *Geophys. Res. Lett., 9,* 921-924, 1982.

Chappell, C. R., T. E. Moore and J. H. Waite Jr., The ionosphere as a fully adequate source of plasma for the Earth's magnetosphere, *J. Geophys. Res., 92,* 5896-5910, 1987.

Chang, S.-W., et al., Cusp energetic ions: a bow shock source, *Geophys. Res. Lett., 25,* 3729-3732, 1998.

Chen, J., et al., Cusp energetic particle events: Implications for a major acceleration region of the magnetosphere, *J. Geophys. Res., 103,* 69-78, 1998.

Crooker, N. U., J. G. Lyon, and J. A. Fedder, MHD model merging with IMF +By: Lobe cells, sunward polar cap convection and overdraped lobes, *J. Geophys. Res., 103,* 9143, 1998.

Curran, D. B., and C. K. Goertz, Particle distributions in a two-dimensional reconnection field geometry, *J. Geophys. Res., 94,* 272, 1989.

Daglis, I. A., W. I. Axford, S. Livi, B. Wilken, M. Grande and F. Søraas, Auroral ionospheric ion feeding of the inner plasma sheet during substorms, *J. Geogmagn. Geoelec., 48,* 729-740, 1996.

Delcourt, D. C., J. A. Sauvaud and T. E. Moore, Cleft contribution to ring current formation, *J. Geophys. Res., 95,* 20937-20943, 1990.

Drakou, E., B. U. Ö. Sonnerup and W. Lotko, Self-consistent steady state model of the low-latitude boundary layer, *J. Geophys. Res., 99,* 2351, 1994.

Erickson, G. M., and R. A. Wolf, Is steady convection possible in the Earth's magnetotail?, *Geophys. Res. Lett., 7,* 897, 1980.

Fuselier, S. A. et al., Bifurcated cusp ion signatures: Evidence for quasi-steady reconnection?, *Geophys. Res. Lett., 24,* 1471, 1997.

Gosling, J. T., M. F. Thomsen, S. J. Bame, R. C. Elphic and C. T. Russell, Observations of reconnection on interplanetary and lobe magnetic field lines at the high-latitude magnetopause, *J. Geophys. Res., 96,* 14097-14106, 1991.

Haerendel, G., G. Paschmann, N. Sckopke, H. Rosenbauer and P. C. Hedgecock, The frontside boundary layer of the magnetosphere and the problem of reconnection, *J. Geophys. Res., 83,* 3195-3216, 1978.

Hardy, D. A., J. W. Freeman, and H. K. Hills, Occurrence of the lobe plasma at lunar distance, *J. Geophys. Res., 84,* 72-78, 1979.

Harel, M., R. W. Wolf, R. W. Spiro, P. H. Reiff, C.-K. Chen, W. J. Burke, F. J. Rich, and M. Smiddy, "Quantitative simulation of a magnetospheric substorm, 2. Comparison with observations," *J. Geophys. Res., 86,* 2242 - 2260, 1981.

Henderson, M. G., et al., First energetic neutral atom images from Polar CEPPAD/IPS, *Geophys. Res. Lett., 24,* 1167, 1997.

Heppner, J. P. and N. C. Maynard, Empirical high-latitude electric field models, *J. Geophys. Res., 92,* 4467-4489, 1987.

Hesse, M., J. Birn, and R. A. Hoffman, On the mapping of ionospheric convection into the magnetosphere, *J. Geophys. Res., 102,* 9543-9551, 1997.

Hill, T. W., Rates of mass, momentum, and energy transfer at the magnetopause, in *Magnetospheric Boundary Layers,* ed. B. Battrick, ESTEC, Nordwijk, 325, 1979.

Hill, T. W., Theoretical models of polar-cap convection under the influence of a northward interplanetary magnetic field, *J. Atmos. Terr. Phys., 56,* 185, 1994.

Hill, T.W., A. J. Dessler, and R. A. Wolf, Mercury and Mars: the role of ionospheric conductivity in the acceleration of magnetospheric particles, *Geophys. Res. Lett., 3,* 429, 1976.

Hill, T. W., and P. H. Reiff, On the cause of plasma sheet thinning during magnetospheric substorms, *Geophys. Res. Lett.*, *7*, 177, 1980.

Hirahara, M., et al., Relationship of topside ionospheric ion outflows to auroral forms and precipitation, plasma waves, and convection observed by Polar, *J. Geophys. Res.*, *103*, 17391-17410, 1998.

Horwitz, J. L. and M. Lockwood, The cleft ion fountain: A two-dimensional kinetic model, *J. Geophys. Res.*, *90*, 9749-9762, 1985.

Jaggi, R. K. and R. A. Wolf, Self-consistent calculation of the motion of a sheet of ions in the magnetosphere, *J. Geophys. Res.*, *78*, 2842, 1973.

Killeen, , T. L., P. B. Hays, G. R. Carignan, R. A. Heelis, W. B. Hanson, N. W. Spencer and L. H. Brace, Ion-neutral coupling in the high-latitude F-region: Evaluation of ion heating terms from Dynamics Explorer-2, *J. Geophys. Res.*, *89*, 7495, 1984.

Killeen, T. L., R. A. Heelis, P. B. Hays, N. W. Spencer, and W. B. Hanson, Neutral motions in the thermosphere for northward interplanetary magnetic field, *Geophys. Res. Lett.*, *12*, 159, 1985.

Lockwood, M., Overlapping cusp ion injections: An explanation invoking magnetopause reconnection, *Geophys. Res. Lett.*, *22*, 1141-1144, 1995

Lu, G., P. H. Reiff, T. E. Moore, and R. A. Heelis, Upflowing ionospheric ions in the auroral region, *J. Geophys. Res.*, *97*, 16,855-16,863, 1992.

Lui, A. T. Y., D. J. Williams, E. C. Roelof, R. W. McEntire and D. C. Mitchell, First composition measurements of energetic neutral atoms, *Geophys. Res. Lett.*, *23*, 2641-2644, 1996.

Lyons, L. R., Discrete aurora as the direct result of an inferred, high-altitude generating potential distribution, *J. Geophys. Res.*, *86*, 1, 1981.

Lyons, L. R., M. Schulz, D. C. Pridmore-Brown and J. L. Roeder, Low-latitude boundary layer near noon: An open field line model, *J. Geophys. Res.*, *99*, 17,367, 1998.

Maynard, N. C., W. J. Burke, D. R. Weimer, F. S. Mozer, J. D. Scudder, C. T. Russell, W. K. Peterson, and R. P. Lepping, Polar observations of convection with northward interplanetary magnetic field at dayside high latitudes, *J. Geophys. Res.*, *103*, 29-46, 1998.

McDiarmid, I. B., J. R. Burrows, and E. E. Budzinski, Particle properties in the dayside cleft, *J. Geophys. Res.*, *81*, 221, 1976.

McFadden, J. P., C. W. Carlson, R. E. Ergun, F. S. Mozer, M. Temerin, W. Peria, D. M. Klumpar, E. G. Shelley, W. K. Peterson, E. Moebius, L. Kistler, R. Elphic, R. Strangeway, C. Cattell, and R. Pfaff, Spatial structure and gradients of ion beams observed by FAST, *Geophys. Res. Lett.*, *25*, 2021-2024, 1998.

Mizera, P. F. and J. F. Fennell, Signatures of electric fields from high and low altitude particle distributions, *Geophys. Res. Lett.*, *4*, 311, 1998.

Möbius, E., L. Tang, et al., Species dependent energies in upward directed ion beams over auroral arcs as oserved with FAST TEAMS, *Geophys. Res. Lett.*, *25*, 2029-2032, 1998.

Moore, T. E., et al., The thermal ion dynamics experiment and plasma source instrument, *Space Sci. Rev.*, *71*, 409-458, 1995.

Moore, T. E., et al., High altitude observations of the polar wind, *Science*, *277*, 349, 1997.

Newell, P. T. and C.-I. Meng, Creation of theta auroras: The isolation of plasma sheet fragments in the polar cap, *Science*, *270*, 1338, 1995.

Nishida, A., Ionospheric signatures of random reconnection on the dayside magnetopause, *J. Atm. Terrest. Phys.*, *53*, 213, 1991.

Onsager, T. G., et al., Low altitude observations and modeling of quasi-steady magnetopause reconnection, *J. Geophys. Res.*, *100*, 11831, 1995.

Peterson, W. K., et al., Simultaneous observations of solar wind plasma entry from FAST and Polar, *Geophys. Res. Lett.*, *25*, 2081-2084, 1998.

Reiff, P. H., J. L. Green, R. F. Benson, D. L. Carpenter, W. Calvert, S. F. Fung, D. L. Gallagher, B. W. Reinisch, M. F. Smith and W. W. L. Taylor, Radio imaging of the magnetosphere, *EOS, Trans. AGU*, **75**, 129-134, 1994.

Reiff, P. H., and J. L. Burch, "By-dependent plasma flow and birkeland currents in the dayside magnetosphere: 2. A global model for southward and northward IMF," *J. Geophys. Res.*, *90*, pp. 1595-1609, 1985.

Reiff, P. H., T. W. Hill, and J. L. Burch, Solar-wind plasma injection at the dayside magnetospheric cusp, *J. Geophys. Res.*, *82*, 479-491,.1977.

Reiff, P. H. and J. G. Luhmann, Solar wind control of the polar-cap voltage, in *Solar-Wind Magnetosphere Coupling*, ed. Y. Kamide and J. Slavin, Terra Publ. Co., Tokyo, pp. 453-476, 1986.

Rich, F. and M. Hairston, Large scale convection patterns observed by DMSP, *J. Geophys. Res.*,.*99*, 3827, 1994.

Rosenbauer, H., H. Grünwaldt, M. D. Montgomery, G. Paschmann and N. Sckopke, Heos 2 plasma observations in the distant polar magnetosphere: The plasma mantle, *J. Geophys. Res.*, *80*, 2723-2737, 1975.

Russell, C. T., The configuration of the magnetosphere, *in Critical Problems in Magnetospheric Physics*, ed. E. R. Dyer, Jr., p. 1, National Academy of Science, Washington, D.C. 1972.

Savin, S. P., S. A. Romanov, et al., The cusp/magnetosheath interface on May 29, 1996: Interball-I and Polar observations, *Geophys. Res. Lett.*, *25*, 2963-2966, 1998.

Shelley, E. G., R. D. Sharp and R. G. Johnson, He^{++} and H^+ flux measurements in the dayside cusp: Estimates of convection electric field, *J. Geophys. Res.*, *81*, 2363, 1976.

Shelley, E. G., H. Balsiger, et al., Initial TIMAS observations of ion conic heating in the cusp, *Adv. Space Res.*, *20*, 841, 1997.

Slavin, J. A., D. H. Fairfield, M. M. Kuznetsova, C. J. Owen, R. P. Lepping, S. Taguchi, T. Mukai, Y. Saito, T. Yamamoto, S. Kokubun, A. T. Y. Lui, and G. D. Reeves, ISTP observations of plasmoid ejection: IMP 8 and Geotail, *J. Geophys. Res.*, *103*, 119-134, 1998.

Spiro, R. W., M. Harel, R. A. Wolf, and P. H. Reiff, Quantitative simulation of a magnetospheric substorm, 3. Plasmaspheric electric fields and evolution of the plasmapause," *J. Geophys. Res.*, *86*, 2261-2272, 1981.

Vasyliunas, V. M., Theoretical models of magnetic field-line merging, 1, *Rev. Geophys. Space Phys.*, *13*, 303-336, 1975.

Weimer, D. R., A flexible, IMF-dependent model of high-latitude electric potentials having "space weather" applications, *Geophys. Res. Lett., 23*, 2549-2552, 1996.

Weiss, L. A., P. H. Reiff, R. Hilmer, J. D. Winningham, and G. Lu, Mapping the aurora into the magnetotail using Dynamics Explorer plasma data, *J. Geodes. Geomagn., 44*, 1121-1144, 1992.

Williams, D. J., E. C. Roelof and D. G. Mitchell, Global magnetospheric imaging, *Rev. Geophys., 30*, 183-208, 1992.

Wing, S., P.T. Newell and T. G. Onsager, Modeling the entry of magnetosheath electrons into the dayside ionosphere, *J. Geophys. Res., 101*, 13,155-13,167, 1996.

Young, D. T., Near-equatorial magnetospheric particles from 1 eV to 1 MeV, *Rev. Geophys. Space Phys., 21*, 402, 1983.

Xue, S., P. H. Reiff, and T. Onsager, Mid-altitude modeling of cusp ion injection under steady and varying conditions, *Geophys. Res. Lett., 24,* 2275-2278, 1997a.

Xue, S., P. H. Reiff, and T. Onsager, Cusp ion injection and number density modeling in realistic electric and magnetic fields, *Phys. Chem. Earth*, 22, pp. 735-740, 1997b.

Patricia H. Reiff, Department of Space Physics and Astronomy, Rice University, 6100 Main St., Mail Stop 108, Houston TX 77005. Phone: (713)527-4634; fax (713)285-5143. Email: reiff@rice.edu.

Cusp Ion Composition as an Indicator of Non-Steady Reconnection

S. A. Fuselier and K. J. Trattner

Lockheed Martin Advanced Technology Center

In situ cusp observations provide a snapshot of magnetopause conditions. For southward IMF, these observations often indicate that reconnection at the magnetopause is occurring in a non-steady manner. The signature for this variability is nearly discontinuous changes in the low energy cutoff of the magnetosheath ions in the cusp. Often this low energy cutoff is difficult to observe. Recently, another indicator for this non-steady reconnection has been identified. Using composition measurements from the POLAR spacecraft, it has been shown that the differences in the magnetosheath velocity distributions leads to a distinctive, relatively slow variation in the solar wind ion composition through the cusp. More rapid changes in this cusp composition indicate changes in reconnection at the magnetopause. All southward IMF cusp intervals sampled here showed these rapid changes in the cusp ion composition. The average period of these oscillations was approximately 2 minutes. These oscillations are consistent with ~20% variations in the reconnection rate and/or the deHoffmann-Teller velocity at the magnetopause.

1. INTRODUCTION

The understanding of the transfer of mass, momentum, and energy from the solar wind into the magnetosphere is an important problem in magnetospheric physics. Spacecraft observations at the Earth's magnetopause [e.g., *Sonnerup et al.*, 1981] have confirmed the prediction by *Dungey* [1961] that this transfer can occur through magnetic reconnection of magnetosheath and magnetospheric field lines. Quantifying the transfer through this process is a more difficult problem. It requires knowledge of the reconnection rate and the spatial extent over which this reconnection is occurring, two quantities that are not easily obtained at the magnetopause. Most in situ observations can be readily related to reconnection as a mechanism for plasma transfer across the magnetopause but few observations relate to its rate. One exception is the magnitude of the normal component of the magnetic field at the magnetopause. However, this component is not easily observed [e.g., *Sonnerup and Ledley*, 1979]. Determining the spatial extent over which reconnection is occurring is very difficult because it requires determining the rate over the entire magnetopause.

Quantifying the transfer of mass, momentum, and energy across the magnetopause through reconnection is further complicated by the fact that reconnection is not steady. Changing conditions in the magnetosheath cause the reconnection x-line (or lines) to move, the field line convection to change, and the reconnection rate to vary. Long term variations in the location of reconnection (and to a lesser extent in the rate of reconnection) have been investigated using statistical methods [e.g., *Phan et al.*, 1996]. Shorter term variations in reconnection (i.e., of the order of minutes) are not easily addressed with in situ observations at the magnetopause.

Magnetospheric field lines at the magnetopause converge in the Earth's cusps. For some time, it has been recognized that the cusp is an excellent place to monitor changes at the magnetopause in general and changes in reconnection

Sun-Earth Plasma Connections
Geophysical Monograph 109
Copyright 1999 by the American Geophysical Union

in particular [e.g., *Lockwood and Smith*, 1992]. This is especially true for southward IMF, where subsolar reconnection produces a relatively simple cusp geometry [e.g., *Rosenbauer et al.*, 1975; *Reiff et al.*, 1977]. In this geometry, the low latitude boundary layer, cusp proper, and mantle [e.g., *Newell and Meng*, 1992] are considered a single region, called the cusp. Poleward convection of reconnected magnetic field lines produces a velocity filter effect of the precipitating magnetosheath plasma. For low altitude spacecraft such DMSP, the spacecraft velocity is rapid compared to this field line convection velocity and a traversal of the cusp yields a snapshot of nearly the entire dayside magnetopause at an "instant" in time. For high altitude spacecraft such as POLAR, the spacecraft velocity can be smaller than the field line convection velocity and the spacecraft can "monitor" a region of the magnetopause at a given location within the cusp for some time.

Relatively simple models of this ion precipitation under southward IMF conditions have been developed and tested. These models have been very successful in reproducing important features of the cusp [e.g., *Lockwood and Smith*, 1992; *Onsager et al.*, 1993].

Specifically, there are observations of energy-latitude dispersion in the cusp whereby the highest energy ions are observed at the equatorial edge of the cusp and successively lower energy ions with a low energy cutoff are observed at higher latitudes. The models have been used to demonstrate that this dispersion is consistent with the velocity filter effect and the finite extent of the reconnection region at the dayside magnetopause.

These models produce a smooth energy-latitude dispersion signature because they have relatively simple, static input conditions. Almost all in situ cusp observations show deviations from the overall smooth energy-latitude dispersion. In particular, fluctuations of the low energy cutoff of the precipitating ions have been interpreted as the result of changes in reconnection at the magnetopause [e.g., *Lockwood and Smith*, 1992]. Unfortunately, using these fluctuations as a monitor of changes in reconnection is complicated by the fact that the low energy cutoff is often difficult to observe.

The purpose of this paper is to present another method for quantifying the changes in reconnection at the magnetopause. Changes in the solar wind ion composition in the cusp are used as a proxy for changes in the precipitation velocity of the ions. Using this method, a survey is conducted of some cusp crossings of the Polar spacecraft under southward IMF conditions. The results from this survey are then related to changes in reconnection at the magnetopause.

2. SAMPLE CUSP ENCOUNTER

Figure 1 shows the Polar orbit for 15 Sept 1997. The spacecraft orbit was nearly in the noon-midnight meridian and Polar encountered the cusp at 6-7 R_E geocentric distance. Over the period of about an hour, the spacecraft moved relatively slowly from the polar cap to the equatorial edge of the cusp. The solar wind conditions were nominal during this interval. The Wind spacecraft observed a solar wind density of ~4 cm^{-3}, a bulk flow velocity of ~425 km/s and a proton thermal speed of 35 km/s. The IMF was southward and had a relatively large $+B_y$ component during the interval. (These data were corrected for the plasma convection time from Wind to Polar.)

The top two panels of Plate 1 show energy time spectrograms of the omni-directional flux of H^+ and He^{2+} observed by the Toroidal Imaging Mass Angle Spectrograph (TIMAS) [*Shelley et al.*, 1995] in the cusp. Only the last 25 minutes of the cusp traversal are shown to emphasize the composition changes discussed below (after 0400 UT, Polar was in the magnetosphere). The time resolution in Plate 1 is 4 spins (24 s) although the TIMAS instrument returns a full energy spectrum for H^+ and He^{2+} every spin (6 s). For H^+, the highest energies are observed near the equatorial edge of the cusp (at ~0400 UT) and lower energies are observed at higher latitudes. H^+ flux below 100 eV/e, especially after 0350 UT, is from the ionosphere. This energy-latitude dispersion of precipitating magnetosheath ions is consistent with magnetic reconnection at the dayside magnetopause. The dispersion is better seen in the He^{2+} spectrogram in the second panel because there is no major source of ionospheric He^{2+}. The peak fluxes for He^{2+} occur at higher energies than the peak fluxes for H^+ because the two distributions precipitate with the same velocity. The third panel shows the He^{2+}/H^+ density ratio. This ratio is initially similar to the average solar wind ratio, it then increases to values well above the solar wind ratio, and finally it decreases again near the equatorial edge of the cusp to values below the average solar wind ratio.

The change in the He^{2+}/H^+ density ratio in the cusp was first noted by *Shelley et al.* [1976]. They interpreted this change as the result of different velocity space distributions for the source H^+ and He^{2+} populations in the magnetosheath. Their interpretation of the magnetosheath source distributions was later confirmed by H^+ and He^{2+} observations in the magnetosheath [e.g., *Peterson et al.*, 1979; *Fuselier et al.*, 1988].

Recently, a direct comparison of cusp observations and models of the magnetosheath H^+ and He^{2+} distributions has been made [*Fuselier et al.*, 1998]. (No comparison of observations has been possible to date because there have been no reported simultaneous composition measurements in the magnetosheath and in the cusp). Figure 2 is an example of this comparison of the model and observations. For the observations in Figure 2, the maximum flux for H^+ and He^{2+} in Plate 1 was converted to phase space density. These H^+ (open squares) and He^{2+} (filled circles) phase space densities are plotted in Figure 2 versus the velocity at which their respective maximum fluxes were measured. Under certain assumptions, this representation of cusp observations is directly comparable to the source distributions in the magnetosheath [*Fuselier et al., 1998*].

The model distributions that were fit to these observations (solid lines in Figure 2) are described in detail elsewhere [*Fuselier et al., 1998*]. Briefly, the H^+ distribution in the magnetosheath is modeled by two maxwellians representing the core component between 0 and 400 km/s and shoulder component above 400 km/s [see e.g., *Sckopke et al., 1983*]. The He^{2+} distribution is modeled by 3 maxwellians. The first 2 represent the shell component between 0 and 500 km/s in Figure 2. Two maxwellians are used to produce the shell by subtracting a lower temperature maxwellian from a higher temperature one to produce a hole in the velocity space distribution. The third maxwellian represents the shoulder above 600 km/s [see e.g., *Fuselier and Schmidt, 1997*]. Parameters which characterize the maxwellian components are obtained from the upstream solar wind conditions measured by the Wind spacecraft. The only free parameter in the fit is the change in the thermal speed for the core H^+ component across the bow shock. This parameter fixes the core H^+ temperature below 400 km/s in Figure 2.

As seen in Figure 2, the model fits and observations compare reasonably well. For He^{2+}, the model and observations deviate from one another below about 200 km/s. This deviation is partly due to a saturation effect in the TIMAS instrument, which results in anomalously high He^{2+} count rates for high H^+ fluxes. This saturation effect produces He^{2+}/H^+ density ratios above ~15% in Plate 1.

Because the two velocity space distributions are different, the He^{2+}/H^+ phase space density (and hence the density ratio in the cusp) is a function of the velocity of the precipitating ions. Thus, as the velocity of the precipitating ions increases in Plate 1, the He^{2+}/H^+ density ratio changes. The third solid line in Figure 2 shows the change in the He^{2+}/H^+ phase space density ratio with velocity and the scale for this ratio is on the right hand side of the figure. When the velocity of the precipitating ions is near zero, the He^{2+} and H^+ phase space densities in Figure 2 are relatively far apart, the phase space density ratio is less than 1%, and the density ratio in the cusp (for example at 0340

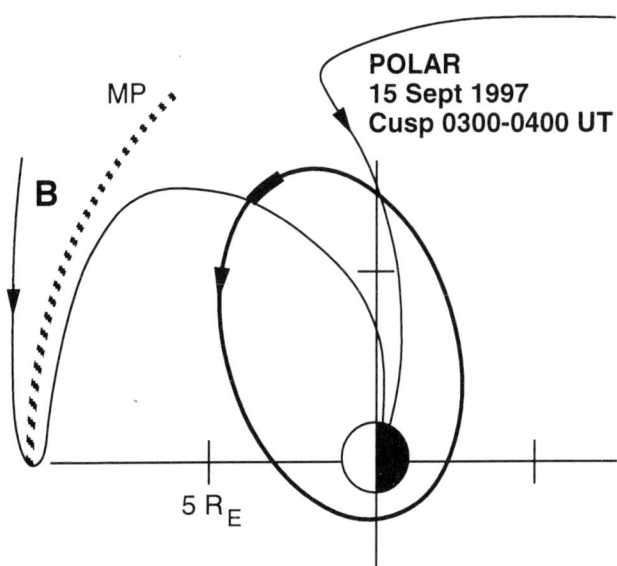

Figure 1. Polar orbit for 15 September 1997. The spacecraft spent almost an hour in the cusp, moving from the poleward edge to the equatorward edge.

UT in Plate 1) is low. As the velocity of the precipitating ions increases to near 350 km/s, the He^{2+} and H^+ phase space densities approach one another in Figure 2, the phase space density ratio is over 10%, and the density ratio in the cusp (for example at 0349 UT in Plate 1) is high. Finally, as the velocity of the precipitating ions increases above 400 km/s, the He^{2+} and H^+ phase space densities become relatively far apart again, the phase space density ratio decreases, and the density ratio in the cusp (for example at 0355 UT) is low again.

Unlike the variation in the phase space density ratio from the model in Figure 2, the variation in the density ratio in the cusp is not smooth. Plate 1 shows several spikes in the density ratio, for example just before and after 0350 UT. Figure 3 shows how these spikes are related to changes in the He^{2+} flux and ultimately changes in the velocity of the precipitating ions. The top panel shows 3 contours of constant He^{2+} flux as a function of time centered on 0350 UT. The bottom panel shows the He^{2+}/H^+ density ratio. The density ratio decreases from a maximum at about 0349:10 UT to a relative minimum at 0350:20 UT and then back to a relative maximum at 0351 UT. The contours of constant flux in the upper panel of Figure 3 decrease in velocity and then increase again in concert with the changes in the density ratio.

Variations in the low energy cutoff of the precipitating ions (represented by the 5×10^4 flux contour in Figure 3) have been directly related to changes in reconnection at the

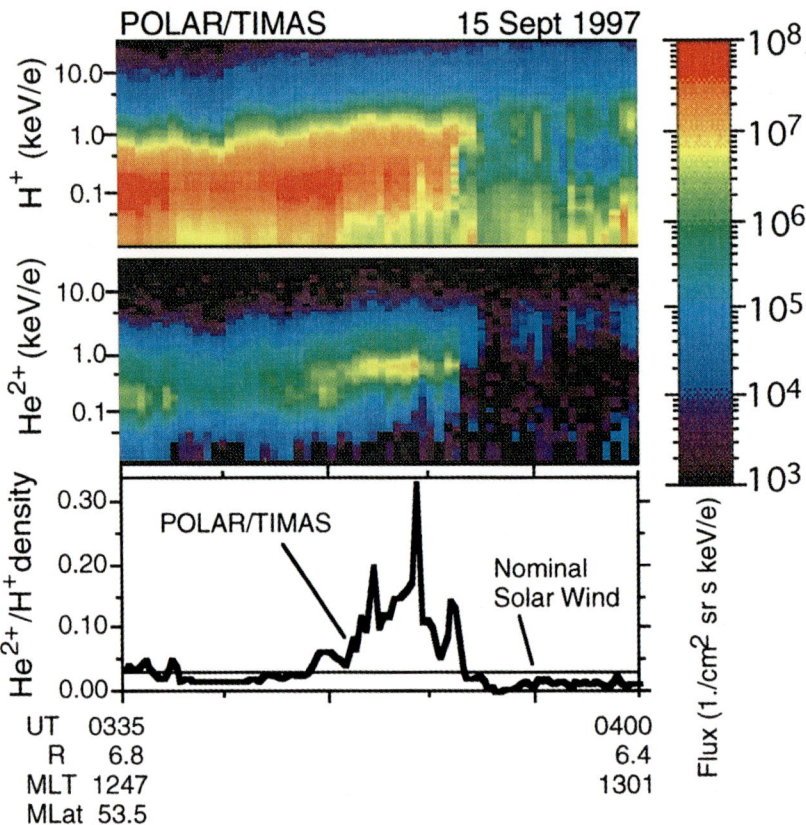

Plate 1. Omni-directional flux of H^+ and He^{2+} and the He^{2+}/H^+ density ratio for a representative traversal of the cusp. H^+ (above ~100 eV/e) and He^{2+} show a characteristic energy-latitude dispersion from high to low energies as the spacecraft moves from high to low latitudes. The He^{2+}/H^+ density ratio changes from below the nominal solar wind ratio to well above it and then back to below it as the spacecraft traverses the cusp.

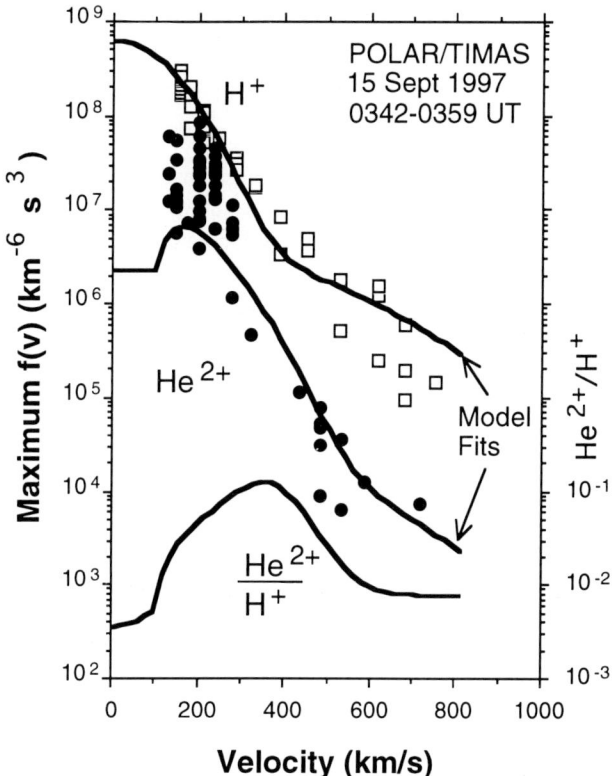

Figure 2. Maximum H^+ and He^{2+} flux from 0342-0359 UT in Plate 1 versus the velocity of the maximum flux. Under certain assumptions, this is directly comparable to the source distributions in the magnetosheath (represented by the model fits). The ratio of the phase space densities has the same profile with velocity as the density ratio in Plate 1.

magnetopause [e.g., *Lockwood and Smith*, 1992]. Thus, Figures 2 and 3 establish a direct link between changes in reconnection at the magnetopause and changes in the He^{2+}/H^+ density ratio in the cusp. With this link, the He^{2+}/H^+ density ratio becomes a proxy for the changes in the energy of the precipitating ions and ultimately the changes in reconnection at the magnetopause.

Detailed consideration of Figures 2 and 3 illustrates how this link results in quantitative determination of the changes in the velocity of the precipitating ions. The ratio of the He^{2+}/H^+ phase space densities from the model distributions in Figure 2 is used to estimate the change in the precipitating ion velocity due to the change in the density ratio observed in Figure 3. In the lower panel of Figure 3 from 0349:10 to 0350:54 UT, the density ratio changes from over 30% to approximately 6% and then back to about 13%. The 30% density ratio is due in part to saturation of the He^{2+} signal. The value of the density ratio at this time corrected for this saturation is approximately 15%. As the density ratio changes from a maximum to a relative minimum and back to a maximum, the precipitation velocity of the ions with the maximum flux changes from about 350 km/s to about 220 km/s and then back to about 350 km/s. Thus, in about 2 minutes, the precipitating ion velocity changes a total of about 130 km/s in one direction and then back by the same amount.

Comparing this total change with the top panel of Figure 3, it is apparent that the estimated change in the precipitating ion velocity is consistent with the change in velocity of the low energy cutoff of the He^{2+} flux represented by the 5×10^4 flux contour. In the 2 minute time period, the low energy cutoff changes from between 160 and 200 km/s to about 90 km/s and then back to 200 km/s. The total change is between 180 and 220 km/s, compared to 260 km/s change in the velocity estimated using the model distributions in Figure 2. The uncertainties in both these estimates are approximately ±30 km/s due to the discreet energy steps in the instrument and, given the fidelity of the comparison, the two numbers compare reasonably well.

The event in Plate 1 was chosen because it had a relatively large change in the precipitating velocity that clearly demonstrated the direct link between changes in the low energy cutoff of He^{2+} and changes in the density ratio. However, the determination of the low energy cutoff is not always easy. Plate 2 shows a cusp event where the changes in the low energy cutoff are less evident. The top 2 panels in the plate are the H^+ and He^{2+} energy-time spectrograms similar to those in Plate 1. As in Plate 1, the H^+ and He^{2+} fluxes in Plate 2 show a relatively smooth energy-latitude (time) dispersion consistent with reconnection at the dayside magnetopause. The energy dispersion is reversed from Plate 1 because the spacecraft was moving toward higher latitudes in Plate 2. Although the energy-latitude dispersion is smooth, the He^{2+}/H^+ density ratio in the third panel shows considerable fluctuations. These fluctuations are correlated with changes in the He^{2+} flux, as seen in the bottom panel of Plate 2. Changes in the H^+ flux and in particular in the low energy cutoff of the flux are much less evident in the top panel.

3. SURVEY OF SELECTED SOUTHWARD IMF EVENTS

Nine cusp events including those in Plates 1 and 2 were chosen to survey the changes in the density ratio and their relation to changes in the velocity of the precipitating ions and changes in reconnection at the magnetopause. All 9 events exhibited good energy-latitude dispersion as in Plates 1 and 2. All events occurred when the IMF was

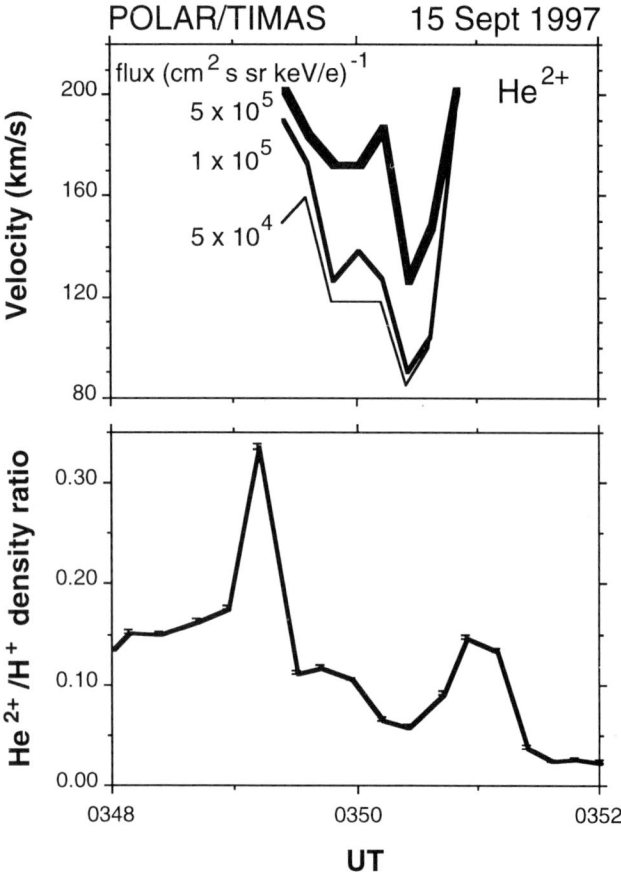

Figure 3. Contours of constant flux and the He^{2+}/H^+ density versus time for part of the event in Plate 1. Changes in the low velocity cutoff of the flux (represented by the 5×10^4 flux contour) are directly related to changes in the density ratio.

southward as observed by the Wind spacecraft. Most significantly, all events exhibited fluctuations in the density ratio similar to the fluctuations in Plates 1 and 2.

For each event, the solar wind conditions from the Wind spacecraft were used to model the magnetosheath distributions as in Figure 2. From these model distributions, the change in the He^{2+}/H^+ density ratio was directly related to the change in the velocity of the precipitating ions as in the discussion of Figures 2 and 3 in the previous section. For all events, the period between maxima in the density ratio was measured. For example, in the third panel of Figure 2, there are 4 peaks in the He^{2+}/H^+ density ratio at about 0345, 0347, 0349, and 0351 UT. For each of these oscillations in the density ratio, the total velocity change was computed using the model distributions. For example, the oscillation isolated in Figure 3 had a total velocity change of the precipitating ions of 260 km/s (as discussed in the previous section). This total change consisted of a 130 km/s excursion to lower velocities followed by a 130 km/s excursion back to higher velocities within a period of about 2 minutes.

Figure 4 shows the total velocity change of the precipitating ions as a function of the period between peaks in the density ratio. There is considerable scatter in the points and no correlation is evident. The average period was about 2 minutes and the total velocity change was 130 km/s. It is significant that the change in the velocity of the precipitating ions in Figure 3 was considerably larger than this average. This made it a good choice for illustrating how the density ratio change and the velocity of the precipitating ions are related.

Figure 5 shows the maximum velocity of the precipitating ions (defined as the velocity where the density ratio was maximum) as a function of the period between maxima in the density ratio. This plot shows that the longer period oscillations in the density ratio occur when the maximum flux is at lower velocities. An example of this trend is seen in Plate 2. The last two oscillations in the density ratio (from 1657 to 1702 UT) occur 5 minutes apart when the precipitating ion flux is at lower energies. Nearer to the equatorial edge of the cusp, where the energies are higher, the periods are of the order of 2 minutes. This may indicate that the cause for precipitating ion velocity changes and reconnection changes are different for plasma in the traditional "cusp" than for plasma in the traditional "mantle". This is a subject for future study.

4. INTERPRETATION

As stated in the introduction, the goal of this study is to relate the changes in the cusp to changes in reconnection at the magnetopause. In the previous section, changes in low energy cutoff of the precipitating ions were related to changes in the density ratio. In Figure 4, the average period between oscillations in the precipitating ion flux was about 2 minutes and the average total change in the velocity of the precipitating ions was 130 km/s. Thus, the low energy cutoff velocity of the precipitating ions changes by about 65 km/s in about 1 minute. In this section, these average changes are related to possible changes in magnetopause reconnection.

In the simple interpretation of the cusp as a velocity filter, a change in the velocity of the precipitating ions occurs because the time-of-flight of the ions that reach the spacecraft has changed. This change can occur by moving the reconnection line, changing the convection velocity of the field line, or changing the rate of reconnection at the magnetopause. Any combination of these changes could also

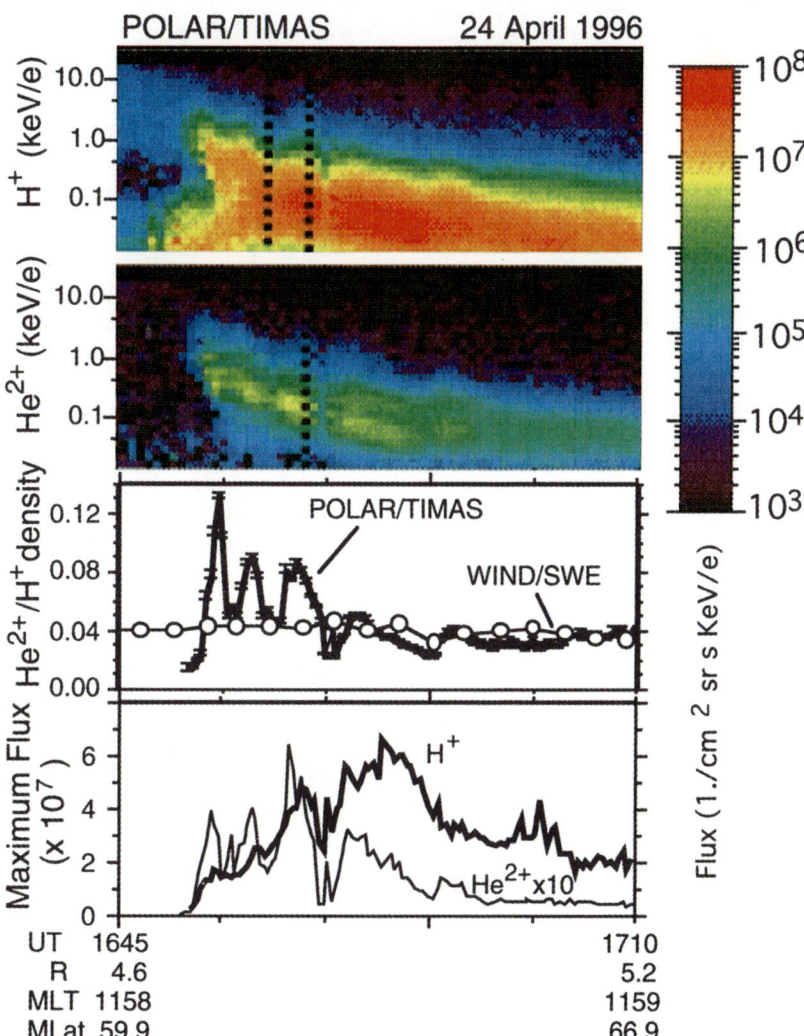

Plate 2. Omni-directional flux of H^+ and He^{2+}, He^{2+}/H^+ density ratio, and the maximum flux for H^+ and He^{2+} for another cusp traversal. The energy-latitude dispersion is similar to Plate 1 but reversed in time because the spacecraft is moving from equatorward to poleward. Oscillations in the He^{2+}/H^+ density ratio are clearly evident but correlation with changes in the lower energy cutoff in H^+ are not evident at all.

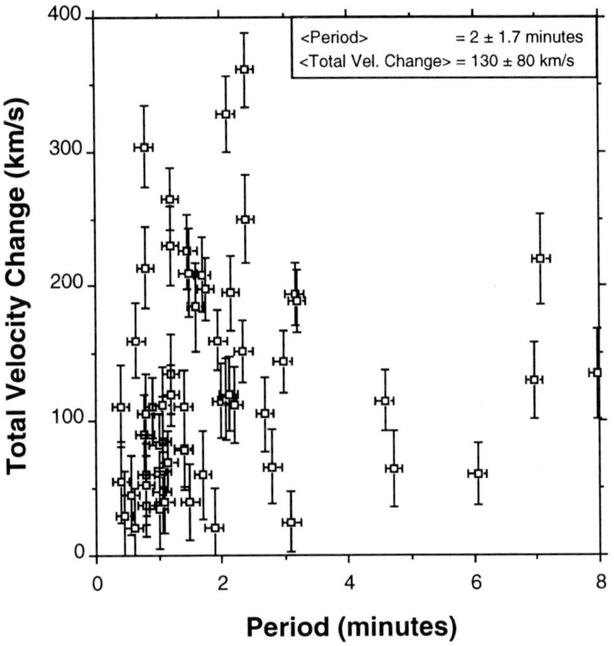

Figure 4. Total velocity change of the precipitating ions versus the period between these changes (determined by the period between successive peaks in the He^{2+}/H^+ density ratio). The average period between these velocity changes from all events was 2 minutes and the average total velocity change was 130 km/s.

occur, but it is important to estimate the magnitude of the required changes for each process individually.

The basic equation for the time of flight of an ion from the reconnection site to the spacecraft is:

$$V_{\|i} = L_i / t \quad (1)$$

where L is the length of the magnetic field line (~10 R_E) and $V_\|$ is the velocity of the ion parallel to the magnetic field.

Moving the Reconnection Line

By moving the reconnection line along the magnetopause, L in (1) is changed. If the time is constant, then the parallel velocity of the precipitating ions must change. For L_1 ~10 R_E (a typical distance from the subsolar point to the Polar spacecraft in the cusp), $V_\|$ ~ 300 km/s (from Figure 5), and the change in $V_\|$ ~65 km/s (from half of the average in Figure 4), L_2 ~ 12 R_E. Thus, to account for the average changes in the velocity of the precipitating ions in the cusp, the reconnection line must move 2 R_E (poleward or equatorward along the magnetopause) and then return 2 R_E to near its original position. These are large changes in the position of the X-line that require accelerations of the X-line of the order of 10 km/s². The consequences of such accelerations would be very obvious in the in situ observations at the magnetopause. The fact that changes in the location of the reconnection site during multiple magnetopause crossings are not typically observed suggests that this is not the dominant means whereby reconnection conditions are changed on 1 minute time scales.

Changing the Field Line Convection Velocity

A reconnected field line convects with the deHofmann-Teller velocity at the magnetopause. If this velocity is increased or decreased, then the parallel velocity of the precipitating ions must increase or decrease by the same amount to keep the time of flight of the ions from the reconnection site to the spacecraft constant. A 65 km/s increase in the parallel velocity from 300 to 365 km/s (from one half the average total change in $V_\|$ in Figure 4 and the average velocity of the precipitating ions in Figure 5) represents about a 20% change in the parallel velocity of the precipitating ions. Thus, in about 1 minute, the deHoffmann-Teller velocity changes by about 65 km/s. The deHoffmann-Teller velocity has been measured for selected in situ observations of magnetopause crossings [e.g., *Sonnerup et al.*, 1990]. The existence of this velocity is a necessary condition for reconnection at the magnetopause. Under certain assumptions, the bulk flow of the plasma on either side of the open magnetopause is at the local Alfven velocity in the frame of reference moving with the deHoffmann-Teller velocity. In a few special cases, it was found that there was better agreement between the observed velocities and the local Alfven speed if the deHoffmann-Teller velocity was not constant. In particular, accelerations of the order of 1 km/s² tangential to the magnetopause were needed to achieve better agreement between theory and observation [*Sonnerup et al.*, 1990]. Typical spacecraft crossings of the magnetopause take about 1 minute and deHoffmann-Teller velocities are typically of the order of 300 km/s. Thus, the ~1 km/s² acceleration of the deHoffmann-Teller frame represents about a 20% change in the deHoffmann-Teller velocity during a magnetopause crossing. Although very few magnetopause crossings have been investigated in such detail, the inferred order of magnitude of the acceleration is consistent with the changes in the cusp ion precipitation in Figure 4.

Changing the Reconnection Rate

If the inflow of magnetic field lines into the reconnection region increases, then the magnetopause moves inward (erosion) and equatorial edge of the cusp moves equatorward. For a near stationary spacecraft such as Polar in the high latitude cusp, this would cause the energy of the precipitating ions that arrive at the spacecraft to decrease. If the tangential electric field and the normal component of the magnetic field at the magnetopause change by equal amounts, then the deHoffmann-Teller velocity remains constant while the reconnection rate changes. Also, the position of the X-line on the magnetopause does not change. Thus, while the field line convection velocity and the X-line position do not change, the movement of the equatorial edge of the cusp to lower latitudes through magnetopause erosion effectively changes the position of the spacecraft in the cusp. This process of changing the reconnection rate without changing the field line convection velocity has been suggested previously [e.g., *Lockwood and Smith*, 1992].

Since this change in the reconnection rate is linearly related to the change in the parallel convection velocity required to reach the spacecraft, the ~20% change in the parallel convection velocity represents a 20% change in the reconnection rate. As stated in the introduction, the reconnection rate is very difficult to measure with in situ observations at the magnetopause. Thus, it is difficult to determine from independent measurements at the magnetopause if the rate varies continuously by about ±20% or more. The only direct signature of this variation is the movement of the magnetopause and the simultaneous shifting of the cusp to higher or lower latitudes. Recent simultaneous observations from the ground and at the magnetopause indicate that this certainly occurs [*Mende et al.*, 1998]. Once again, there are few simultaneous observations of the magnetopause and cusp positions so there is only enough information to conclude that erosion may be a mechanism for producing the changes in the precipitation velocity in Figure 4.

5. CONCLUSIONS

In this paper, an additional method for quantitative investigation of short term (~minute) variations in reconnection at the magnetopause was presented. Previously, these changes were investigated by directly observing changes in the low energy cutoff velocity of the precipitating ions in the cusp. Here, the change in the He^{2+}/H^+ density ratio was introduced as a proxy for this change in the cutoff velocity.

Oscillations in the He^{2+}/H^+ density ratio were observed in all events surveyed here. The average change in the precipitating ion velocities deduced from these oscillations was about 130 km/s (65 km/s in one direction and then 65 km/s in the other) over an average period of 2 minutes.

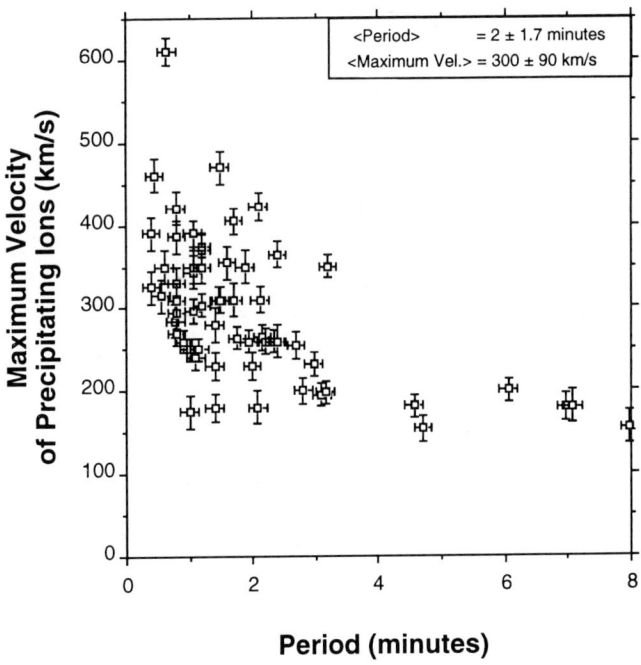

Figure 5. Maximum velocity (defined as the velocity at which the He^{2+}/H^+ density ratio is maximum) versus the period. Longer period oscillations in the precipitating ion velocity occur when the maximum velocity is low. This may be an indication that there are longer period changes in reconnection for field lines connected to the mantle.

Such changes maybe due to movement of the reconnection line along (tangential to) the magnetopause, changes in the deHoffmann-Teller velocity at the magnetopause, and/or changes in the reconnection rate. The amplitude of the changes in the velocity of the precipitating ions in the cusp is large enough to rule out movement of the reconnection line along the magnetopause as the prime reason for the changes. This amplitude is consistent with ~20% changes in either the deHoffmann-Teller velocity or the reconnection rate or some combination of both possibilities. Changes in the deHoffmann-Teller velocity of that magnitude have been observed at the magnetopause for some magnetopause crossings. The signature of a 20% change in the reconnection rate would be a simultaneous inward (outward) motion of the magnetopause and equatorward (poleward) motion of the cusp. Such simultaneous motion has been observed. Distinguishing between these two possibilities requires simultaneous observations at the magnetopause and in the cusp.

Acknowledgments. The TIMAS investigation is the result of more than a decade of work by many dedicated engineers and scientists at several institutions. Until his retirement in 1998, E. G. Shelley was the PI of the TIMAS instrument. The PI is now W. K. Peterson. Solar wind data were obtained from the Wind Magnetometer Experiment (R. Lepping, PI) and the Wind Solar Wind Experiment (K. Ogilvie, PI).

REFERENCES

Dungey, J. W., Interplanetary field and the auroral zones, *Phys. Rev. Lett., 6*, 47, 1961.

Fuselier, S. A., E. G. Shelley, and D. M. Klumpar, AMPTE/CCE observations of shell-like He^{2+} and O^{6+} distributions in the magnetosheath, *Geophys. Res. Lett., 15*, 1333, 1988.

Fuselier, S. A., and W. K. H. Schmidt, Solar wind He^{2+} ring beam distributions downstream from the Earth's bow shock, *J. Geophys. Res., 102*, 11,273, 1997.

Fuselier, S. A., E. G. Shelley, W. K. Peterson, and O. W. Lennartsson, Solar wind He^{2+} and H^+ distributions in the cusp for southward IMF, in Polar Cap Boundary Phenomena, ed. J. Moen et al., p. 63, Kluwer Academic, Netherlands, 1998.

Lockwood, M. and M. F. Smith, The variation of reconnection rate at the dayside magnetopause and cups ion precipitation, *J. Geophys. Res., 97*, 14,841, 1992.

Mende, S. B., D. M. Klumpar, S. A. Fuselier, and B. J. Anderson, Dayside auroral dynamics: South Pole - AMPTE/CCE observations, *J. Geophys. Res., 103*, 6891, 1998.

Newell, P., and C.-I. Meng, Mapping the dayside ionosphere to the magnetosphere according to particle precipitation characteristics, *Geophys. Res., Lett., 19*, 609, 1992.

Onsager, T. G., C. A. Kletzing, J. B. Austin, and H. MacKiernan, Model of magnetosheath plasma in the magnetosphere: Cusp and mantle particles at low altitudes, *Geophys. Res. Lett., 20*, 479, 1993.

Peterson, W. K., E. G. Shelley, R. D. Sharp, R. G. Johnson J. Geiss, and H. Rosenbauer, H^+ and He^{++} in the dawnside magnetosheath, *Geophys. Res. Lett., 6*, 667, 1979.

Phan, T.-D., G. Paschmann, B. U. Ö. Sonnerup, Low latitude dayside magnetopause and boundary layer for high magnetic shear 2. Occurrence of magnetic reconnection, *J. Geophys. Res., 101*, 7817, 1996.

Reiff, P., T. H. Hill, and J. L. Burch, Solar wind injection at the dayside magnetospheric cusp, *J. Geophys. Res., 82*, 479, 1977.

Rosenbauer, H., H. Grünwaldt, M. D. Montgomery, G. Paschmann, and N. Sckopke, HEOS 2 plasma observations in the distant polar magnetosphere: The plasma mantle, *J. Geophys. Res., 80*, 2723, 1975.

Sckopke, N., G. Paschmann, S. J. Bame, J. T. Gosling, and C. T. Russell, Evolution of ion distributions across a nearly perpendicular bow shock: Specularly and non-specularly reflected-gyrating ions, *J. Geophys. Res., 88*, 6,121, 1983.

Shelley, E. G., R. D. Sharp, and R. G. Johnson, He^{++} and H^+ flux measurements in the dayside cusp: Estimates of convection electric field, *J. Geophys. Res., 81*, 2363, 1976.

Shelley, E. G., et al., The toroidal imaging mass-angle spectrograph (TIMAS) for the Polar mission, in *The Global Geospace Mission*, ed. C. T. Russell, p 497 Kluwer Academic, Netherlands, 1995.

Sonnerup, B. U. Ö., and B. G. Ledley, Electromagnetic structure of the magnetopause and boundary layer, *in Magnetospheric Boundary Layers*, B. Battrick, ed., p. 401, ESA SP 148, European Space Agency, Paris, 1979.

Sonnerup, B. U. Ö., G. Paschmann, I. Papamastorakis, N. Sckopke, G. Haerendel, S. J. Bame, J. R. Asbridge, J. T. Gosling, and C. T. Russell, Evidence for magnetic field reconnection at the Earth's magnetopause, *J. Geophys. Res., 86*, 10,049, 1981.

Sonnerup, B. U. Ö., I. Papamastorakis, G. Paschmann, and H. Lühr, The magnetopause for large magnetic shear: Analysis of convection electric fields from AMPTE/IRM, *J. Geophys. Res., 95*, 10,541, 1990.

Stephen A. Fuselier and Karlheinz J. Trattner, Dept H1-11 Bldg 255, Lockheed Martin Advanced Technology Center, 3251 Hanover St., Palo Alto, CA 94304, USA.

Simulation of Radiation Belt Dynamics Driven by Solar Wind Variations

M. K. Hudson, S. R. Elkington and J. G. Lyon

Physics and Astronomy Department, Dartmouth College, Hanover

C. C. Goodrich

Astronomy Department, University of Maryland, College Park

T. J. Rosenberg

Institute for Physical Science and Technology, University of Maryland, College Park

The rapid rise of relativistic electron fluxes inside geosynchronous orbit during the January 10-11, 1997, CME-driven magnetic cloud event has been simulated using a relativistic guiding center test particle code driven by output from a 3D global MHD simulation of the event. A comparison can be made of this event class, characterized by a moderate solar wind speed (< 600 km/s), and those commonly observed at the last solar maximum with a higher solar wind speed and shock accelerated solar energetic proton component. Relativistic electron flux increase occurred over several hours for the January event, during a period of prolonged southward IMF B_z, more rapidly than the 1-2 day delay typical of flux increases driven by solar wind high speed stream interactions. Simulations of the January event captured the flux increase around L = 4 observed by GPS satellites, following the flux decrease associated with build up of the ring current. Analysis of ULF oscillations in the simulation data shows toroidal mode structure commensurate with electron drift periods in the 0.2 - 3.2 MeV energy range between L = 3 - 9. Oscillations in the same frequency range seen in riometer and magnetometer data suggest that resonance with ULF oscillations may play a role in energizing relativistic electrons. The radial electric field component of toroidal oscillations at the electron drift period provides a mechanism for continuous acceleration of relativistic electrons in the absence of a large inductive electric field impulse.

INTRODUCTION

The interaction of a CME-driven magnetic cloud with the earth's magnetosphere on January 10 - 11, 1997, produced an increase in outer zone relativistic electron fluxes by several orders of magnitude, depending on energy and radial location [*Li et al.*, 1998; *Reeves et al.*, 1998a; *Selesnick and Blake*, 1998]. The expanding magnetic cloud, embedded in nominal solar wind flow speed ~400 km/s, produced an interplanetary shock which crossed the WIND spacecraft ~01:00 UT on January 10. The shock impacted the magnetosphere 20 minutes later, followed by a period of average southward IMF beginning ~04:40 UT, which lasted until 17:30 UT on January 10 [*Burlaga et al.*, 1998]. There was substantial buildup of the ring current to Dst ~ -85 nT prior to northward turning of the IMF, with substorm activity indicated by a maximum three hour average $K_p = 6$, using preliminary index data from the ISTP web site (www-spof.gsfc.nasa.gov/istp/cloud_jan97/).

Here we will focus on simulations of the rise in relativistic electron flux occurring first around L = 4.5 [*Li et al.*, 1998; *Reeves et al.*, 1998a], as seen by GPS satellites in circular (L = 4.2, 55 deg inclination) orbit, which map flux at L > 4.2 extrapolated from measurements off-equator, and provide a relatively continuous determination of the rise in flux vs. L within geosynchronous orbit at 0.2 - 0.4, 0.4 - 0.8, 0.8 - 1.6 and 1.6 - 3.2 MeV. In addition, POLAR provides cuts in L every 17.5 hours which show a jump in flux of >1.6 MeV electrons between outer zone crossings at 04:00 - 07:00 UT on January 10, and 20:00 - 01:00 UT on January 10 - 11, by a factor of $10^3 - 10^4$, peaking around L = 4.3 [*Selesnick and Blake*, 1998]. Geosynchronous data, on the other hand, show a rise in > 1.6 - 2 MeV electron fluxes somewhat later, with peak values and rise times sensitive to local time [*Reeves et al.*, 1998b].

SIMULATION

The January 1997 magnetic cloud event has provided an opportunity to combine a very complete satellite and groundbased data set with modelling tools developed in support of the ISTP program. The model described here employs field output from a global MHD simulation code as input to a guiding center test particle code used to advance radiation belt particle trajectories [*Hudson et al.*, 1996; 1997]. These results in turn were an outgrowth of a simpler analytic model for the effect of a CME-produced storm sudden commencement (SSC) which generates a magnetosonic wave compression of the dayside magnetosphere. This model, applied to the March 24, 1991 great storm (Dst = -300), was very effective at reproducing the observed transport and acceleration of outer zone electrons into L = 2.5 on the time scale of the electron drift period [*Li et al.*, 1993]. The rise in relativistic electron flux occurred several hours after the SSC in January 1997, and was precipitated by a much slower interplanetary shock moving at a nominal solar wind speed of 400 km/s [*Burlaga et al.*, 1998], vs. the 1000 - 1400 km/s estimates for the March 1991 event. Another significant difference from the diagnostic point of view was the absence of an upstream solar wind monitor for the March 1991 event, in contrast to the current era of WIND and other L1 spacecraft measurements.

A magnetic and electric field time series has been obtained from a 3D global MHD simulation of the January 1997 magnetic cloud event. The evolving solar wind parameters measured by WIND are used as input to the Lyon-Fedder-Mobarry code [*Fedder and Lyon*, 1987]. The solar wind conditions were taken from WIND key parameter data interpolated to a constant one minute time resolution. Only the transverse components – B_x and B_y – of the IMF were used. The data were input in GSM coordinates with the Earth's dipole at a fixed tilt of 23 degrees away from the sun in the Northern hemisphere (corresponding roughly to 0 UT). The inner boundary of the simulation was at 2 R_E to allow for low-latitude convection during the magnetic cloud passage and to allow for the extreme compression inside geosynchronous orbit observed later in the event. The simulation covered the entire time period from 0 UT on January 10 to 13:00 UT on January 11, with field results recorded approximately every 75 seconds. Other details of the MHD simulation are described elsewhere [*Goodrich et al.*, 1998].

A guiding center test particle code was implemented to follow electron trajectories in the equatorial plane with interpolation of output from the MHD field time series [*Hudson et al.*, 1996; 1997]. A first adiabatic invariant conservation criterion is well satisfied for > 99.9% of simulation electrons [*Hudson et al.*, 1997]. The specific criterion imposed here is to remove from the simulation domain electrons whose gyroradius equals the magnetic gradient scale length (field variation on the gyrofrequency time scale is not an issue in these MHD-driven simulations). An initial AE8MIN energy - flux profile is used as input [*Vette*, 1991], and relative flux in selected energy ranges at fixed L and varying longitude ϕ can be plotted, with ϕ variation corresponding to an equatorial satellite period at the specified L value. The entire (L - ϕ) flux time series is available, and can be extrapolated to other latitudes assuming that flux dynamics is dominated by behavior in the equatorial plane.

Plate 1a shows two components of the electric field time series, azimuthal E_ϕ and radial E_r, along with total magnetic field strength B for a virtual satellite flown

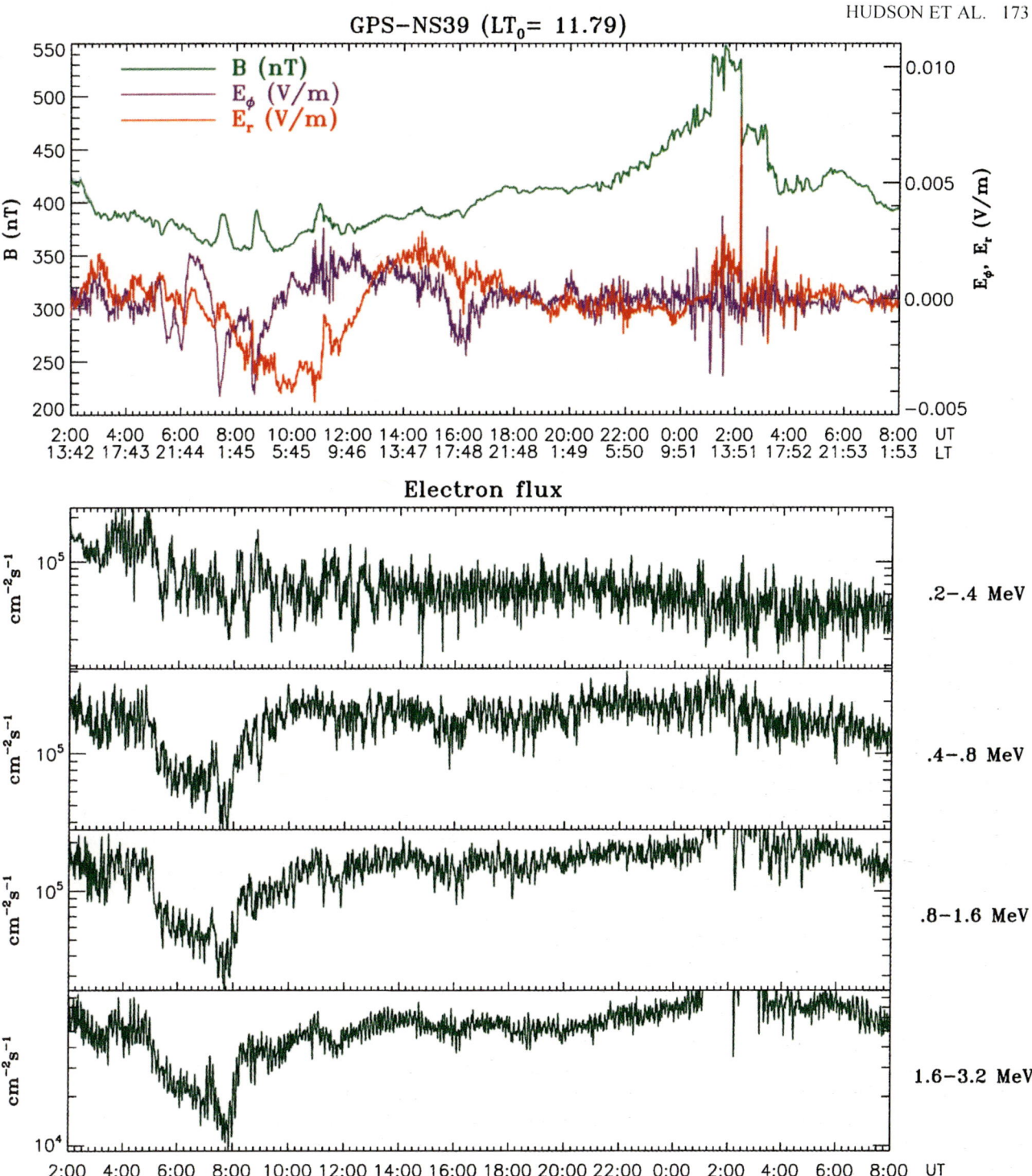

Plate 1. (a) Total **B** (nT) and two components of **E** (mV/m), azimuthal E_ϕ and radial E_r, at L = 4.2 in the equatorial plane at LT of GPS satellite NS39, from 02:00 UT Jan 10 to 08:00 UT Jan 11. Initial GPS NS39 local time was $LT_0 = 12.79$, moving roughly two hours eastward in LT for each hour UT. (b)-(e) Relative flux in four energy channels measured by GPS, 0.2 - 0.4, 0.4 - 0.8, 0.8 - 1.6 and 1.6 - 3.2 MeV, with same trajectory format as (a).

through the MHD simulation data in an equatorial orbit at L = 4.2. The particle simulations are begun at 02:02 UT on January 10 when GPS satellite NS39 was at 12.8 hours LT, with little change occurring during the first hour. Since GPS data has been used to infer time evolution of equatorial plane fluxes [*Li et al.*, 1998; *Reeves et al.*, 1998a], and since the present particle simulation is restricted to the equatorial plane, we do not attempt here to reconstruct fluxes at the GPS orbit inclination. Relative electron flux in four energy channels measured by GPS NS39 is shown in Plates 1b - 1e for the same trajectory and time period, encompassing a rapid rise in flux around 09:00 UT by half an order of magnitude in the simulation output from a value initially depressed by buildup of the ring current (Dst effect discussed below). Field signatures characteristic of substorm dipolarization (compression in B and large negative, westward E_{phi}) are seen before and after this flux increase, while it continues to rise gradually and flatten out by the time the IMF turns northward ~ 17:30 UT. Plates 2a - 2e, in the same format, show simulation data at the location of geosynchronous satellite 1994-084. Here, ULF oscillations are evident in the field data, beginning at the time of substorm field signatures in Plate 1a, ~ 006:00 UT. The dropout in flux associated with the passage of the satellite into the magnetosheath during the period following a high density solar wind pressure pulse [*Burlaga et al.*, 1998] is in good agreement with the geosynchronous measurements [*Reeves et al.*, 1998b].

Plate 3a shows a plot of flux vs. energy and L for the initial AE8MIN model, peaked between L = 5 - 6 at low energies. By 09:46 UT, Plate 3b shows that enhanced convection during several hours of southward IMF B_z has transported the outer boundary of the AE8MIN profile radially inward in the simulations, along with the flux peak at low energies, which by this time lies between L = 4 - 5. The flux stays peaked around L = 4, as the outer boundary expands outward during the ensuing period of northward IMF, with the end simulation result shown in Plate 3e at 12:02 UT on January 11, eleven hours after arrival of the high density solar wind impulse which caused the magnetopause to move inside geosynchronous orbit.

ULF OSCILLATIONS

In both observations and simulations of the January 1997 event, very limited particle acceleration at MeV energies was seen during the initial phases of the storm. The magnetic cloud was embedded in a solar wind of moderate velocity ~ 400 km/s, compared to estimates as great as 1400 km/s for the March 24, 1991 event, where significant radiation belt flux enhancement occurred on the time scale of an MeV electron drift period following the SSC. Instead of a high speed interplanetary shock, the organizing feature of the event studied here was an extended period of southward IMF characterized by substorm activity preceding the rise in relativistic electron fluxes after 09:00 UT on January 10. In the period between about 9 and 12 UT, ground magnetometers located at College and Gakona, Alaska, recorded large amplitude (several hundred nT) oscillations in the magnetic field coincident with the rise in electron flux observed by GPS spacecraft between 4.2 and 4.5 R_E. These waves had periods of around 10 minutes, also seen in riometer and scanning meridian photometer data, corresponding to the drift frequency of, e.g. 1.6 MeV electrons at L = 4.2. The riometer data shown in Figure 1 indicates further enhancement in activity around 11 UT when a moderate solar wind pressure pulse impacted the magnetosphere [*Li et al.*, 1998]. However, this enhancement was clearly embedded in ULF wave activity in the same frequency range over a three hour period at Gakona, during which significant electron flux increase was seen in the simulations at L = 4.

To analyze the effect of ULF oscillations in the simulations, a power spectrum of the model MHD fields has been produced in the various field components. A frequency analysis of the simulated radial electric field at the azimuthal location of Gakona (214.85 deg E) is shown in Plate 4, taken over 10:00 - 12:16 UT on January 10. Assuming the oscillations seen are Alfvénic in nature, this component of the electric field corresponds to a toroidal-mode field line resonance [*Southwood*, 1974]. The ascending black lines in the figure indicate the drift frequency of a particle at constant energy, while the descending lines between L = 4.2 and L = 6.6 indicate the drift frequency seen by a particle moving through this range of L-shells at constant M, where M is the relativistic first adiabatic invariant [*Schulz and Lanzerotti*, 1974]. Clearly, a particle moving from just inside geosynchronous orbit to the radial distances covered by the GPS spacecraft would encounter significant power in the spectral range matching its drift frequency. This suggests that a drift resonance between the particles and fields might be responsible for the energization observed in both in-situ measurements and in the simulation.

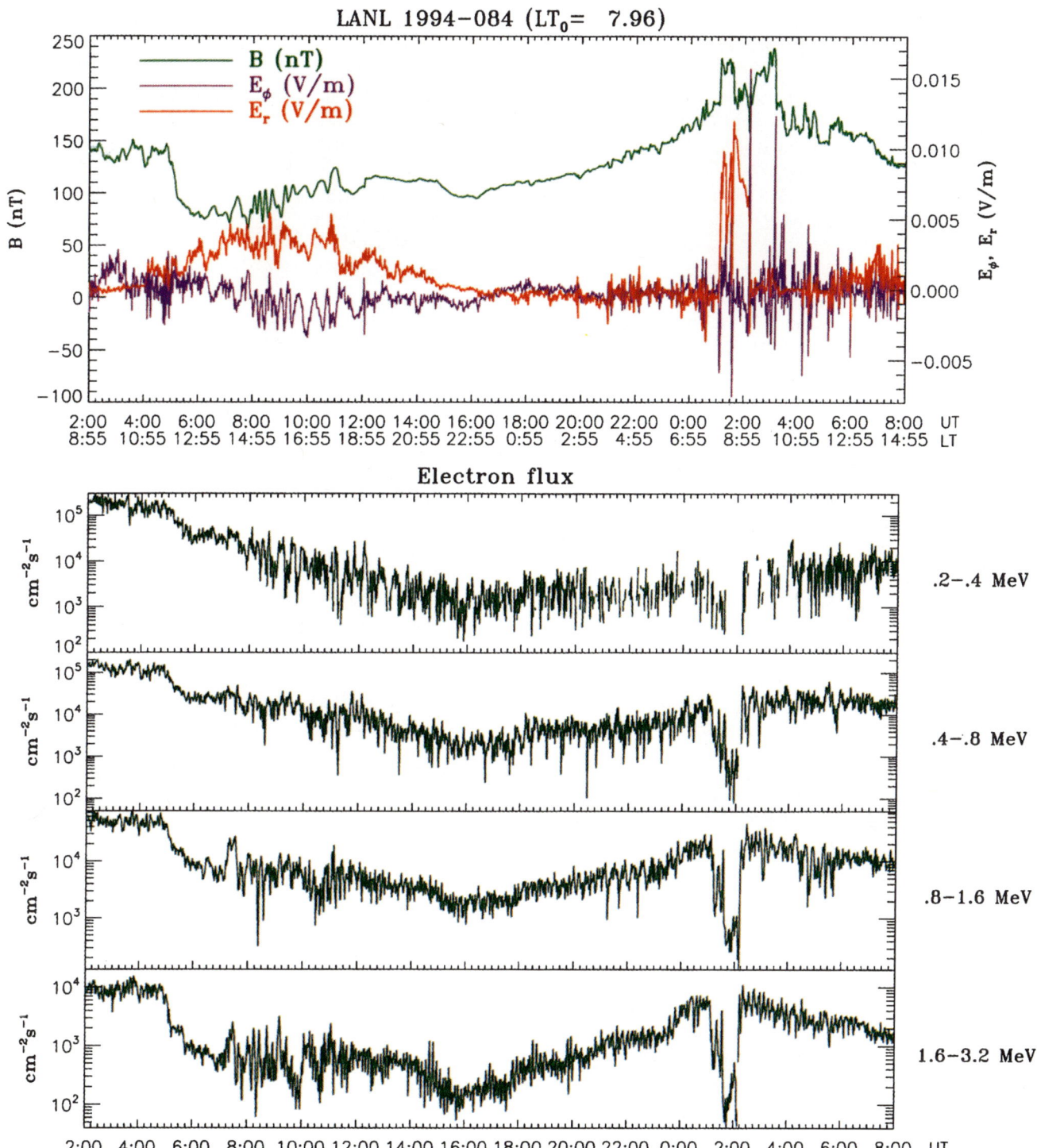

Plate 2. Same as Plate 1 at location of geosynchronous satellite 1994-084.

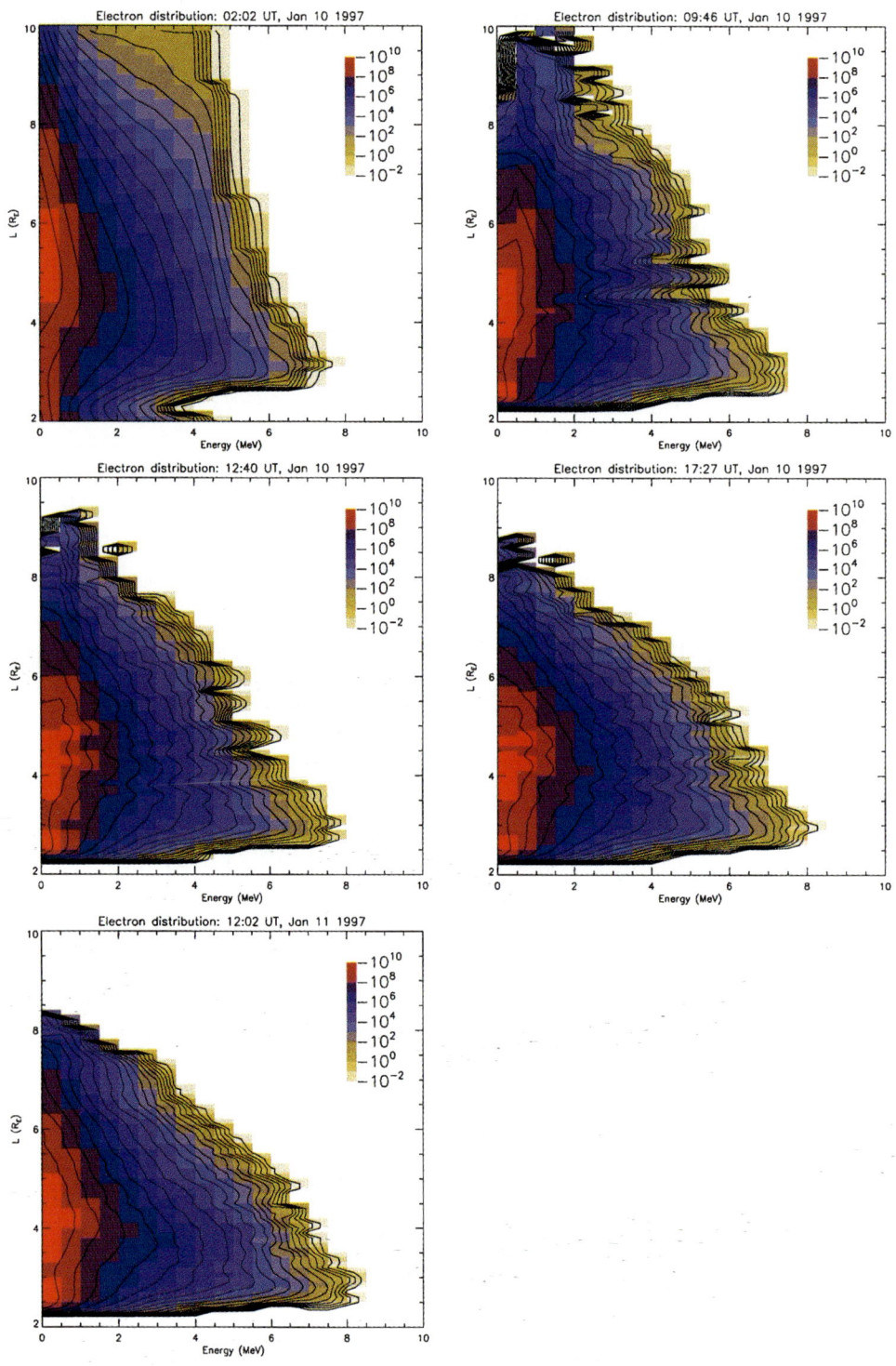

Plate 3. (a)Flux vs. energy and L for the initial, AE8MIN model electron population, whose drift trajectories are then advanced in time with MHD code field output. (b)Same as (a) after advancing electron trajectories to 09:46 UT on 10 Jan 98. (c)Same at 12:40 UT. (d) Same at 17:27 UT. (e)Same at 12:02 UT on 11 Jan 98.

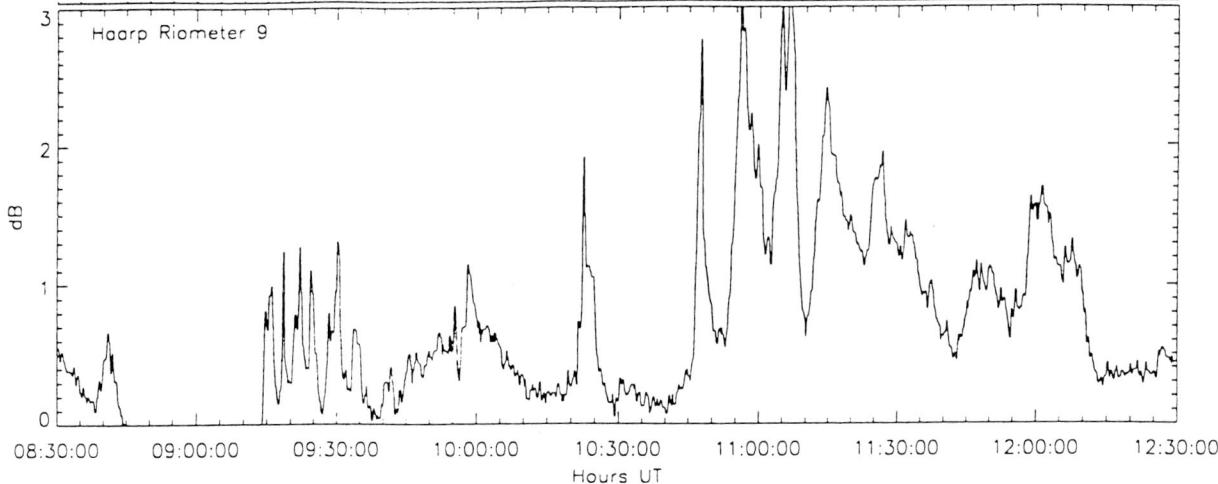

January 10, 1997 Event

Figure 1. Auroral absorption at 38.6 MHz recorded at Gakona, Alaska by one of the 16 beams of a phased-array riometer antenna. The antenna array is phased only in the meridional direction and covers a range of magnetic latitudes from 61 to 65 degrees. The trace shown is for beam 9 corresponding to the nearly vertical (63 deg) direction. A sequence of long-period (10-15 min) absorption oscillations, reflecting similar modulation of energetic electron precipitation, is evident from 10:30-12:30 UT.

DISCUSSION

The following features are noteworthy in the simulated GPS and geosynchronous flux levels (Plates 1 and 2):

1. There is an initial decrease in electron flux at $L = 4.2$ due to buildup of the ring current, captured by the MHD simulations as an increased pressure in the inner magnetosphere due to enhanced convection during the period of southward IMF. The increased ring current causes a local decrease in total B which is more evident at geosynchronous orbit than at $L = 4.2$. Electrons move radially outward conserving the third adiabatic invariant in response to the decrease in local magnetic field strength due to build up of the ring current, which occurs on a time scale long compared to radiation belt electron drift periods [Schulz, 1997] (10.35 min at $L = 4.2$ and 6.38 min at $L = 6.6$ at 1.6 MeV, using $B_0 = .305G$ at the Earth's equator). With conservation of their first invariant, energy decreases as does flux as electrons move radially outward. This decrease has been referred to as the Dst effect [Li et al., 1998; Kim and Chan, 1997], and is observed to be greater with increasing energy, as borne out in Plates 1 and 2, simply due to the steepness of the power law energy spectrum.

2. There is an abrupt rise in flux $\sim$ 08:00 UT, by half an order of magnitude in the most energetic simulated GPS channel (1.6 - 3.2 MeV), and a correspondingly smaller rise with decreasing energy, evident in Plate 1.

3. There is a dropout in flux at geosynchronous orbit in Plate 2 $\sim$ 01:00 - 02:00 UT on January 11, due to compression of the magnetopause inside geosynchronous orbit and recovery following arrival of the high density solar wind impulse at the magnetosphere [Reeves et al., 1998b].

The flux vs. energy and L plots (Plate 3) show that the flux peak moved inward from $L = 5$ to $L = 4$ on a timescale of several hours. Inward transport of the source population resulted in an increase in flux at higher energies as well, due to conservation of the first invariant. The spectrum continued to harden slightly after northward turning of the IMF at 17:30 UT, but there was no further inward radial transport of the flux peak, in particular associated with the high density solar wind pressure pulse at $\sim$ 01:00 UT on January 10.

The drift resonance between > 200 keV electrons and the toroidal oscillations apparent in the radial electric field component of the simulation data suggests the following coherent acceleration mechanism. Electron fluxes at geosynchronous were amplified by successive substorm injections throughout the period of steady

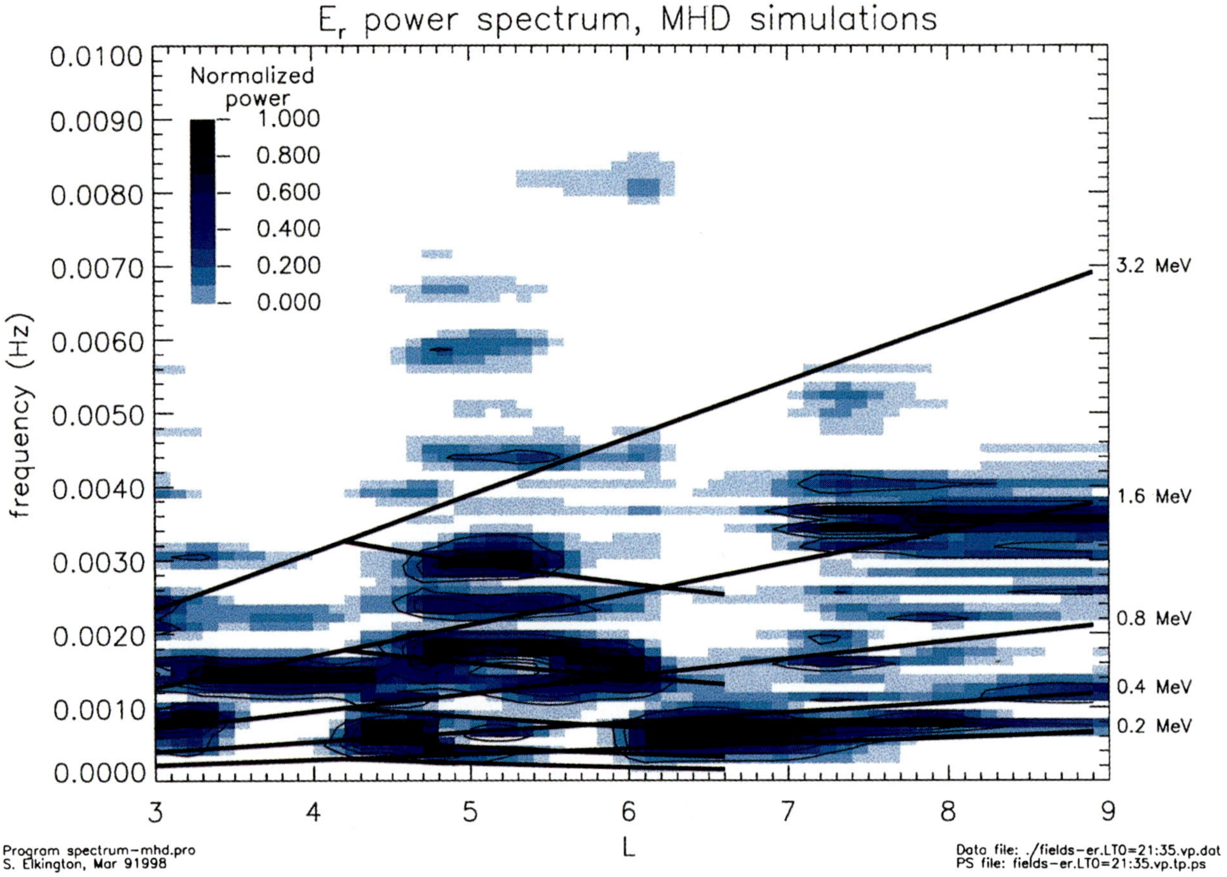

Plate 4. Power spectrum of the radial component of the MHD simulation electric field in the period between 10:00 UT and 12:16 UT on 10 Jan 98, at the local time of Gakona, AK. The ascending black lines indicate dipole drift frequency as a function of L for particles with energies shown in Figure 1. The descending lines between L = 4.2 and L = 6.6 give the dipole drift frequency of a particle moving radially through this range while conserving the relativistic adiabatic invariant.

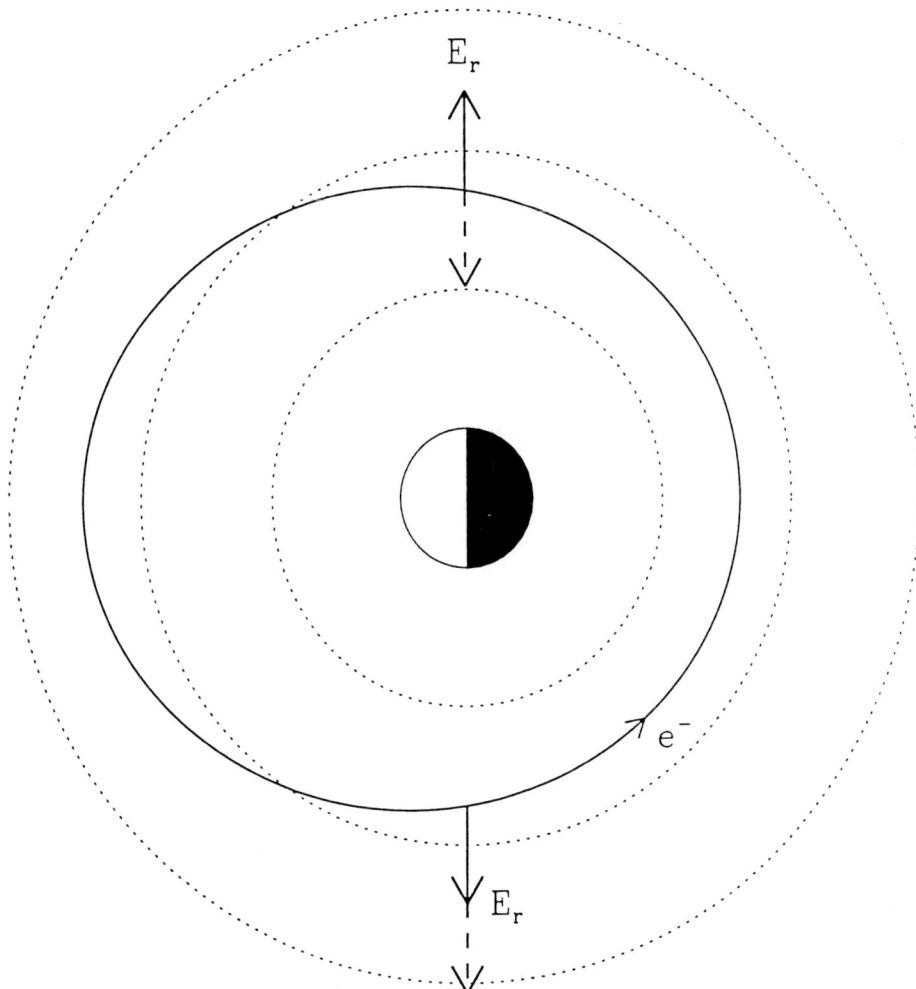

Figure 2. Sketch of electron drift path and radial electric field orientation in a toroidal mode oscillation cycle. Solid arrows indicate electric field at t = 0, and dashed as seen by an electron with wave frequency $\omega = \omega_D$, the electron drift frequency, starting at dusk for an m = 2 azimuthal mode number.

southward IMF B_z beginning at 04:40 UT [*Reeves et al.*, 1998b]. Electrons with the right drift phase are subjected to continuous acceleration by a radial electric field over a non-azimuthally symmetric drift path, given the electric field reversal on the timescale of half an electron drift period, as sketched in Figure 2. Electrons with the opposite drift phase are continuously decelerated. Consider, for example, E_r with azimuthal mode number m = 2, indicated by solid arrows in Figure 2. An electron with drift frequency $\omega_D = \omega$, the ULF wave frequency, drifting eastward from dusk (bottom) sees an electric field indicated by dashed arrows as it moves along its drift orbit. By the time it drifts around to dawn (top), half a drift period later, the electric field has reversed sign, but now $dr/dt > 0$ as it moves towards the dayside, vs. $dr/dt < 0$ as it moves towards the nightside, so such an electron experiences continuous $-\int E_r dr > 0$, and is accelerated over its drift orbit. Examining an m = 1 mode, there would be acceleration all the way around the drift path if the wave frequency is twice the drift frequency, with other resonance conditions possible, depending on the m-spectrum. An estimate of the magnitude of the acceleration has been made using the maximum radial electric field strength seen in the simulations, ~ 8 mV/m, assuming a radial drift path of $\sim 1.5 R_E$. A 100 keV electron at geosynchronous takes approximately one hour to drift around the earth and increase its energy to 200 keV for the assumed parameters, in another hour it increases its energy to 400 keV, and in two more hours it exceeds 1.6 MeV. Starting at 200 keV, its energy exceeds 1.6 MeV in less than 3 hours. These estimates, which assume continuous acceleration over a drift path, may be optimistic by factors of two, but are supported by the relativistic test particle results. Thus, it seems reasonable to conclude that the observed increase in relativistic electron flux in the simulations, and measured for the January 1997 event, can be explained in part by drift resonant acceleration in the radial electric field of toroidal eigenmodes which show enhanced power during the period of rise in electron fluxes seen by GPS.

The GPS observations suggest that the most rapid rise in flux at lower energies, 0.8 - 1.6 MeV, occurred around 11:00 UT, which coincided with the arrival of a moderate solar wind pressure pulse at the magnetosphere as seen by WIND, GEOTAIL and GOES spacecraft, as well as preliminary Dst [*Li et al.*, 1998]. Figure 1 shows that ULF wave power in the same frequency range as seen in the power spectral analysis of the MHD simulation data (Plate 4) is also enhanced at this time. However, it is embedded in a pre-existing enhancement in ULF wave power in the same frequency range, following substorm activity which produced a peak in average $K_p = 6$ during the three hour intervals 06:00 - 09:00 and 09:00 - 12:00 UT. This activity interval is readily apparent in the Canopus magnetometer data as well [*Baker et al.*, 1998], which provides a preliminary indication of longitudinal extent of the ULF oscillations in ground magnetometer data, under analysis. The MHD simulation data has been analyzed at local times shifted by 6, 12 and 18 hours relative to Plate 4, and similar but less well defined structure is apparent. Figure 3 shows a snapshot of the full electric field vectors in the equatorial plane of the simulations at $\sim$ 09:00 UT, and one can see a dominant m = 2 azimuthal mode number, and large scale coherence of the toroidal oscillation superimposed on the convection electric field and magnetopause signatures.

CONCLUSIONS

The rise in relativistic electron flux observed by numerous spacecraft within the magnetosphere on January 10, 1997 has been simulated with a guiding center test particle code using output field data from a 3D global MHD simulation of the event driven by upstream solar wind parameters measured by the WIND spacecraft. The rise in flux is greatest in the 1.6 - 3.2 MeV energy range of the four GPS detector channels simulated, and greatest around the radial location of GPS (L = 4.2) in the equatorial plane of the simulations. The January 1997 event was characterized by an extended period of steady southward IMF which produced a sequence of substorms leading to a peak in average $K_p = 6$ just prior to and during the rise in relativistic electron flux seen by GPS spacecraft. Simultaneous groundbased measurements in Alaska and Canada, which were situated postmidnight in local time, indicated enhanced ULF wave power coincident with the rise in relativistic electron flux measured inside geosynchronous orbit [*Baker et al.*, 1998].

The acceleration mechanism proposed here is distinct from that attributed to rapid compressions of the magnetopause on the electron drift timescale, such as occurred for the March 24, 1991 CME-driven SSC [*Li et al.*, 1993; *Hudson et al.*, 1996; 1997]. In the latter case, it was shown that electrons drift eastward synchronously with a magnetosonic pulse launched within the magnetosphere by the dayside SSC shock compression. It is the azimuthal electric field component which transports electrons (and protons) radially inward in this case, increasing their energy as the first adiabatic invariant is conserved. The pulse spreads around the flanks of the magnetosphere at a characteristic magnetosonic speed determined by the plasma density and magnetic field, with maximum effect on those electrons drifting eastward at a comparable velocity. Lower energy electrons are not efficiently transported inward by such a pulse, which is mainly bipolar after reflection from the ionosphere (see *Hudson et al.*, [1997], Figure 3). Electrons in drift coherence with the magnetosonic pulse as it spreads around the flanks of the magnetosphere become drift-phase bunched in less than one drift period, as observed for the March 1991 event by energetic particle detectors on the CRRES satellite [*Blake et al.*, 1992]. By contrast, the ULF-wave drift-coherent mechanism proposed here requires multiple drift orbits to increase electron energies from the hundred keV to MeV range, and drift echo features are not expected. Nonetheless, because it is a coherent process, it is much more efficient and rapid than acceleration by standard radial diffusion, based on incoherent electric or magnetic fluctuations.

In the broader context of so-called "killer electron" flux increase, the January 1997 event is noteworthy for the lack of 1 - 2 day delay typically observed for outer zone electron flux buildup associated with high speed solar wind stream interactions [*Blake et al.*, 1997], as well as for moderate solar wind velocity (< 500 km/s), until after passage of the magnetic cloud on January 11 [*Burlaga et al.*, 1998]. While the correlation between solar wind speed and outer zone electron fluxes is well

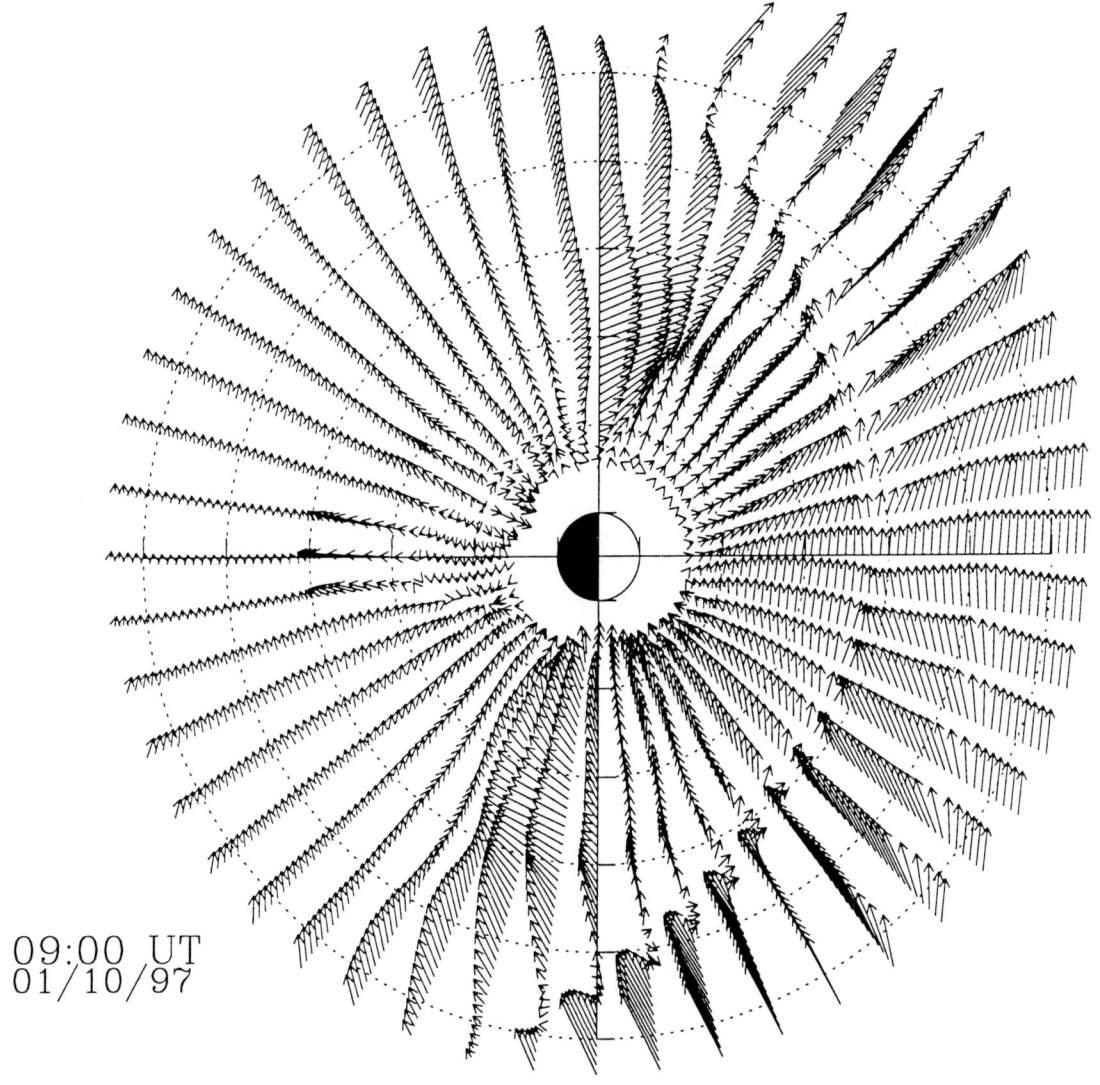

Figure 3. Snapshot of MHD simulation electric field vectors in the equatorial plane at 09:00 UT, showing low azimuthal mode number, large scale coherence of toroidal oscillation, superimposed on the convection electric field pattern.

documented [*Paulikas and Blake*, 1979], the additional factor of steady southward IMF B_z [*Blake et al.*, 1997] clearly played a role in the January 1997 case. It appears that such steady southward IMF B_z, well known to facilitate substorms, may affect the relativistic electron population in two ways. First, it provides an enhanced seed population in the hundred keV energy range due to substorm injections. Second, the ULF wave activity in the 10 minute period range is enhanced. Further investigation of what we have interpreted as toroidal oscillations, based on analysis of the MHD simulation data, will be pursued elsewhere.

In summary, we have simulated an impulsive rise in relativistic electron flux on January 10, 1997, correlated with an observed increase in ULF wave activity on the nightside in a frequency range commensurate with electron drift periods. Solar wind parameters from the WIND spacecraft were used to drive the longest 3D global MHD simulation to date, and the output fields were used to push guiding center relativistic electron trajectories in the equatorial plane. Power enhancement in the radial electric field component of the simulations coincident with the electron drift period suggests drift-resonant acceleration over multiple electron drift periods. As electrons move inward, gaining energy from this radial electric field while conserving the first invariant, the flux peak moves inward to L = 4, during a time interval both of enhanced ULF wave activity as seen by ground instrumentation on January 10, and the rise in relativistic electron fluxes seen by GPS [*Reeves et al.*, 1998a]. A continuous and time-dependent electron injection scheme is under development, which will increase flux levels above those seen in Plates 1 and 2, and should improve direct comparisons with spacecraft measurements over the present results, which simply apply the time evolving MHD fields to an initial AE8MIN electron model.

Acknowledgments. We thank J. B. Blake for providing Polar as well as HEO data, which motivated initial work on this event, G. D. Reeves for preprints and discussion of this event and for providing GPS and geosynchronous data for comparison with the simulations, V. Marchenko, I. Roth and M. Temerin for helpful discussions, and X. Li and R. Selesnick for preprints and discussion of their work on this event. This work is supported by NASA grants NAG5-1098, NAGW 4728 and NAG5-2252 to Dartmouth College. The work at the University of Maryland has been carried out with support received from NASA grant NAG5-101, APTI Subcontract No. 1011-91-01 and NRL Contract N0001495C2088.

REFERENCES

Baker, D. N., T. I. Pulkinen, X. Li, S. G. Kanekal, J. B. Blake, R. S. Selesnick, M. G. Henderson, G. D. Reeves, H. E. Spence and G. Rostoker, Coronal mass ejections, magnetic clouds, and relativistic magnetospheric electron events: ISTP, *J. Geophys. Res.*, *103*, 17229, 1998.

Blake, J. B., W. A. Kolasinski, R. W. Filius and E. G. Mullen, Injection of electrons and protons with energies of tens of MeV into L , 3 on March 24, 1991, *Geophys. Res. Lett.*, *19*, 821, 1992.

Blake, J. B., D. N. Baker, N. Turner, K. W. Ogilvie, and R. P. Lepping, Correlation of changes in the outer-zone relativistic-electron population with upstream solar wind and magnetic field measurements, *Geophys. Res. Lett.*, *24*, 927, 1997.

Burlaga, L., R. Fitzenreiter, R. Lepping, K. Ogilvie, A. Szabo, A. Lazarus, J. Steinberg, G. Gloeckler, R. Howard, D. Michels, C. Farrugia, R. P. Lin, and D. E. Larson, A magnetic cloud containing prominence material: January, 1997, *J. Geophys. Res.*, *103*, 277, 1998.

Fedder, J. A. and J. G. Lyon, The magnetospheric current-voltage curve, *Geophys. Res. Lett.*, 880, 1987.

Goodrich, C. C., M. W. Wiltberger, R. E. Lopez, K. Papadopoulos and J. G. Lyon, An overview of the impact of the January 10 - 11, 1997 magnetic cloud on the magnetosphere via global MHD simulation, *Geophys. Res. Lett.*, *25*, 2537, 1998.

Hudson, M. K., S. R. Elkington, J. G. Lyon, V. A. Marchenko, I. Roth, M. Temerin and M. S. Gussenhoven, MHD/particle simulations of radiation belt formation during a storm sudden commencement, in *Radiation Belts: Models and Standards*, edited by J. F. Lemaire, D. Heynderickx, and D. N. Baker, Geophys. Momogr. Ser., vol. 97, p. 57, AGU, Washington, D. C., 1996.

Hudson, M. K., S. R. Elkington, J. G. Lyon, V. A. Marchenko, I. Roth, M. Temerin, J. B. Blake, M. S. Gussenhoven, and J. R. Wygant, Simulations of radiation belt formation during storm sudden commencements, *J. Geophys. Res.*, *102*, 14,087, 1997.

Kim, H. -J. and A. A. Chan, Fully adiabatic changes in storm-time relativistic electron fluxes, *J. Geophys. Res.*, *102*, 22107, 1997.

Li, X., I. Roth, M. Temerin, J. R. Wygant, M. K. Hudson and J. B. Blake, Simulation of the prompt energization and transport of radiation belt particles during the March 24, 1991, SSC, *Geophys. Res. Lett.*, *20*, 2423, 1993.

Li, X., D. N. Baker, M. Temerin, T. Cayton, G. D. Reeves, T. Araki, H. Singer, D. Larson, R. P. Lin and S. G. Kanekal, Energetic electron injections into the inner magnetosphere during the January 10-11, 1997, magnetic storm, *Geophys. Res. Lett.*, *25*, 2561, 1998.

Paulikas, G. A., and J B. Blake, Effects of the solar wind on magnetospheric dynamics: Energetic electrons at the synchronous orbit, *Quantitative Modeling of Magnetospheric Processes*, 21, Geophys. Monograph Series, 1979.

Reeves, G. D., D. N. Baker, R. D. Belian, J. B. Blake, T. E. Cayton, J. F. Fennell, R. H. W. Friedel, M. M. Meier, R. S. Selesnick and H. E. Spence, The global response of relativistic radiation belt electrons to the January 1997 magnetic cloud, *Geophys. Res. Lett.*, *25*, 3265, 1998a.

Reeves, G. D., R. H. W. Friedel, R. D. Belian, M. M. Meier, M. G. Henderson, T. Onsager, H. J. Singer, D. N. Baker and X. Li, The relativistic electron response at geosynchronous orbit during the January 10, 1997, magnetic storm, *J. Geophys. Res.*, *103*, 17559, 1998b.

Schulz, M., Direct influence of ring current on auroral oval diameter, *J. Geophys. Res.*, *102*, 14,149, 1997.

Schulz M. and L. J. Lanzerotti, *Particle Diffusion in the Radiation Belts*, Springer-Verlag, Berlin, 1974.

Selesnick, R. S. and J. B. Blake, Radiation belt electron observations from January 6 to 20, 1997, *Geophys. Res. Lett.*, *25*, 2253, 1998.

Vette, J. I., The AE8 trapped electron model environment, NSSDC/WDC-A-R&S 91-24, 1991.

M. K. Hudson, S. R. Elkington and J. G. Lyon, Physics and Astronomy Dept., Dartmouth College, Hanover, NH 03755. (e-mail: mary.hudson@dartmouth.edu)

C. C. Goodrich, Department of Astronomy, University of Maryland, College Park, MD 20742.

T. J. Rosenberg, Institute for Physical Science and Technology, University of Maryland, College Park, MD 20742.

Origins and Transport of Ions During Magnetospheric Substorms

M. Ashour-Abdalla, M. El-Alaoui, V. Peroomian, J. Raeder, and R. J. Walker

Institute of Geophysics and Planetary Physics, UCLA, Los Angeles, California

L. A. Frank and W. R. Paterson

Department of Physics and Astronomy, The University of Iowa, Iowa City, Iowa

We investigate the origins and the transport of ions observed in the near-Earth plasma sheet during the growth and expansion phases of a magnetospheric substorm that occurred on November 24, 1996. Ions observed at Geotail were traced backward in time in time-dependent magnetic and electric fields to determine their origins and the acceleration mechanisms responsible for their energization. Results from this investigation indicate that, during the growth phase of the substorm, most of the ions reaching Geotail had origins in the low latitude boundary layer (LLBL) and had already entered the magnetosphere when the growth phase began. Late in the growth phase and in the expansion phase a higher proportion of the ions reaching Geotail had their origin in the plasma mantle. Indeed, during the expansion phase more than 90% of the ions seen by Geotail were from the mantle. The ions were accelerated enroute to the spacecraft; however, most of the ions' energy gain was achieved by non-adiabatic acceleration while crossing the equatorial current sheet just prior to their detection by Geotail. In general, the plasma mantle from both southern and northern hemispheres supplied non-adiabatic ions to Geotail, whereas the LLBL supplied mostly adiabatic ions to the distributions measured by the spacecraft.

1. INTRODUCTION

The dynamic releases of energy and the dramatic energization of particles during geomagnetic substorms is an intriguing and yet unresolved problem in magnetospheric physics. The Earth's magnetosphere is known to be strongly affected by the dynamics of the interplanetary magnetic field (IMF) and the solar wind, which are the ultimate energy sources for magnetospheric substorms. Less well known, though, are the plasma energization that results from substorms and the composition and sources of magnetospheric plasmas during this dynamic process.

A fundamental goal of magnetospheric physics is to understand the transport of plasma through the solar wind-magnetosphere-ionosphere system. To attain such an understanding, we must determine the sources of the plasma, the trajectories of the particles through the magnetospheric electric and magnetic fields to the point of observation, and the acceleration processes they undergo enroute. At any given time the number of spacecraft that are in the magnetosphere is insufficient for the use of observations alone to be an effective method of studying transport through the entire system. Fortunately, theory and modeling can be used to augment the available observations.

Sun-Earth Plasma Connections
Geophysical Monograph 109
Copyright 1999 by the American Geophysical Union

Studies of the population of the magnetotail from either the solar wind or the ionosphere, or a combination of both, are numerous. Following the observations of energetic O^+ ions precipitating into the ionosphere [*Shelley et al.*, 1972], the auroral zone and the cleft ion fountain were found to be robust sources of ions for the near-Earth magnetotail [e.g., *Yau et al.*, 1985; *Lockwood et al.*, 1985a, b]. Solar wind ions, on the other hand, were found to enter the magnetosphere through the plasma mantle and the low latitude boundary layer (LLBL). Many studies have confirmed that both the solar wind and ionosphere contribute to the magnetospheric plasma population [*Eastman et al.*, 1976; *Lennartsson and Shelley*, 1986; *Chappell et al.*, 1987]. However, the role each source plays, the transport of plasma from each source to the near-Earth tail, and the acceleration mechanisms responsible for particle heating remain unclear.

In the past, studies of plasma transport and acceleration were carried out using empirical models of magnetospheric fields. These studies investigated the magnetotail plasma population using sources in the plasma mantle [*Ashour-Abdalla et al.*, 1993], the auroral ionosphere [*Peroomian and Ashour-Abdalla*, 1996], and the cleft ion fountain [*Delcourt et al.*, 1989, 1990; *Cladis and Francis*, 1992]. More importantly, these studies successfully showed that non-adiabatic ion acceleration plays a significant role in the energization of particles in the magnetotail [*Büchner and Zelenyi*, 1986; *Ashour-Abdalla et al.*, 1993].

The launch of the Geotail spacecraft has allowed us to take these studies one step further, and to use observations and theory together to obtain a quantitative picture of transport and acceleration in the magnetotail. For instance, *Ashour-Abdalla et al.* [1997, 1998] and *El-Alaoui et al.* [1998] used theory and three-dimensional distribution functions observed by Geotail during geomagnetically quiet intervals [*Frank et al.*, 1996] to trace structures in the distributions to specific plasma sources.

This study employs plasma distributions observed in the near-Earth plasma sheet and theory to investigate the ion sources and the transport of plasma during the growth and expansion phases of a substorm on November 24, 1996. In section 2 we describe our approach in investigating this problem, while in section 3 we discuss the observations of this substorm along with the results of our magnetohydrodynamic (MHD) simulation based on these observations. In section 4 we determine the sources of the ions, how the sources vary as a function of time and how the ions are accelerated en route to Geotail, after which in section 5 we discuss the implications of these findings.

2. METHODOLOGY

This study investigates ion transport through the magnetosphere by tracing the trajectories of thousands of particles in time-dependent electric and magnetic fields obtained from a global MHD simulation [*Raeder et al.*, 1995] of the November 24, 1996 substorm. The MHD model essentially solves the ideal MHD equations that are modified to include an anomalous resistivity term for the magnetosphere and a potential equation for the ionosphere. A few numerical effects, such as diffusion, viscosity, and resistivity, are necessarily introduced by the numerical methods. These permit viscous interactions and, to a limited extent, magnetic field reconnection. However, the numerical scheme is optimized to minimize numerical effects. For instance, numerical resistivity is so low that it is necessary to introduce an anomalous resistivity term in order to model substorms correctly [see *Raeder et al.*, 1996]. The ionospheric part of the model takes into account three sources of ionospheric conductance: solar EUV ionization is modeled using the empirical model of *Moen and Brekke* [1993], diffuse auroral precipitation is modeled by assuming full pitch angle scattering at the inner boundary of the MHD simulation (at 3.7 R_E), and accelerated electron precipitation associated with upward field-aligned currents is modeled in accordance with the approach of *Knight* [1972] and *Lyons et al.* [1979]. The empirical formulas of *Robinson et al.* [1987] are used to calculate the ionospheric conductances from the electron mean energies and the energy fluxes. A detailed description of the MHD model, including initial and boundary conditions, can be found in *Raeder et al.* [1996, 1997].

The electric field used in this calculation is given by $\mathbf{E} = -\mathbf{V} \times \mathbf{B} + \eta \mathbf{J}$, where $\mathbf{V}$ is the velocity from the MHD model, $\mathbf{B}$ is the magnetic field, η is the resistivity, and $\mathbf{J}$ is the current density. The electric field has both a convective ($-\mathbf{V} \times \mathbf{B}$) and a resistive ($\eta \mathbf{J}$) term. The resistive term becomes important near the magnetopause and near x-lines but is negligible elsewhere. We construct three-dimensional distribution functions in velocity space by placing ~75,000 ions in V_x-V_y-V_z bins (100 km/s × 100 km/s × 100 km/s) such that the number in each bin is proportional to the phase space density observed at the Geotail spacecraft. One particle in the computational distribution function corresponds to 2×10^{-26} $s^3 cm^{-6}$ in the Geotail distribution function. Particles in each velocity bin are randomized in phase space to avoid the possibility of launching identical particles. For each particle we integrate the equation of motion ($d\mathbf{V}/dt = q\mathbf{V} \times \mathbf{B} + q\mathbf{E}$) backward in

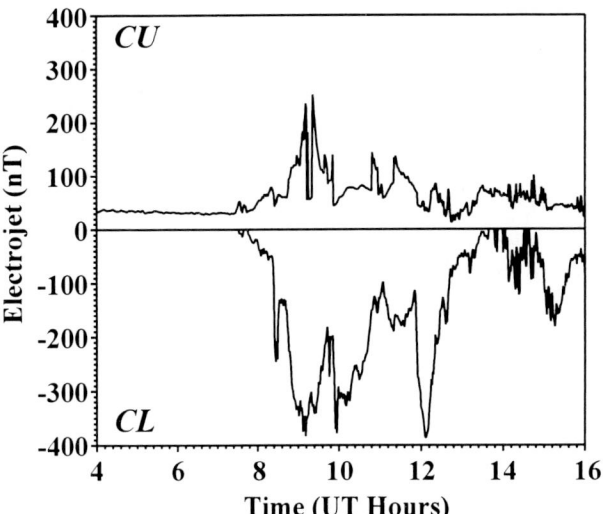

Figure 1. CANOPUS magnetometer data for November 24, 1996.

time until the particle encounters the magnetopause (as defined by the current distribution in the MHD model) or the ionosphere (taken as the inner boundary of the MHD simulation at $r = 3.7\ R_E$). Since the minimum grid spacing in the global MHD model is relatively large (~0.5 R_E) and the simulation data are saved at four-minute time intervals, we use linear interpolation in both space and time to determine the instantaneous values of the MHD fields on scales smaller than the grid spacing. We use a fourth-order Runge-Kutta method to calculate the ion trajectories in the evolving magnetic and electric fields. The time step for the particle trajectory calculation is nominally set at 0.002 times the local ion gyro-period, with an upper limit imposed to ensure that the time step does not get too large in weak field regions. This ensures that all the particles in the simulation conserve energy (to 6 significant figures) and that the trajectory is calculated correctly in the model.

Because the magnetospheric topology, especially near the magnetopause and the magnetotail current sheet, causes ions to behave nonadiabatically and to violate the conservation of the first adiabatic invariant [*Speiser*, 1965; *Lyons and Speiser*, 1982; *Chen and Palmadesso*, 1986; *Büchner and Zelenyi*, 1986, 1989] it is necessary to follow the exact motion of ions (rather than use Liouville's Theorem). Using Liouville's Theorem under the conditions we are simulating would require the numerical calculation of particle trajectories in a manner similar to the one carried out

in this study, and, additionally, the instantaneous measurements of the source distributions (at the magnetopause and in the ionosphere) for all times during the time interval being studied; these measurements are not available.

3. OBSERVATIONS AND MHD CALCULATIONS

November 24, 1996 was a relatively quiet day in the magnetosphere. For the first 6 hours K_p was 0^+ and 1^-. Figure 1 shows the auroral electrojet indices calculated from the CANOPUS chain, which was ideally situated to capture the onset of the substorm. Although the CANOPUS indices are not necessarily representative of the global auroral indices, they provide useful information about the local auroral dynamics and electrojet activity during the period of interest. The upper (lower) panel of Figure 1 shows the CU (CL) index, which is the highest (lowest) values of the horizontal magnetic field at one minute resolution. The CANOPUS data indicate that the growth phase of the substorm began at ~0730 UT and the expansion phase at ~0820 UT.

Figure 2 shows the solar wind and interplanetary magnetic field (IMF) observations from the WIND spacecraft from 0300 UT until 1000 UT. During this interval WIND was located upstream of the Earth at (72.4, –20.7, 8.06) R_E. The panels in this plot are, from top to bottom, the three components of the interplanetary magnetic field (IMF) (in nT), the three components of solar wind velocity (in km/s), the solar wind plasma density (in cm^{-3}) and plasma pressure (in pPa). The IMF data show that B_z was predominantly northward prior to substorm onset and turned southward at ~ 0700 UT. The B_x and B_y components of the IMF were both small during this interval. However, the B_y component showed a significant negative turning that coincided with the southward turning of the B_z component. The solar wind velocity was steady and predominantly in the x direction with an average speed of ~310 km/s, placing the WIND spacecraft approximately 25 minutes upstream of Earth. The plasma density at the beginning of the interval was higher than average, ~20 cm^{-3}, and increased steadily, reaching ~35 cm^{-3} during the substorm. The increase in density was mirrored in the solar wind plasma pressure profile, which increased from ~ 6 pPa to over 30 pPa during the substorm. We note that the increases in density and pressure occurred at the same time as the southward turning of the IMF, suggesting that one or more of these changes precipitated the onset of the substorm.

186 ORIGINS AND TRANSPORT OF IONS DURING SUBSTORMS

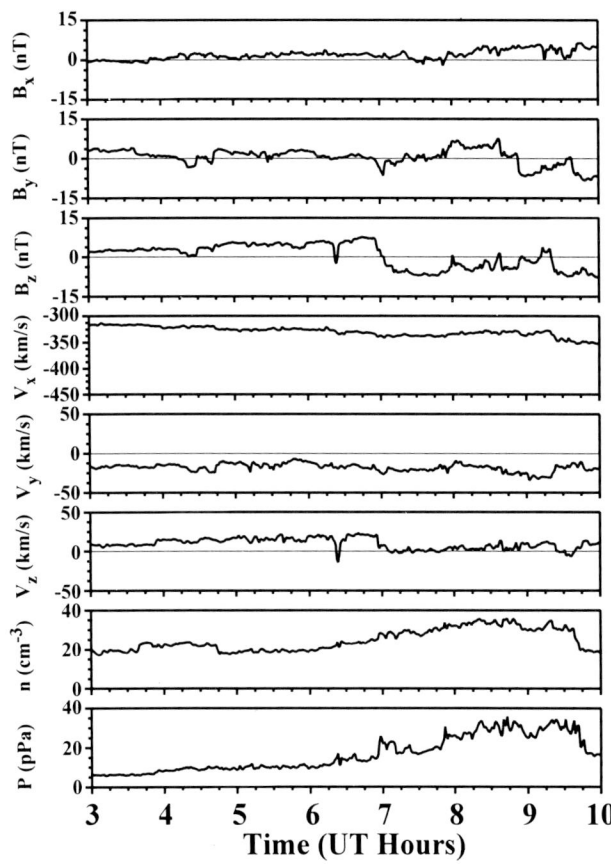

Figure 2. WIND data for November 24, 1996.

During the November 24, 1996 substorm the Geotail spacecraft was located in the near-midnight magnetotail about 21 R_E from the Earth. (Geotail moved from (–20.7, 2.8, –2.3) R_E to (–21.5, 1.7, –2.4) R_E during the substorm). Three-dimensional ion distribution functions (Figure 3) from the Comprehensive Plasma Instrumentation (CPI) on Geotail [*Frank et al.*, 1994] reveal a very complex structure. In Figure 3 the distribution function is plotted in V_y - V_z, V_x - V_z, and V_x - V_y projections in GSE coordinates. The four rows give the distribution functions at 0740 UT, 0750 UT and 0800 UT during the growth phase and at 0840 UT during the expansion phase. The projection of the average magnetic field vector has been superimposed on each of the distribution functions.

At 0740 UT the magnetic field was mainly tailward and strong earthward field-aligned flows existed, signs that Geotail was located in the southern plasma sheet boundary layer (PSBL). Thereafter, Geotail moved deeper into the plasma sheet, as indicated by the larger B_z and the earthward convection found ten minutes later (at 0750 UT). By 0800 UT near the end of the growth phase the distribution function became much colder and the magnetic field was pointed predominantly earthward, from which we conclude that Geotail was near the northern boundary of the plasma sheet. Finally at 0840 UT Geotail moved to the outer edge of the northern plasma sheet boundary layer and observed mostly cold plasma convecting equatorward into the plasma sheet.

The solar wind observations in Figure 2 were used as input to the global MHD code. The southward IMF observed by WIND at 0700 UT reached the magnetopause at about 0725 UT. The predominantly duskward B_y component of the IMF during the four hours prior to the southward turning caused the magnetotail to rotate such that the plasma sheet was north of the $z = 0$ plane on the dusk side of the tail and south of it on the dawn side. This placed the Geotail location in the southern PSBL in the simulation during the early growth phase of the substorm, as was observed. By 0750 UT the duskside plasma sheet near Geotail began to rotate in response to the dawnward rotation of the IMF, and the spacecraft moved toward the center of the plasma sheet in both the observations and the simulation. Figure 4 shows the pressure in the x-z plane through the location of Geotail during the substorm. At 0800 UT (top panel), during the growth phase of the substorm, Geotail was in the plasma sheet. A thin current sheet formed and extended tailward of Geotail. At 0830 UT (middle panel), early in the expansion phase, reconnection began tailward of the spacecraft. Following the onset of reconnection, the plasma sheet earthward of Geotail thickened (compare the 0830 UT plasma sheet distribution with that at 0902 UT in the lower panel). At the same time the entire duskside plasma sheet continued to move southward in response to the dawnward turning of the IMF ($B_y < 0$). The net effect of these movements was to place Geotail in the northern plasma sheet boundary layer during the expansion phase of the substorm.

4. RESULTS

We followed trajectories of the ions observed at Geotail backwards in time to their points of entry into the magnetosphere (Plate 1). Each color-coded dot in the plate represents the number of ions originating from a 1 R_E × 1 R_E × 1 R_E region centered at the point. The green sphere shows the location of Geotail, and the contour plot in each panel represents a profile of the plasma pressure at $x = -150$ R_E at the time of measurement. Most (84%) of the ions observed at Geotail at 0740 (Plate 1a) during the growth phase of the substorm enter the magnetosphere at low latitudes in the dusk flank. These ions enter the magnetosphere prior to the beginning of the growth phase. About

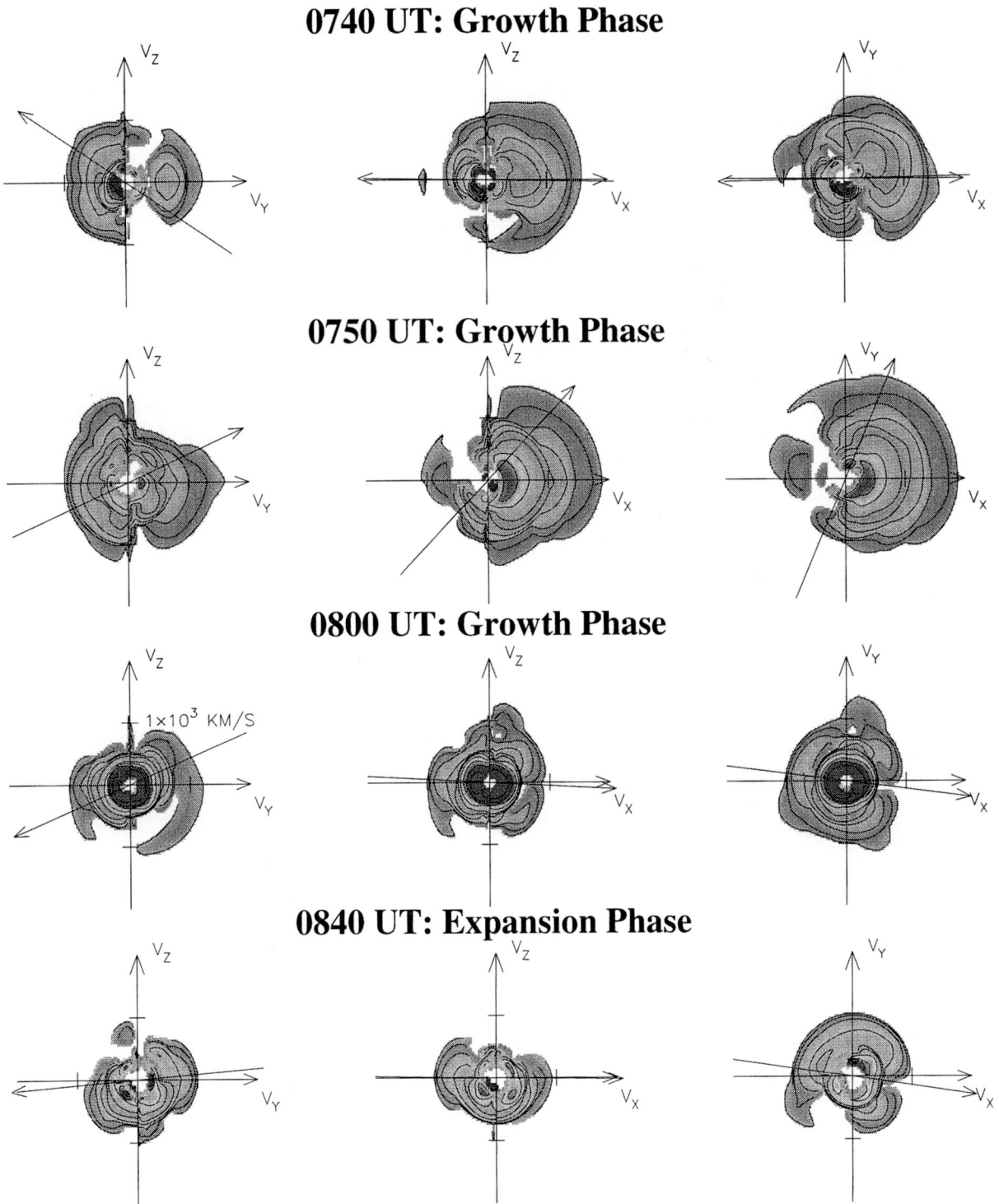

Figure 3. V_y - V_z, V_x - V_z, and V_x - V_y cuts of the Geotail distribution for four time intervals.

188 ORIGINS AND TRANSPORT OF IONS DURING SUBSTORMS

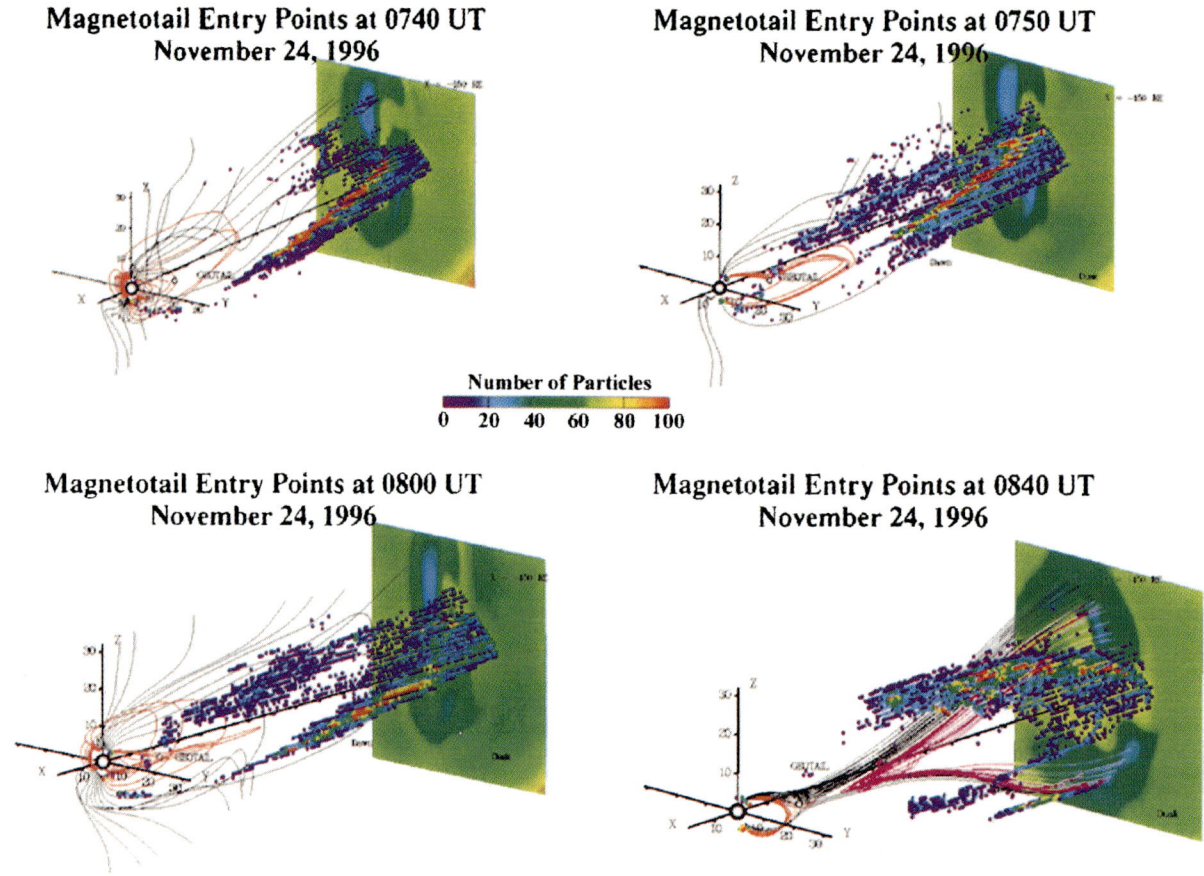

Plate 1. Magnetotail entry points for the ions in the Geotail distribution functions observed at (a) 0740 UT, (b) 0750 UT, (c) 0800 UT, and (d) 0840 UT.

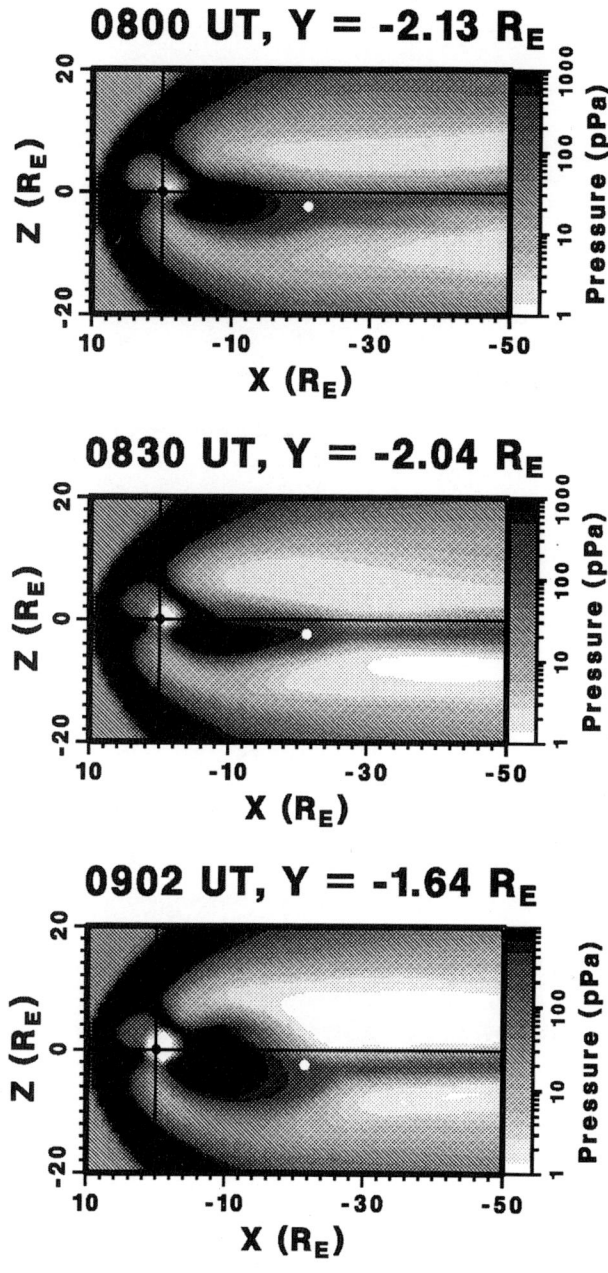

Figure 4. MHD pressure profile in the meridional plane intersecting the Geotail spacecraft for three time intervals during the substorm.

14% of the ions enter the magnetosphere in the distant ($x < -100\ R_E$) plasma mantle. These ions are primarily field aligned and are responsible for the earthward streaming component seen at Geotail. Ionospheric ions make up < 2% of the population at Geotail.

Ten minutes later, at 0750 UT (Plate 1b), the LLBL is still the dominant source of ions (82%), with the bulk of this plasma flowing in the direction perpendicular to the magnetic field. Dayside reconnection has enhanced the region from which mantle ions can reach Geotail. The plasma mantle now supplies ~17% of the ions at Geotail, most of which are concentrated in a heated earthward flowing beam (Figure 3). The ionospheric contribution at this time is still negligible. At 0800 UT the plasma mantle has become a significant (~37%) source of ions in the near-Earth tail.

The formation of a near-Earth neutral line and the ejection of a plasmoid downtail causes dramatic changes in the ability of ions to reach Geotail. In Plate 1d there is a large region along the near-Earth flank magnetopause from which LLBL ions can no longer reach Geotail. Particles entering the magnetosphere in this region are swept downtail by a tailward flow generated by the near-Earth neutral line. In fact, only LLBL ions that are inside the magnetosphere prior to the end of the growth phase make it to the spacecraft. These ions constitute 10% of the particles seen at Geotail (Plate 1d).

At 0840 UT more than two-thirds of the ions reaching Geotail are from the northern plasma mantle. The changes in the magnetospheric configuration that occurred during the expansion phase now for the first time allow ions from the southern plasma mantle to reach the spacecraft. The trajectory of one such ion is plotted in Figure 5. The top panel of this figure shows a three-dimensional view of the trajectory of the ion, which originates in the southern plasma mantle. The particle trajectory is shown in black. The field lines encountered by the particle as a function of time during its trajectory are also shown. The ion shown in Figure 6 travels earthward from its entry point on open field lines, mirrors in the ionosphere and bounces back onto the closed field line connected to Geotail. The lower panel of Figure 5 is a plot of the particle's kinetic energy (solid curve, scale on the left) and parameter of adiabaticity κ [*Büchner and Zelenyi*, 1986] (gray dots, scale on the right). The energy gain of the ion occurs in the current sheet, where $\kappa \ll 1$.

Plate 2 shows the contribution to the observed distribution function by each of the plasma sources at 0840 UT. Ions from the northern plasma mantle make up 69% of the particles observed at Geotail, while 21% come from the southern mantle. The northern mantle ions are moving southward toward the equator (see the middle and right-hand columns) while the southern mantle ions are moving upward past Geotail. As noted above, by this stage the LLBL particles can no longer reach the spacecraft in large

190 ORIGINS AND TRANSPORT OF IONS DURING SUBSTORMS

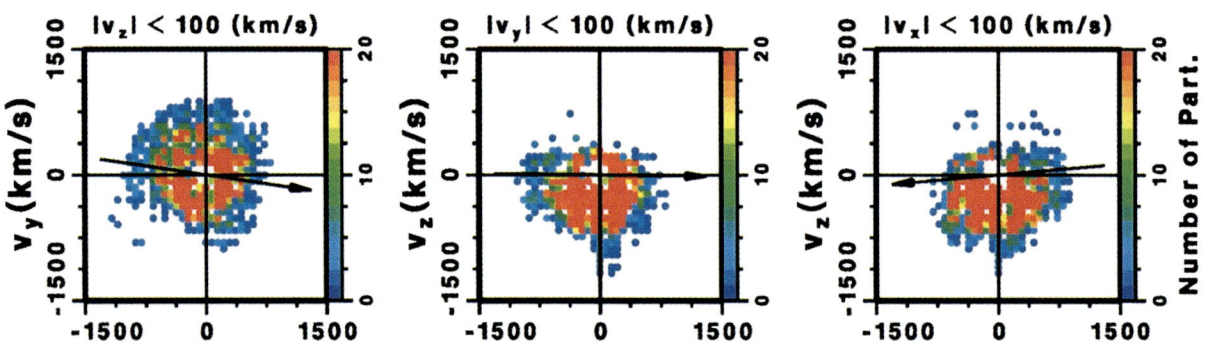

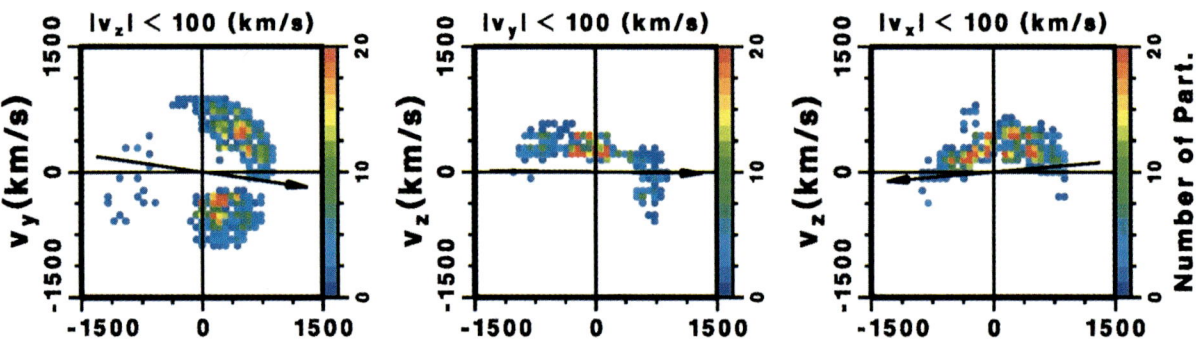

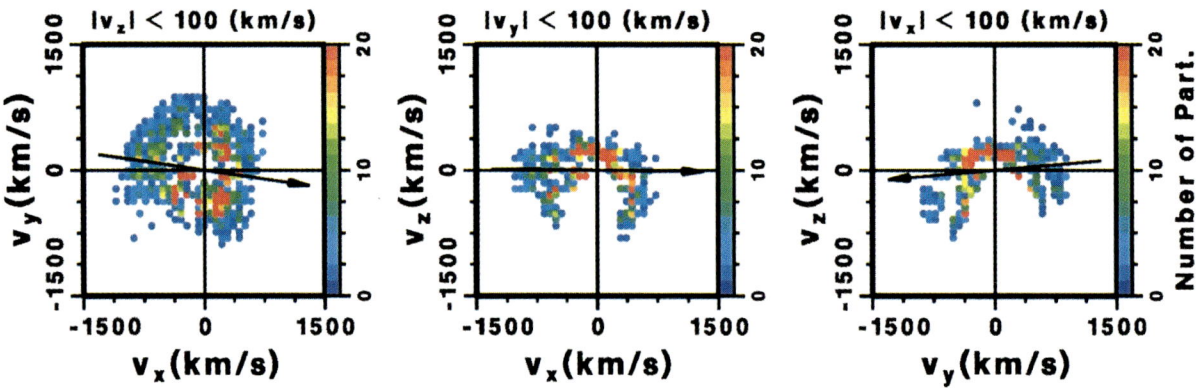

Plate 2. Contribution of individual sources to the Geotail distribution function observed at 0840 UT. (Top row) Northern plasma mantle, (middle row) Southern plasma mantle, (bottom row) LLBL.

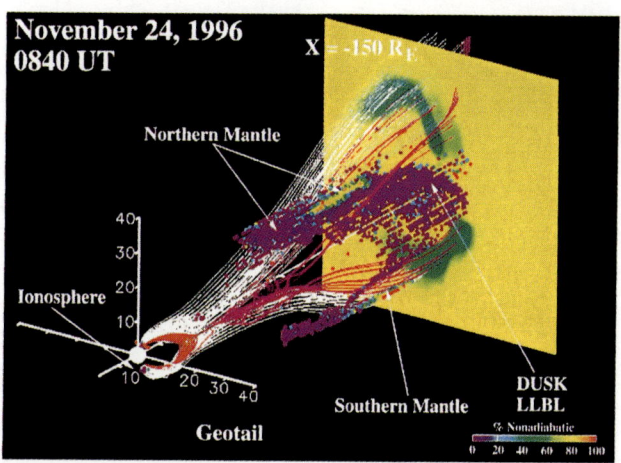

Plate 3. Magnetotail entry points for ions in the 0840 UT Geotail distribution function color-coded according to the percentage of ions with $\kappa < 2$.

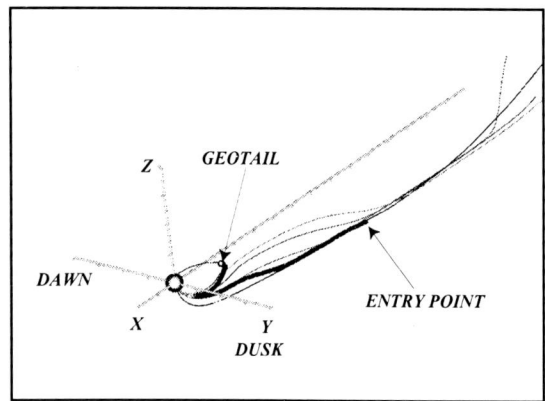

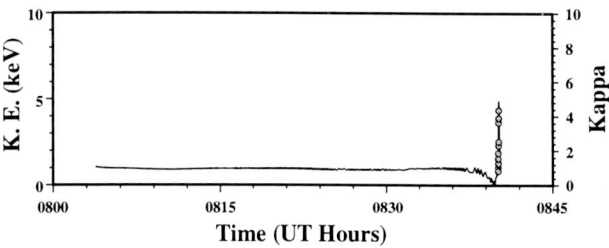

Figure 5. (Top panel) Trajectory of a southern mantle particle reaching Geotail at 0840 UT. (Lower Panel) Kinetic energy (solid curve) and κ (gray dots) for the particle.

numbers, and those that do are less energetic than the plasma mantle ions.

In order to ascertain the importance of non-adiabatic acceleration on the particle distributions at this time, we plot in Plate 3 the lowest κ experienced by ions entering the magnetosphere at 0840 UT. Plate 3 is similar in format to the panels in Plate 1 and shows the locations of particle entry into the magnetosphere. The color coding of Plate 3, however, differs; it corresponds to the percentage of ions from each 1 R_E^3 bin that have κ < 2. Plate 3 shows two regions in the northern and southern plasma mantles that supply non-adiabatic ions to Geotail. Comparison of Plate 3 to Plate 1d shows that these locations coincide with the place of origin of the bulk of the ions from these two sources. Thus, non-adiabatic acceleration affects a large portion of the Geotail distribution functions observed at 0840 UT.

5. SUMMARY AND DISCUSSION

We used ion trajectory calculations in magnetic and electric fields from a global MHD simulation of the November 24, 1996, magnetospheric substorm to investigate the sources and transport of the ions observed in the near-Earth magnetotail during the growth and expansion phases of the substorm. We found that

1) Early in the growth phase the LLBL was the main source of ions reaching the Geotail spacecraft. These ions were in the magnetosphere before the growth phase began and were slowly drifting towards Geotail on adiabatic trajectories.

2) Later in time, the plasma mantle ions became more important, and in the expansion phase, the mantle ions became dominant (90%).

3) Plasma mantle ions from both the northern and southern hemispheres reached the spacecraft primarily on non-adiabatic orbits. These ions were rapidly energized in the current sheet just prior to reaching Geotail.

Both the observations and the results from the global MHD model are consistent with Geotail's moving northward through the southern plasma sheet boundary layer and the plasma sheet during the substorm. During the expansion phase Geotail was nearly in the northern tail lobes. It is interesting to consider what can be learned about ion dynamics from observations taken far from the central plasma sheet. Ions from both the northern and the southern mantle reach Geotail (Plate 1). Significantly, many of the ions from the southern mantle are like the ion in Figure 5 in that they moved through the equatorial current sheet before reaching Geotail. Since reconnection was occurring just tailward of the spacecraft (Figure 4) when these ions reached Geotail, the ions passed through a region where the radius of curvature of the field lines was small. Here the parameter of adiabaticity κ [*Büchner and Zelenyi*, 1986] is less than one, and the ions gained energy from non-adiabatic motion across the electric field. This is the dominant form of acceleration for these ions. Thus, even though Geotail was not in the central plasma sheet during the period in question the region's distribution function revealed information about the processes occurring there.

Acknowledgments. We would like to thank R. L. Richard, L. M. Zelenyi, and J. M. Bosqued for helpful discussions, and G. Rostoker for supplying CANOPUS magnetometer data. This work was supported by NASA ISTP grants NAG5-1100 and NAG5-6689 and The University of Iowa subcontract PO V50494. Computing support was provided by the San Diego Supercomputer Center and by the Office of Academic Computing at UCLA.

REFERENCES

Ashour-Abdalla, M., Ashour-Abdalla, M., J. Berchem, J. Büchner, and L. M. Zelenyi, Shaping of the magnetotail from the mantle: Global and local structuring, *J. Geophys. Res.*, 98, 5651, 1993.

Ashour-Abdalla, et al., Ion sources and acceleration mechanisms inferred from local distribution functions *Geophys. Res. Lett., 24*, 955, 1997.

Ashour-Abdalla, M., J. Raeder, M. El-Alaoui, V. Peroomian, Magnetotail structure and its internal particle dynamics during northward IMF, in *New Perspectives of the Earth's Magnetotail, Geophys. Monogr. Ser.*, AGU, Washington, DC, in press 1998.

Büchner, J., and L. M. Zelenyi, Deterministic chaos in the dynamics of charged particles near a magnetic field reversal, *Physics Letters A, 118*, 395, 1986.

Büchner, J. and L. Zelenyi, Regular and chaotic charged particle motion in magnetotail-like field reversals, 1. Basic theory of trapped motion, *J. Geophys. Res., 94*, 11,821, 1989.

Chappel, C. R., R. E. Moore, and J. H. Waite, Jr., The ionosphere as a fully adequate source of plasma for the Earth's magnetosphere, *J. Geophys. Res., 92*, 5896, 1987.

Cladis, J. B. and W. E. Francis, Distribution in magnetotail of O^+ ions from cusp/cleft ionosphere: a possible substorm trigger, *J. Geophys. Res., 97*, 123, 1992.

Delcourt, D. C., C. R. Chappell, T. E. Moore, and J. H. Waite, Jr., A three-dimensional numerical model of ionospheric plasma in the magnetosphere, *J. Geophys. Res., 94*, 11,893, 1989.

Delcourt, D. C., J. A. Sauvaud, and T. E. Moore, Cleft Contribution to Ring Current Formation, *J. Geophys. Res., 95*, 20937, 1990.

Eastman, T.E., E.W. Hones, Jr., S.J. Bame, and J. R. Asbridge, The magnetospheric boundary layer: Site of plasma, momentum and energy transfer from the magnetosheath to the magnetosphere, *Geophys. Res. Lett., 3*, 685, 1976.

El-Alaoui, M., et al., Modeling magnetotail ion distributions with global magnetohydrodynamic and ion trajectory calculations, in *Geospace Mass and Energy Flow: Results from the International Solar-Terrestrial Physics Program, Geophys. Monogr. Ser., 104*, edited by J. Horwitz and D. Gallagher, p. 291, AGU, Washington, DC, 1998.

Frank, L. A et al., The Comprehensive Plasma Instrumentation (CPI) for the Geotail spacecraft, *J. Geomag. Geoelec., 46*, 23, 1994.

Frank, L. A., et al., The plasma velocity distributions in the near-Earth plasma sheet: A first look with the Geotail spacecraft, *J. Geophys. Res., 101*, 10,627, 1996.

Knight, S., Parallel electric fields, *Planet. Space Sci., 21*, 741, 1972.

Lennartsson, W. and E. G. Shelley, Survey of 0.1- to 16-keV/e plasma sheet ion compositions, *J. Geophys. Res., 91*, 3061, 1986.

Lockwood, M., J. H. Waite, Jr., T. E. Moore, J. F. E. Johnson, and C. R. Chappell, A new source of suprathermal O^+ ions near the dayside polar cap boundary, *J. Geophys. Res., 90*, 4099, 1985a.

Lockwood, M., M. O. Chandler, J. L. Horwitz, J. H. Waite, Jr., T. E. Moore, and C. R. Chappell, The cleft ion fountain, *J. Geophys. Res., 90*, 9376, 1985b.

Lyons, L. R., D. Evans, and R. Lundin, An observed relation between magnetic field aligned electric fields and downward electron energy fluxes in the vicinity of auroral forms, *J. Geophys. Res., 84*, 457, 1979.

Moen, J., and A. Brekke, The solar flux influence on quiet time conductances in the auroral ionosphere, *Geophys. Res. Lett., 20*, 971, 1993.

Peroomian, V., and M. Ashour-Abdalla, Population of the near-Earth magnetotail from the auroral zone, *J. Geophys. Res., 101*, 15,387, 1996.

Raeder, J., R. J. Walker, and M. Ashour-Abdalla, The structure of the distant geomagnetic tail during long periods of northward IMF, *Geophys. Res. Lett., 22*, 349, 1995.

Raeder, J., J. Berchem, and M. Ashour-Abdalla, The importance of small-scale processes in global MHD simulations: Some numerical experiments, in *The Physics of Space Plasmas*, vol. 14, T. Chang and J. R. Jasperse, eds., p. 403, MIT Center for Theoretical Geo/Cosmo Plasma Physics, Cambridge, MA, 1996.

Raeder, J., J. Berchem, M. Ashour-Abdalla, L. A. Frank, W. R. Paterson, K. L. Ackerson, J. M. Bosqued, R. P. Lepping, S. Kokubun, T. Yamamoto, S. A. Slavin, Boundary layer formation in the magnetotail: Geotail observations and comparisons with a global MHD model, *Geophys. Res. Lett.*, in press, 1997.

Robinson, R. M., R. R. Vondrak, K. Miller, T. Dabbs, and D. Hardy, On calculating ionospheric conductances from the flux and energy of precipitating electrons, *J. Geophys. Res., 92*, 2565, 1987.

Scholer, M., and R. A. Treumann, The low-latitude boundary layer at the flanks of the magnetopause, in *Transport Across the Boundaries of the Magnetosphere*, edited by B. Hultqvist and M. Øieroset, p. 341-367, Kluwer Academic Publishers, Dordrecht, 1997.

Shelley, E. G., R. G. Johnson, and R. D. Sharp, Satellite Observations of Energetic Heavy Ions during a Geomagnetic Storm, *J. Geophys. Res., 77*, 6104, 1972.

Williams, D. J., Considerations of source, transport, acceleration/heating and loss processes responsible for geomagnetic tail particle populations, in *Transport Across the Boundaries of the Magnetosphere*, edited by B. Hultqvist and M. Øieroset, p. 369-389, Kluwer Academic Publishers, Dordrecht, 1997.

Yau, A. W., E. G. Shelley, W. K. Peterson, and L. Lenchyshyn, Energetic auroral and polar ion outflow at DE 1 altitudes: magnitude, composition, magnetic activity dependence, and long-term variations, *J. Geophys. Res., 90*, 8417, 1985.

Maha Ashour-Abdalla, Mostafa El-Alaoui, Vahé Peroomian, Joachim Raeder, and Ray J. Walker, Institute of Geophysics and Planetary Physics, University of California, Los Angeles, CA 90095-1567 (e-mail madalla@igpp.ucla.edu).

L. A. Frank, and W. R. Paterson, Department of Physics and Astronomy, The University of Iowa, Iowa City, IA 52242.

Ionospheric Outflow

R. W. Schunk

Center for Atmospheric and Space Sciences, Utah State University, Logan

Upflowing ions of ionospheric origin are an important source of magnetospheric plasma at all latitudes, particularly during geomagnetic storms and substorms. Upflowing ions from the cusp populate the mantle/ boundary layer and plasma sheet regions. Polar wind ions escaping from the polar cap can populate both the plasma sheet and magnetotail lobes. Ionospheric ions energized in the nocturnal auroral oval can populate the plasma sheet and ring current regions. Unfortunately, at the present time, there are no time-dependent global models that can properly account for all of the sources of escaping ionospheric plasma. However, we have recently conducted several time-dependent global simulations of the coupled ionosphere-polar wind system that show how this system reacts to changing geomagnetic activity. The system's response is found to be nonlinear, and the temporal response of O^+ can be opposite to that of H^+. Also, the temporal response of the ions at high altitudes can be opposite to that at low altitudes, depending on the seasonal and solar cycle conditions. In general, the simulation results are in good agreement with the available measurements. These global ionosphere-polar wind simulations, and related results, are reviewed in this paper.

1. INTRODUCTION

Shortly after the configuration of the magnetosphere was deduced [*Axford and Hines*, 1961; *Dungey*, 1961], it was suggested that there should be a continual outflow of ionospheric plasma (H^+ and He^+) along the 'open' magnetic field lines in the polar cap, because they extend deep into space. The early models used to support this suggestion were based on a continual escape of light ions via a thermal evaporation process [*Dessler and Michel*, 1966; *Bauer*, 1966], while the later models were based on a supersonic hydrodynamic formulation [*Axford*, 1968; *Banks and Holzer*, 1968]. Now, after thirty years of intensive study, it is well known that the 'classical' polar wind is an ambipolar outflow of thermal plasma from the topside ionosphere at high latitudes. As the plasma escapes the ionosphere along diverging magnetic field lines, it undergoes four major transitions, including a transition from chemical to diffusion dominance, a transition from subsonic to supersonic flow, a transition from collision-dominated to collisionless regimes, and a transition from a heavy to a light ion. It is also well known that the polar wind is penetrated by energetic ion beams and conics that are created on auroral field lines and then convect into the polar cap [*Shelley et al.*, 1982].

Over the years, a myriad of polar wind studies have been conducted. Also, numerous studies have been conducted of the so-called 'nonclassical' polar wind, which may contain nonthermal components such as ion beams or hot electrons. Polar wind studies have been conducted of its supersonic nature, its anisotropic thermal structure, its evolution through the collision-dominated to collisionless transition region, its inherent stability, and the maximum ion escape fluxes that are possible for different seasonal and solar cycle conditions. Simulations have also been conducted of the extent to which various processes can affect the polar wind, including charge exchange, photoelectrons, elevated thermal electron temperatures, elevated thermal ion temperatures (at low, middle and high altitudes), imposed transverse ion heating, hot electrons and ions of magnetospheric origin, centrifugal acceleration, wave-particle interactions in the polar cap, and field-aligned auroral currents [cf. *Ganguli*, 1996; *Schunk and Sojka*, 1997; and references therein]. The different processes may or may not be important, depending on the geophysical conditions, the location in the polar cap, the altitude, and the magnetic activity level. Nevertheless, the large number of possible polar wind processes must be kept in mind when results from relatively simple simulations, containing only one or two of the processes, are compared to satellite measurements.

Nearly all of the polar wind studies conducted to date were either for steady state conditions or for simple time-dependent scenarios in which 'one-dimensional' models were applied to single 'fixed' locations, with the temporal variations driven by imposed energy inputs. Three-dimensional simulations were conducted by *Ganguli* [1994] that self-consistently take into account cross-field transport due to plasma instabilities, but they had a restricted spatial domain in the direction perpendicular to **B**. On the other hand, global polar wind simulations were conducted by *Schunk and Sojka* [1989, 1997], but they did not self-consistently include the cross-field transport due to plasma instabilities. Nevertheless, the latter simulations were realistic in that they were based on 'convecting' flux tubes of plasma, and they provided the first global view of the polar wind. This latter approach is the focus of this paper, because the horizontal motion of plasma into and out of the sunlit ionosphere, auroral oval, polar cap, and main trough must be taken into account if one is to properly model the dynamic polar wind. Specifically, this paper provides a review of the recent and on-going work in this area.

2. GLOBAL IONOSPHERE-POLAR WIND MODEL

The global ionosphere-polar wind model developed by *Schunk and Sojka* [1989] covers the altitude range from 90 to 9000 km for latitudes greater than 50° magnetic in either the northern or southern hemispheres. At altitudes between 90 and 800 km, the model calculates 3-dimensional time-dependent distributions for the electron and ion (NO^+, O_2^+, N_2^+, O^+, N^+) densities and temperatures (T_e, $T_{i\parallel}$, $T_{i\perp}$) from diffusion and heat conduction equations. At altitudes above 500 km, the time-dependent nonlinear hydrodynamic equations for O^+ and H^+ are solved self-consistently with the ionospheric equations. The model is an Euler-Lagrange hybrid model in that the transport equations are solved as a function of altitude for convecting flux tubes of plasma. The 3-dimensional nature of the model is obtained by following many plasma flux tubes, while keeping track of their positions at all times. The coupled ionosphere-polar wind model takes account of ion-neutral frictional (Joule) heating, anisotropic ion heating, the magnetic mirror force, thermal electron heating due to precipitating electrons, supersonic ion outflow, shock formation, and ion energization during a plasma expansion event.

The main inputs required for a model run include the atmospheric densities, temperatures, and winds, and the magnetospheric convection and precipitation patterns. Typically, the atmospheric parameters are obtained from empirical models, such as those developed by *Hedin et al.* [1977]. In the first global simulation [*Schunk and Sojka*, 1989], a *Heelis et al.* [1982] 'symmetric' two-cell convection pattern and a *Spiro et al.* [1982] auroral oval were adopted. This global simulation, which was for winter/solar minimum conditions, was based on 59 convecting plasma flux tubes. In the second study [*Schunk and Sojka*, 1997], the 'BC' two-cell convection pattern of *Heppner and Maynard* [1987] and the *Hardy et al.* [1985] auroral oval were adopted. The BC convection pattern is asymmetric, with enhanced convection in the dawn cell. It is appropriate for southward IMF and $B_y > 0$ in the northern hemisphere. This second study involved four geophysical cases (summer and winter for solar maximum and minimum), and each global simulation was based on 159 convecting plasma flux tubes. Our most recent global ionosphere-polar wind simulations contain 1000 convecting plasma flux tubes, which yield a 130-200 km horizontal spatial resolution in the polar cap (Figure 1).

All of the global simulations conducted to date were for an idealized geomagnetic storm that contains growth, main, and decay phases. As magnetic activity increases, the plasma convection and auroral precipitation patterns expand, the convection speeds increase, and the particle precipitation becomes more intense. The reverse occurs during declining magnetic activity. Figure 2 shows the assumed variation of K_p for the idealized storm. At 0400 UT, K_p increases exponentially over a 1-hour period from 1 to 6. The high K_p is maintained for a 1-hour period, and then K_p decreases exponentially to 1 over a 4-hour period. Subsequently, K_p is held fixed at 1. The dependence of the convection pattern on K_p is not well known, but in our global simulations, K_p is only used as a guide in producing a time-varying convection pattern. During changing magnetic activity, the size of the pattern and the cross polar cap potential (Φ) are varied continuously, following the K_p variation shown in Figure 2. The simple linear relationship proposed by *Heppner* [1973] is used to relate K_p to Φ; $\Phi = 20 + 13\ K_p$ (keV). In a similar manner, the precipitation pattern is varied continuously according to the K_p variation

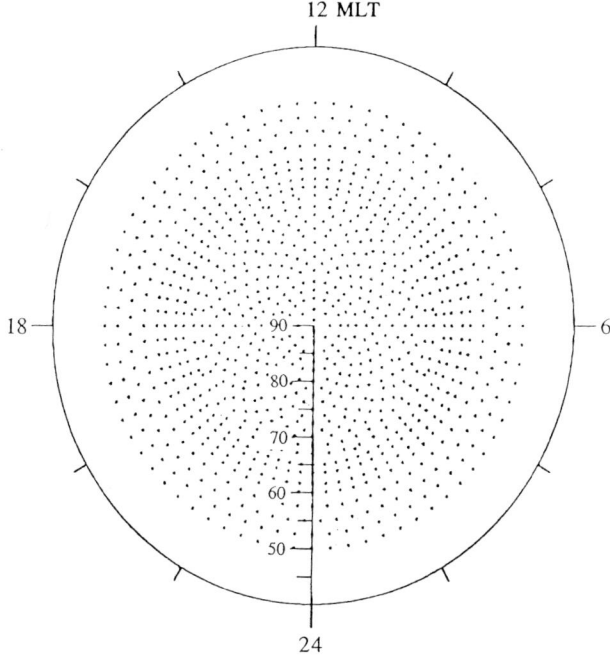

Figure 1. Locations of the 1000 plasma flux tubes, at 0300 UT, that are used in simulating a geomagnetic storm.

shown in Figure 2. During changing magnetic activity, both the size of the oval and the intensity of precipitation change.

3. IONOSPHERE-POLAR WIND SIMULATIONS

The global simulations of the ionosphere-polar wind system led to some general conclusions about the system's response to changing magnetic activity. It was found that: (1) Plasma pressure changes in the ionosphere due to N_e, T_e, or T_i variations produce disturbances in the polar wind. Specifically, the horizontal convection of the plasma through the auroral oval and regions of large electric fields produces transient large-scale upflows and downflows; (2) The density structure in the polar wind tends to be more complicated than that in the underlying ionosphere because the vertical propagation speeds change as the plasma convects into different high-latitude regions; (3) During increasing magnetic activity, there is an overall increase in the polar wind outflow rate, while the reverse occurs during declining magnetic activity; and (4) There are significant time delays with respect to when disturbances in the ionosphere create disturbances at high altitudes in the polar wind. This occurs because the convection time across the polar cap is comparable to the time it takes plasma to flow from the ionosphere (500 km) to high altitudes (9000 km).

Plates 1a and 1b show, respectively, the temporal variations of the O$^+$ and H$^+$ densities during the storm for winter, solar minimum conditions. There are 3 rows in both of the figures, corresponding to altitudes of 500, 1000, and 4000 km. Each row contains 13 snapshots of the entire polar region at 1/2-hour intervals from 0400 to 1000 UT. Each snapshot shows a color-coded ion density distribution from 50°-90° magnetic latitude and 0-24 MLT, with pink corresponding to high densities and blue to low densities. Each row has a fixed color scale, but there are different color scales at different altitudes. This format allows one to get a quick global view of the system's response to the storm, by simply observing the change in color with time and altitude.

When the storm commences, there is an O$^+$ upwelling throughout the bulk of the polar region in response to the storm-enhanced electron and ion temperatures, but there is a half-hour time delay in the O$^+$ density increase at 4000 km that is associated with the transit time required for O$^+$ to flow from low to high altitudes. During the storm's main phase (0500 to 0600 UT), there is an increase in the O$^+$ density throughout the polar region at both 1000 and 4000 km as the upwelling continues, and then there is a decrease in the O$^+$ density as the storm subsides and the temperatures decrease (t > 0630 UT). The O$^+$ flow is generally downward during the declining phase of the storm at 1000 and 4000 km. However, at 500 km, the O$^+$ density basically continues to increase, from about 0500 to 0800 UT, due to the production of ionization at low altitudes as a result of particle precipitation and its subsequent upward

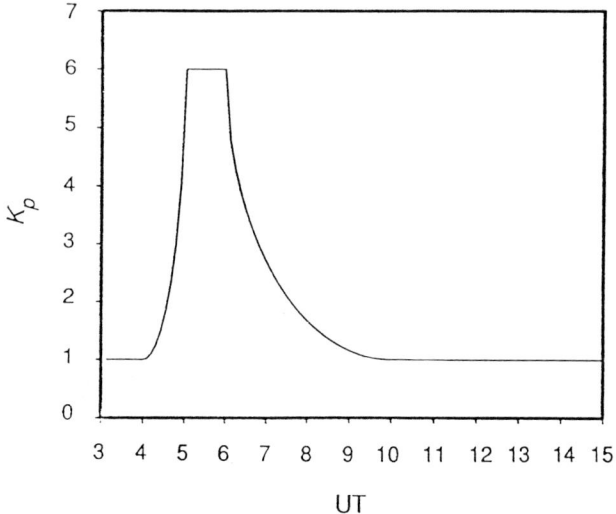

Figure 2. Variation of K_p with time for an idealized geomagnetic storm. The storm lasts six hours, from 0400 to 1000 UT. From *Schunk and Sojka* [1997].

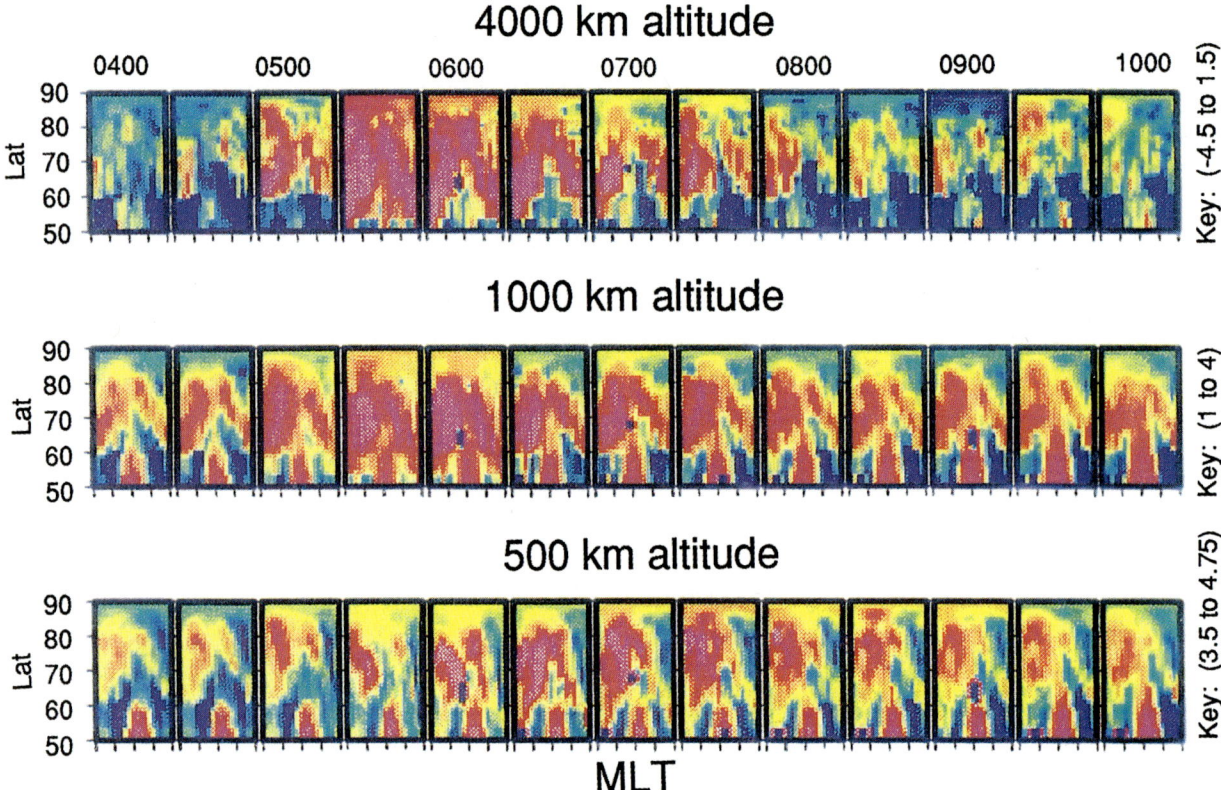

Plate 1a. Snapshots of the O⁺ density distribution at 500 km (bottom), 1000 km (middle), and 4000 km (top) at half-hour intervals from 0400 to 1000 UT for winter, solar minimum conditions. Each rectangular plot shows a color coded distribution of $\log_{10}[n(O^+)$ in cm$^{-3}]$ over the polar region from 50° to 90° magnetic latitude and 0-24 MLT. The key at the right only shows the color coding range. From *Schunk and Sojka* [1997].

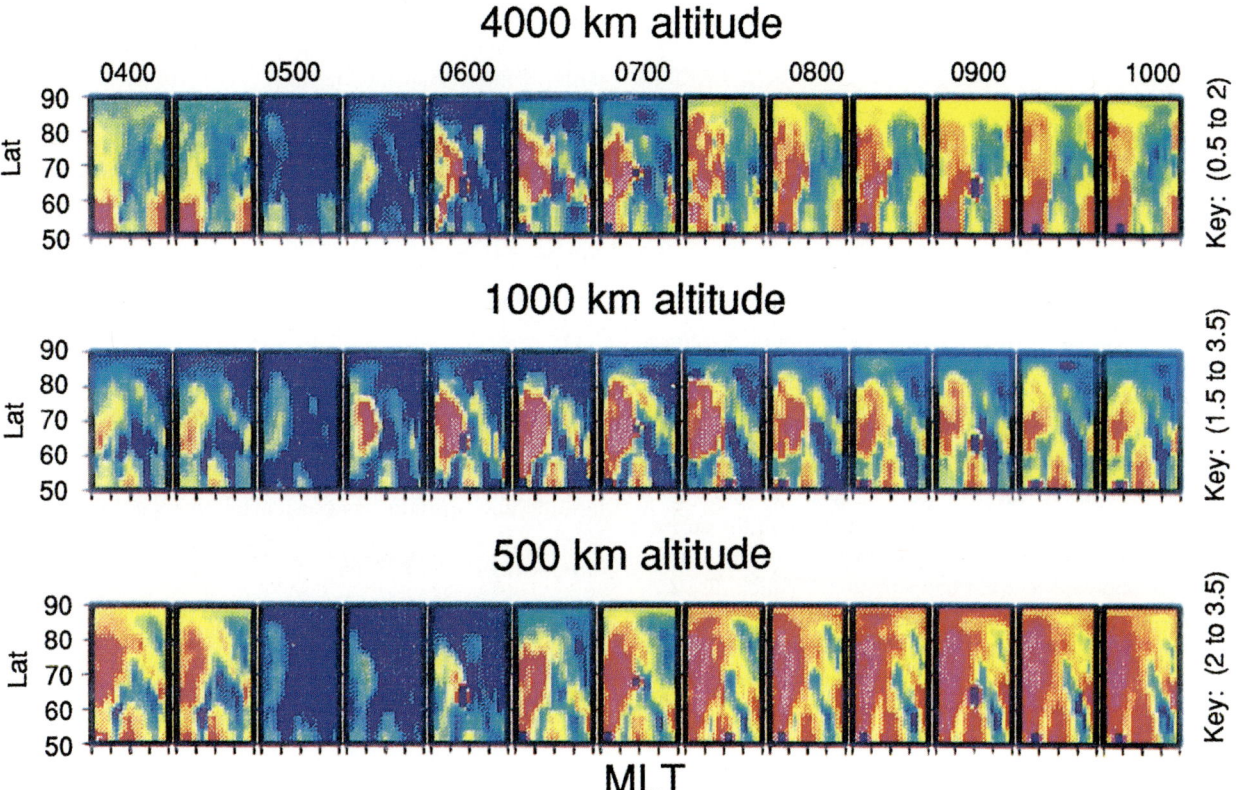

Plate 1b. Snapshots of the H$^+$ density distribution for winter, solar minimum conditions. The plotting format is the same as for Plate 1a. From *Schunk and Sojka* [1997].

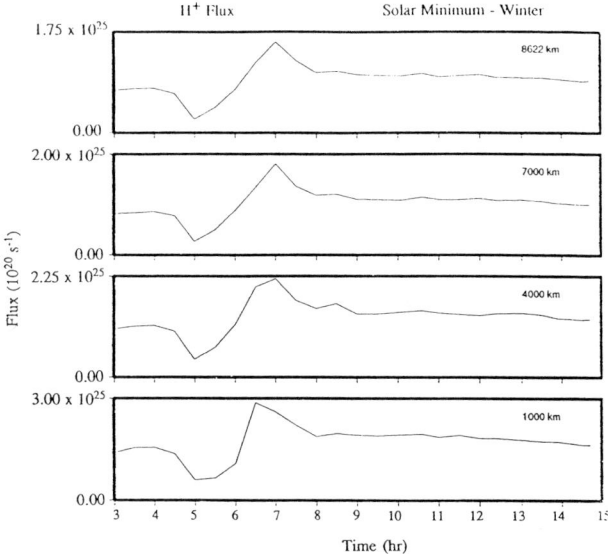

Figure 3. Total H⁺ outflow rate (ions s⁻¹) versus time at selected altitudes. The outflow rate is obtained by integrating the H⁺ flux over the entire polar region at each altitude. From *Schunk and Sojka* [1997].

diffusion to 500 km. Consequently, the temporal variation of the O⁺ density is opposite at high and low altitudes during the storm's recovery phase (0600 to 0800 UT).

The H⁺ behavior during the storm is different than that of O⁺. Shortly after the storm commences, there is an 'H⁺ blowout' throughout the polar region at all altitudes as a consequence of the upward flow associated with the storm-enhanced electron and ion temperatures (Plate 1b). This behavior is opposite to what happens to O⁺, which displays a density increase during the storm's growth and main phases (Plate 1a). The net result is that O⁺ is the dominant ion at high altitudes (4000 km) throughout the polar region during enhanced magnetic activity (0500 to 0630 UT). Subsequently, as the O⁺ density slowly increases at low altitudes, the H⁺ density also increases due to the O⁺ + H ⇔ H⁺ + O reaction. However, the H⁺ recovery does not occur at the same rate at all altitudes.

For an analysis of the global polar wind, a useful quantity is the total ion outflow rate (ions s⁻¹) across a given altitude integrated over latitudes greater than 50° magnetic. The variation of this quantity versus time is shown in Figure 3 for H⁺ and winter, solar minimum conditions. The net flow rate is upward at all altitudes and times, and at 8622 km the flow rate corresponds to the global H⁺ escape rate. During the storm's growth phase, the H⁺ flow rate decreases at all altitudes as the initial reservoir of H⁺ is depleted (the H⁺ blowout). It reaches a minimum at about 0500 UT and then increases throughout the main phase (0500-0600 UT). The maximum H⁺ escape rate of 1.7×10^{25} ions s⁻¹ occurs at 0700 UT, which is one hour after the main phase of the storm. This occurs because the O⁺ density at low altitudes undergoes a gradual increase during the storm's main phase, and this increasing source of H⁺ accounts for the gradual increase in the H⁺ escape rate. For O⁺, on the other hand, the maximum upward flow rate, at each altitude, occurs during the storm's main phase (not shown). However, the magnitude of the O⁺ flow rate decreases with altitude, from about 5×10^{25} ions s⁻¹ at 500 km to 3×10^{23} ions s⁻¹ at 8622 km. This decrease in the O⁺ flow rate with altitude indicates that most of the O⁺ ions do not have sufficient energy to escape and return to the ionosphere.

The effect of a solar cycle change on the polar wind's response to a storm can be determined by comparing the winter, solar minimum ($F_{10.7} = 70$) case discussed above with a winter, solar maximum ($F_{10.7} = 210$) case. This latter case has been run with 1000 plasma flux tubes, which yield a horizontal spatial resolution of better than 200 km in the polar region. Plates 2a and 2b show, respectively, snapshots of O⁺ and H⁺ density distributions as a function of altitude and latitude at the end of the storm's main phase (0600 UT) for the winter, solar maximum case. The altitude range extends from 500 to 8000 km and the latitude range is from 50° on the dayside to 50° on the nightside along the noon-midnight meridian. The color scale is the same in both figures and is chosen such that densities greater than 10^3 cm⁻³ are pink and those below 10^0 cm⁻³ are dark blue. The most evident feature in both figures is the spatial structure in the density distributions, both with latitude and altitude. In particular, note the biteout in the H⁺ density in the 1400-2000 km region on the nightside at latitudes between 80° and 65°. Another H⁺ density biteout occurs on the dayside near 80° latitude in the same altitude region. These occur because of the complex and time-dependent interplay between chemistry and both vertical and horizontal plasma transport. Only by studying the past history of the plasma can the causative mechanism be elucidated. Hence, it would be impossible to establish the physics governing the biteouts, or any other density feature, by using data from just one satellite or one ground-based observing site. Another interesting feature to note is the extensive O⁺ upwelling, which results in O⁺ densities at high altitudes that are generally greater than the H⁺ densities.

With regard to a general comparison of the solar maximum and minimum cases, the H⁺ density distribution for winter, solar maximum displays a variation with time and altitude that is 'qualitatively' very similar to that shown in Plate 1b for winter, solar minimum conditions. However, there are important quantitative differences between the two cases. For O⁺, on the other hand, the solar maximum and minimum cases are 'qualitatively' different.

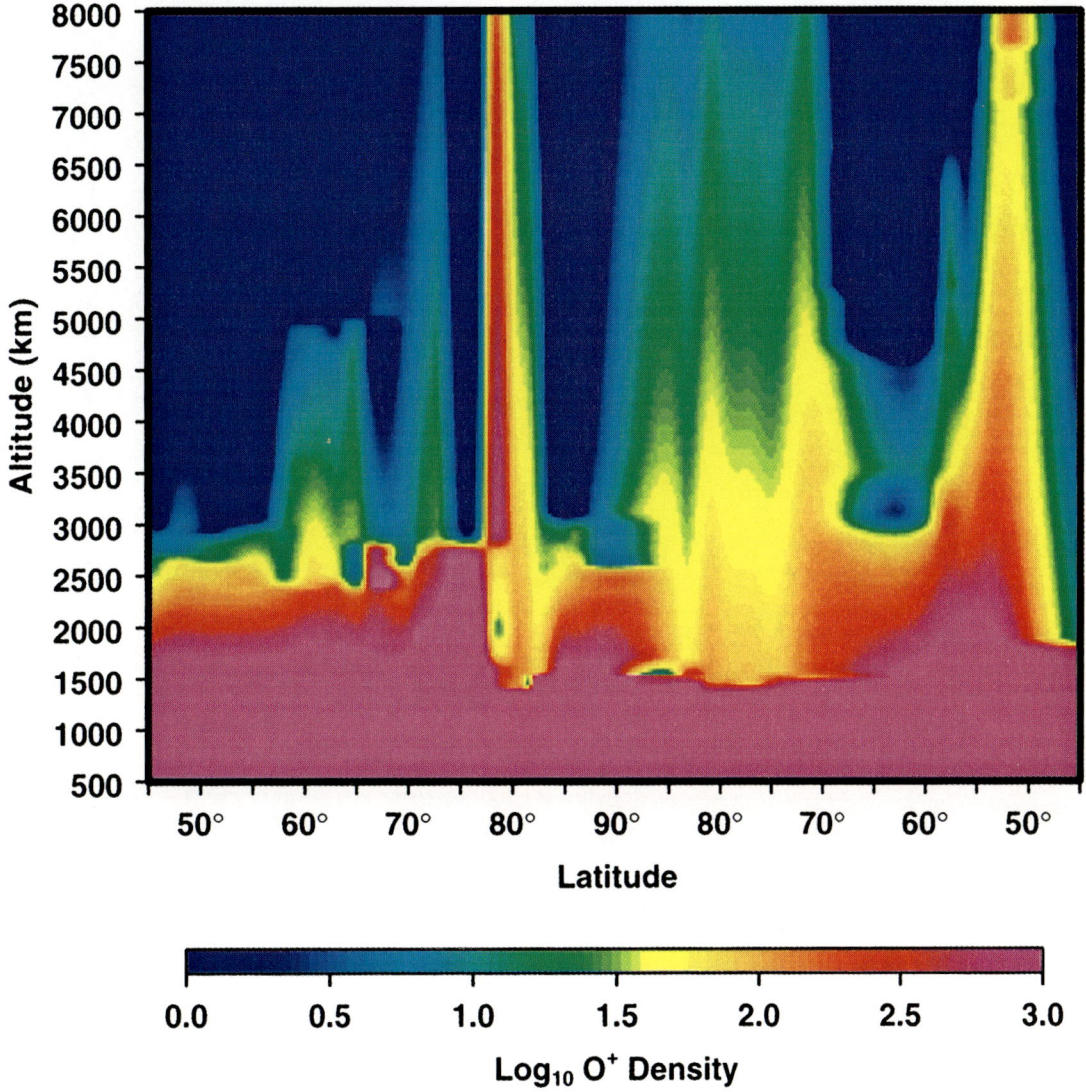

Plate 2a. Snapshot of the O⁺ density distribution as a function of latitude and altitude at 0600 UT for winter, solar maximum conditions. The latitude range is from 50° on the dayside to 50° on the nightside along the noon-midnight meridian. Densities greater than 10^3 cm^{-3} are colored pink and those below 10^0 cm^{-3} are dark blue.

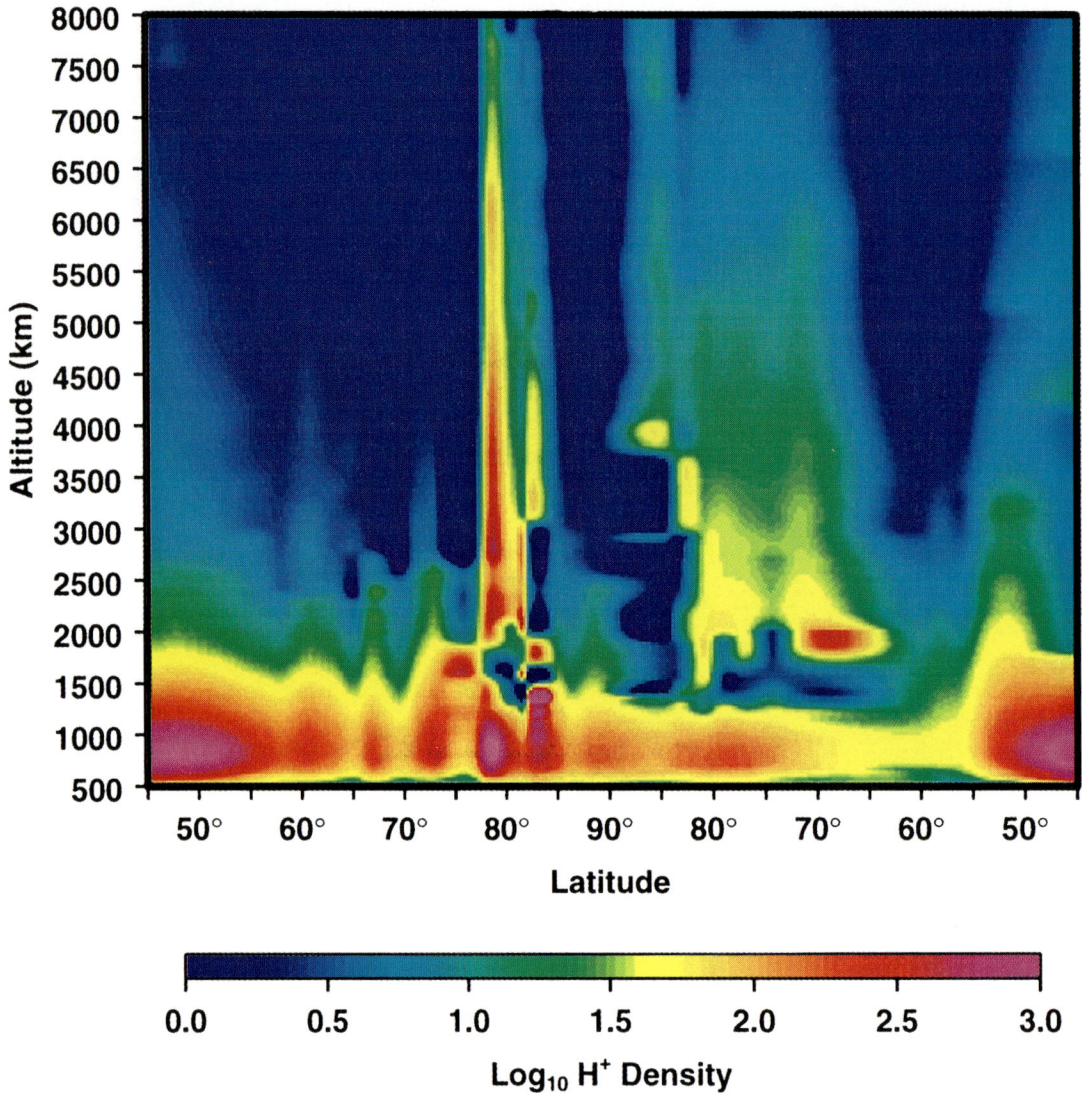

Plate 2b. Snapshot of the H⁺ density distribution as a function of latitude and altitude at 0600 UT for winter, solar maximum conditions. The plotting format is the same as for Plate 2a.

For solar maximum, there is an increase in the O$^+$ density at altitudes between 1000 and 8000 km during enhanced magnetic activity (0500 to 0630 UT), while there is a decrease at 500 km. The decrease at 500 km is opposite to what occurs at solar minimum (see Plate 1a). However, 'quantitatively' the total O$^+$ flow rates at solar maximum are similar to those obtained at solar minimum.

4. GENERAL COMPARISON WITH MEASUREMENTS

During the last two decades, numerous experimental databases have been collected that pertain to thermal ion upflows and the polar wind. Incoherent scatter radar measurements have clearly shown that enhanced upward flows of thermal ions are common in the cusp, nocturnal oval, and polar cap [*Wahlund et al.*, 1992; *Blelly et al.*, 1996; *Keating et al.*, 1990]. *Tsunoda et al.* [1989] studied thermal ion upwellings in the cusp/cleft region using Hilat satellite data taken at 800 km, and *Pollock et al.* [1990] analyzed DE-1 satellite data for 39 thermal ion upwelling events at high altitudes in the vicinity of the cleft. *Loranc et al.* [1991] conducted a statistical study of thermal ion upwellings using DE-2 satellite data taken at altitudes between 200 and 1000 km. *Chandler* [1995] conducted a statistical study of polar wind measurements made by the DE-1 satellite in the polar cap at altitudes between 1000 and 4000 km. *Abe et al.* [1993] presented field-aligned ion drift measurements in the polar cap. The velocities were measured by the Akebono satellite at altitudes from 2000 to 10,000 km. Finally, *Persoon et al.* [1983] published DE-1 satellite measurements of the local electron density in the polar cap that covered the altitude range from 6378 top 23,343 km.

A careful comparison of the various satellite and ground-based measurements with our global ionosphere-polar wind simulations indicates a generally good agreement. Specifically, the model and measurements agree on the following issues: (1) Most of the transient O$^+$ upflows are subsonic; (2) The upflows can be generated by T_e, T_i, and n_e enhancements; (3) O$^+$ upflows typically occur in the cusp and auroral zone at all local times, and downflows occur in the polar cap. However, during increasing magnetic activity, O$^+$ upflows can also occur in the polar cap; (4) T_e is the main driver of the large-scale ion upflow events, but the magnitudes of the upward O$^+$ fluxes and velocities are enhanced if T_i is also elevated; (5) The upward H$^+$ and O$^+$ velocities increase with T_e, and this results in both seasonal and day-night asymmetries in the ion velocities; (6) During increasing magnetic activity, O$^+$ is the dominant ion at all altitudes throughout the bulk of the polar region; (7) Upward O$^+$ fluxes in the range of from 10^7 to 10^{10} cm^{-2} s^{-1} can occur; (8) Between 2000 and 4000 km, the H$^+$ velocity typically displays its greatest increase with altitude, and it exhibits a large variability in this altitude range; (9) The calculated H$^+$ and O$^+$ velocities are in good quantitative agreement with the measured velocities, particularly those of *Chandler* [1995]; and (10) The calculated H$^+$ and O$^+$ densities generally lie within the scatter of the measurements, which increases with altitude.

The only real disagreement between the model and measurements occurs after the storm at altitudes above about 4000 km, where the calculated O$^+$ densities can be orders of magnitude lower than the 'average' measured values. When this occurs, H$^+$ is the dominant ion at these altitudes and times. Also, as noted earlier, the bulk of the upflowing O$^+$ ions do not have sufficient energy to escape and they simply return to the ionosphere. These shortcomings may be related to our neglect of nonclassical polar wind processes, which generally operate at high altitudes.

5. NONCLASSICAL PROCESSES

The global ionosphere-polar wind simulations presented above correspond to the behavior of the 'classical' polar wind, which is driven by temperature and density perturbations in the underlying ionosphere. However, the polar wind may be affected by other processes not included in the classical picture. Figure 4 is a schematic diagram of some possible nonclassical processes. In sunlit regions, escaping photoelectrons may provide an additional acceleration of the polar wind at high altitudes (>7000 km) as they drag the thermal ions with them [*Lemaire*, 1972; *Khazanov et al.*, 1997]. Also, the polar wind can be driven to be unstable by cusp ion beams and conics that have convected into the polar cap and then pass through the polar wind at high altitudes [*Barakat and Schunk*, 1989; *Chen and Ashour-Abdallah*, 1990]. The interaction of the cold, outflowing, polar wind electrons with the hot magnetospheric electrons (polar rain, showers and squall) can result in a double layer electric field over the polar cap, which can energize the O$^+$ and H$^+$ ions [*Barakat and Schunk*, 1984; *Winningham and Gurgiolo*, 1982]. At altitudes above 6000 km, electromagnetic turbulence can affect the ion outflow via perpendicular heating through wave-particle interactions [*Ludin et al.*, 1990; *Barghouthi*, 1997]. In addition, centrifugal acceleration will act to increase the ion outflow velocities above 3000 km as the plasma flux tubes convect across the polar cap [*Cladis*, 1986]. Finally, anomalous resistivity on auroral field lines can affect the polar wind as the plasma flux tubes convect through the oval [*Ganguli*, 1996].

To date, when the various nonclassical processes were deduced to be important, the deductions were based on 'one-dimensional' steady state or time-dependent simulations applied at a 'fixed' location. Also, it was

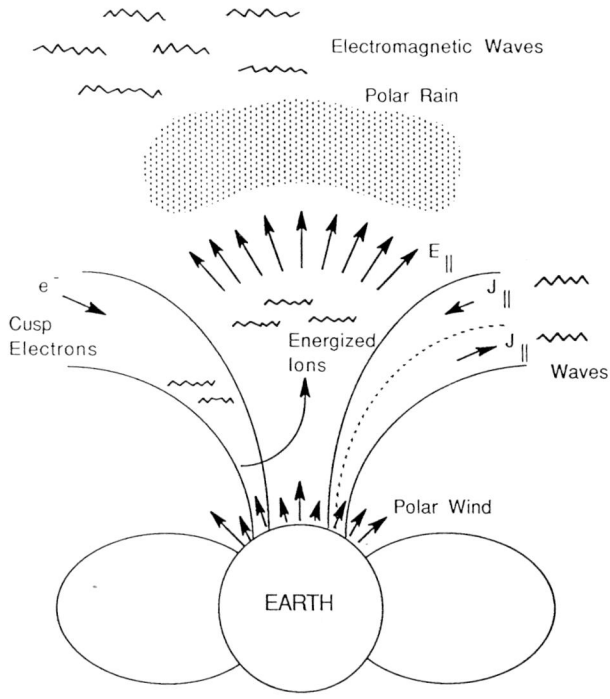

Figure 4. Schematic diagram showing the nonclassical processes that may affect the polar wind. From *Schunk and Sojka* [1997].

typical to include only one nonclassical process at a time. However, when all of the processes are included and when the plasma is allowed to convect through the different high-latitude regions in a realistic manner, the conclusion as to the importance of a given process may change. Nevertheless, it is important to include nonclassical processes in polar wind models.

As an example of how a nonclassical process can affect the polar wind, we consider the interaction of the cold, upflowing, ionospheric electrons with the hot magnetospheric electrons. Similar to what was done in all previous studies, only one nonclassical process is considered, but unlike the previous studies, its effect will be shown for a convecting plasma flux tube. The trajectory that the selected plasma flux tube follows is shown in Figure 5. This flux tube is one of the 159 flux tubes considered in the storm simulation of *Schunk and Sojka* [1997]. The plasma following this trajectory starts at location 'a' on the nightside (4:20 MLT, 65° magnetic) at 0300 UT. The plasma then convects sunward, turns antisunward, convects across the polar cap (segment b-c), through the nocturnal oval, and finally convects sunward again. Note that the auroral oval and convection pattern change continuously with time as the plasma follows this trajectory because of the storm (see Figure 2).

When nonclassical polar wind processes are considered, it is more convenient to use a macroscopic particle-in-cell (PIC) model. The PIC model we used covered the altitude range from 2000 km to 8 Earth radii (R_E) and contained 2 million ions [*Barakat et al.*, 1998]. Both H^+ and O^+ ions were considered, as were both hot magnetospheric and cold ionospheric electrons. In this model, the ions were kinetic and were subjected to electrostatic, gravitational, mirror, centrifugal, and collisional forces, while both electron populations were assumed to obey the Boltzmann relation. The time-dependent boundary conditions at 2000 km that were needed for the macroscopic PIC simulation were taken from the storm simulation of *Schunk and Sojka* [1997] for the selected trajectory shown in Figure 5. As the PIC plasma followed this trajectory, it was only subjected to magnetospheric electrons in the polar cap (along the segment b-c). In the polar cap, the hot electron density and temperature at 2000 km were held constant, while the values above this altitude were calculated with the model. Three different hot electron cases were considered, corresponding to three different hot electron temperatures and one value for the hot electron density. These hot electron values were obtained by using the ionospheric electron density and temperature values at the starting location 'a' as a reference. The hot electron density was

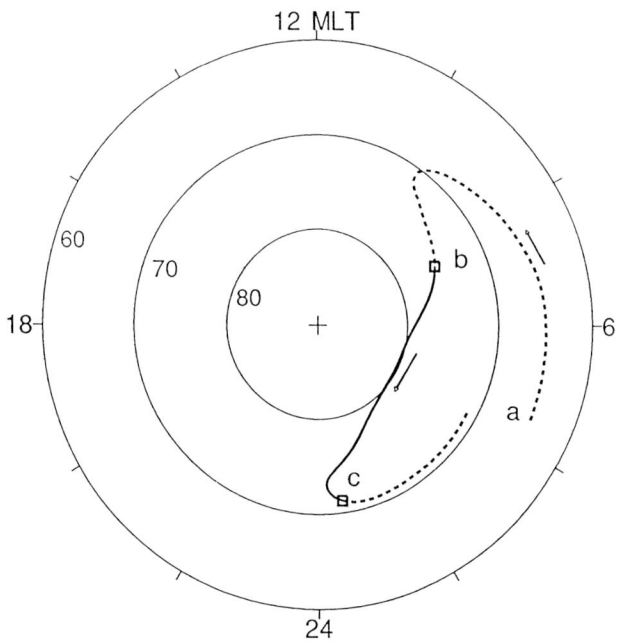

Figure 5. Selected convection trajectory for a flux tube of plasma in the dawn sector of the polar region. When the flux tube is in the polar cap along the segment marked b-c, it is subjected to hot magnetospheric electrons. From *Barakat et al.* [1998].

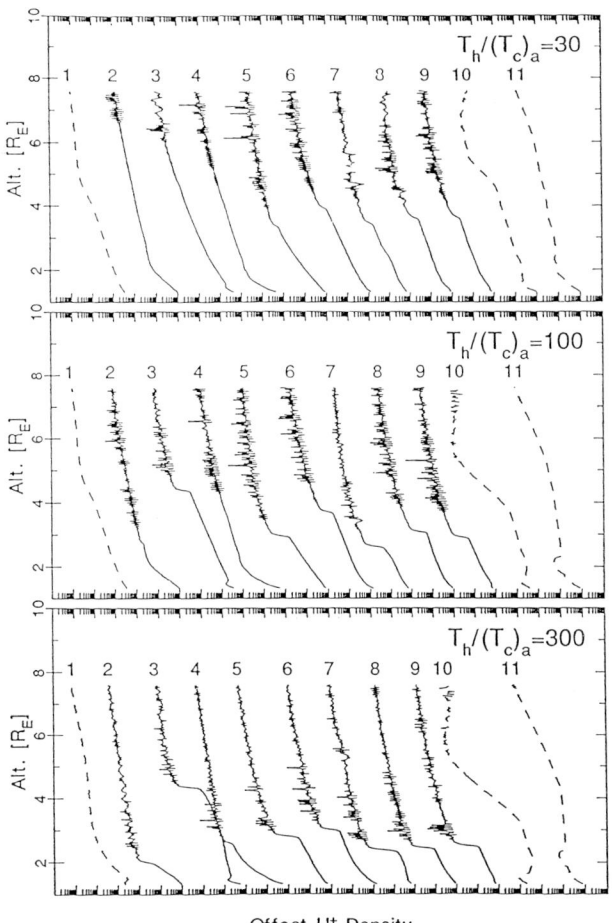

Figure 6. Offset H+ density profiles for hot/cold electron temperature ratios of 30 (top), 100 (middle), and 300 (bottom), and for a hot/cold density ratio of 0.01 at 2000 km altitude using the cold electron density at location "a" in Figure 5. The time interval between successive curves is 750 seconds. The dashed curves correspond to the times when the plasma flux tube is outside of the polar cap. From *Barakat et al.* [1998].

assumed to be 1% of the cold electron density at point a, and the three temperature values were 30, 100, and 300 times the cold electron temperature at point a. As noted above, once these hot electron values at 2000 km were determined, they were only applied along the trajectory segment b-c and they were held constant.

Figure 6 shows the results of the three PIC simulations, where offset H+ density profiles are plotted at 750 second intervals as the plasma follows the trajectory in Figure 5 [*Barakat et al.*, 1998]. The interaction of the hot magnetospheric electrons and the cold ionospheric electrons results in the formation of a 'double layer' electric field, which is parallel to **B** and points upward. This electric field acts to abruptly enhance the upward ion velocities and, hence, reduce their densities. However, the amplitude and altitude of the parallel electric field vary appreciably as the plasma flux tube convects across the polar cap. The amplitude and altitude of the electric field are primarily determined by a balance of the magnetospheric electron pressure and the ionospheric electron pressure (kinetic and dynamic). In each of the three simulations in Figure 6, the magnetospheric electron pressure is constant, so the variability shown in this figure is due entirely to the variability in the underlying ionosphere. With regard to the energy gain as H+ crosses the double layer, the H+ velocity above the double layer varies from 10 to 110 km/s for the three cases shown in Figure 6. These values are in agreement with the recent POLAR satellite measurements [*Moore et al.*, 1997].

6. FUTURE DIRECTION

The global ionosphere-polar wind simulations we conducted have shown that it is important to include the horizontal convection of the high-latitude plasma. Specifically, the ionosphere-polar wind plasma continually convects into and out of the sunlit hemisphere, cusp, polar cap, nocturnal oval, and main trough, and this convection determines how long a region-specific plasma process (precipitation, Joule heating, photoionization, etc.) operates on the ionosphere-polar wind plasma elements. To date, the relevant ionospheric processes have been included in a reasonable manner, but the effect of magnetospheric (nonclassical) processes on the high-altitude polar wind needs improvement. To accomplish this, we are constructing a hybrid fluid-PIC model of the global ionosphere-polar wind system that will cover the altitude range from 90 km to 8 R_E and that will be based on 1000 convecting plasma flux tubes, with 5-10 million particles per flux tube. When completed, the model will provide 3-dimensional time-dependent distributions of the ionosphere and polar wind parameters, including ion velocity distributions. The numerical datasets can then be used to more rigorously calculate ionospheric plasma escape rates and height-integrated conductivities, which can be used as inputs to global MHD models of the magnetosphere. The 3-dimensional time-dependent numerical data can also be probed the way satellites probe the real ionosphere-polar wind system, which will allow for direct model-measurement comparisons. Hopefully, this will lead to an improved understanding of ionosphere-polar wind-magnetosphere coupling processes.

Acknowledgement. This research was supported by NASA grant NAG5-1484 and NSF grant ATM-9612638 to Utah State University. We thank H. G. Demars for producing Figures 5a and 5b.

REFERENCES

Abe, T., B. A. Whalen, A. W. Yau, R. E. Horita, S. Watanabe, and E. Sagawa, EXOS D (Akebono) suprathermal mass spectrometer observations of the polar wind, *J. Geophys. Res.*, 98, 11191-11203, 1993.

Axford, W. I., The polar wind and the terrestrial helium budget, *J. Geophys. Res.*, 73, 6855-6859, 1968.

Axford, W. I., and C. O. Hines, A unifying theory of high-latitude geophysical phenomena and geomagnetic storms, *Can. J. Phys.*, 39, 1433-1464, 1961.

Banks, P. M. and T. E. Holzer, The polar wind, *J. Geophys. Res.*, 73, 6846-6854, 1968.

Barakat, A. R. and R. W. Schunk, Effect of hot electrons on the polar wind, *J. Geophys. Res.*, 88, 9771-9784, 1984.

Barakat, A. R. and R. W. Schunk, Stability of H^+ beams in the polar wind, *J. Geophys. Res.*, 94, 1487-1494, 1989.

Barakat, A. R., H. G. Demars, and R. W. Schunk, Dynamic features of the polar wind in the presence of hot electrons, *J. Geophys. Res.*, submitted, 1998.

Barghouthi, I. A., Effects of wave-particle interactions on H^+ and O^+ outflow at high latitude: A comparative study, *J. Geophys. Res.*, 102, 22065-22075, 1997.

Bauer, S. J., The structure of the topside ionosphere, in *Electron Density Profiles in Ionosphere and Exosphere* (Edited by Frihagen, J.), p. 387, North-Holland Publishing Company, Amsterdam, 1966.

Blelly, P.-L., A. Robineau, and D. Alcaydé, Numerical modeling of intermittent ion outflow events above EISCAT, *J. Atmos. Terr. Phys.*, 58, 273-285, 1996.

Chandler, M. O., Observations of downward moving O^+ in the polar topside ionosphere, *J. Geophys. Res.*, 100, 5795-5800, 1995.

Chen, M. W. and M. Ashour-Abdallah, Heating of the polar wind due to ion beam instabilities, *J. Geophys. Res.*, 95, 18949-18968, 1990.

Cladis, J. B., Parallel acceleration and transport of ions from polar ionosphere to plasma sheet, *Geophys. Res. Lett.*, 13, 893-896, 1986.

Dessler, A. J. and F. C. Michel, Plasma in the geomagnetic tail, *J. Geophys. Res.*, 71, 1421-1426, 1966.

Dungey, J. W., Interplanetary magnetic field and the auroral zones, *Phys. Rev. Lett.*, 6, 47-48, 1961.

Ganguli, S. B., 3-D simulations of the polar wind, *STEP SIMPO Newsletter*, 4, 17, Kyoto University, Uji, Kyoto, Japan, 1994.

Ganguli, S. B., The polar wind, *Rev. Geophys.*, 34, 311-348, 1996.

Hardy, D. A., M. S. Gussenhoven, and E. Holeman, A statistical model of auroral electron precipitation, *J. Geophys. Res.*, 90, 4229-4248, 1985.

Hedin, A. E., J. E. Salah, J. V. Evans, C. A. Reber, G. P. Newton, N. W. Spencer, D. C. Kayser, D. Alcaydé, P. Bauer, L. Cogger, and J. P. McClure, A global thermospheric model based on mass spectrometer and incoherent scatter data, MSIS 1, N_2 density and temperature, *J. Geophys. Res.*, 82, 2139-2147, 1977.

Heelis, R. A., J. K. Lowell and R. W. Spiro, A model of the high-latitude ionospheric convection pattern, *J. Geophys. Res.*, 87, 6339-6345, 1982.

Heppner, J. P., High latitude electric fields and the modulations related to interplanetary magnetic field parameters, *Radio Sci.*, 8, 933-948, 1973.

Heppner, J. P. and N. C. Maynard, Empirical high-latitude electric field models, *J. Geophys. Res.*, 92, 4467-4489, 1987.

Keating, J. G., F. J. Mulligan, D. B. Doyle, K. J. Winser, and M. Lockwood, *Planet. Space Sci.*, 38, 1187-1201, 1990.

Khazanov, G. V., M. W. Liemohn, and T. E. Moore, Photoelectron effects on the self-consistent potential in the collisionless polar wind, *J. Geophys. Res.*, 102, 7509, 1997.

Lemaire, J., Effect of escaping photoelectrons in a polar exospheric model, *Space Res.*, 12, 1413-1416, 1972.

Loranc, M. and J.-P. St.-Maurice, A time-dependent gyro-kinetic model of thermal ion upflows in the high-latitude F region, *J. Geophys. Res.*, 99, 17429-17451, 1994.

Ludin, R., G. Gustafsson, A. I. Eriksson, and G. Marklund, On the importance of high-altitude low-frequency electric fluctuations for the escape of ionospheric ions, *J. Geophys. Res.*, 95, 5905-5919, 1990.

Moore, T. E., et al., High-altitude observations of the polar wind, *Science*, 277, 349-351, 1997.

Persoon, A. M., D. A. Gurnett, and S. D. Shawhan, Polar cap electron densities from DE 1 plasma wave observations, *J. Geophys. Res.*, 88, 10,123-10,136, 1983.

Pollock, C. J., M. O. Chandler, T. E. Moore, J. H. Waite, Jr., C. R. Chappell, and D. A. Gurnett, A survey of upwelling ion events characteristics, *J. Geophys. Res.*, 95, 18969, 1990.

Schunk, R. W. and J. J. Sojka, A three-dimensional time-dependent model of the polar wind, *J. Geophys. Res.*, 94, 8973-8991, 1989.

Schunk, R. W., and J. J. Sojka, Global ionosphere-polar wind system during changing magnetic activity, *J. Geophys. Res.*, 102, 11625-11651, 1997.

Shelley, E. G., W. K. Peterson, A. G. Ghielmetti, and J. Geiss, The polar ionosphere as a source of energetic magnetospheric plasma, *Geophys. Res. Lett.*, 9, 941-944, 1982.

Spiro, R. W., P. H. Reiff, and L. J. Maher, Precipitating electron energy flux and auroral zone conductances: An empirical model, *J. Geophys. Res.*, 87, 8215-8227, 1982.

Tsunoda, R. T., R. C. Livingston, J. F. Vickrey, R. A. Heelis, W. B. Hanson, F. J. Rich, and P. F. Bythrow, Dayside observations of thermal-ion upwellings at 800 km altitude: An ionospheric signature of the cleft ion fountain, *J. Geophys. Res.*, 94, 15277-15290, 1989.

Wahlund, J.-E., H. J. Opgenoorth, I. Häggström, K. J. Winser, and G. O. L. Jones, EISCAT observations of topside ionospheric ion outflows during auroral activity: Revisited, *J. Geophys. Res.*, 97, 3019-3037, 1992.

Winningham, J. D. and C. Gurgiolo, DE-2 photoelectron measurements consistent with a large scale parallel electric field over the polar cap, *Geophys. Res. Lett.*, 9, 977, 1982.

R. W. Schunk, Center for Atmospheric and Space Sciences, Utah State University, Logan, UT 84322-4405.

The Science of Solar-B

Spiro K. Antiochos

Naval Research Laboratory, Washington, D. C.

Solar-B is an ISAS mission designed as a follow-on to the highly successful Japan/US/UK *Yohkoh* (Solar-A) colloboration. The mission consists of a coordinated set of optical, EUV and X-ray instruments that will observe the direct response of the Sun's corona to changes in the photospheric magnetic and velocity fields. Solar-B will reveal the mechanisms that give rise to solar variability and determine how this variability modulates the solar output and creates the drivers of space weather. Hence, Solar-B can be considered as the vanguard of the generation of missions post-ISTP that will study the Sun-Earth Connection. We present some of the science objectives of Solar-B, focusing on the creation and destruction of the Sun's magnetic field and on solar eruptions.

1. OVERVIEW

Solar variability is due to the energy that is generated deep in the Sun's interior by nuclear burning, carried to the surface by convective motions, and finally transported to the outer atmosphere and heliosphere by the Sun's magnetic field. This variable transport of energy is the origin of all the many manifestations of solar activity that drive the Sun-Earth connection. Magnetic fields play the central role in the variable transport of the Sun's energy. The solar magnetic field modulates the thermal energy that escapes the Sun's surface, the photosphere, in the form of black-body radiation, and the magnetic field carries directly the non-thermal energy that produces the dynamic corona and heliosphere. Variability in the Sun's nonthermal emissions drive space weather and variability in the thermal radiation is an important driver of global change.

Our view of solar variability has been revolutionized by Yohkoh, an ISAS mission with major NASA participation. Yohkoh has shown that the hot corona is extremely dynamic, with magnetic reconnection, rapid heating and mass acceleration being common phenomena. SOHO, a joint ESA/NASA mission and a major component of the ISTP, has shown in detail the physical parameters of plasma heating and acceleration. The next vital step is to understand the magnetic origins of variability. Solar-B, the next ISAS mission, is designed to address this fundamental question of how magnetic fields interact with plasma to produce solar variablity.

The mission has a number of unique capabilities that will enable us to answer the outstanding questions on solar magnetism. First, by escaping atmospheric seeing, it will deliver continuous observations of the solar surface with unprecedented spatial resolution, approximately 150 km on the solar surface. Solar-B will allow us for the first time to observe the dynamics of the elemental, discrete magnetic flux tubes that form the photospheric magnetic field. It is the dynamics of these flux tubes that is thought to be responsible for the activity observed in the corona by Yohkoh.

Second, Solar-B will deliver the first accurate measurements of all three components of the photospheric magnetic field, with sensitivity better than 100 G for the transverse field components. In order to have the free energy necessary to power solar activity, the magnetic field must contain electric currents, It is vital,

Sun-Earth Plasma Connections
Geophysical Monograph 109
This paper not subject to U.S. copyright
Published in 1999 by the American Geophysical Union

therefore, that all components of the field be observed so that the currents can be calculated. However, the magnetic components transverse to the line of sight are difficult to observe, and cannot be measured with any degree of accuracy if the field is not spatially resolved. By resolving the magnetic structures, Solar-B will yield measurements of the transverse field that will be truly revolutionary.

Finally, Solar-B will measure both the magnetic energy driving at the high-beta photosphere and, simultaneously, its effects in the low beta corona. Solar-B will image the detailed structure and dynamics of coronal plasma, and deliver spectra so that key physical parameters such as temperature and density can be determined. The mission consists of a complement of instruments that will observe the solar surface and atmosphere as one coupled system. This instrument complement contains an optical vector magnetograph for the photosphere and coordinated X-ray/XUV imaging telescopes and spectrographs for the corona.

The overarching goal for the Solar-B mission is to understand comprehensively the solar photosphere and corona, as a system. To address this broad goal, the Solar-B Science Definition Team [*Antiochos et al.*, 1997] has identified four promising areas of exploration:

- Magnetic field generation and transport
- Magnetic modulation of the Sun's luminosity
- Heating of the upper solar atmosphere
- Eruptive events: coronal mass ejections and flares

Taken together, these topics encompass much of the effort by the US and international solar physics communities, and are at the heart of solar variability.

Due to page limitations, only the first and last topic are discussed in this brief paper. These demonstrate the great breadth of the science of Solar-B. Discussion of all four topics can be found in the report of the Solar-B Science Definition Team [*Antiochos et al.*, 1997], from which the material in this paper was derived.

2. MAGNETIC FIELD GENERATION AND TRANSPORT

Generation of magnetic flux is believed to take place through the interaction of solar rotation with the convecting, highly conductive plasma deep within the Sun. Once generated, magnetic flux rises through the convection zone to the visible solar surface. Recent results show that the topology of the magnetic fields, as they emerge into the visible atmosphere, reflects the workings of both the dynamo and the passage through the convection zone. The processes by which solar flux reconnects, erupts, and leaves the Sun to produce a solar cycle are unknown. Solar-B will observe the topological changes associated with magnetic flux emergence, reconnection, and eruption, through simultaneous X-ray and vector magnetic field observations.

Understanding the generation of magnetic fields, their emergence into the solar atmosphere, and their energetic consequences is a challenging and fascinating problem in its own right. Not only does the study of the physics of the magnetized solar plasma provide a cornerstone for magnetized plasmas occurring in other astronomical contexts, it also has broader implications for society. Ultimately, the prediction of solar variability, in the form of coronal eruptions and luminosity variations, will most likely hinge on a firm physical understanding of the solar processes involved.

Three aspects of solar magnetism have highest priority for Solar-B:

- The dynamo process within the Sun which is the origin of the solar magnetic cycle.
- The processes which govern the transport and subsurface evolution of magnetic fields as they rise to and penetrate the surface.
- The dispersal, decay, and ejection of magnetic energy and flux once the fields have emerged.

To address these challenges, the key strength of Solar-B is its capability to simultaneously observe coronal structure and photospheric vector magnetic field.

2.1. The Solar Dynamo

Magnetic fields of lower main-sequence stars such as the Sun are believed to arise from a dynamo process operating near the base of their convective envelopes. The magnetic field is generated through the interaction of solar rotation with the convecting, highly electrically conductive plasma. Evidence of this process is seen in the hemispheric dependence of the twist of active region magnetic fields, which recent studies have shown cannot simply be explained in terms of differential rotation or the dynamics of flux ropes rising through the convection zone. It is therefore believed that twisting of the magnetic field into flux ropes by subsurface flows is a principal physical action of the dynamo process. This process is understood only in broad outline, as the best extant models of the solar and stellar dynamos lack the ability to predict fundamental observed properties of the solar magnetic cycle. New observational inputs are now possible, through vector magnetic field observations from space, which will enable advancement in this important area. A major goal of Solar-B is to strengthen our understanding of the solar dynamo.

To date we have only very limited knowledge of plasma motions and magnetic fields deep within the Sun where the dynamo is believed to operate, but our knowledge is rapidly improving as a result of recent developments in helioseismology: the use of oscillations generated by the Sun itself to probe the physical conditions within, in an analogous fashion to methods developed by geophysicists to probe the Earth's interior. New observational programs of the coming decade, GONG for the ground-based effort and SOHO from space, will widen vastly our understanding of the solar interior. However, the magnetic fields measured at the surface, and their large-scale structure in the solar corona provide us with the most directly observable manifestation of the dynamo process. In particular, it is important to understand not only the spatial distribution of the polarity of magnetic fields at the solar surface, but also their geometric structure as they penetrate the solar surface and how they fill the coronal volume above. Thus, the combination of vector magnetic field structure at the solar surface observed simultaneously with the accompanying structure of the plasma in the solar corona (Solar-B), and the insight about subsurface flows derived from helioseismology (GONG and SOHO), will provide a unified observational picture of the dynamo process heretofore unattainable.

The ways in which Solar-B will most effectively contribute to this goal, through its novel combination of vector magnetic field measurements and X-ray images, are:

- measurement of the global patterns of the topology of the emerging magnetic field (handedness and density of twist),

- measurement of the distribution of size scales for the twist of the magnetic field, an observable quantity fundamental to the dynamo process,

- observational description of the modes of evolution and expulsion of magnetic flux and helicity from the Sun in relation to flux emergence, dispersal, and reconnection.

- observational characterization of the role of reconnection and expansion of the flux into the corona on the fields below.

- quantification, for the first time, of the flux history of individual solar active regions, through measurements of the vector magnetic field as an active region rotates across the solar disk.

2.2. Subsurface Transport and Flux Emergence

The evolution of magnetic flux as it is transported from its dynamo origin at the base of the convection zone, upward through the convecting and rotating solar envelope is surely geometrically complex. However, large-scale magnetic features such as prominences, active region complexes, and coronal holes (as well as overall active-region magnetic field twist), suggest that the fields retain some of their twisted topology in spite of their journey through the turbulent envelope of the Sun. Furthermore, recent theoretical work has indicated that it may not be possible for the Sun to transfer large-scale helicity to smaller and smaller scales, so that the solar magnetic field retains its large-scale helicity even after it penetrates the surface.

SOHO observations of the emergence of magnetic fields at the smallest scales yet observed show that the process is a complex one. As well, it is a continuous one, in which new flux emerges in the form of very small bipoles (ephemeral regions). These bipoles, on average, bring to the surface a total flux of about 10^{19} Mx. The poles separate with an initial velocity of about 4 km/s to a distance of 7000 km, then more slowly to a separation speed of about 0.5 km/s to a distance of 15,000 km. On a time scale of about 4 hours both poles are in the supergranulation boundaries. These bipoles bring enough flux to the surface to replace all of the magnetic field on the Sun on a time scale of 20 to 100 hours. The total field in the quiet Sun is 10 to 100 times more than the net flux. The observed distribution of flux concentrations, the number of a given flux versus flux at the SOHO resolution, can be predicted on the basis of a statistical mechanical model based on the assumptions that fields moving along intergranular boundaries have a given probability of merging and canceling, and that they spontaneously fragment in proportion to the amount of flux in the concentration. The merging, cancellation, and fragmentation rates have been measured and it has been possible to predict the flux distributions of both polarities in very quiet Sun and dense plage. When the flux concentrations observed at SOHO resolution are examined at higher resolution with ground-based magnetograms, the concentrations are observed to break up into smaller fragments that are embedded in the intergranular lane pattern just as the concentrations at the MDI resolution are embedded in the supergranular pattern.

A new generation of instrument capable of providing vector magnetic field measurements is needed to quantitatively describe the state of the magnetic field as it emerges. The magnetic field is a vector quantity, and it is not sufficient to measure only one component of this vector to answer the questions we pose. This information is crucial not only to understanding the large-scale evolution of the field which concern the dynamo process, but also the topology of structures of the outer atmosphere of the Sun.

Several important questions regarding subsurface flux transport and emergence may be addressed by Solar-B:

- Does the natural buoyancy of magnetized plasmas drive the emergence of magnetic flux, or are the inductive effects of flows such as differential solar rotation, meridional, supergranular, and granular flows important, as well?

- What is the evolution of the local helicity density during the process of flux emergence? Is most of the twist resident in the fields as they emerge, or is significant twist imparted to the fields afterwards?

- What is the magnetic flux history of individual solar active regions, during the process of flux emergence? Does flux submergence play a significant role?

- Does submergence of magnetic fields commonly occur, presumably as a result of magnetic tension forces in curved subsurface magnetic fields, or is this phenomenon rare due to both buoyancy and the rapid expansion of field structures into the corona?

2.3. *Dispersal, Decay, and Ejection of Magnetic Fields*

The processes by which the magnetic flux of solar active regions disappears from the solar surface remain largely unknown, yet we know such processes must exist because the net polarity of the magnetic field in each hemisphere reverses with each 22-year magnetic cycle. It is suspected that sunspot magnetic fields are dispersed, through the action of turbulent convection, into the quiet Sun in the form of many very small, intense magnetic flux tubes, which comprise the magnetic network of the quiet Sun. Their fate from that point on is even more speculative, but either they undergo reconnection and dissipation in the form of micro-flares and atmospheric heating, or they are ejected into the corona and solar wind. As yet, these speculations have been neither refuted nor supported because we lack the high resolution, continuous observations of the photospheric magnetic field which Solar-B will provide.

In order for the solar magnetic polarity to reverse itself every 11 years, it is probably necessary for the Sun to expel the large-scale twisted field structures, in the form of coronal mass ejections. There is no other obvious avenue for the Sun to dissipate the large-scale helicity which clearly appears to be present in solar magnetic fields in the form of the long chromospheric filaments which commonly reside at high latitudes. Interestingly, the extremely high resolution magnetic observations from Solar-B are absolutely necessary to define the nature of even the largest scale structures of the solar magnetic field and its resulting coronal structure.

The most important objectives of Solar-B magnetic field measurements for elucidating the processes by which magnetic flux evolves are:

- to identify the modes of decay and dispersal of active regions in quantitative terms, allowing for the contributions to the flux loss by reconnection, ejection, and dissipation into the quiet magnetic network,

- to explore the physical properties of the intense, but very small magnetic flux tubes which apparently are responsible for most of the 11-year periodic variability of the net radiative output of the Sun,

- to determine the fate of quiet region magnetic flux by in-situ dissipation of the accompanying electric currents (leading to atmospheric heating, either impulsive or steady), by reconnection (leading to small scale ejection of magnetic flux – a process not yet observed but within the capability of Solax-ray instruments), or even by submergence, and

- to determine the nature and importance of turbulent, weak internetwork magnetic fields in the context of the apparently dominant intense fields.

Whenever quantitative magnetic flux measurements are called for, as in the objectives stated above, measurement precision is essential. Solar-B will provide the first opportunity to explore vector magnetic fields with sufficient precision and temporal continuity to address these fundamental objectives.

3. ERUPTIVE EVENTS AND FLARES

Some of the most important effects that force changes in our local space environment originate in eruptive events on the Sun. Solar eruptions range over orders of magnitude in size, duration and energy output. Examples of such events vary in size from spectacular flares and coronal mass ejections (CMEs), which have direct and measurable effects on the Earth, to innumerable tiny spicules and coronal jets, which may only impact the Earth indirectly – through their coronal heating and subsequent solar wind generation. This leads us to the basic question:

- Are these highly diverse eruptive phenomena caused by the same fundamental physical processes? If so, what are they?

Eruptive events can be very rapid. This requires, for the larger events at least, that the energy be stored in the corona and released in a catastrophic manner. Hence, we must understand:

- How is a critically unstable state produced by energy buildup?
- How is the eruption triggered?
- How does the eruption propagate?

To accomplish these ambitious goals we have to be able to probe the solar photosphere with high spatial resolution to determine the evolution of the vector magnetic field, which is an indicator of free energy buildup. In addition we need to determine the topology, location and timing of the energy release. The determination of these boundary conditions will require high-resolution and high-cadence coronal imaging. To understand the physical and dynamic environment of the energy release process to help constrain MHD models, we will also need coronal spectroscopy with high spectral resolution. Solar-B will supply the broad new observational view needed for this purpose.

The recent results from SOHO EIT show clearly that even at solar minimum the Sun is almost continuously producing eruptive events over a wide range of spatial and temporal scales. Solar-B will be launched during the declining phase of the current cycle and therefore should see a full range of solar activity levels from intense flaring to quiet solar-minimum conditions. Consequently, Solar-B will be an ideal tool to investigate the evolution of the magnetic field that leads to plasma erupting from the Sun.

Eruptions can happen rapidly (in seconds) so there is insufficient time to transport energy from remote sources and have a mechanism capable of dissipating it efficiently and quickly enough. Hence it is generally accepted that the energy is stored in nonpotential (sheared) coronal fields prior to the event. But the question remains as to how the fields become sheared. Possible explanations for the origin of these magnetic stresses include:

- footpoint motion caused by the convection flows visible in the photosphere;
- proper motions of flux tubes in the photosphere due to the rotation or relative motion of sunspots;
- the emergence of magnetic field stressed far below the photosphere, perhaps by the solar dynamo itself.

The answer, of course, could be a combination of these physically different mechanisms.

Solar-B will be able to follow the build up of energy in the magnetic field because of its unique vector magnetogram capability and its orbit. Such measurements are not possible from the ground, at the high spatial resolution available to Solar-B, because of the image distortion due to the Earth's atmosphere. Equally important, Solar-B will have continuous coverage of the Sun because it is in a Sun-synchronous orbit. In contrast ground based observatories can observe only 6-8 hours each day assuming the best of conditions. The Solar-B view of the changing magnetic fields will produce a revolution in our understanding of the emergence and dissipation of the strong solar magnetic fields.

3.1. Triggering Solar Eruptions

A question that has troubled solar observers for years concerns the existence of coronal signatures prior to an eruption. Solar-B with its high cadence, continuous coverage and the high spatial resolution of its coronal images and spectra should be able to identify any coronal indicator of the build up of energy, e.g., an increasing shear of the coronal loops with respect to the neutral line, changing energy dissipation in the loops or increased dynamic activity (turbulence or flows). The identification of such a signature could, for the first time, allow us to make reliable predictions of solar eruptions, one of the primary goals of the space weather program.

There are several possible mechanisms to explain the triggering of an eruptive flare or coronal mass ejection.

It can be driven by an essentially ideal process such as a kink instability or a resistive instability (e.g., tearing). Yohkoh has already demonstrated the importance of reconnection in the flare process by showing the dramatic evolution of the coronal fields as the result of such events. While this can be accounted for by the relaxation of the field after the event (i.e., after the explosive eruption of the magnetic field) there is still the intriguing idea that reconnecting magnetic field may also be the origin of the instability.

Solar-B can attack the problem of the triggering mechanism in a number of ways using combinations of the vector magnetic field and coronal imaging data. For example, the new observations can:

- locate the time and the site of the initiation of the energy release

- determine the critical state of the coronal and photospheric fields at the time that the magnetic configuration becomes unstable and erupts

- determine at what stage of the event the coronal fields reconfigure

- derive changes in the 3-D field configuration, including topology changes, throughout the event

- determine physical parameters in the eruptive and pre-eruptive plasma such as temperature and turbulence

3.2. Theory and Modeling of Solar Eruptions

The combination of Yohkoh and SOHO is proving to be a very powerful tool in understanding the propagation of solar eruptions, particularly CMEs. However, Solar-B will determine the vector magnetic field during these phenomena. It will also bring much superior (continuous) coverage, plus higher spatial and temporal resolution. Solar-B will definitively follow the changes in the magnetic configuration of a region during an eruption. There is also the tantalizing prospect of seeing a relaxation directly in the photospheric fields and consequently being able to derive a self-consistent energy budget for such an event.

The study of eruptions that followed the discovery of coronal jets by the Yohkoh team naturally led to attempts to understand the process theoretically, and to model it numerically. The Solar-B data would equivalently revolutionize our view of eruptive events by providing accurate boundary conditions and evolving physical parameters to feed into the models. The 3-D magnetic field data are crucial for producing realistic 3-D simulations of the overall eruptive process. It is only when we can fully simulate the observations in their entirety that we can claim to have answered the above questions.

In the last decade there has been enormous progress in the development of 2-D and 3-D numerical simulation capabilities due to the development of massively-parallel computers coupled with the development of highly sophisticated numerical techniques. The goal of the National High Performance Computing and Communications Program is to develop machines that by the turn of the century will be able to operate at the teraflop rate. With these speeds it will be possible to perform fully time-dependent simulations with grids of order 1024^3, so that we can finally achieve closure between data and theory. Solar-B provides the high resolution observations necessary to test and refine this next generation of numerical models.

Acknowledgments.

I am very pleased to have this opportunity to express my sincerest gratitude to the members of the Solar-B Science Definition Team for all their excellent and successful work on behalf of Solar-B, a small part of which consisted of preparing the report which is the source of the material in this paper. The members of this team were: S. Antiochos (*Chair*), L. Acton, R. Canfield, J. Davila, J. Davis, K. Dere, G. Doschek, L. Golub, J. Harvey, D. Hathaway, H. Hudson, R. Moore, B. Lites, D. Rust, K. Strong, and A. Title.

REFERENCES

Antiochos, S. K. et al., The Solar-B Mission: Final Report of the Science Definition Team,, NASA Report, June 19, 1997, *available at:*
http://wwwssl.msfc.nasa.gov/ssl/pad/solar/solar-b.htm

S. K. Antiochos, Code 7675, Naval Research Lab, Washington, DC 20375-5352. (e-mail: antiochos@nrl.navy.mil

The Solar Stereo Mission

D. M. Rust

Applied Physics Laboratory, Johns Hopkins University, Laurel, Maryland 20723

The principal scientific objective of the Solar-Terrestrial Relations Observatory (STEREO) is to understand the origin and consequences of coronal mass ejections (CMEs). CMEs are the most energetic eruptions on the Sun. They are responsible for essentially all of the largest solar energetic particle events and are the primary cause of major geomagnetic storms. They may be a critical element in the solar dynamo because they remove the dynamo-generated magnetic flux from the Sun. Two spacecraft at 1 AU from the Sun, one drifting ahead of Earth and one behind, will image CMEs. They will also map the distribution of magnetic fields and plasmas in the heliosphere and accomplish a variety of science goals described in the 1997 report of the NASA Science Definition Team for the STEREO Mission. Current plans call for the two STEREO launches in early 2003. Simultaneous image pairs will be obtained by the STEREO telescopes at gradually increasing spacecraft separations in the course of the mission. Additionally, *in-situ* measurements will provide accurate information about the state of the ambient solar wind and energetic particle populations ahead of and behind CMEs. These measurements will allow definitive tests of CME and interplanetary shock models. The mission will include a "beacon mode" to warn of either coronal or interplanetary conditions indicative of impending disturbances at Earth.

INTRODUCTION

NASA's Sun-Earth Connections program aims to improve mankind's understanding of the origins of solar variability, how that variability transforms the interplanetary medium, how eruptive events on the Sun impact geospace, and how they might affect climate and weather.

STEREO is one of five Solar-Terrestrial Probes called for in NASA's Space Science Enterprise Strategic Plan to accomplish the goals of the Sun-Earth Connections program. The other missions are *TIMED*, the Thermosphere-Ionosphere-Mesosphere Energetics and Dynamics mission, *Solar-B*, which will obtain high-resolution images of the solar magnetic field, *Magnetospheric Multiscale*, which will provide a network of *in situ* measurements of Earth's magnetosphere, and *Global Electrodynamics*, which will probe Earth's upper atmosphere to determine how variations in particle flux and solar electromagnetic radiation affect it.

The Solar-Terrestrial Relations Observatory (STEREO) will focus on coronal mass ejections (CME) and their influence in the heliosphere. Unfortunately, the CMEs that most affect Earth are also the least likely to be detected with ground-based or Earth-orbiting telescopes. To understand CMEs better and to forecast their arrival and effects at Earth, a totally new perspective is needed. Achieving this perspective requires two spacecraft moving away from our customary lookout point. This report on such a mission, STEREO, is based on the work of the NASA Science Definition Team [*Rust et al.*, 1997]. The mission is planned for a 2003 launch.

SCIENTIFIC OBJECTIVES

The principal science objectives to be addressed by the STEREO mission are as follows:

- Understand the origin and consequences of CMEs.
- Determine the processes that control CME evolution in the heliosphere by tracking CME-driven disturbances from the Sun to Earth's orbit.
- Discover the mechanisms and sites of solar energetic particle acceleration.
- Determine the 3-D structure and dynamics of coronal and interplanetary plasmas and magnetic fields.
- Probe the solar dynamo through its effects on the corona and heliosphere.

The scientific literature dealing with these topics is extensive. Several recent conference proceedings are particularly helpful in explaining the scientific issues motivating the STEREO mission [*Crooker et al.*, 1997, *Bentley and Mariska*, 1996, *Winterhalter et al.*, 1996, *Hunt*, 1994].

Coronal Mass Ejections

A primary scientific motivation for studying CMEs stems from their enormous and difficult-to-explain spatial scales, masses, speeds, and energies. CMEs appear to be the means by which the corona evolves through the solar cycle. They may be the means of removing dynamo-generated magnetic flux from the Sun [*Bieber and Rust*, 1995]. They appear thus to be a crucial link to Earth from the solar dynamo. Further, the striking effects of CMEs on planetary magnetospheres, comets, and cosmic rays extend the interest in mass ejections well beyond the traditional realm of solar physics, as emphasized in the Sun-Earth Connection Roadmap. Finally, there may be astrophysical analogues of mass ejections, perhaps in accretion disks and active galactic nuclei, that will be better understood when we understand CMEs.

Explaining the sudden expulsion of a highly conducting plasma from the magnetized Sun presents a major challenge to space physics. The spectacular nature of these large mass ejections is illustrated in Figure 1 by a time sequence of images obtained with the Large Angle Spectroscopic Coronagraph (LASCO) [*Brueckner et al.*, 1995] flown on the SOHO mission. The bright, loop-like feature contains more than 10^{15} g of plasma. The energy required to lift the material off the Sun may be as high as 4×10^{32} ergs.

Here are some of the models that have been proposed for the origins of CMEs:

- Magnetic shear by surface motions, causing loss of equilibrium in the corona
- Magnetic helicity charging from beneath the surface, causing a kink instability in the corona
- Emerging magnetic flux, causing loss of equilibrium in a coronal arcade
- Magnetic helicity charging of the corona by flares, causing loss of equilibrium in a coronal arcade
- Thermally driven blast wave from a large flare, blowing the corona open
- Buoyancy, due to a low-density cavity in the corona

One should be able to distinguish among the models by careful examination of the structure of the pre-CME corona. CMEs frequently follow several days of "swelling" of a coronal helmet streamer. There may be corresponding changes in the low corona: magnetic shear [*Karpen et al.*, 1998, *Mikic and Linker*, 1994] (by differential rotation's effect on emerged coronal loops) or magnetic helicity charging (loop twisting by subsurface flows) [*Rust and Kumar*, 1994]. Three-dimensional reconstructions by triangulation on coronal features should reveal the key signatures of these processes and even allow us to specify the density, temperature, and magnetic fields of the pre-event structures.

As the list of models suggests, several fundamental questions must be answered if we are to understand the physical causes of CME eruption:

- Are CMEs driven primarily by magnetic or nonmagnetic forces?
- What is the geometry and magnetic topology of CMEs?
- What key coronal phenomena accompany CME onset?
- What initiates CMEs?
- What is the role of magnetic reconnection?
- What is the role of evolving surface features?

These questions cannot be satisfactorily addressed with single-vantage-point observations of the type currently available. The corona is optically thin, both in the emissions seen by X-ray and UV imagers and in the Thompson-scattered photospheric light seen by coronagraphs. Line-of-sight integration effects are a major source of ambiguity and confusion. Stereoscopic observations can sort out the overlapping 3-D structures.

CME geometry and onset signatures. A CME frequently starts with a sinuous brightening in the low corona and an outward movement of coronal structures on many scales. Observations with the Extreme-ultraviolet Imaging Telescope (EIT) on SOHO [*Delaboudiniere et al.*, 1995] show that CMEs are often accompanied by a wave front in the corona. An example is shown in Figure 2, courtesy of B. Thompson of the EIT team. It was once thought that such waves are triggered only by flares, but now the whole issue must be reexamined. While the waves, variously called Moreton waves or EIT waves, do not appear to be "thermal blast waves," they seem to be intimately involved with CMEs. What is the relationship

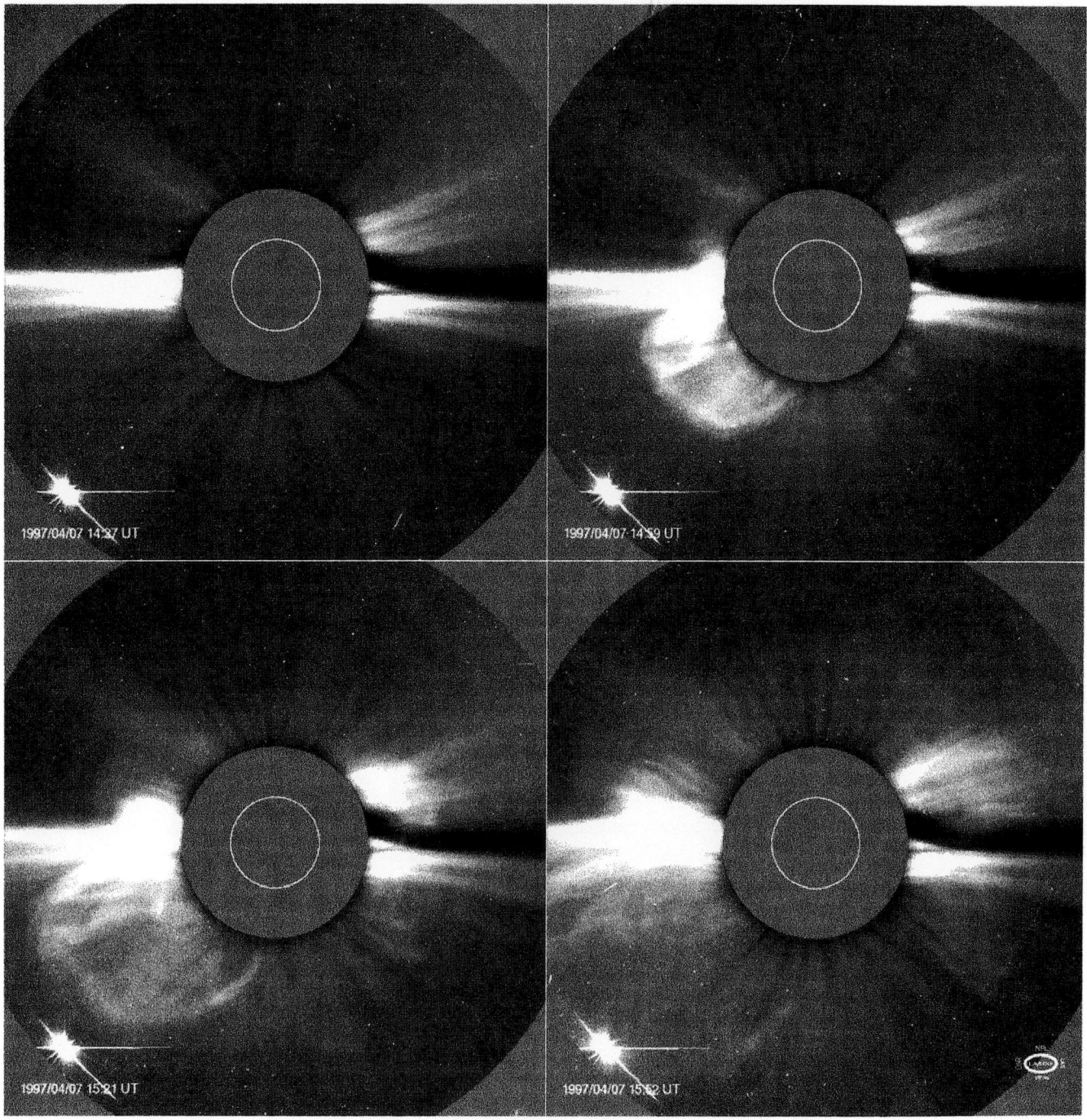

Figure 1. Four images from the LASCO coronagraph on SOHO, showing a CME on 7 April 1997.

of the wave to the CME? Which is the trigger? In order to resolve these questions, STEREO will be designed to provide images with a much higher cadence than SOHO does.

Reconnection. In many CME models, magnetic reconnection is necessary for the eruption to begin or to proceed. The physical role of reconnection varies from model to model, however, and it is possible that

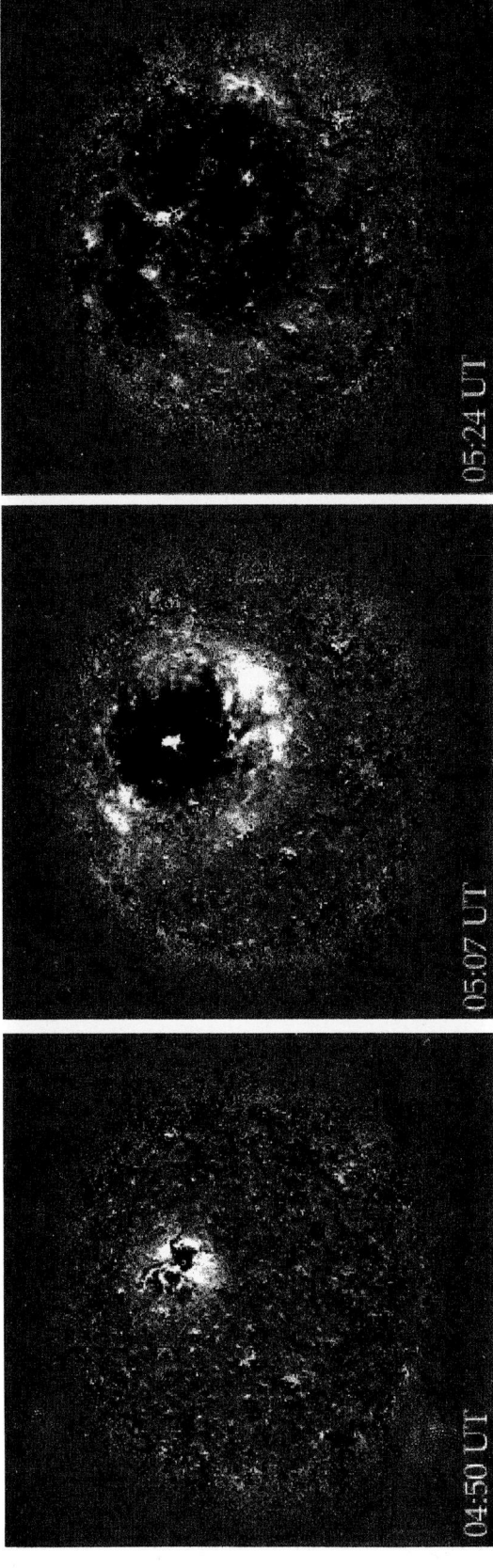

Figure 2. SOHO-EIT observations of a coronal wave spreading from a CME initiation site on 12 May 1997. The images are successive differences of images in Fe XII 195 Å.

reconnection plays no active role whatsoever. In the dipolar arcade and suspended flux rope models, the stretched fields reconnect beneath the CME at the same time that the CME is lifting off. Without reconnection, a full eruption is not possible. In contrast, the quadrupolar model involves reconnection high in the corona above the erupting arcade, and no reconnection is necessary at low altitudes.

These different scenarios will be tested with STEREO observations. For example, reconnection in the dipole arcade and flux rope models produces closed magnetic loops under the CME that should be visible at the time of the eruption. If no loops are seen, then the models must be either rejected or modified.

The Heliosphere Between The Sun And Earth

The heliosphere extending from 30 R_{Sun} to 1 AU (215 R_{Sun}), i.e., from the edge of the widest-field LASCO coronagraph to Earth orbit, contains nearly 400 times the volume of the currently imaged region close to the Sun. This volume has remained unexplored, except during the Helios mission 20 years ago. The two Helios spacecraft carried solar wind analyzers and low-resolution photometers that mapped the solar wind density distribution in the heliosphere and proved that CMEs can be detected well beyond 30 R_{Sun}, even well into the heliosphere [*Webb and Jackson*, 1990]. No such observations are available now, and determination of the instantaneous distribution of matter in the heliosphere is an important goal for STEREO. Solar wind density and velocity measured *in situ* at 1 AU can be related to 3-D reconstructions from STEREO heliosphere imagers and coronagraphs and traced almost all the way down to the solar surface.

Maps derived from STEREO data will allow tests of a new generation of advanced models of interplanetary propagation of solar disturbances [*Riley et al.*, 1997]. The input requirements are relatively straightforward but also impossible to obtain with present capabilities. The crucial elements are the time and location of CME launch, the initial direction, the speed, the spatial extent, the magnetic configuration, and the mass.

Tracking Disturbances From The Sun To Earth

Aboard STEREO, interplanetary disturbances will be detected remotely not only by heliosphere imagers but also by radio telescopes. With only one spacecraft, parameters such as the electron density at the Type II and Type III radio emission sites and the path of the disturbance through interplanetary space can be inferred but they are model dependent because it is not possible to determine exactly where along the measured line of sight the radio source lies [*Bougeret et al.*, 1995]. The planned radio direction-finding capabilities, together with the wide separation between the STEREO spacecraft, will permit the type II radio source, at a given frequency, to be located by triangulation. A single triangulated source position is sufficient to establish the density scale and, therefore, determine the CME shock speed through the interplanetary medium. Once the shock speed and density scale are obtained, one can readily predict, to within about 2 hours, when Earth will encounter the disturbance.

With the stereoscopic observations, trajectories of kilometric type III radio bursts will be constructed and studied in a systematic way for the first time. The type III radio burst trajectory can be constructed from measurements made at a number of different frequencies. Stereoscopic observations will allow interplanetary densities along the radio burst trajectory and electron exciter speeds to be remotely measured and the average interplanetary magnetic field topology to be mapped. STEREO observations will also allow intrinsic properties of the radio source region, such as the brightness temperature, the source size, and the source effective beam width, to be derived and studied in an unambiguous manner.

Particle Acceleration

Solar energetic particle studies with STEREO have two main objectives: to understand how and where CMEs accelerate charged particles, and to develop tools for greatly improved forecasts of large solar energetic particle (SEP) events and/or to warn of their onset.

More than 95% of the largest solar energetic particle events are associated with CMEs, but only about one-third of CMEs produce shocks, and not all shocks result in large events. In the largest SEP events, the particle fluxes spread out over 180° in longitude. Since the particle flux at a given spacecraft depends on how well it is connected to the shock [*Reames et al.*, 1997], the objectives above are best addressed by observations of particles and fields at several heliospheric longitudes.

SEP events can be classified into: *impulsive* events, which have minor increases in particle flux, are rich in He^3, heavy ions, and electrons, and last from minutes to hours; and *gradual* events, which are major proton flux increases on time scales of hours to days. Impulsive events are associated with solar flares, and gradual events are associated with fast CMEs that drive interplanetary shocks. In both types of events the propagation and properties of the charged particles depend crucially on the structure of the coronal and interplanetary magnetic fields and plasmas, so particle flux measurements will allow a new kind of remote sensing of the acceleration and propagation regions, especially when the measurements are combined with stereoscopic images of the corona and heliosphere.

Compression, plasma turbulence, and shock acceleration of particles are expected to be strongest near the western face of fast CMEs, and the duration of the particle events

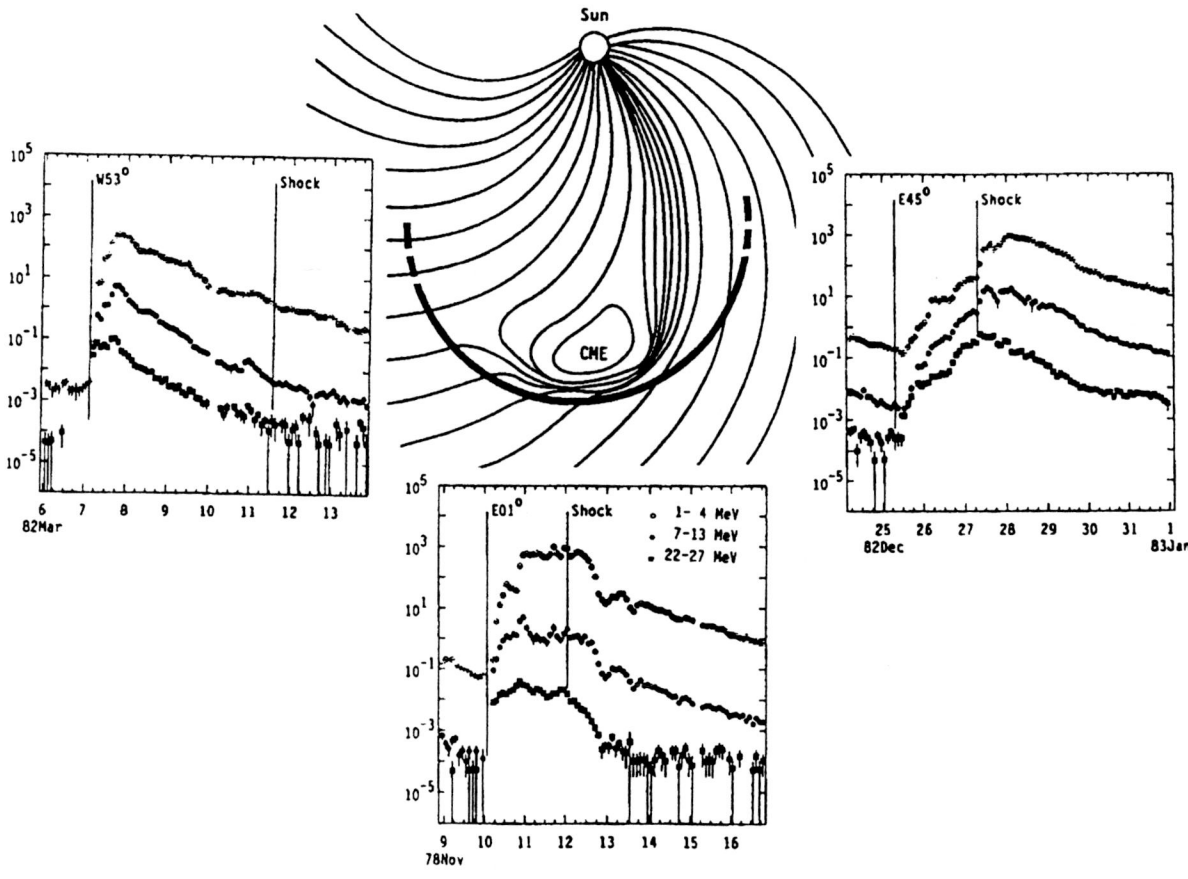

Figure 3. Typical intensity-time profiles for protons from a CME-driven shock as seen from three longitudes. The observer seeing a "western" event (left panel) is well-connected to the nose of the shock early on and sees a rapid rise and decline. The observer near central meridian is well-connected until the shock passes, and he sees a flat profile. The observer viewing an "eastern" event is poorly connected until after the shock passes [*Reames*, 1994].

should depend on the time that the shock and CME affect the field lines that connect the observer with the Sun. Thus, the particle experiments on the two STEREO spacecraft will provide stereoscopic observations of the large-scale structure of CMEs, their effects on the ambient interplanetary medium, and their evolution in interplanetary space. The effects of shocks and CMEs can be sensed from CME onset up to and beyond the time when the CMEs or shocks pass over the spacecraft. Model calculations can then determine the large-scale structure and the position of the CME's center in the heliosphere. To test the models, it is important to be able to sense the different regions around CMEs and shocks on a large scale.

Figure 3 shows an example of how particle measurements at different energies provide information from the onset of the CME at the Sun up to its arrival at Earth. Early in the event, the particle intensity peaks at MeV energies, followed by increases at keV energies up to and beyond the time the shock passes the observer. The MeV electrons and protons provide information on the dynamics of the corona and CME close to the Sun, and the keV particle measurements can be used to track the shock and CME on the way to Earth. In addition to helping profile CME-associated particles, the trailing spacecraft will detect steady interplanetary particle streams, such as those in co-rotating interaction regions, before they pass Earth.

Magnetic Clouds

Many CMEs are associated with erupting prominences (called filaments when seen against the bright solar disk). They often appear to be twisted strands, like ropes. Interplanetary magnetic clouds are also flux ropes, as determined from fitting their *in situ* fields to flux rope models [*Lepping et al.*, 1990]. Many, if not all, CMEs are associated with magnetic clouds. This may be a vital clue to their possible origins in helicity-charged features in the corona. It is expected that magnetic helicity is conserved in flux ropes once they leave the Sun. The magnetic helicity of a twisted flux rope is TF^2, where T is the total twist in radians and F is the magnetic flux in the rope. The STEREO chromospheric and low corona imagers will have enough spatial resolving power to allow determination of T in eruptive prominences. Magnetograms from ground-based observatories or Solar B will allow reasonable estimates of the flux. The imagers should also be able to test models that attribute CME onset to a helical kink instability.

Helicity-conserving flux-rope models have been constructed under the assumption that twisted filaments and their surrounding loops become the magnetic clouds seen in interplanetary space. The models fit the average thermodynamic and magnetic properties of magnetic clouds [*Kumar and Rust, 1996*]. If the helicity of eruptive prominences can be determined with STEREO observations, one should be able to predict the magnetic field structure and strength of magnetic clouds at 1 AU. This would represent a major advance in estimating the geomagnetic effects of the most potent CMEs.

Escape Of Magnetic Flux And The Solar Dynamo

Two mechanisms can facilitate net flux escape from the Sun: helicity charging to push open the fields with reconnection to close them off. Measurements of the solar wind magnetic fields at 1 AU appear to show that 10^{24} Mx of azimuthal flux is ejected by the Sun in each solar cycle [*Bieber and Rust, 1995*]. This rate is the same as the expected rate of toroidal flux generation by the solar dynamo. This measured flux ejection rate is also consistent with estimates of flux escaping in CMEs and prominence eruptions and with the apparent rate of flux emergence at the solar surface, as measured by ground-based magnetographs. It appears that escaping toroids remove at least 20%, and possibly 100%, of the emerging flux in each cycle. Flux escape can be checked with STEREO data, and it may prove to be the key to understanding the cyclic behavior of the Sun.

Coronal Magnetic Fields

"Moreton waves," once also known as flare blast waves, were discovered by Gale Moreton, an observer at the Lockheed Solar Observatory in the 1960s. These waves propagate horizontally across the disk of the Sun at velocities up to 1000 km s^{-1}. They are fast-mode magnetosonic waves associated with large flares [*Uchida et al., 1973*]. Moreton waves are visible in the wings of the Hα line, and until the SOHO mission, they were known only by their effect on the chromosphere and by their correlation with type II radio bursts.

As Figure 2 shows, there are distinctive waves in coronal emission line images from the EIT instrument on SOHO. These waves sweep across almost the entire corona, and the EIT observations make it clear that two coronal emission-line imagers operating at higher cadence than is possible with SOHO will be able to specify the wave fronts in three dimensions.

The motion and distortions of the wave fronts reflect the conditions for wave propagation in the corona. According to Uchida's theory, propagation of the slow mode and the Alfvén mode wave packets is confined to local magnetic field lines, but the propagation of the fast mode wave packets can reveal the distribution of the field strength. The field strength distribution in the corona can be inferred by entering a field distribution and computing the paths of the wave packets, then adjusting the field distribution until there is agreement with the observed wave fronts. Thus, STEREO observations of the wave fronts can achieve a dramatic advance in measuring the coronal magnetic field. *This "seismology of the corona" may finally achieve what has been impossible with older approaches: a complete specification of coronal magnetic field strength.*

Collateral Research

The STEREO platforms offer opportunities for many unique kinds of observation in areas not directly related to the solar activity that affects Earth. Coronal loop heating, reconnections among loops, solar wind sources, and coronal streamer evolution are a few of the research areas that will benefit uniquely from stereoscopic observations. Also, there is another category of observations that should be accommodated to the extent that resources permit, providing that this does not compromise the primary mission objectives. Examples are photospheric magnetic field and velocity observations for helioseismology, solar irradiance measurements, and X-ray and gamma ray burst spectroscopy. Also, unique studies of faint sources in the sky other than heliospheric plasmas can be undertaken with the STEREO coronagraphs and heliosphere imager. Examples are the zodiacal light, asteroids, and comets. Images of comets and of the distribution of dust down to the level of the zodiacal cloud brightness will provide fundamental information about the dust replenishment of the zodiacal cloud. Finally, stellar light curves with ~0.1% photometric precision and 1-day time resolution can be obtained for the 1000 brightest stars.

MAKING THE BEST USE OF STEREO IMAGES

Other research fields have long had the benefit of stereo data, and some of their developed analysis techniques can probably be carried over to space physics. One example is automatic feature tracking in which patterns within many sections, or "patches," in one of the images are searched for and identified automatically in the other image. In contrast to manual methods, the relative offsets of matching points are computed with cross correlations, usually to sub-pixel accuracy. Then, using ray intersection techniques, a sophisticated algorithm determines the coordinates for conjugate points in the two images in three dimensions. While this is a standard technique in producing digital terrain models, much development remains before we will understand its full potential and limitations in interpreting the optically thin features of the corona.

Resolving Line-of-Sight Ambiguities with Stereo Observations

Many coronal images show loops apparently interacting with adjacent loops. Without a stereo view, however, it is not possible to resolve the ambiguity of whether the brightenings of the loops are a result of summing intensities along the line of sight or if the loops physically interact. In some eruptive event scenarios, the energy release is triggered by the interaction of neighboring flux systems, but a close neighbor in a 2-D view may be quite distant when the third dimension is considered. Through visual examination of image pairs and by triangulation on loop features, stereoscopic observations of the X-ray/EUV corona can be used to resolve ambiguities in the interpretation of changes in the coronal structure.

Magnetic-Field-Constrained Tomographic Reconstruction of the Corona

Tomography can be used to directly determine the 3-D structure of the optically thin corona if one has many viewing angles [*Davila*, 1994]. STEREO will provide images from only two angles. However, it is possible to make a tomographic-like reconstruction of the corona from only two views by assuming a magnetic field configuration *a priori*. In this approach, the spatial distribution of coronal emissivity is determined by constraining the stereo reconstruction with a 3-D magnetic field model. The technique is a modification of the multiplicative algebraic reconstruction technique [*Gary et al.* 1998]. In it, the constraint is applied by assuming that emitting plasma only exists within a loose volume defined by the magnetic field model. Figure 4 illustrates the technique and shows results of a tomographic reconstruction both with and without a magnetic field constraint. The magnetic-field-constrained tomographic reconstruction has reproduced the original loops with little smearing (considerably less than the range of the envelope), illustrating the importance of using *a priori* knowledge of the magnetic field.

Visual Evaluation of Stereo Images

Human beings are equipped with an exquisite computer that quickly evaluates stereo image pairs and develops an intense image in three dimensions. Just viewing stereo image pairs and time sequences of stereo pairs will provide valuable insights on the structure and dynamics of the phenomena we seek to understand. Examination of image pairs with stereo viewers may be enough to eliminate some models. For example, models of CME initiation involving buoyancy require that there be a cavity, but the absence of a cavity in a single image may be due to a line-of-sight effect. However, if stereo observations show some CME initiations with no cavity, buoyancy models can be eliminated.

BEACON MODE

Successful integration of STEREO into the national space weather forecast effort hinges on the implementation of simple but robust onboard processing schemes to automatically identify events of interest, broadcast an alert, and trigger the transmission of a pre-stored, high-cadence image and ancillary data stream necessary to sharpen the warning and maximize its utility.

A "beacon" mode of operation can be particularly useful in warning of dangerous solar particle activity. For example, a microprocessor could make real-time classifications of gradual or impulsive events based on the measured particle composition and other characteristics. The microprocessor would also determine the maximum particle flux, rate of rise, proton and helium energy spectra, and elemental composition. If these parameters exceed pre-determined threshold levels, an alert could be sent to Earth.

If suitably designed, STEREO will provide a real-time capability for warning of Earthward-directed CMEs. For example, if an onboard microprocessor identifies a coronal transient, significant particle fluxes, or a strong interplanetary shock, an alert could be sent to Earth at a low bit rate. The immediate alert will need to provide positive identification of CME launch time and direction. Estimates of speed, mass, and relation to structures in the lower atmosphere (to provide an idea of the magnetic content of the CME) would be desirable but perhaps too difficult to include in a simple algorithm. Most likely, preliminary values will have to be derived from the first few images sent down, and more accurate ones would follow from analysis of the full series of event images.

The value of the STEREO mission in pioneering and developing the use of deep space monitors at large angles to the Sun–Earth line cannot be overemphasized. The work here is truly exploratory, since although we now have some idea of what is involved in gathering observations relevant to space weather applications, the full scope of what is required can only be determined by direct experience.

MISSION OVERVIEW

STEREO must lead to a depth of understanding of solar activity that is incisive enough to predict solar eruptions and their effects throughout the heliosphere. To accomplish this, each STEREO spacecraft will carry a cluster of state-of-the-art telescopes and environmental sensors. Images from STEREO's solar telescopes will be combined with solar magnetograms and other data from ground-based or Earth-orbiting observatories to document in detail both the buildup of magnetic energy and CME liftoffs. Other STEREO telescopes will track CMEs and their shocks through interplanetary space. Onboard sensors will sample particles accelerated by the shocks as well as the disturbed plasmas and magnetic fields themselves.

The NASA Science Definition Team recommended that the STEREO mission consist of two identically instrumented Sun-pointed spacecraft at 1 AU. The spacecraft should slowly drift away from Earth, so that after 2 years, STEREO #1 will lead Earth by 45° and STEREO #2 will lag by 60°. Each spacecraft will generate at least 250 images per day plus *in situ* magnetic field and particle data. The solar images should be simultaneous ± 1 s. Science data should be transmitted once a day, and both spacecraft should provide real-time alerts (beacon mode).

The mission is divided into four phases, as described in Section 6 of the SDT report. Primary science operations will occupy the first two years. The goal for total mission lifetime is five years. The schedule, with a launch in 2003, is based on the Solar Terrestrial Probe strategic plan developed for the Sun-Earth Connection Roadmap. The scientific program does not depend on the phase of the solar cycle because CMEs and the other phenomena to be studied are common to all phases of the cycle.

Angular Spacing Between the Two Spacecraft

There is no single angular spacing that is best for all instruments and science goals. The coronagraphs effectively detect only the corona within ± 60° of the plane of the sky. This implies that for triangulation on CMEs aimed at Earth, the spacecraft should be at least 60° apart. Other CMEs will be detectable by both coronagraphs for spacecraft separations ranging between 0° and 120°. On the other hand, it is best to have the high-resolution chromosphere and low corona imagers separated by only 15° – 60° so that features can be identified in the images from both spacecraft. Triangulation on shock fronts with the radio receivers is likely to be most accurate when the spacecraft are separated by ~ 60°. If ACE or WIND or other near-Earth spacecraft are not available, then a STEREO spacecraft near Earth would be desirable to monitor the fields and particles input to the magnetosphere. The Science Definition Team's solution was a four-phase plan, which focuses on different mission objectives at different times. Thus, they recommend that the two spacecraft be launched into slightly elliptical orbits at 1 AU, one leading Earth and one lagging (see Figure 5, so that the angles between the spacecrafts and the Sun–Earth line increase gradually with dwells at selected angles.

PHASES OF THE STEREO MISSION

The studied STEREO mission will have four distinct phases corresponding to different scientific and practical applications of the data and to the angle α separating the two spacecraft.

Phase 1: The 3-D Structure of the Corona (first 400 days, $\alpha \leq 50°$)

While the angular separation a is small and the satellites are close to Earth, telemetry is hardly restricted, and the STEREO satellite configuration will be optimum for making rapid-cadence high-resolution 3-D images of coronal structures. The coronal imagers will be able unambiguously to determine the important physical properties of coronal loops and to determine whether coronal loop interactions include reconnection. Stereoscopic image pairs and sequences will capture the three-dimensional structure of the corona before, during, and after CMEs. They will also allow one to delineate the subtle swelling and the sigmoid features that often foreshadow CME onset. The period when the STEREO spacecraft are close together will also be used to intercalibrate the instruments.

Phase 2: The Physics of CMEs (days 400 to 800, $50° \leq \alpha \leq 110°$)

As the two spacecraft drift farther apart, they become ideally placed to triangulate on CMEs to determine their true dimensions and trajectory. Each spacecraft will be able to image CMEs directed toward the other. Detectors on each spacecraft will measure the magnetic field and plasma properties of CMEs tracked by the other spacecraft, thereby linking the characteristics of a CME (composition, magnetic field orientation, density, and velocity at 1 AU) with its launch and propagation parameters (size, velocity, and source region characteristics).

Phase 3: Earth-Directed CMEs (days 800 to 1100, $110° \leq \alpha \leq 180°$)

In Phase 3, the viewing angles become ideal for observing CMEs aimed at Earth. The coronagraphs,

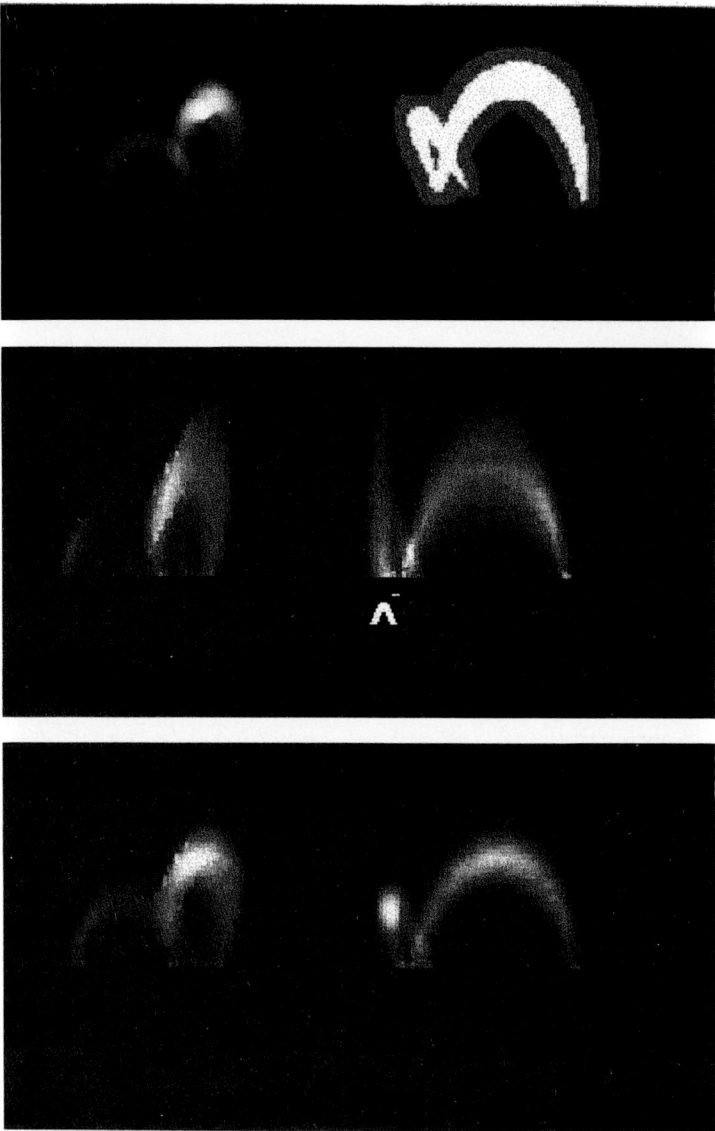

Figure 4. Magnetic-field-constrained tomographic reconstruction [see *Gary et al.* 1998]. (Top left) Original simulated X-ray loops. (Top right) The envelope is the magnetic field constraint applied. The view here is orthogonal to that on the left. (Middle left and right) Two views of the tomographic reconstruction of the simulated loops from a simulated pair of stereo X-ray images (28° separation angle) with no magnetic field constraint; the white arrow head points to the more badly smeared loop. (Bottom left and right) Reconstruction from the same image pair but with the added constraint that the loops are within the loose magnetic envelope shown in the top right frame.

heliosphere imagers, and radio receivers will track the development of CMEs and their shocks as they propagate to Earth, where the Magnetospheric Multiscale and Global Electrodynamics missions will measure their geoeffectiveness.

At this phase of the STEREO mission, the spacecraft will have nearly a 360° view of the Sun, allowing the longitudinal extent of CMEs and other activity to be determined. There have been tantalizing suggestions from Yohkoh soft X-ray images and from the SOHO/LASCO experiment that CMEs can stretch over more than 180° of longitude. STEREO will not only test this suggestion but will also provide global maps of the coronal structures that participate in the activity.

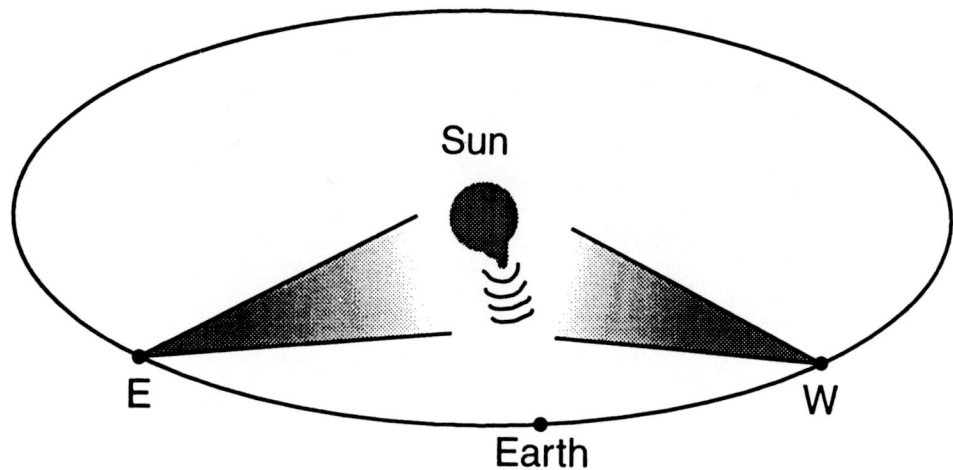

Figure 5. Position of the STEREO spacecraft after about one year. STEREO #1, leading Earth, can see around the west limb; STEREO #2, lagging Earth, can see around the east limb.

Phase 4: Global Solar Evolution and Space Weather (after day 1100, $\alpha > 180°$)

When the separation of each STEREO spacecraft from the Sun–Earth line becomes greater than 90°, events on the far side of the Sun that launch particles toward Earth will be visible for the first time. Active regions can be tracked and studied for their eruptive potential from their emergence, wherever it occurs on the Sun. The results will have a tremendous impact on our ability to anticipate changes in solar activity and to predict changes in space weather conditions. Such a predictive capability is vital if we are to build permanent lunar bases or send astronauts to Mars.

INSTRUMENTATION

The SDT recommended that the baseline instrument complement for each of the two STEREO spacecraft consist of seven instruments as summarized below. The principal restriction on added investigations is the Solar Terrestrial Probe cost cap. Hence, instruments provided by non-NASA-supported institutions may be included to strengthen the overall science program.

- *Chromosphere and low corona imager:* an extreme ultraviolet (EUV) and/or X-ray telescope that images 1 R_{Sun} to 1.5 R_{Sun}
- *Coronagraph:* a white-light coronagraph that images 1.5 R_{Sun} to 30 R_{Sun}
- *Radio burst tracker:* a radio receiver that tracks shocks from the outer corona to beyond Earth
- *Heliosphere imager:* a visible-light telescope that images 30 R_{Sun} to beyond Earth
- *Solar wind analyzer:* a plasma analyzer that samples CME and ambient plasmas at 1 AU
- *Magnetometer:* a sensor that detects magnetic fields inside and outside CMEs
- *Solar energetic particle detector:* detectors of prompt and delayed electrons and ions from 0.1 to 50 MeV

CONCLUSIONS

I have presented a summary of the work of the Science Definition Team for the STEREO mission. The SDT reviewed recent progress in understanding CMEs and identified the major scientific questions to be answered by the STEREO mission. They concluded that two spacecraft at 1 AU, one drifting well ahead of Earth and one well behind, will serve the objectives of NASA's Sun-Earth Connection Initiative by (1) enabling fundamental research on the three-dimensional structure and dynamical processes of CMEs, (2) providing the science base for greatly improved forecasts of disturbances at Earth, and (3) providing comprehensive measurements of the interplanetary environment in support of follow-on Solar Terrestrial Probes. Readers should access the STEREO mission web page at sd-www.jhuapl.edu/STEREO where the full report of the SDT can be obtained, with color illustrations.

The SDT recommended that the STEREO spacecraft carry identical complements of instruments, including chromosphere and coronal imagers, a heliosphere imager, a radio telescope, and sensors of interplanetary particles and magnetic fields The recommended complement of instruments will accomplish the science goals of the STEREO mission.

The needed technologies are available now and the mission can be launched in mid-2003 within the cost restrictions of the Solar Terrestrial Probe line of missions. In order to maximize the scientific return from the unique opportunity provided by STEREO, further studies should be conducted to maximize the information that can be extracted from stereo observations. Such studies, which should include simulated observations of prescribed structures (e.g., CMEs, streamers, loops), will help assure the optimum design and selection of STEREO instrumentation. And, studies of various instruments not on the strawman list, including magnetographs, should be pursued vigorously to increase the potential scientific return.

Acknowledgements. This paper is based on the Report of the NASA Science Definition Team for the STEREO Mission: J. Davila, V. Bothmer, L. Culhane, R. Fisher, J. Gosling, L. Guhathakurta, H. Hudson, M. Kaiser, J. Klimchuk, P. Liewer, R. Mewaldt, M. Neugebauer, V. Pizzo, D. Socker, K. Strong, D. Rust and J. Watzin. D.M.R's work was supported by NASA grant NAG5-4399.

REFERENCES

Bentley, R. D. and J. T. Mariska (Ed.), *Magnetic reconnection in the solar atmosphere*, Astron. Soc. Pacific Conf. Ser. *111*, 1996.

Bieber, J. W. and D. M. Rust, The Escape of Magnetic Flux from the Sun, *Astrophys. J., 453*, 911, 1995.

Bougeret, J. L. et al., Waves: The Radio and Plasma Wave Investigation on the Wind Spacecraft, *Space Sci. Rev. 71*, 231-263, 1995

Brueckner, G. E. et al., The Large Angle Spectroscopic Coronagraph (LASCO), *Solar Phys. 162*, 357-402, 1995.

Crooker, N., J. A. Joselyn and J. Feynman (Ed.), *Coronal mass ejections, Geophys. Monogr. Ser. 99*, AGU, 1997.

Davila, J. M., Solar Tomography, *Astrophys. J. 423*, 871, 1994.

Delaboudiniere, J. P. et al., EIT: Extreme-Ultraviolet Imaging Telescope for the SOHO Mission, *Solar Phys. 162*, 291-312., 1995.

Gary, G. A., J. M. Davis and R. Moore, On analysis of dual spacecraft stereoscopic observations todetermine the three-dimensional morphology and plasma properties of solar coronal flux tubes, *Solar Phys.* (submitted), 1998.

Hunt, J. J. (Ed.), *Solar Dynamic Phenomena and Solar Wind Consequences*, SP-373, European Space Agency, 1994.

Karpen, J. T., S. K. Antiochos, C. R. Devore and L. Golub, Dynamic Responses to Magnetic Reconnection in Solar Arcades, *Astrophys. J., 495*, 491, 1998.

Kumar, A. and D. M. Rust, Interplanetary magnetic clouds, helicity conservation and intrinsic-scale flux ropes, *J. Geophys. Res., 101*, 15667, 1996.

Lepping, R. P., J. J. A. and L. F. Burlaga, Magnetic field structure of interplanetary magnetic clouds at 1 AU, *J. Geophys. Res., 95*, 11957, 1990.

Mikic, Z. and J. A. Linker, Disruption of coronal magnetic field arcades, *The Astrophys. J. 430*, 898-912, 1994.

Reames, D. V., Acceleration of energetic particles which accompany coronal mass ejections, in *Workshop—Solar Dynamic Phenomena and Solar Wind Consequences*, edited by J. J. Hunt, pp 107–116, ESA, 1994.

Reames, D. V., S. W. Kahler and C. K. Ng, Spatial and Temporal Invariance in the Spectra of Energetic Particles in Gradual Solar Events, *Astrophys. J. 491*, 414, 1997.

Riley, P., J. T. Gosling and V. J. Pizzo, A two-dimensional simulation of the radial and latitudinal evolution of a solar wind disturbance driven by a fast, high-pressure coronal mass ejection, *J. Geophys. Res., 102*, 14,677-14,686, 1997.

Rust, D. M. et al., The Sun and Heliosphere in Three Dimensions: Report of the NASA Science Definition Team for the STEREO Mission, Johns Hopkins University Applied Physics Laboratory, Laurel, Maryland, 1997.

Rust, D. M. and A. Kumar, Helicity charging and eruption of magnetic flux from the sun, in *Proc. 3rd SOHO Workshop - Solar Dynamic Phenomena and Solar Wind Consequences*, pp 39-43, European Space Agency, 1994.

Uchida, Y., M. D. Altschuler and G. Newkirk, Flare-Produced Coronal MHD-Fast-Mode Wavefronts and Moreton's Wave Phenomenon, *Solar Phys. 28*, 495-516, 1973.

Webb, D. F. and B. V. Jackson, The identification and characteristics of solar mass ejections observed in the heliosphere by the Helios 2 photometers, *J. Geophys. Res. 95*, 20641-20661., 1990.

Winterhalter, D., J. T. Gosling, S. R. Habbal, W. S. Kurth and M. Neugebauer (Ed.), *Solar Wind Eight, AIP Conf. Proc. 182*, AIP Press, 1996.

D. M. Rust, Applied Physics Laboratory, 11100 Johns Hopkins Road, Laurel, MD 20723

Magnetospheric Multiscale and Global Electrodynamics Missions

Barry H. Mauk and Richard W. McEntire

The Johns Hopkins University, Applied Physics Laboratory, Laurel, MD

Roderick A. Heelis

University of Texas at Dallas

Robert F. Pfaff, Jr.

NASA/Goddard Space Flight Center, Greenbelt, MD

The resolution of many fundamental scientific questions regarding the behavior of the Earth's upper atmosphere and space environment (termed "geospace") will require the use of multiple diagnostic satellites operating in a highly coordinated fashion. This "multiprobe" approach, increasingly viable given technology advances, allows one to: (1) discriminate spatial and temporal variations, (2) expose the prevalent cross-scale coupling processes, and (3) provide spatial and temporal sampling (revealing gradients and other differential characteristics) commensurate with theoretical models and data assimilation procedures. Given cost constraints, a principal trade-off that the community must contemplate is whether and when one chooses to fly a few (~4 to 6) spacecraft that are well instrumented or many (10 to 10s) spacecraft that are minimally instrumented. We document here two missions that take the first approach. They were defined as candidate Solar-Terrestrial Probes by NASA's Geospace Multiprobes Science Definition Team. The Magnetospheric Multiscale Mission uses six spacecraft to focus on understanding the fundamental plasma processes that operate at space plasma boundaries and current sheets at positions ranging from the subsolar magnetopause to the distant magnetic tail. *In situ* multipoint measurements are combined with global imaging, exposing localized processes and the global response. The Global Electrodynamics Mission employs up to five spacecraft to provide systematic multipoint measurements in the ionosphere/thermosphere system to determine how the ionized and neutral gases exchange energy and how energy is injected and dissipated in the lower thermosphere.

INTRODUCTION

Earth's upper atmosphere (>60-km altitude) and space environment, termed "geospace," are fascinating from the perspective of fundamental astrophysical processes, but they are also becoming increasingly important for their practical effects on space-based and globally distributed engineering infrastructures (e.g., the effects on communications, navigation, power distribution, etc.) [*Maynard*, 1995]. Many fundamental scientific questions remain regarding the behaviors of these environments, in part because of the severe observational impediments associated with previous space missions. Because important aspects of the geospace medium are invisible to remote sensing techniques (electric fields, magnetic fields, etc.), certain critical parameters can only be measured by *in situ* techniques, that is, by means of spacecraft that fly through the regions of interest and measure the environments in the immediate vicinity of the spacecraft. However, a single satellite is severely constrained in its ability to make the needed measurements: it cannot separate spatial from temporal effects, measure certain critical parameters that depend on spatial gradients and other differential quantities, and relate phenomena acting over one particular spatial or temporal scale to phenomena acting over very different scales. To remove these observational difficulties, new approaches are needed.

One new approach that is necessary and becoming increasingly viable given technological advances is the use of multiple diagnostic satellites that are operated in highly controlled and coordinated fashions with respect to each other. Missions utilizing such "multiprobe" approaches can diminish substantially the observational constraints highlighted above. In particular, they can (1) separate spatial from temporal effects in uniquely capable ways, (2) measure gradients and other differential characteristics of important parameters, and (3) expose cross-scale coupling processes that are prevalent in the geospace and related astrophysical environments. These unique features of the multiprobe approach will be critical factors in moving from a phenomenological characterization of geospace to an understanding of the fundamental processes that control the behavior of geospace.

Recognizing the potential power of the multiprobe approach, NASA organized the Geospace Multiprobes Science Definition Team (SDT; Table 1) in late 1996 to examine the potential applications of this approach to specific problems and regions of the Earth's magnetosphere and its boundaries. The committee delivered its final report to NASA in December 1997 [*Heelis et al.*, 1997]. The present paper draws heavily on that report.

The implementation of multiprobe missions is difficult in today's political environment because of severe cost constraints. Given these constraints and the present state of technological development, the space science community must consider an important trade-off as it promotes the multiprobes concept: Do we fly a few spacecraft (4 to 6) that are well instrumented, or do we consider the use of many spacecraft (10 to 10s) that are minimally instrumented? The answer lies in the nature of the science questions that are being addressed. The diagnosis of fundamental transport processes requires that a full complement of electromagnetic parameters be measured (the first approach). On the other hand, in some regions the connectivity of different phenomenological features must be understood before fundamental transport processes can be attacked. Ascertaining such connectivities may require the use of more spacecraft, purchased at the expense of measurements of key parameters (the second approach). Both approaches were examined by the *Heelis et al.* [1977] SDT report. In this paper we summarize two proposed missions that utilize the first approach: the Magnetospheric Multiscale (MMS) Mission and the Global Electrodynamics (GED) Mission. Missions that take the second approach (Magnetospheric Constellation Missions) will be described elsewhere.

MAGNETOSPHERIC MULTISCALE MISSION

Mission Summary

This mission focuses on understanding the fundamental plasma processes that operate in the vicinity of space plasma boundaries and current sheets. Plasma processes at boundaries transport, accelerate, and energize plasmas, and by doing so help control the structure and dynamics of the Earth's magnetosphere and related astrophysical plasma environments. MMS will (1) measure three-dimensional (3D) fields and particle distributions; (2) measure and discriminate their temporal variations and 3D spatial gradients, with high

Table 1. Full membership of the Geospace Multiprobes Science Definition Team.

Member	Affiliation
R. A. Heelis (Chair)	University of Texas at Dallas
S. A. Curtis (Study Scientist)	NASA/Goddard Space Flight Center
V. E. Angelopoulous	University of California, Berkeley
J. H. Clemmons	Aerospace Corporation
R. Ergun	University of California, Berkeley
T. L. Killeen	University of Michigan
D. M. Klumpar	Lockheed Palo Alto Research Laboratory
K. A. Lynch	University of New Hampshire
R. W. McEntire	The Johns Hopkins University Applied Physics Laboratory
B. H. Mauk	The Johns Hopkins University Applied Physics Laboratory
R. F. Pfaff	NASA/Goddard Space Flight Center
C. T. Russell	University of California, Los Angeles
S. C. Solomon	University of Colorado
H. E. Spence	Boston University
R. R. Vondrak	NASA/Goddard Space Flight Center

resolution, while dwelling in the key magnetospheric boundary regions (from the subsolar magnetopause to the distant tail; Plate 1 scale lengths appropriate to the processes being studied—connecting the small-scale kinetic regime to the larger-scale regimes appropriate for magnetohydrodynamic (MHD) calculations. The MMS ability to separate spatial from temporal effects and to measure gradients is the critical missing link of previous missions that will allow this mission to dramatically enhance our understandings of the fundamental processes that operate in the vicinity of high-altitude magnetospheric boundary regions.

The MMS Mission consists of both a "telescope" and a "microscope." The telescope consists of two small microsats in elliptical orbit that will yield global stereo energetic neutral atom (ENA) imaging to provide the context of large-scale dynamics. The heart of the MMS Mission, its microscope, is a boundary layer probe consisting of four identical spinning spacecraft, flying approximately in a tetrahedral configuration (Plate 2), with spacing variable from greater than 1 km (the kinetic regime) to several R_E. Each spacecraft will contain an identical set of 3D instruments with high angular and temporal resolution (plasma electron and ion composition, energetic electron and ion composition, magnetometer, electric fields and waves). Interspacecraft VHF ranging will determine spacecraft spacing and allow trigger-mode burst, high-rate data recording on all spacecraft for maximum resolution of boundaries or events. Onboard propulsion will allow the orbits of the MMS "probe" to have four separate mission phases covering almost the entire magnetosphere, from the near-Earth equator to the magnetotail. In each phase, this four-spacecraft probe will dwell at apogee in key boundary regions of magnetic reconnection and energy conversion.

Background and Science Objectives

The Earth's magnetic canopy or magnetosphere acts both as a shield against the solar particle flux and as a storage reservoir for energy extracted from the solar wind flow. The coupling between the solar wind flow and the magnetosphere, as well as the transport of energy within the magnetosphere, appears to be controlled by processes occurring in a number of thin boundary regions that separate vast and topologically distinct volumes of magnetospheric plasma (Plate 1). Models (sometimes conflicting) exist for some of these controlling processes [*Walker and Ashour-Abdalla*, 1995], but they fail to uniquely identify the kinetic processes responsible for dissipation, fail to connect the small-scale (kinetic) to the larger-scale (MHD) processes, and are not adequately testable using existing data sets. The key science questions for the MMS Mission are:

- How do microscale processes near plasma boundaries couple to larger-scale dynamics and structures?
- What controls the transport of magnetic fields, and thus energy, across plasma boundaries?
- How are electric currents generated at boundaries to influence distant magnetospheric regions?
- How do processes at plasma boundaries accelerate charged particles?

Key goals of the mission are to understand how the magnetosphere is energized by the solar wind and to help in the development of quantitative models based on first principles that predict the energy transfer. We wish to provide a detailed understanding of magnetic reconnection in all of its guises. Magnetic reconnection is the primary process that couples solar wind plasmas and fields into the Earth's magnetosphere and converts magnetic field energy density to particle energization inside the magnetosphere. Magnetic reconnection will be directly observed in each of the regimes where it is thought to occur around the Earth, including both the magnetopause and the magnetotail. Magnetic reconnection is central to many astrophysical theories, yet its operation in collisionless space plasmas is very poorly understood. As with other boundary processes, the kinetic processes responsible for the dissipation and the connection between kinetic and larger-scale processes are not known. Our goal is to quantitatively characterize the acceleration of charged particles to high energies in different distinct magnetic geometries. We expect to specify the spatial and temporal properties of plasma waves and turbulence, which are of significance in many cosmic settings. Through the detailed, highly coordinated measurement of particles and fields, it is the goal of MMS to determine with unprecedented resolution the microscale and mesoscale spatial properties of space plasmas in the key magnetospheric boundary regions while connecting the effect of those processes to the larger-scale activities of the magnetosphere.

Required Measurements

Table 2 provides a description of the measurements required at four separate, closely spaced points to determine the 3D structure and dynamics of the key thin magnetospheric boundaries. The data must be properly synchronized among each spacecraft and must be sampled at a rate consistent with resolving spatial scales down to a few kilometers to measure the kinetic scales associated with some regions.

The measurement of differential quantities is essential for discriminating among different transport theories (Plate 2). For example, theories espoused to explain the current disruptions associated with substorm dipolarizations and reconnection include the tearing instability, the cross-field current instability, the drift-kink instability, and the ballooning instability. Critical parameters associated with these various theories include the electric current (measurable using

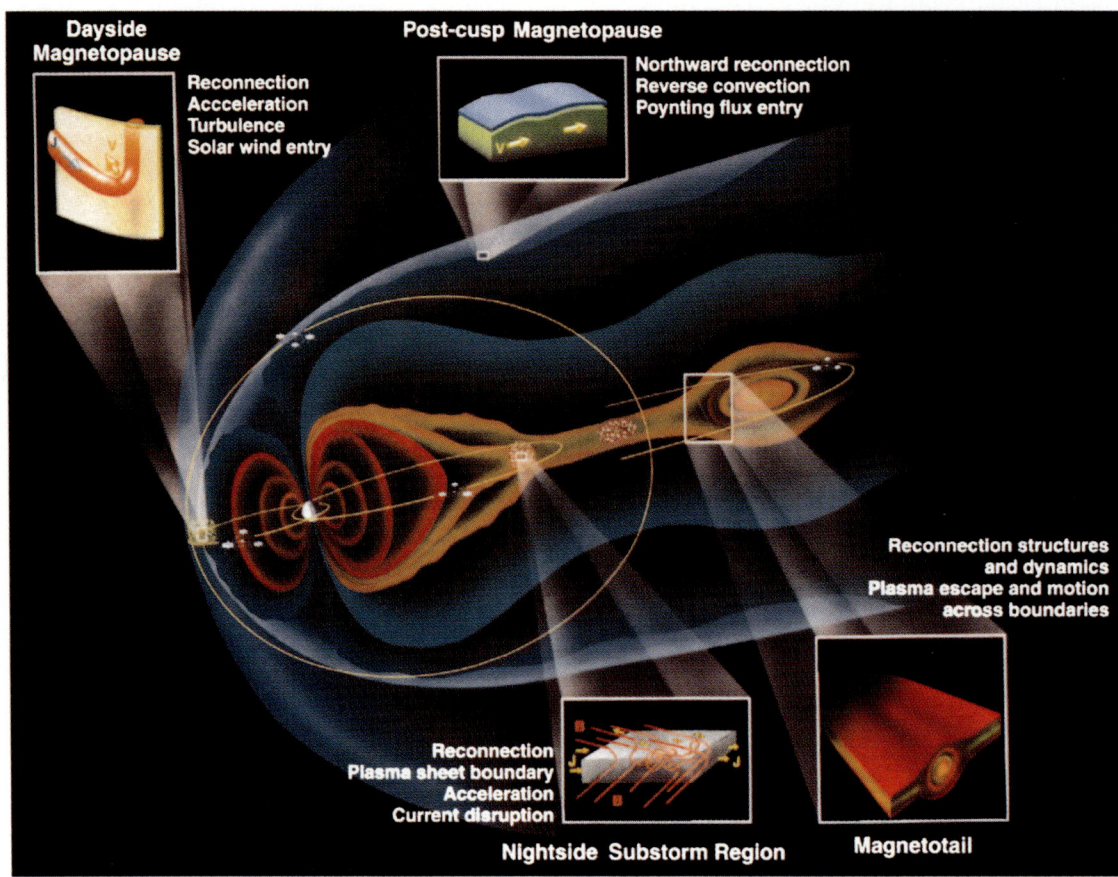

Plate 1. Regions and processes to be addressed by the Magnetospheric Multiscale (MMS) Mission. MMS will focus on understanding fundamental processes at space plasma boundaries.

Table 2. Required Measurements for the Magnetospheric MultiScale Mission.

Parameter	Range	Accuracy
Electric Field Vector	−500 to 500 mV m^{-1}	±0.5 mV m^{-1}
Magnetic Field Vector	−1000 to 1000 nT	±0.1 nT
Magnetospheric Plasma	1 eV to 30 keV	±10%
Energetic Particle Distribution	30 keV to 3 MeV	±5%
AC Electric Field (100 Hz to 1 MHz)	10^{-6} (V/m)^2/Hz to 10^{-16} (V/m)^2/Hz	±5%
AC Magnetic Field (100 Hz to 100 kHz)	10^{-6} (nT)^2/Hz to 10^{-16} (nT)^2/Hz	±5%

Curl(B)), the pressure gradient (**Grad(P)**), the magnetic field gradient (**Grad(B)**), the field-line curvature (**B·Grad)B**), etc. From the measured gradients and curls of the fields and particle distributions, spatial variations in currents, densities, velocities, pressures, and heat fluxes can be calculated. Finally, identification of particle acceleration processes and regions where anomalous transport of charged particles is taking place will be uncovered through measurements of AC electric and magnetic field emissions.

In addition to these *in situ* measurements of essential plasma parameters, a global context for these measurements is required to assess the degree to which the mesoscale and microscale processes might regulate or be regulated by the global scale configuration and dynamics of the magnetosphere. Such a context will be obtained by two small microsatellites dedicated to ENA imaging of the magnetosphere and by access to other space- and ground-based imagers and monitors. The ENA imagers, measuring remotely the spatial, energy, and compositional distributions of the >5-keV ion populations that generate the ENAs Earthward of 10 to12 R_E, will complement the existing ENA imaging missions (e.g., IMAGE) by viewing the plasma dynamics from a near-equatorial perspective. This perspective, for example, will increase the probability of viewing topology changes associated with reconnection and/or related processes in the near-Earth magnetotail. The ENA imaging satellites will remain in their insertion orbits of ~12 R_E apogee and ~28° inclination throughout the mission. Stereo-imaging is achieved by the natural random positioning of the two spacecraft on the same orbit trajectory. The ENA imagers will be sensitive to ENA intensities <1 (cm^2·s·sr·keV)$^{-1}$ for accumulations of ~10 min (strong magnetic storm intensities have intensities >1000 at lower altitudes) with angular resolution capabilities of <4°.

Measurement Strategy

The measurement strategy of MMS has four basic components:

1. Four spacecraft flying in close proximity to form a single probe that can uniquely separate space and time
2. Interspacecraft ranging and communication to explore a wide range of scales down to less than 10 km and up to several R_E
3. Four different orbital phases to cover the entire magnetosphere
4. Two microsatellites in separate orbits providing stereo-connection to global scales

The spacecraft complement will be launched from a single vehicle. The two imaging spacecraft will be placed in elliptical orbits with apogee of 12 R_E and inclinations of ~28° (the small ellipses in Plate 3). They will provide continuous imaging of the inner and mid-magnetosphere. The four remaining identical spacecraft will initially be placed in similar orbits, but in the equatorial plane. During later phases, the orbit apogee will be varied so that, with the rotation of the line of apsides, the tetrahedral satellite configuration can pass through many critical boundary regions in the equatorial boundary layers, the dayside magnetopause, and the distant tail. The final phase of the mission will involve changing the orbital inclination to polar so that the satellite configuration may pass through the high-latitude cusp and magnetosheath boundaries. During all phases of the mission, the spacecraft separation will be varied from <10 km to several R_E. Such separations will allow space and time variations to be examined for processes operating on micro- and mesoscales.

The initial orbit of the four-spacecraft cluster is 1.2 × 12 R_E (perigee × apogee), equatorial (inclination < 10°), with initial apogee in the nightside. This orbit allows measurements with long dwell times in the near tail region, where dynamic changes in the magnetic field occur during substorms, and in the dayside magnetopause, where reconnection is inferred to occur. Phase 2 extends the apogee of the orbit from 12 to 30 R_E. Orbit adjustment will be phased so that the apogee remains in the dawnside magnetopause as the magnetic local time of the apogee changes from ~1000 to ~0400. At that time, the spacecraft will make one traversal in local time of the midrange magnetotail, dwelling during each orbit near 30 R_E. Plasmoid formation, x-line motion, turbulent current disruptions, and plasmoid disconnections are the key processes to study in this distance range. Phase 3 is a lunar swingby, with mid-tail apogees. Here, MMS will be able to answer definitively questions about the distant x line, plasma entry, and dynamic plasma tail disconnection and transport characteristics.

Phase 4 (Plate 4) is a $10 \times 30\ R_E$ polar orbit, with the apogee in the equatorial plane. Although this phase appears superficially similar to the ESA Cluster orbit, this low-eccentricity orbit is the first spacecraft track that can skim the entire dayside magnetopause from cusp to cusp and cover the high-latitude transcusp magnetopause. Thus, spatial scales and repetition rates of flux transfer events can be analyzed. This orbit also allows north–south cuts through the magnetotail at $30\ R_E$, complementing the east–west equatorial plane cuts done in phase 2.

Spacecraft Description

The overall satellite configuration consists of six spacecraft. Four are identically instrumented to perform the measurements outlined in Table 2. They have onboard propulsion for apogee adjustment and interspacecraft ranging and alerts to constitute a closely spaced configuration that will enable space and time variations to be separated over scales ranging from a few kilometers to several R_E. Sun sensors and a star tracker provide attitude determination. In addition, two identically instrumented microsatellites deployed in separate orbits provide global ENA imaging data.

GLOBAL ELECTRODYNAMICS MISSION

Mission Summary

The target of the GED Mission is the ionosphere/thermosphere system that constitutes the lower boundary of the Earth's space environment (Plate 5). In this region, a very substantial fraction of the solar wind energies captured by the magnetosphere are dissipated. It is also substantially affected by the dynamics of the lower atmospheric regions: the troposphere, stratosphere, and mesosphere. This boundary provides a partially ionized collisional plasma that allows electric currents to be generated and dissipated within it. The medium behaves nonlinearly, since the presence of electric fields within it can change the conductivity and currents. Changes in the conductivity at large scales will affect the electrodynamics at smaller scales. Furthermore, the magnitudes of the changes resulting from external inputs will depend on the present dynamical state of the medium. To ascertain the most important factors determining the dynamics and dynamical response of the lower regions of geospace, measurements of changes within an already dynamic system are required. This long-standing problem of space and time ambiguity can only be removed by successive measurements in the same volume of geospace (the multiprobes approach).

The GED Mission will provide a suite of up to four spacecraft platforms making *in situ* measurements that can be used to (1) distinguish spatial and temporal effects and (2) determine how nonlinear interactions of neutral and plasma gases affect the distributions of each species and thus the electrical coupling that they mediate. The spacecraft will be parked in 3000×350 km polar orbits, but will have sufficient fuel to allow for many transient excursion to low (<200-km) perigee altitudes, where the energy dissipation associated with magnetospheric forcing maximizes. A possible fifth spacecraft will be considered to perform remote sensing of mesosphere/thermosphere neutral winds and auroral emission patterns. The GED measurements will constitute a global database of the upper atmosphere so that it may be evaluated as a system, one in which the effects of localized energy transport and dissipation may be examined in all regions and at all local times. Multipoint measurements will allow space/time ambiguities to be resolved and thus enable the distance scales of the most effective energy coupling to be ascertained. These measurements will also provide a means to distinguish between the relative importance of turbulent mixing vs. instabilities in several smaller-scale dissipation processes. Following the legacy that will be provided by the TIMED Mission to uncover the energy balance of the lower thermosphere and mesosphere, the GED Mission will elucidate the dynamic response of the region to the energy inputs.

Background and Science Goals

Decades of highly successful experimental programs have identified the critical energy dissipation mechanisms that take place in the Earth's upper atmosphere, and often in localized regions. However, we still do not understand how these mechanisms actually redistribute and channel energy or the relative importance of many of the competing mechanisms. This state of affairs is due to several factors: the highly nonlinear nature of many of these processes over a wide range of spatial and temporal scales, the fact that electric fields affect neighboring regions nonlocally, and the fact that response times for the neutral and ionized gases differ significantly. Figure 1 shows a sampling of the highly interacting elements of the ionosphere/thermosphere system that must be understood as an interacting whole in order to understand the response of the system to magnetospheric forcing. Not included in Figure 1 are the forcings from the lower atmospheric, for example, in the form of gravity and tidal waves and thunderstorm electric fields (Plate 5). The key science questions for the GED Mission are:

- How are energy and momentum exchanged between ion and neutral gases in the upper atmosphere?

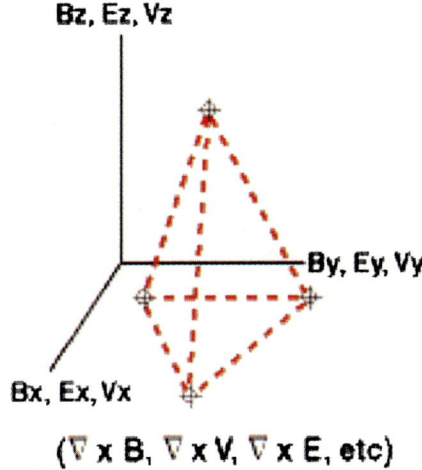

Plate 2. Ideal MMS spacecraft cluster configuration.

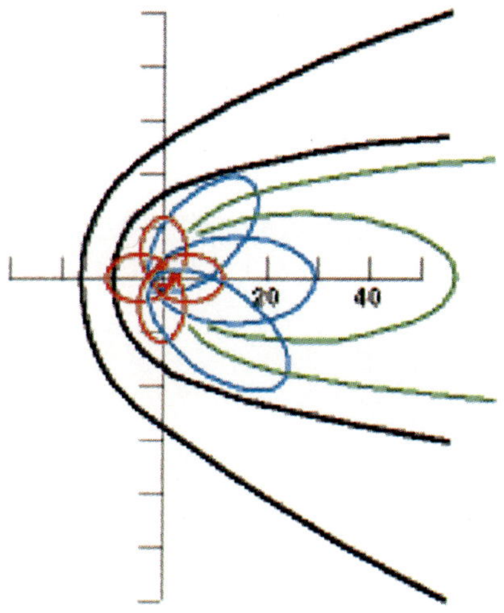

Plate 3. MMS Mission phases 1–3 within the equatorial plane.

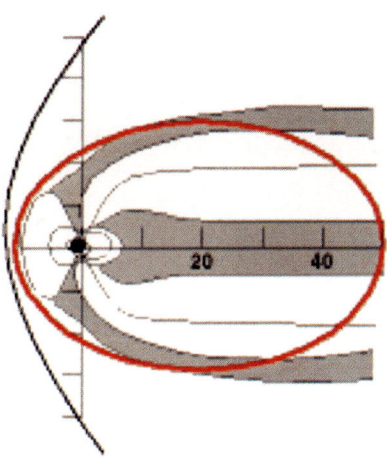

Plate 4. MMS Mission phase 4, a polar orbit.

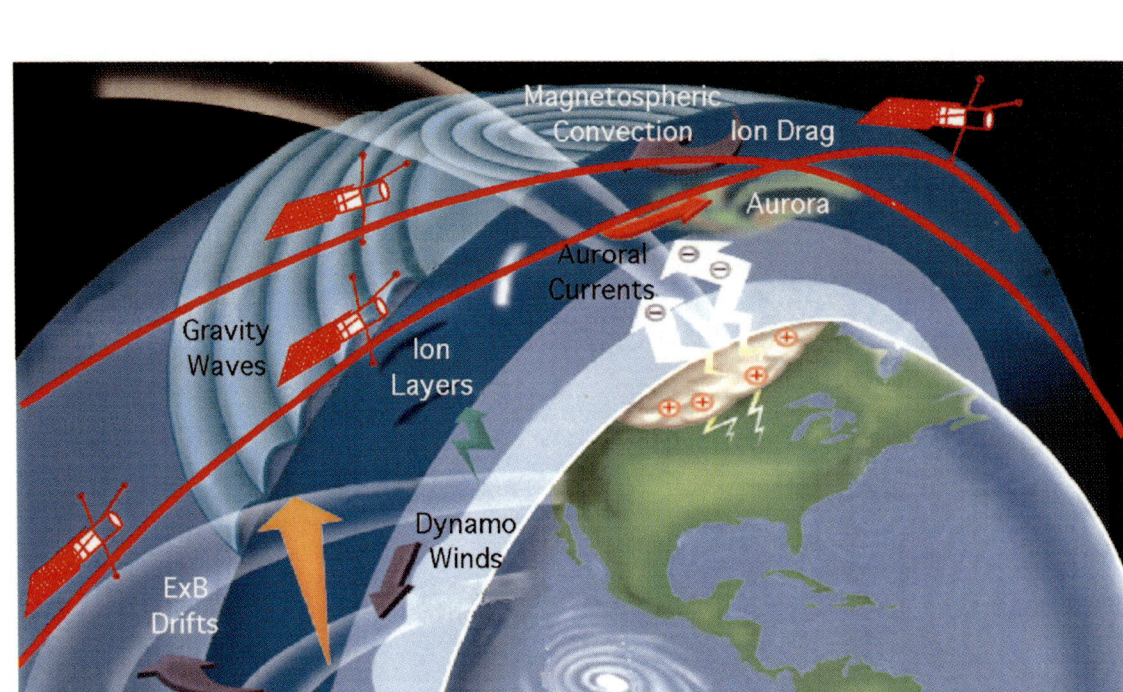

Plate 5. Schematic of the ionosphere/thermosphere regions, illustrating some processes that are the focus of the Global Electrodynamics Mission (GED).

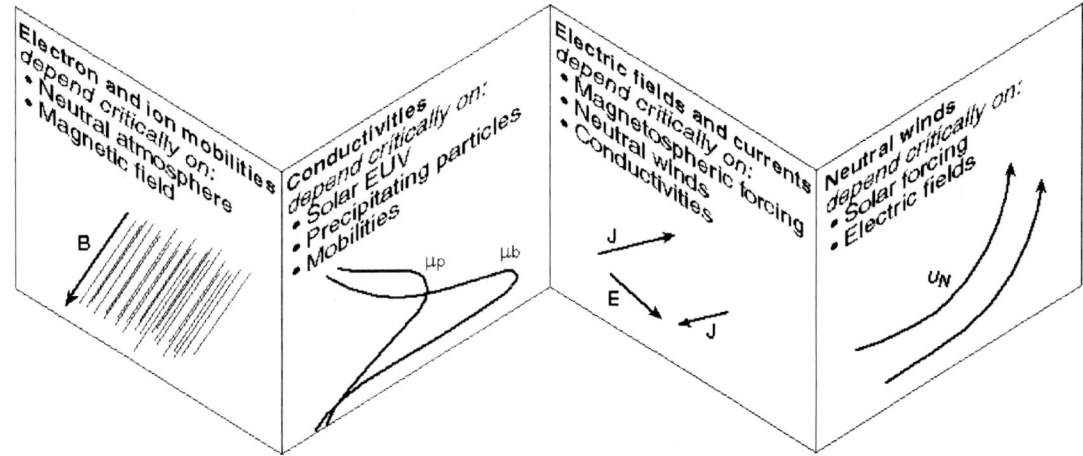

Figure 1. Key elements of the ionosphere/thermosphere regions that must be understood as part of an interaction in order to understand the response of the system to magnetospheric and upper atmospheric forcing.

- How does the state of the ionosphere determine its response to a given external forcing?
- How are the ionosphere and magnetosphere actively coupled?
- What are the roles of turbulence and coherent flows in energy exchange processes?

Required Measurements

To accomplish the stated objectives, the key characteristics of the charged and neutral gases in the upper atmosphere must be measured. These measurements must be accompanied by knowledge of the magnetic field perturbations and electric fields that are internally generated and imposed from the magnetosphere. Energetic particle precipitation represents a substantial energy source from the magnetosphere that must be quantified at high latitudes. *In situ* measurements must be made at altitudes below 200 km. In this region, the dynamics of the lower thermosphere and ionosphere is strongly influenced by winds and waves that propagate from lower altitudes. Thus some merit is attached to remotely sensing these wind fields. Similarly, at high latitudes, the dynamics of the region is largely controlled by the global energy inputs from the magnetosphere. Global imaging of the auroral could be effectively accomplished from high altitudes and may be usefully combined with *in situ* measurements at low altitudes. Table 3 summarizes the measurement requirements that should be considered.

Measurement Strategy

Measurements of the basic state variables describing the ionosphere and thermosphere are required in a region where the interaction between the ionosphere and thermosphere dominates the energy and momentum balance of the region. In this region below a 350-km altitude, the time scales for interactions vary from seconds to a few hours and over spatial scales of a few kilometers to thousands of kilometers. Separation of space and time variations over these scales can only be achieved by ensuring that observations are made sequentially in the same volume. Resolution of time and spatial gradients along the track of the spacecraft required that three satellites be in nominally the same orbit, with separation distances varying from a small fraction of an orbit to 1/2 orbit. A measure of spatial gradients across the spacecraft track requires that one spacecraft make similar measurements in the same orbit but displaced in local time. A requirement to launch all satellites from a single vehicle limits the local time displacement of a fourth vehicle to that achievable from launch maneuvers.

Three spacecraft will be placed into the same elliptical orbit (3000 × 350 km); a fourth will be placed into the same eccentricity orbit at a local time separated from the previous satellites orbit by about 1 h. Figure 2 shows a notional view of the nominal spacecraft orbits. The orbit inclination will be 78°, thus allowing access to the highest magnetic latitudes at some longitudes while retaining the ability to access all local times over a relatively short period (3 months).

Table 3. Required Measurements for the Global Electrodynamics Mission.

Parameter	Range	Accuracy
Constituent Atmospheric Density	1 to 60 amu	±10%
Neutral Gas Temperature	100 to 5000 K	±10%
Neutral Gas Velocity	−3 to +3 km s^{-1}	±10 m s^{-1}
Constituent Ionospheric Density	1 to 56 amu	±10%
Ion and Electron Temperature	100 to 10,000 K	±10%
Ion Velocity Vector	−4 to +4 km s^{-1}	±10 m s^{-1}
Electric Field Vector	−200 to 200 mV m^{-1}	±0.5 mV m^{-1}
Magnetic Field Vector	−60,000 to 60,000 nT	±5 nT
Energetic Particle Distribution	0 to 50 keV	±1 eV
Doppler Imaging	To be determined	
Global UV Imaging	To be determined	

Each spacecraft will also be equipped with propulsion, allowing excursions and retreats to and from perigee altitudes as low as 140 km. Propulsion will also be used to adjust individual orbit periods so that variable spacing along the orbit can be obtained. Identically equipped spacecraft allow a good deal of flexibility in the conduct of the mission. For example, perigee excursions by two or three spacecraft may be combined with remote sensing observations taken over the same location by optical instrumentation. These options will need further investigation to optimize the use of fuel and the science investigations that are proposed in response to an opportunity.

Spacecraft Description

The requirement for excursions into a high-drag region of the atmosphere is met by providing an elongated geometry along the spacecraft velocity vector, with solar cells deployed edge-on in the wake (Plate 5). A spring-loaded ejection mechanism is used to deploy the spacecraft that are subsequently three-axis stabilized with the use of Sun and horizon sensors. The spacecraft will be maintained with one axis pointing toward the Earth, thus performing one revolution per orbit. Each of the four spacecraft will carry significant propellant for small adjustments to the orbit period and

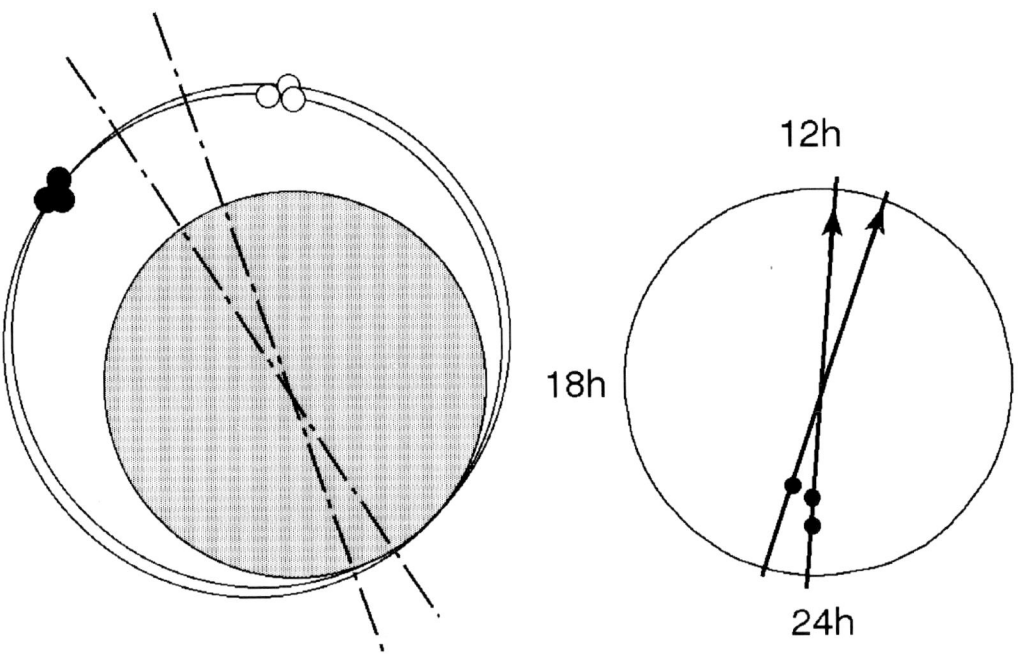

Figure 2. Notational orbit configuration of the GED spacecraft.

to support multiple perigee excursions to altitudes near or below 150 km during the targeted 2-year mission.

The requirement for remote sensing to provide 2D views of the aurora and a measure of low-altitude winds would most sensibly be achieved by incorporating an additional spacecraft in a circular orbit optimized for such measurements. In such a case, careful study is required to ensure that data obtained from both perigee excursions and remote sensing are available in the same regions of space at the same time.

CONCLUSIONS

The MMS and GED missions have been configured by the *Heelis et al.* [1997] Committee to focus on the understanding of fundamental transport processes within (1) the high-altitude boundaries of the Earth's magnetosphere and (2) the ionosphere/thermosphere boundary of the Earth's space environment, where much of the magnetospheric energies are dissipated. The diagnosis and closure of fundamental transport processes require the measurement of the values and differential characteristics of a relatively complete set of key plasma and electromagnetic parameters. Thus, satellites that are configured to address such processes must be well instrumented and operated in a well coordinated fashion that allows for the exploration of scale sizes and times from the microscale to the macroscale. Given cost constraints, the approach is presently limited to the use of a few (4 to 6) spacecraft. Missions that are focused on establishing the connectivity between different phenomena that are as yet poorly diagnosed may require larger numbers of spacecraft. The advantages accrued by the larger numbers must be weighed against the loss of the measurement of parameters that are critical to the diagnosis of the transport processes.

Acknowledgments. Two of us (Barry H. Mauk and Richard W. McEntire) appreciate valuable discussions with A T. Y. Lui, particularly regarding experimental tests of various current disruption and diversion mechanisms.

REFERENCES

Heelis, R., et al., *Geospace Multiprobes; Report from the Science Definition Team*, NASA Office of Space Science, December, 1997.

Maynard, N. C., Space weather prediction, *Reviews of Geospace, Supplement, 33*, 1995.

Walker, R. J., and M. Ashour-Abdalla, The magnetosphere in the machine. Large scale theoretical models of the magnetosphere, *Reviews of Geophysics, Supplement, 33*, 1995.

Roderick A. Heelis, University of Texas at Dallas, P.O. Box 830688 MS FO22, Richardson, TX, 75083-0688. Phone: 214.883.2851. E-mail: Heelis@utdallas.edu

Barry H. Mauk, The Johns Hopkins University Applied Physics Laboratory, Laurel, MD. Phone: 301.953.6023; fax: 301.953.6670. E-mail: Barry.Mauk@jhuapl.edu

Richard E. McEntire, The Johns Hopkins University Applied Physics Laboratory, Laurel, MD. Phone: 301.953.5410; fax: 301.953.6670. E-mail: Richard.McEntire@jhuapl.edu

Robert F. Pfaff, Jr., NASA/Goddard Space Flight Center, Greenbelt, MD 20771. Phone: 301.286.1648. E-mail: Robert.Pfaff@gsfc.nasa.gov

Solar Probe: A Mission to the Sun and the Inner Core of the Heliosphere

G. Gloeckler[1], S. T. Suess[2], S. R. Habbal[3], R. L. McNutt[4],
J. E. Randolph[5], A. M. Title[6] and B. T. Tsurutani[5]

Solar Probe, the first mission to the Sun, is a voyage of exploration, discovery and comprehension. It will address the basic questions of the origin of the slow and fast solar winds by flying through the acceleration region where these winds are born, and it will determine solar surface properties over the poles. Driven by the extraordinary observations from Ulysses and SOHO, Solar Probe measurements made close to the Sun can provide closure to these fundamental problems. Solar Probe, scheduled for launch in February 2007, will travel along a polar trajectory to the Sun, where it will arrive in 2010 flying as close as 3 solar radii above its surface. The second closest approach will be near solar minimum in 2015. Imaging and in-situ miniaturized instruments will provide the first three-dimensional view of the corona, high-resolution spatial and temporal observations of the magnetic fields and helioseismic measurements of the solar polar regions, and local sampling of plasmas and fields at all latitudes. We describe here the prime scientific objectives and the strawman instrument payload of the baseline Solar Probe mission.

1. INTRODUCTION

One of the last unexplored regions of the solar system is the innermost portion of the heliosphere, the region inside the orbit of Mercury, the region that reaches to within a few solar radii (R_S) from the Sun's surface. We have

[1]University of Maryland, College Park, Maryland
[2]NASA Marshall Space Flight Center, Huntsville, Alabama
[3]Harvard-Smithsonian Center for Astrophysics, Cambridge, Massachusetts
[4]Applied Physics Laboratory, Johns Hopkins University, Laurel, Maryland
[5]Jet Propulsion Laboratory, Pasadena, California
[6]Lockheed Research, Palo Alto, California

Sun-Earth Plasma Connections
Geophysical Monograph 109
Copyright 1999 by the American Geophysical Union

flown by many planets. Even now, Galileo is orbiting Jupiter and Cassini is on its way to circle Saturn. With Ulysses we are exploring the high latitude heliosphere and the Voyagers are soon expected to reach and report on the distant boundary of the solar system. From its 1 AU orbit SOHO is imaging the Sun and its atmosphere far better than ever before, and the solar wind and solar energetic particles at 1 AU are being measured in unsurpassed detail and precision with Wind and ACE. Yet we have never encountered the Sun. The inner heliosphere, the solar corona, and the polar photosphere remain essentially unexplored. Ulysses and SOHO have shown us that we do not understand how energy flows into the solar atmosphere, heating the corona and driving the solar wind which affects the Earth and all other planets and determines the size and shape of the heliosphere. Now it is clear that in-situ measurements offer the opportunity to achieve that understanding. We have the technology to send a well-instrumented and affordable spacecraft close to the Sun's surface to explore for the first time this last frontier, the inner heliosphere, from a few to ~100 R_S. Solar Probe is this mission.

Solar Probe is a mission of exploration, of discovery and of comprehension. Flying from pole to pole through the solar atmosphere, as close as 3 R_S above the solar surface, Solar Probe will perform the first close-up exploration of the only star accessible to humankind, the Sun. This pioneering mission will sample the slow and fast solar winds directly where their acceleration takes place, and will image the solar atmosphere with spatial resolution higher than is currently possible and take the best images of the polar regions of the Sun's surface. This missing "ground truth" picture will link the enormous wealth of existing solar and coronal observations to the actual physical state and dynamics of the solar corona. Solar Probe will determine the origin and acceleration of the fast and slow solar wind which engulfs the entire solar system, creates magnetospheres, produces auroras, modulates the penetrating cosmic rays from the galaxy into the solar system and onto earth, and controls interplanetary space from the Sun to the local interstellar medium far beyond the outermost planets. Solar Probe is the third of three missions in NASA's Outer Solar System/Solar Probe Program.

2. CURRENT KNOWLEDGE OF THE INITIAL AND TERMINAL SOLAR WIND AND OF THE SOLAR SURFACE

Recent measurements with state-of-the-art instruments on Ulysses, SOHO, Wind and ACE have advanced our current understanding of the initial and terminal solar wind and of the characteristics of the solar surface. These new results, highlighted below, point to unresolved questions that only Solar Probe can properly address.

2.1 Fast and Slow Solar Wind

Ulysses, with its near polar 1.4 by 5.4 AU orbit, has given us graphic evidence (figure 1) that solar wind comes in two distinct states: an irregular slow wind with typical speeds of 400 km/s and a smooth fast wind with a speed of ~750 km/s [*McComas et al.*, 1998]. This "bimodality" is most apparent around solar minimum when the data in figure 1 were taken. Fast wind flows everywhere in the heliosphere above ~30° latitude and comes most likely from the large polar coronal holes. The slow wind is believed to originate from the boundaries or interior of streamers. Solar Probe will encounter streamers in both 2010 and 2015 and will pass through the large polar coronal holes at 5-10 R_S in 2015.

2.2 Fast Wind is Steady and Simple

The fast wind is relatively steady (figure 1) and its charge state distribution is characterized by a single, low freezing-in coronal temperature for each element, as shown

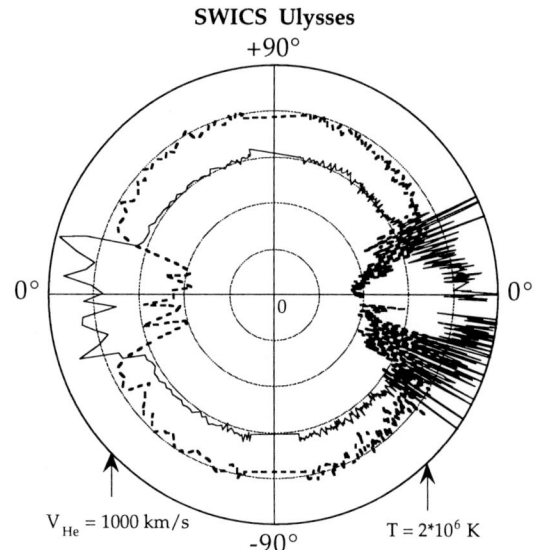

Figure 1. Polar plot of the solar wind speed and freezing-in temperature versus heliographic latitude. The 3-day averages of the solar wind helium speed (dashed curve) and freezing-in temperature based on the O^{7+}/O^{6+} ratio (solid curve) are derived from data of the SWICS instrument on Ulysses between 1.4 and 5.4 AU. It is quite evident that during solar minimum the steady, low temperature (~$1.2 \cdot 10^6$ K), high speed wind (~750 km/s) fills most of the heliosphere.

in figure 2 [*Geiss et al.*, 1995a]. The elemental composition is least biased and most closely resembles the photospheric composition; the overabundance of low FIP (first ionization potential) elements is at most weak, as shown in figure 3 [*Geiss et al.*, 1995b]. The isotopic ratio of $^3\text{He}^{++}/^4\text{He}^{++}$ has its lowest and least variable values in the fast wind [*Gloeckler and Geiss*, 1998]. Fast wind is permeated by an evolving field of magnetohydrodynamic (MHD) turbulence which is presumed to be a remnant or imprint of the coronal acceleration process [*Marsch*, 1991].

2.3 Slow Wind is Variable and Complicated

The slow wind is highly variable in speed and freezing-in temperature (figure 1) and more complicated than the fast wind in its other characteristics. Figure 4 shows that the charge state distribution can no longer be characterized by a single freezing-in temperature [*von Steiger et al.*, 1997]. The FIP effect is far more pronounced (figure 3) and the $^3\text{He}^{++}/^4\text{He}^{++}$ ratio is both higher and more variable in the slow wind compared to the fast wind [*Gloeckler and Geiss*, 1998]. MHD turbulence in the slow wind is less evolved and more intermittent than in fast wind [*Marsch*, 1991].

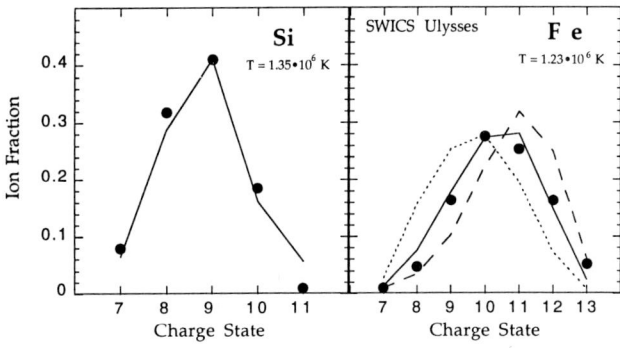

Figure 2. Ionization fractions (filled circles) of Si (left panel) and Fe (right panel) observed in the high speed solar wind with SWICS/Ulysses. The solid curves are equilibrium freeze-in distributions for temperatures of $1.35 \cdot 10^6$ K and $1.23 \cdot 10^6$ K, respectively. For iron curves for $1.13 \cdot 10^6$ K (dotted curve) and $1.33 \cdot 10^6$ K (dashed curves) are also shown. The narrowness of the observed charge distributions indicates a single (but different) freeze-in temperature for each element and implies that temperatures at the freeze-in altitudes are homogeneous down to the smallest scale. (Adapted from Geiss et al., [1995a].)

2.4 Sharp Boundary Between the Fast and Slow Wind

Figure 5 [*Geiss et al.*, 1995b] shows that the sharp speed-boundary between the fast and slow wind is also sharp in freezing-in temperature (measured by the O^{6+}/O^{7+} ratio) and FIP strength (measured by the Mg/O ratio). This boundary thus extends to the lower corona (see also *Habbal et al.*, [1997]) where the charge states freeze-in, and even down to the chromosphere, where the composition is established.

2.5 Coronal Structure and the Solar Cycle

The corona changes dramatically over the solar cycle, with coronal holes dominating at sunspot minimum and streamers likely to be predominant at solar maximum. Solar Probe will pass through the corona at both solar maximum and minimum to provide good data on both streamers and coronal holes.

2.6 Characteristics of the Initial Solar Wind in Polar Coronal Holes

SPARTAN and interplanetary scintillation (IPS) results indicate that fast wind reaches its terminal speed at about 10 R_S and that the strongest acceleration occurs between 1 and 4 R_S (figure 6). SOHO/UVCS observations in the initial fast wind between ~1 and ~4 R_S [*Kohl et al.*, 1997; *Li et al.*, 1998] support the IPS results and suggest (figure 7) that the temperature of heavy ions is much larger than that of protons. In the fast terminal wind this temperature difference is considerably smaller. The proton temperature between ~1 and ~4 R_S in coronal holes is found to be two to three times higher than the electron temperature inferred from charge state measurements in the terminal wind while they differ by less than a factor of two at 1 AU. Inferred ion temperature anisotropies are enormous between 2 and 10 R_S and are believed to be due to an Alfvén or ion-cyclotron wave field contributing to the perpendicular temperature. A true proton temperature anisotropy exists in the 1 AU fast solar wind, but is smaller than inferred from the coronal observations.

Plumes (rays of dense material) permeate all coronal holes and are visible to more than 10 R_S. While possible remnants of these structures have been detected using in-situ plasma measurements [*Thieme et al.*, 1989; *Neugebauer et al.*, 1995], they are generally invisible in the terminal solar wind. How this variable, filamented flow becomes the uniform fast wind is unknown. Whether this is related to the source and evolution of plasma waves in the solar wind will be answered by Solar Probe.

2.7 Characteristics of the Initial Solar Wind in and above Streamers

SOHO and other observations indicate that flow speeds in and around streamers are consistent with the origin of slow wind. However, how this happens has not been determined. One problem is that the standard concept of a streamer is a magnetostatic structure which releases no wind in the steady state. Conversely, SOHO/LASCO has clearly shown sporadic escapes of mass from the tops of streamers which seem to ride on a preexisting slow flow [*Sheeley et al.*, 1997]. Solar Probe will pass through and take measurements at the tops of streamers, precisely where this process is occurring.

Measurements imply that the proton temperature in the initial slow solar wind is comparable to or lower than the inferred electron temperature, while it is distinctly less in the terminal solar wind. Inferred ion temperature anisotropies are less than they are in coronal holes. Composition measurements in streamers show a difference in core and boundary composition which is consistent with the core being essentially static. Solar Probe will measure how these differences map out into the solar wind.

2.8 Properties of the Polar Photosphere

SOHO observations have hinted at some remarkable features for the polar photosphere. But, since the poles cannot

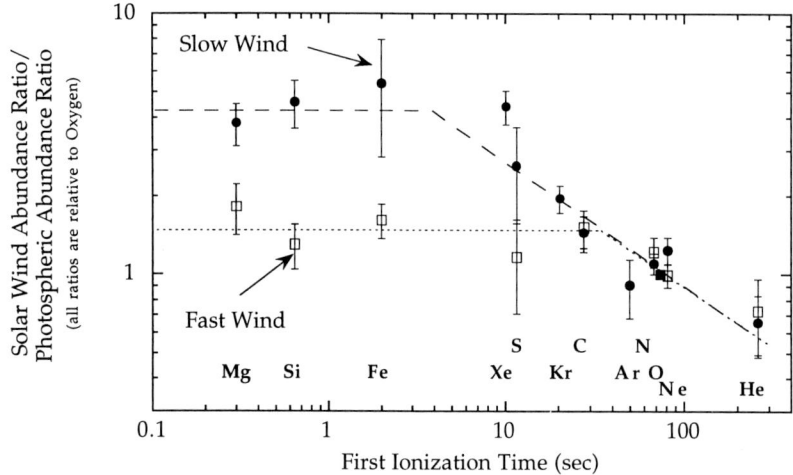

Figure 3. Abundance of indicated elements relative to Oxygen in the high speed stream (open squares) and in the slow solar wind (filled circles) versus First Ionization Time (FIT). FIT, the average time it takes to ionize a given atom in the chromosphere, is closely related to its First Ionization Potential (FIP). The composition in the fast solar wind resembles most closely the photospheric abundances, indicating that the solar wind from these regions has experienced little compositional changes. (Adapted from Geiss et al., [1995b].)

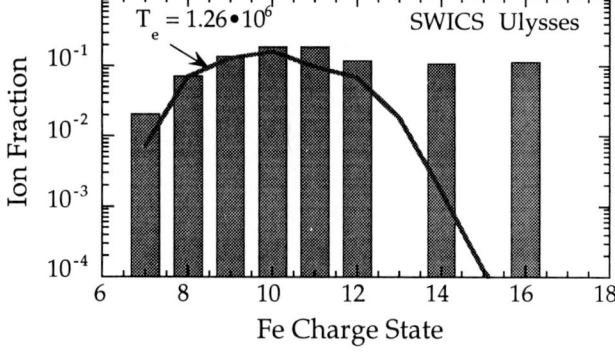

Figure 4. Ionization fraction of iron in the slow interstream wind. In contrast to the charge distributions in the fast wind (Fig. 2) the slow wind charge distributions are broad and cannot be modeled using a single freeze-in coronal electron temperature. These observations imply that the slow wind is a mixture from different sources with different freeze-in temperatures. (Adapted from von Steiger et al., [1997].)

be viewed effectively by SOHO (or any spacecraft confined to near the ecliptic plane), these features remain poorly defined. These findings are: (1) the rotation rate of the solar surface at higher latitudes is 10% to 20% lower than expected; (2) there is good evidence for a polar vortex; (3) there is some evidence of a polar concentration of magnetic flux; (4) measurements of surface and subsurface motion indicate meridional flows that are a factor of more than two higher than previously estimated; (5) there are indications that small and large scale magnetic fields on the sun are rooted at different depths in the convection zone. These results, combined with the SOHO/MDI observations that magnetic flux is replaced very rapidly everywhere on the surface of the Sun - approximately every 40 hours [*Title and Schrijver*, 1997], suggest the importance of a close examination of the photospheric dynamics and magnetic fields with Solar Probe to extend our understanding of how these relate to the flow of energy into the corona.

2.9 Unresolved Questions

The recent findings summarized above have not resolved many fundamental questions regarding solar wind origin and mechanisms for its acceleration, or of coronal heating mechanisms and flow of energy from the solar surface to the corona. We do not know magnetic field topology and surface and subsurface flow patterns in the polar regions and how they differ from those at lower latitudes. We have no direct information on the nature of wave turbulence and of wave-plasma interactions in the acceleration region. We have no direct information on the initial energetic particle populations, their production and acceleration. Plasma waves in the upper corona and transient events at lower altitudes provide appropriate conditions for particle acceleration. Identification of the active mechanisms will depend

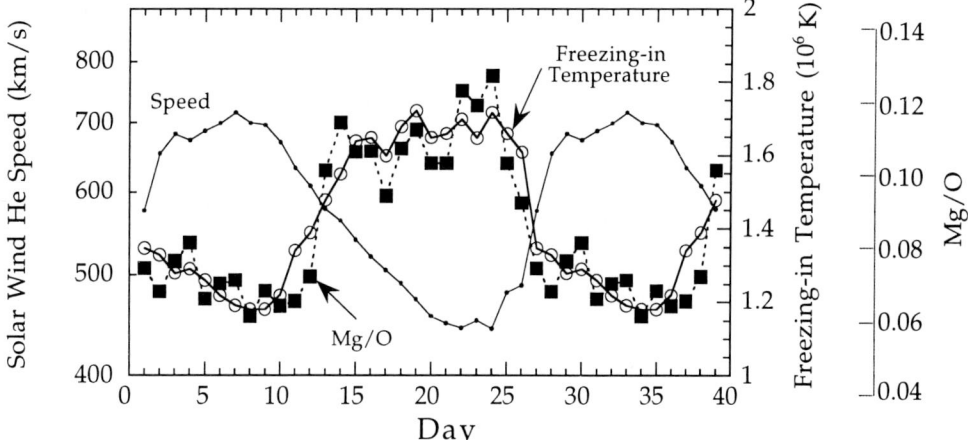

Figure 5. Superposed epoch plot of the SWICS/Ulysses data (day 191, 1992 to day 98, 1993) when Ulysses repeatedly crossed the boundary of the high speed stream once every solar rotation (26 days). Shown are the solar wind He speed (solid curve), the coronal freeze-in temperature of Oxygen (open circles), and the Mg/O abundance ratio (solid squares). The data are repeated for one half cycle to reveal the entire pattern. The steep changes in both the freeze-in temperature and abundance ratio demonstrate the existence of a sharp boundary for the high-speed stream reaching into the corona and chromosphere. (Adapted from Geiss et al., [1995b].)

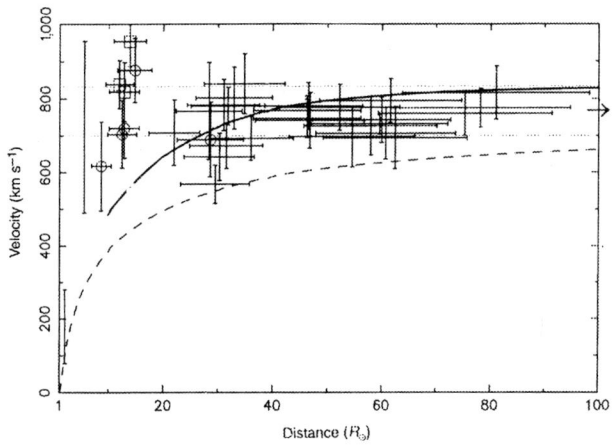

Figure 6. Solar wind speed in coronal holes versus radius with 90% confidence limits. Also shown are SPARTAN 201-01 speeds at 2 and 5.5 R_S. The curves are model solutions (dashed) and models plus wave bias (solid). It is concluded that: (i) the mean apparent speed is already 800 km/s at 10 R_S and probably even at 5 R_S, (ii) the apparent radial speed of the polar wind exhibits great "spatio-temporal fine structure" and is not well described as a smooth, spherically diverging flow. The vertical spread in points around a given radius represents the true flow speed dispersion. The dotted horizontal lines are the upper and lower bounds of Ulysses measurements over the polar regions. (Adapted from Grall et al., [1996].)

on knowing the underlying particle population and wave environments, their spatial extent and dynamical evolution. All of these questions will remain unanswered until in-situ measurements in the solar wind acceleration region near the sun are made and high-resolution images of the polar regions of the Sun are taken. These questions are the basis for the Solar Probe Mission.

3. SOLAR PROBE: THE FIRST CLOSE ENCOUNTER WITH THE SUN

The Solar Probe Science Definition Team (SDT) listed in Table 1 was charged with providing the prime scientific objectives for the Solar Probe mission and core strawman instruments to address these objectives. The science objectives were further prioritized by the SDT into three categories: (A) irreducible objectives to be fulfilled with the core payload, (B) objectives that would require minimum enhancement to the payload, and (C) objectives that could be addressed with additions to the core payload. The SDT also identified payload and measurement requirements, including nadir viewing for the plasma and remote sensing instruments viewing the solar surface. The core payload requirements were then used for the baseline spacecraft and mission design.

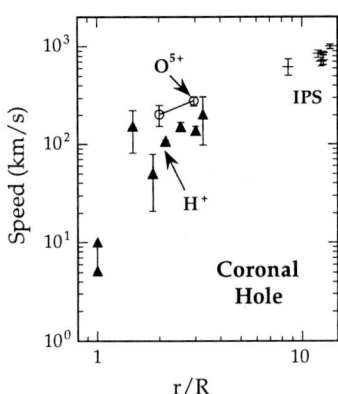
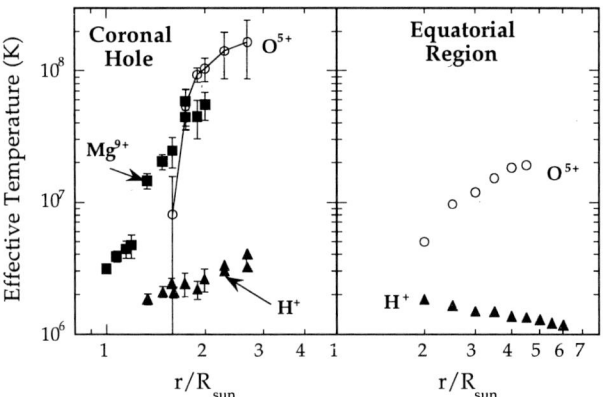

Figure 7. Parameters of the initial solar wind inferred from model calculations and remote-sensing 1 AU observation of the coronal hole (left and center panels) and the equatorial region (right panel) of the Sun. In the left panel, the speeds of protons (filled triangles) and of Oxygen (open circles) are based on UVCS/SOHO measurements [Kohl et al, 1997], as are the effective temperatures for the indicated ions (center panel). The observations, in spite of significant errors, show clearly that the coronal hole solar wind is still being accelerated at 4 R_S and is a factor of ~5 lower than its terminal speed of ~750 km/s, and that the effective temperature of heavy ions (Mg and O) is far greater than that of protons in both the coronal hole wind as well as in the equatorial wind.

Table 1. Solar Probe Science Definition Team.

Member	Affiliation
William Feldman	Los Alamos National Labs
George Gloeckler, Chairman	University of Maryland
Shadia Habbal	Harvard Smithsonian
Clarence Korendyke	Naval Research Laboratory
Paulett Liewer	Jet Propulsion Laboratory
Ralph McNutt, Deputy Chairman	Johns Hopkins University/APL
Eberhard Möbius	University of New Hampshire
Thomas Moore	Goddard Space Flight Center
Stewart Moses	TRW Space & Technology Group
James Randolph, Study Manager	Jet Propulsion Laboratory
Robert Rosner	University of Chicago
James Slavin	Goddard Space Flight Center
Steven T. Suess	NASA Marshall Space Flight Center
Bruce Tsurutani, Study Scientist	Jet Propulsion Laboratory
Alan Title	Lockheed Martin

3.1 Prime Scientific Objectives

The three categories of the Solar Probe prime scientific objectives are given below:

Category A Objectives:

- Determine the acceleration processes and find the source regions of the fast and slow solar wind at maximum and minimum solar activity

- Locate the source and trace the flow of energy that heats the corona

- Construct the three-dimensional density configuration from pole to pole, and determine the subsurface flow pattern, the structure of the polar magnetic field and its relationship with the overlying corona

- Identify the acceleration mechanisms and locate the source regions of energetic particles, and determine the role of plasma turbulence in the production of solar wind and energetic particles

Category B Objectives:

- Investigate dust rings and particulates in the near-sun environment

- Determine the outflow of atoms from the Sun and their relationship to the solar wind

- Establish the relationship between remote sensing, near-earth observations at 1 AU and plasma structures near the Sun

Category C Objectives:

- Determine the role of X-ray microflares in the dynamics of the corona

- Probe nuclear processes near the solar surface using measurements of solar gamma-rays and slow neutrons

Table 2 Solar Probe Strawman Instruments: Measurement Requirements

Strawman Instrument	Parameters or Measured Quantities	Sensitivity *Dynamic Range*	Spectral Range *Resolution*	Angular Range *Resolution*	Time or Spatial Resolution
Remote Sensing Package					
Visible Magnetograph - Helioseismograph	Magnetic Field, Line-of-Sight Velocity Field, Intensity	10 G *300* 20 m/s *400* 1% *400*	3 Å Visible 70 mÅ	1024 arc sec 2 arc sec	2 sec 32 km
XUV Imager	Intensity @ Entrance aperture	100 ergs/cm^{2}sr *400*	EUV Band provides Coronal Imaging 8Å	2560 arc sec 5 arc sec	< 1 sec
All-sky, 3-D Coronagraph Imager	White light	Signal to Noise > 100 *> 1000*	400-700 nm	20-180° from S/C-Sun line < 1°	< 1 min
In-Situ Package					
Magnetometer	Vector DC Magnetic Field	±0.05 nT *10^3*	---	---	10 ms 3 km
Solar Wind Ion Composition and Electron Spectrometer	Distribution functions of dominant charge states of H, He, C, O, Ne, Si and Fe; electrons.	10^5/cm^{2}s *2•10^7*	0.05 < E < 50 keV/e $\Delta E/E < 0.07$	Nadir ±20° and 135° x 300° 10 x 10°	1 sec for H, He,e$^-$: 10 sec for heavy ions
Energetic Particle Composition Spectrometer	Differential fluxes of H, ^{3}He, ^{4}He, C, O, Si, Fe, and electrons	10/cm^{2}s-sr keV *10^7*	0.02 < E < 2 MeV/n e$^-$: 0.02 - 1.0 MeV $\Delta E/E < 0.07$	135° x 300° 20° x 20°	1 sec for e$^-$: 5 sec for H 30 sec for heavy ions
Plasma Wave Sensor	AC Electric and Magnetic Fields	10^{-5} V/m 10^{-9} nT/Hz *10^6*	0.05 - 150 kHz $\Delta\omega/\omega = 0.05$	---	1 ms (wave cap) 1 sec (spectral)
Fast Solar Wind Ion Detector	Distribution function of ions	10^6/cm^{2}s *10^6*	0.02 < E < 50 keV/e $\Delta E/E < 0.07$	90° x 300° 10 x 10	1 ms

3.2 Core Science Payload

The strawman payload consists of five in-situ and three remote-sensing miniaturized instruments. The measurements required to address the Category A science objectives are listed for each of these instruments in Table 2. The spacecraft resources required to accommodate each of the instruments are given in Table 3. The allocation for the Solar Wind Ion Composition and Electron Spectrometer includes some allowance for a nadir viewing deflector. The most economical use of these resources is achieved by configuring the core instruments into two instrument packages - Remote Sensing and In-Situ - each with its own common data interface unit (DIU). With this dual configuration the total mass and power for the strawman payload is 18.8 kg and 15.5 W respectively. The data rate at the time of closest approach is over 112,000 bits/s. During the first perihelion flyby, roughly half of the data will be transmitted in real time with the rest stored on board for transmission after perihelion.

3.3 Solar Probe Spacecraft

The unique feature of the Solar Probe spacecraft is the large but low mass carbon-carbon parabolic heat shield to

Table 3. Solar Probe Strawman Payload: Instrument Requirements

Strawman Instruments	Mass (kg)	Power (W)	Data Rate (kbps)
Remote Sensing Package			
Visible Magnetograph - Helioseismograph	3.0	1.2	30
XUV Imager	3.0	1.2	30
All-sky, 3-D Coronagraph Imager	2.8	2.0	2
Data Interface Unit (DIU) for Remote Sensing Instruments	0.3	0.8	--
In-Situ Package			
Magnetometer (with boom cables)	0.8	0.5	1.2
Solar Wind Ion Composition and Electron Spectrometer (with nadir viewing)	4.4	4.4	15.6
Energetic Particle Composition Spectrometer	0.7	0.6	4.8
Plasma Wave Sensor (with boom cables)	2.5	2.5	9.6
Fast Solar Wind Ion Detector	1.0	1.5	19.2
DIU for In-situ Instruments	0.3	0.8	--
Total	18.8	15.5	112.4

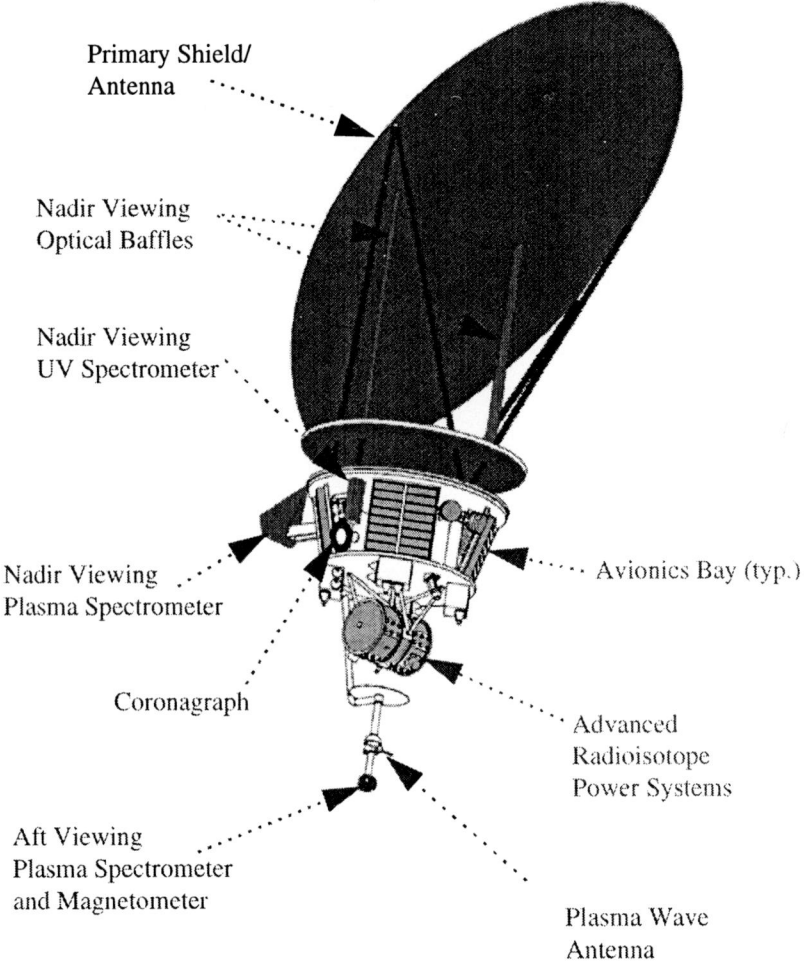

Figure 8. View of a possible Solar Probe spacecraft showing the parabolic side-mounted heat shield/high-gain antenna, the two conical secondary heat shields and the spacecraft bus. Instruments are mounted on the sides and inside the bus and on the boom. The size and shape of the bus and the length of the boom are constrained to fit inside the umbra of the heat shield. Nadir viewing for the plasma instrument is achieved by means of an ion deflection system. A possible configuration of a plasma deflector is shown on the left side of the spacecraft. Also shown are the nadir viewing light baffles for the imaging instruments.

provide thermal protection for the payload and spacecraft (figure 8). This heat shield, serving also as the high-gain antenna, has undergone extensive development and testing. In particular, tests of the carbon-carbon material to be used for the heat shield indicate that the mass loss rate is insignificant. Nadir viewing for the visible and XUV imagers is accomplished by means of carbon-carbon tubes that penetrate the heat shield and spacecraft bus. It is recognized that the baseline mission with two perihelion passes requires an Advanced Radioisotope Power Source (ARPS). ARPSs will be used in each of the other two missions (Europa and Pluto) in the Outer Solar System/Solar Probe Program.

3.4 Baseline Mission

Solar Probe is planned to be launched by a Delta III in February 2007 on a direct trajectory to Jupiter to minimize flight time. Gravity assist by Jupiter ~ 1.5 years after launch places Solar Probe in a highly elliptic polar orbit around the Sun. Solar Probe's closest approach takes place in late 2010, ~3.6 years after launch. At perihelion the orbital plane will be perpendicular to the earth-spacecraft line which permits use of the parabolic heat shield as the high gain antenna and allows real-time high rate data transmission. Because the first encounter is at solar maximum, which will produce excellent data on the active corona but

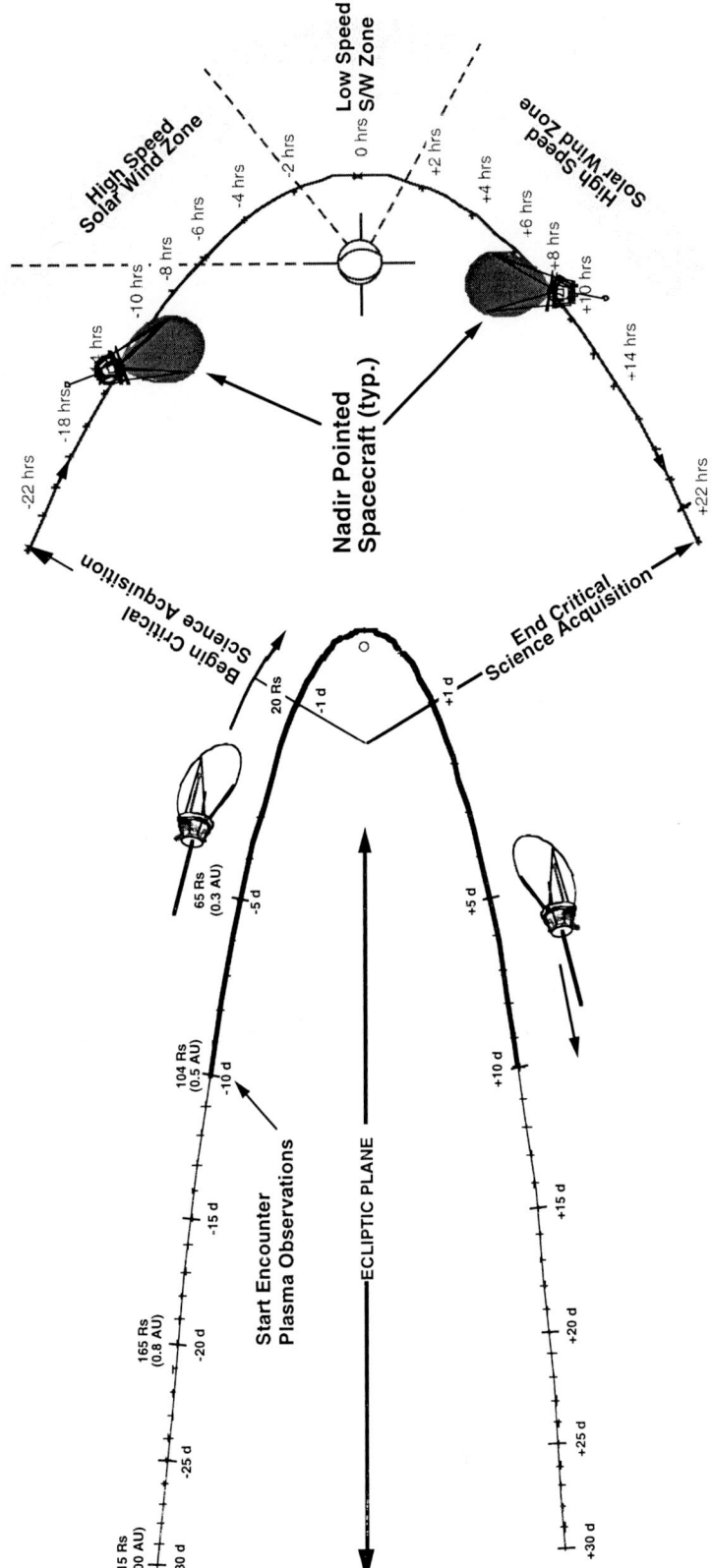

Figure 9. The Solar Probe trajectory as seen from Earth at ± 30 days from perihelion. An enlarged view of the critical science acquisition phase for ± ~1 day is shown on the right. Two perihelion passes, one at solar maximum and one at solar minimum, are planned for the baseline Solar Probe Mission which is planned to be launched in February 2007.

limit chances of encountering a coronal hole inside 8 R_S, a second perihelion pass is now baselined. This will take place in early 2015 at solar minimum when the presence of large polar coronal holes is assured.

The near perihelion trajectory and activities are shown in figure 9. For each of the two passes, encounter measurements by the in-situ instruments start at 10 days before closest approach and end 10 days after perihelion passage. During this twenty day period, the inner heliosphere (< ~0.5 AU) and the corona will be observed in-situ for the first time. Helioseismology observations begin at -4 days (~0.2 AU) from closest approach. The most intense observations by all instruments will take place in the two-day period at distances of < 20 R_S from the Sun. During this period Solar Probe will make high time resolution in-situ measurements in the inner corona and high spatial resolution pole to equator to pole observations of the solar surface. Three-dimensional pictures of the solar corona will be transmitted as Solar Probe flies through it.

4. SUMMARY

Solar Probe, one of the first 3 missions in the Outer Solar System/Solar Probe Program of NASA, is scheduled to be launched in February 2007. A highly capable payload of three remote sensing and five in-situ miniaturized instruments will make unprecedented measurements of the inner heliosphere and polar regions of the Sun both at solar maximum and solar minimum. Thus, Solar Probe, a mission of exploration, discovery and comprehension, offers an unparalleled opportunity to explore the last uncharted regions of the Sun and inner solar system. Progress to make substantial advances in our understanding of solar wind origin and the dynamics of the Sun is not possible without the critical in-situ measurements in the solar wind acceleration region, and the high-resolution spatial observations of the polar regions of the Sun made with Solar Probe.

Acknowledgments. We wish to thank all members of the current and previous Solar Probe Science Definition teams for critical contributions in defining the Solar Probe Mission. The progress made in making this challenging mission technically feasible and ready for development and launch would not have been possible without the support of the talented JPL engineering staff. Portions of this work were performed at the Jet Propulsion Laboratory, California Institute of Technology, under contract with NASA.

REFERENCES

Geiss, J., and 10 others, "The Southern High Speed Stream: Results from SWICS/Ulysses", *Science*, 268, 1033-1036, 1995a.

Geiss, J., G. Gloeckler and R. von Steiger, "Origin of the Solar Wind from Composition Measurements", *Space Science Reviews*, 72, 49, 1995b.

Gloeckler, G. and J. Geiss, "Measurement of the abundance of Helium-3 in the Sun and in the Local Interstellar Cloud with SWICS on Ulysses", *Space Science Reviews*, 84, 275, 1998.

Grall, R. R., and 6 others, "Rapid acceleration of the polar solar wind", *Nature*, 379, 429, 1996.

Habbal, S. R., and 6 others, "Origins of the slow and ubiquitous fast solar wind", *Ap. J.*, 489, L103-L106, 1997.

Kohl, J. L., and 25 others, "First results from the SOHO Ultraviolet Coronagraph Spectrometer", *Solar Phys.*, 175(2), 613-644, 1997.

Li, X, S. R. Habbal, J. L. Kohl, G. Noci, "The effect of temperature anisotropy on observations of Doppler dimming and pumping in the inner corona", *Ap. J. Lett.*, 501, L133-L136, 1998.

Marsch, E., "MHD turbulence in the solar wind", in *Physics of the Inner Heliosphere*, (eds. R. Schwenn & E. Marsch), 159, Springer-Verlag, 1991.

McComas, D. J., and 11 others, "Ulysses' return to the slow solar wind", *Geophys. Res. Lett.*, 25, 1-4, 1998.

Neugebauer, M., B. E. Goldstein, D. J. McComas, S. T. Suess, and A. Balogh. "Ulysses observations of microstreams in the solar wind from coronal holes", *J. Geophys Res.*, 100, 23389, 1995.

Sheeley, N. R., and 18 others, "Measurements of flow speeds in the corona between 2 and 30 RS", *Ap. J.*, 484(1), 472 1997.

Thieme, K. M., R. Schwenn, and E. Marsch, "Are structures in high speed streams signatures of coronal fine structures?", *Adv. Space Res.*, 9, 127, 1989.

Title, A. M. and C. J. Schrijver, in "Cool stars, stellar systems and the Sun", (eds. R. Donahue and J. A. Bookbinder), *Astron. Soc. of the Pacific Conf. Ser.*, 1997.

von Steiger, R., J. Geiss, and G. Gloeckler, "Composition of the Solar Wind", in *Cosmic Winds and the Heliosphere*, (eds. J. R. Jokipii, C. P. Sonnett, and M. S. Giampapa), University of Arizona Press, 581-616, 1997.

George Gloeckler, University of Maryland, Dept. of Physics & Astronomy, Space Science Building, College Park, Maryland 20742 (email: gg10@umail.umd.edu)

Shadia Habbal, Harvard Smithsonian, Center for Astrophysics, 60 Garden Street, Cambridge, Massachusetts 02138 (email: habbal@cfassp29.harvard.edu)

Ralph McNutt, Applied Physics Laboratory, Johns Hopkins University, Laurel, Maryland 20723-4638 (email: ralph_mcnutt@jhuapl.edu)

James E. Randolph, MS 301-170U, Jet Propulsion Laboratory, 4800 Oak Grove Drive, Pasadena, California 91109 (email: James.E.Randolph@jpl.nasa.gov)

Steven T. Suess, NASA Marshall Space Flight Center, Solar Physics/ES82, Huntsville, Alabama 35812-9999 (email: steve.suess@msfc.nasa.gov)

Bruce Tsurutani, MS 169-506, Jet Propulsion Laboratory, 4800 Oak Grove Drive, Pasadena, California 91109 (email: Bruce.T.Tsurutani@jpl.nasa.gov)

Alan Title, Lockheed Research, Bldg 252, Dept H1AL, 3251 Hanover Street, Palo Alto, California 94304 (email: title@sag.lmsal.com)

Magnetospheric Constellation: Past, Present and Future

V. Angelopoulos

Space Sciences Laboratory, University of California, Berkeley, CA

H. E. Spence

Boston University Center for Space Physics, Boston, MA

Essential information regarding the magnetosphere's dynamic evolution and mapping resides in the coupling between global and local processes. This information, pertaining to the instantaneous wavenumber spectrum of the magnetospheric system remains today virtually untapped, mostly due to the enormity of the system's volume and the scarcity of fortuitous spatial conjunctions from previous platforms. The Constellation mission aspires to deploy a series of distributed observational platforms with varying degrees of sophistication, with the goal to understand the interchange of information and energy throughout the magnetospheric system. This paper summarizes proposed approaches and their scientific and technical justification. A phased deployment is proposed, whereby each platform will contribute significantly to the Constellation goal but building on its predecessors ensures an increasingly more comprehensive sampling of the medium. Prioritization of science goals, definition of data ingestion methods and manifestation of science closure are essential next steps towards the realization of the mission.

"Looking to the future I believe that progress requires bunches of satellites, though these are as yet in no published program. One is continually conscious of this need for reasons which have a direct analogue on the ground... [S]ince satellites are being launched singly, the scientific returns are less than they could be." Jim Dungey [1966]

"By believing passionately in something that still does not exist, we create it. The non-existent is whatever we have not sufficiently desired." Nikos Kazantzakis

1. INTRODUCTION

Over the last 30 years much knowledge has been accumulated on the Sun-Earth connection through single or paired spacecraft that were exploratory in nature: They characterized each region under a variety of solar inputs and exlored inter-relationships in a statistical manner. From such studies it has become clear that the main driver of magnetospheric activity and energy circulation is the flow of solar plasma when it is carrying solar magnetic-flux pointing antiparallel to Earth's magnetic dipole. Previous satellites have provided excellent information regarding wave-particle interactions and plasma processes in the time domain. The recently fully-implemented International Solar-Terrestrial Physics (ISTP) program, which combines satellite data acquired simultaneously from different regions is the epitome of the single satellite approach, as it allows us to test the global interactions, something that previously was only possible in a statistical sense or in few fortuitous cases (Plate 1).

As the space physics discipline moves to bridge the gap between the microphysics learned from single point measurements and the global evolution obtained from ISTP studies, it is becoming increasingly evident that sampling in the temporal domain alone is illuminating only a part of the physics at play (Figure 1). The conditions governing the growth of a local instability depend on and, in turn, affect the topological evolution of the system. Collective behavior results from the long-range nature of the electromagnetic force and the effectiveness of electrical currents in line-tying disparate regions. The interaction between neighboring sites is, by definition, the means by which the system reconfigures.

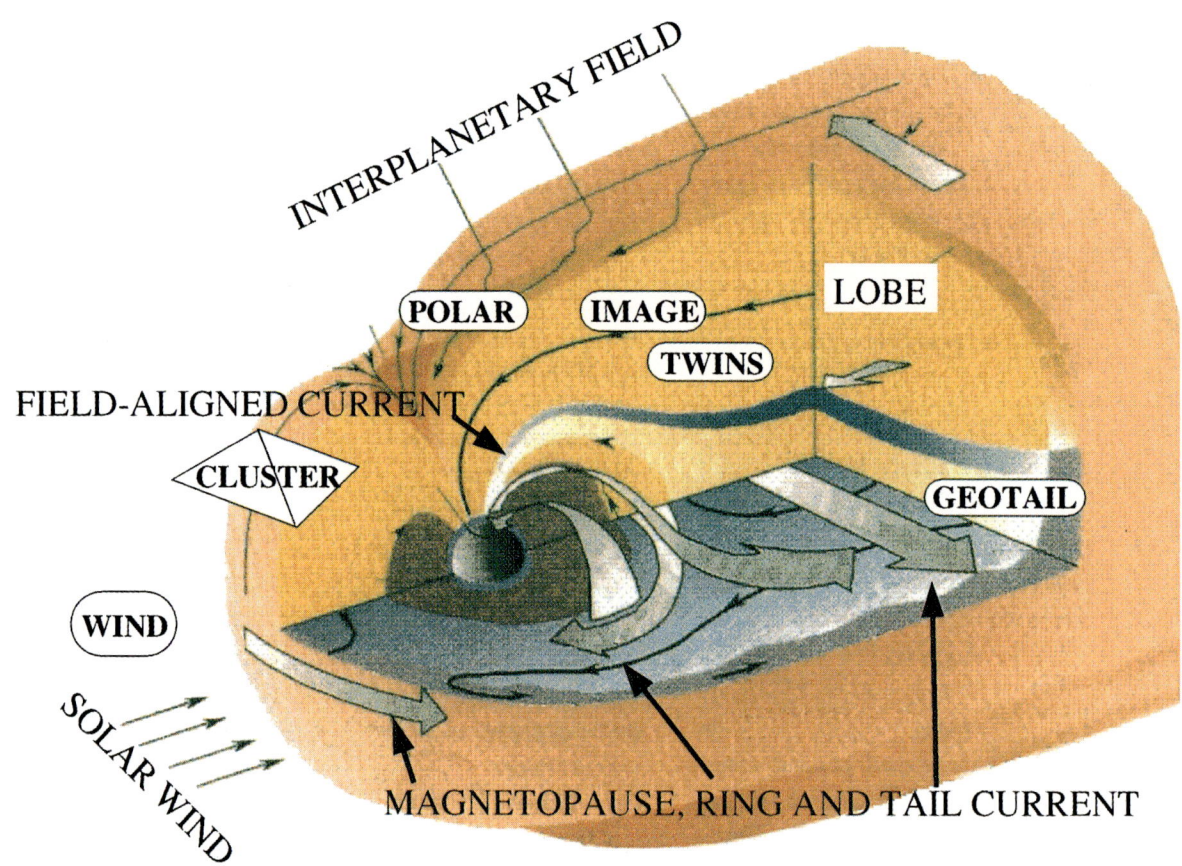

Plate 1. Current missions and average magnetosphere.

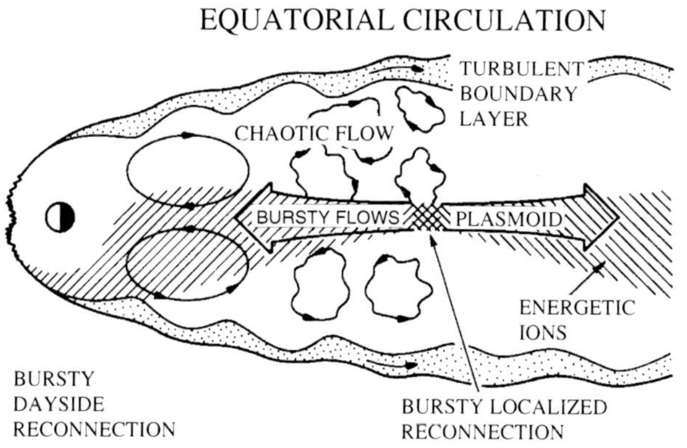

EQUATORIAL CIRCULATION

Figure 1. Cross-scale coupling and localization characterize the instantaneous magnetospheric circulation at intermediate scales, as depicted in this schematic [*Kennel*, 1995].

Progress necessitates multiple measurements at different scale sizes and spanning a variety of k-vectors. Coverage in the spatial domain is as essential to space physics as multispectral imaging is to astrophysics or assembly of pressure and temperature maps is to meteorology.

Carrying the latter analogy further [*Siscoe*, 1997a], there are two ways in which new information can come about: The first is a network of ground meteorological stations that can provide maps of pressure, temperature, velocity etc. Combining information from different locations, altitudes and measurement techniques, advances knowledge on the physical mechanisms that drive weather. The second is imaging networks, which together with in situ information and models obtain predictions of the system's state some time in the future, once the current state is established and the basic equations of motion are known. In meteorology, imaging of cloud coverage and motion is folded into operational models. For magnetospheric physics various proposed methods of global imaging are possible and promise to provide initially advancement in understanding and, eventually, operational functionality.

The Magnetospheric Constellation mission aspires to study the electrodynamic coupling of microphysical processes through a variety of distances, propagation speeds and synergistic behaviors, both in the temporal and in the spatial domains. The mission's primary goal is to understand the instantaneous magnetospheric circulation pattern. Reaching computer-aided, data-restricted, predictive capability of the magnetospheric response to the external, solar wind conditions, is the mission's confirmation of success. Thus, by design, Magnetospheric Constellation will provide an understanding of the basic energy and plasma circulation processes in the closest astrophysical laboratory available for in situ observations, and the capability to predict space weather that affects the sphere of human activity in space.

The mission opens a new observational domain in the exploration of the magnetosphere. In this domain spatio-temporal ambiguities will be curtailed and cause-and-effect relationships will be delineated. Data from previous missions, which may be better equipped to study microphysical processes will be revisited in light of the new understanding of how those processes partake in the system's evolution.

The new approach necessitates large numbers of observational points. Since launch costs represent a significant part of the total mission expenditure, keeping it at bay necessitates instrument and spacecraft component miniaturization. This, in turn, will stimulate technical innovation and efficiency, with technology spin-offs for planetary and solar exploration.

2. HISTORY

The beginnings of the Constellation mission date back to the early years of space physics [*Dungey*, 1966]. McIlwain [1968] proposed that "experiments of large scale are needed more than large spacecraft with many instruments". The limited accessibility to space that could be tackled only by a national agency, NASA, mandated a centralized management structure. Increasing instrument sophistication, experiment size and quality assurance requirements favored the opposite approach. Although the need for global multipoint sampling was continuously re-affirmed through National Academy of Sciences reports [e.g., NAS 1988] such ideas were either considered impractical or futuristic.

With the full implementation of ISTP, the next step in space physics research has to be defined. This need places a clear and immediate focus on the multiprobe approach, which was recognized during the first community-wide Sun-Earth connection roadmap workshop at JHU/APL to hold the promise for a "major leap" in magnetospheric research. In response to the clear need for direction in the post-ISTP era, NASA is supporting studies of future mission concepts (NRA 96-OSS 03) and has sponsored a science definition team on "geospace multiprobes" which discussed multiprobe missions in both ionospheric and magnetospheric physics. The team, chaired by R. Heelis, concluded its work in the summer of 1997 and the outcome of its study was released to the science community prior to the 1997 AGU meeting. The "Heelis Report" concludes that future "multiprobe" missions in the magnetosphere are the next step towards final understanding and development of predictive capability of geospace processes. Preliminary technical studies of tentative mission implementations by NASA have suggested that some versions of the proposed Constellation missions are achievable under the Solar Terrestrial Probe Line budgetary constraints ($110M) and with a single Delta 7920 launch.

Community-wide response to a call for papers on the "Science closure and enabling technologies for constellation missions" at the 1997 AGU meeting was overwhelming. Forty papers were presented and the lively discussions during and after it reinforced the grass-roots support for the proposed concept. A publication assembling the papers resulting from the AGU meeting is expected to reach the scientific community by December of 1998. A science definition team on Constellation class missions is expected to be assembled by NASA within 1998.

This paper's plan is to summarize Constellation-related efforts currently under way. An implicit prioritization of the science quests is unavoidable and in fact imperative, as is a tentative implementation strategy driven by this prioritization.

3. SCIENCE QUESTS

By exploring a new dimension of the magnetosphere, Constellation will provide data that will leave few space physics disciplines untouched. It is promising an advancement in our understanding of space physics processes comparable to the one that resulted in the late 19th century, when a sufficiently dense network of weather stations allowed meteorologists to track the growth, motion and decay of storm-producing pressure fronts [Siscoe, 1997a]. The revelation of the extra-tropical cyclone in assimilative reconstruction of atmospheric observables is analogous to the anticipated revelation of the modes of magnetospheric circulation through assimilative reconstruction of magnetospheric observables.

Meta-cognition will undoubtedly raise questions asked to a level transcending current vernacular. Serendipitous discoveries will surely re-order the importance of the scientific themes attacked. Nevertheless, impasses justified by analysis of current datasets point towards specific families of questions and objectives.

3.1 Topology Reconstruction

Ionospheric current, conductivity and electric field maps can be obtained from a combination of low altitude satellites (e.g. DMSP), radar chains (e.g. SUPERDARN), and auroral images (e.g. POLAR). The magnetospheric counterparts of those maps remain theoretical constructs. Since the magnetosphere is the main conduit of solar wind energy to the auroral ionosphere, the relevant driving processes are, therefore, still unresolved. The magnetospheric topology is primarily determined by the magnetopause, ring and tail currents but it is also significantly influenced by quasi-permanent field aligned currents (Region 1 and 2, cusp, NBz and mantle) as well as by substorm currents. The presence and temporal variation of these currents modifies statistical models so dramatically [Kaufmann and Larson, 1989; Stern and Tsyganenko, 1992], that it has been to-date impossible to reach consensus on which magnetospheric process is responsible for a large number of ionospheric phenomena. Most notably, delineating the mapping of the Harang discontinuity, substorm initiation arc, westward traveling surge position and motion and travelling convection vortices can revitalize our understanding of magnetospheric dynamics and circulation.

Field aligned currents can be most naturally monitored using low altitude magnetometer-bearing satellites whose density and sensitivity requirements remain to be studied. Modeling magnetospheric currents requires a network of high altitude satellites and possibly ground stations [Sergeev et al., 1996]. Magnetometer data from such platforms when inverted using techniques available from statistical [Tsyganenko and Usmanov, 1982], simulation [Berchem et al., 1995] and atmospheric [Ghil and Malanotte-Rizzoli, 1991] models, an instantaneous "image" of currents and fields consistent with the average tail structure but representative the instantaneous fields.

The Tsyganenko assimilation technique (Plate 2) can be used to reproduce the model magnetospheric currents at a given instant from model Constellation data [Tsyganenko, 1997]. This can be a powerful tool for Constellation mission orbit planning. The model data can arise by flying the satellites through a model (statistical or MHD) magnetosphere (Plate 3). The goodness of the topological reconstruction can be evaluated by comparing the Constellation field values within the initial, model magnetosphere and the values within the reconstructed magnetosphere. In such a way Tsyganenko found that using magnetometer-only bearing satellites on the order of 80-100 probes are necessary for self-consistent magnetotail topology reconstruction. The probes should be placed in the tail lobes at random but well-distributed locations as far as 40 R_E downtail and between -15 R_E and +15 R_E in the dawn-dusk direction. The combined usage of plasma and field data will add information (such as total pressure) that may allow utilization of data from probes inside the high beta plasma sheet.

Flying constellation probes through global MHD models will produce realistic Constellation data whose study can enhance confidence in the mission's reconstruction methods and will allow development of the necessary suite of tools for constellation mission analysis [Raeder, 1997].

3.2 Energy and Flux Circulation

3.2.1 Substorm science. Substorms are a primary mode of magnetospheric variability at times of moderate solar wind power input (~10^{11} Watts), and modulate storm-time magnetospheric circulation at times of intense solar wind power input [Siscoe, 1997b]. Statistical studies or multi-spacecraft studies utilizing fortuitous (and thus unoptimized) conjunctions have resulted in contention rather than convergence on the cause-and-effect relationship between current disruption and reconnection. Although a small number of satellites strategically positioned at the magnetic equator may resolve this issue, the study of the boundary conditions, effects and evolution of the magnetospheric activation(s) responsible for a substorm requires a series of satellites that sample simultaneously over a wide range of scale-sizes.

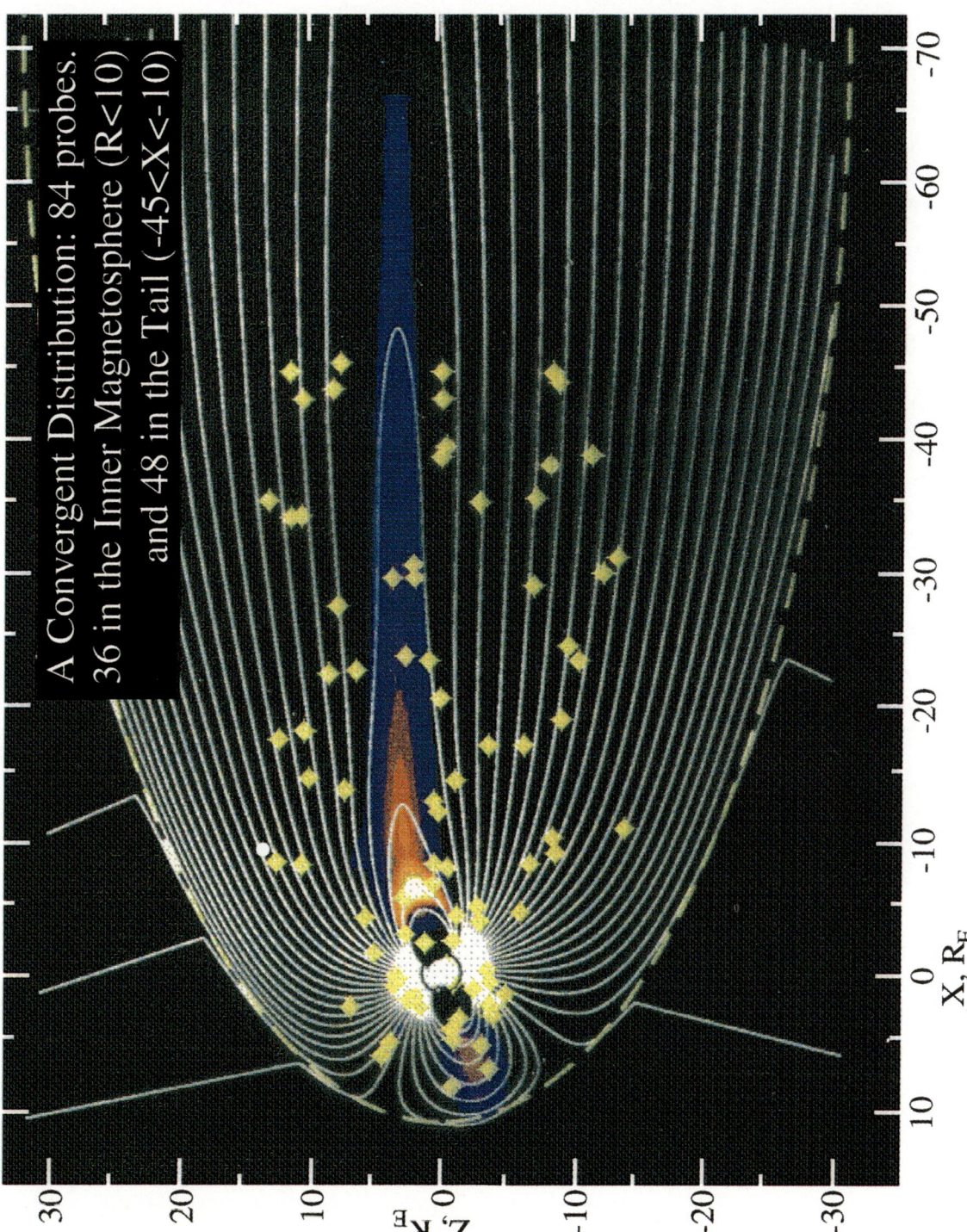

Plate 2. Usage of Tsyganenko methods for assimilative reconstruction of the instantaneous magnetospheric currents.

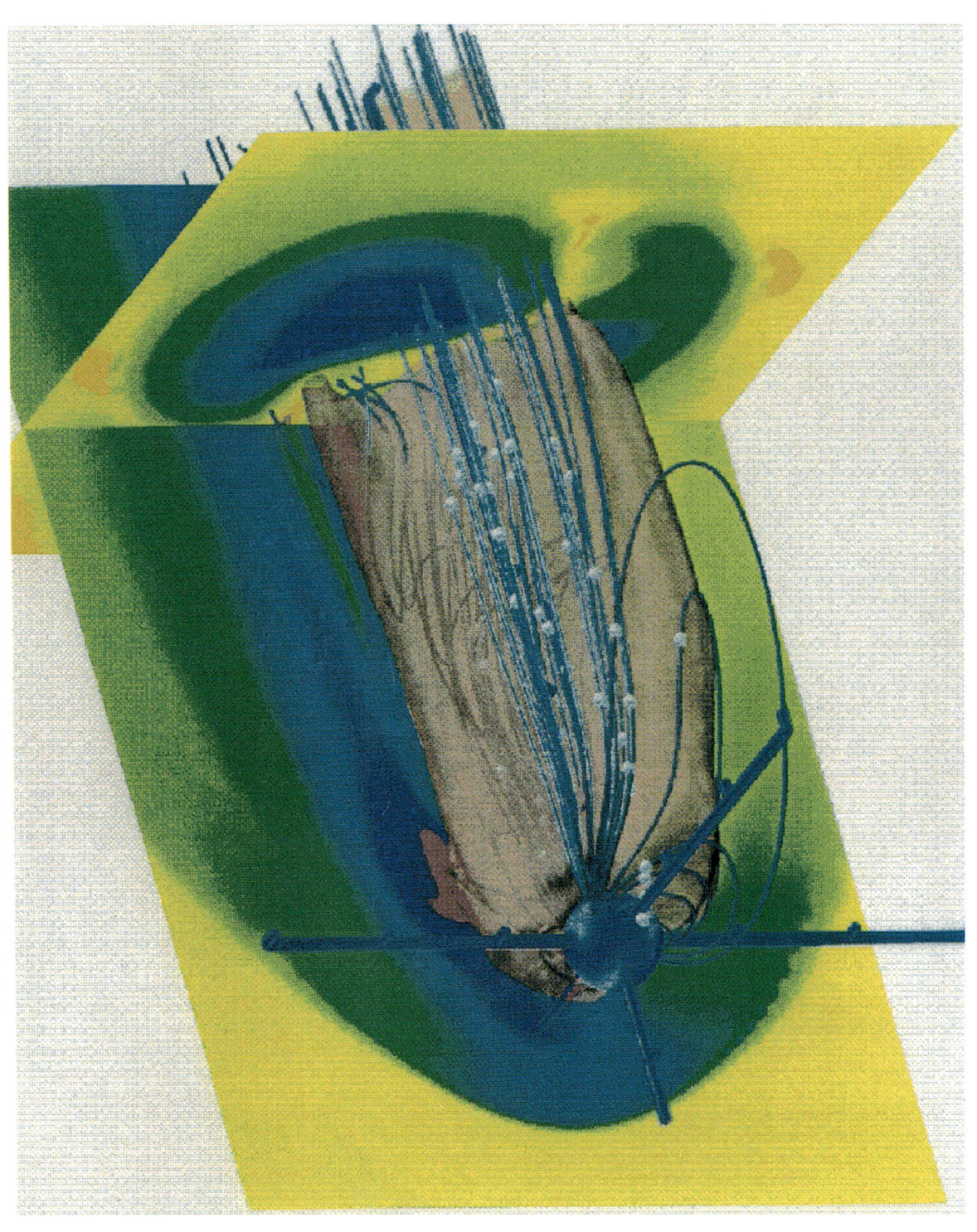

Plate 3. MHD models as mission-planning tools.

High speed magnetotail flows observed during substorms are thought to be important means of energy and flux circulation in the magnetosphere. They redistribute energy and flux and modify local particle pressure. Pressure gradients are responsible for field-aligned currents that, in turn, power the aurora and induce particle outflow from the ionosphere. The magnetospheric pressure system responsible for the aurora has yet to be measured in the magnetosphere.

The scale size of the initial magnetotail activation region is on the order of 1 R_E or less [Ohtani et al., 1991, Angelopoulos et al., 1997]. The active region expands across and along the magnetotail at an expansion speed of ~200 km/s [Nagai, 1982; Jacquey et al., 1993]. The transport-efficient flows at the edge of the expanding region reach a large magnetotail region [Slavin et al., 1997]. The correlation length of individual flow bursts within the expanding boundary is also small (order of 1 R_E or less). The wave-particle interactions leading to local energy transformation and flow burst production is the objective of other missions, such as CLUSTERII. Magnetospheric Constellation aspires to unravel the expansion mechanism of, and the topological reconfiguration induced by these bursts. Equatorial satellites, at separations of 0.5-10 R_E and sampling every ~10s are required to characterize this process.

The scale length of the active boundary has not yet been measured but CLUSTERII will provide that information. Based on few fortuitous satellite conjunction observations, 3 R_E is a reasonable value for the activity region. Sampling the entire plasma sheet between 10 and 40 R_E downtail and +-10 R_E in Y, necessitates ~60 probes. Scales from 0.1-20 R_E are sampled due to multiple conjunctions from such a deployment. Shorter scale structures are measured more frequently by reducing the area sampled to 12X20 R_E^2. Should the separation be required to be smaller, a series of clusters can be deployed at strategic locations.

3.2.2 Instantaneous circulation pattern. When flow bursts (associated with substorm activations) are excised from the ensemble of magnetotail states the average flow pattern is surprisingly consistent with the one expected from particle tracing or two-fluid motion. However, the peak-to-peak flow variability is typically many times larger than the average, rendering the classical picture of Earthward, steady-state magnetotail convection questionable (Figure 2).

When particle populations are advanced in a computer model of a slowly convecting magnetotail a large ion pressure and temperature is expected in the dusk plasma sheet as far down-tail as 20 R_E. This is not observed at regions tailward of ~12 R_E (Plate 4). Earthward of that distance, there is qualitative agreement only at active times [Wing and Newell, 1998].

Finally, if a flux tube is followed in a realistic model field in response to an external electric field, it will convect inwards and increase adiabatically in pressure. If the pressure increase is kept consistent with a realistic field topology it violates total magnetotail pressure balance. Following a flux tube along its path and establishing the means by which the

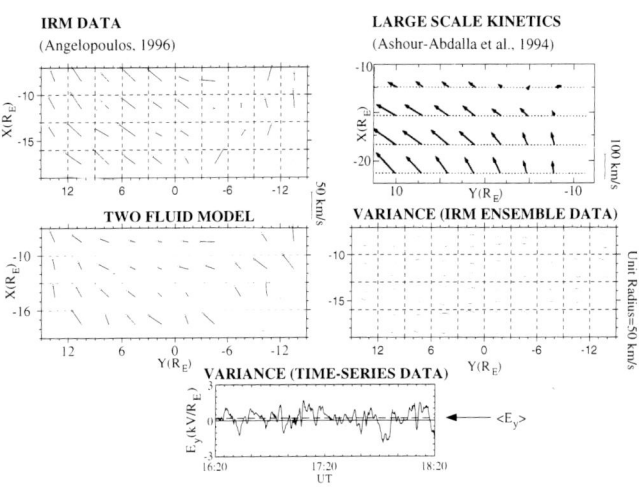

Figure 2. Instantaneous equatorial circulation is elusive.

distribution function evolves in response to measured external fields is the only way to resolve this paradox. Particle losses through precipitation, gradient drifts, drift-shell splitting and confirmation of laminar (versus turbulent) convection can then be studied.

The above discussion exemplifies the types of questions that phase magnetospheric physics in the post-ISTP era. These questions are fundamentally different from the ones that space physics missions attacked in the last 30 years: They call for taking one step back from exploration of space microphysics, and extending our domain of knowledge in the realm of k-space, exploring maps of particle spectra, and DC fields over a large volume of space.

This can be achieved by: (1) In situ sampling of plasmas using classical sensors, and (2) Remote sensing techniques combined with in situ measurements. Mission-specific development of miniaturized sensors and electronics to obtain more refined sampling or reduce weight, power and total mission cost can help both approaches. Dual operational and scientific use of existing commercial or government platforms represents an efficient means of implementing parts of the Constellation plan. Prioritization of missions should depend on their promise to definitively answer clear and major outstanding science questions, and their effectiveness in maximizing science return for a given cost.

4. APPROACHES

4.1 In Situ Measurements

4.1.1 Topology reconstruction. A network of high altitude, low data accumulation rate (1 min), magnetometer-only-bearing microsats that satisfies the science quest of topology reconstruction is achievable with current technologies, at a small fraction of the cost of the STP line [Petscheck

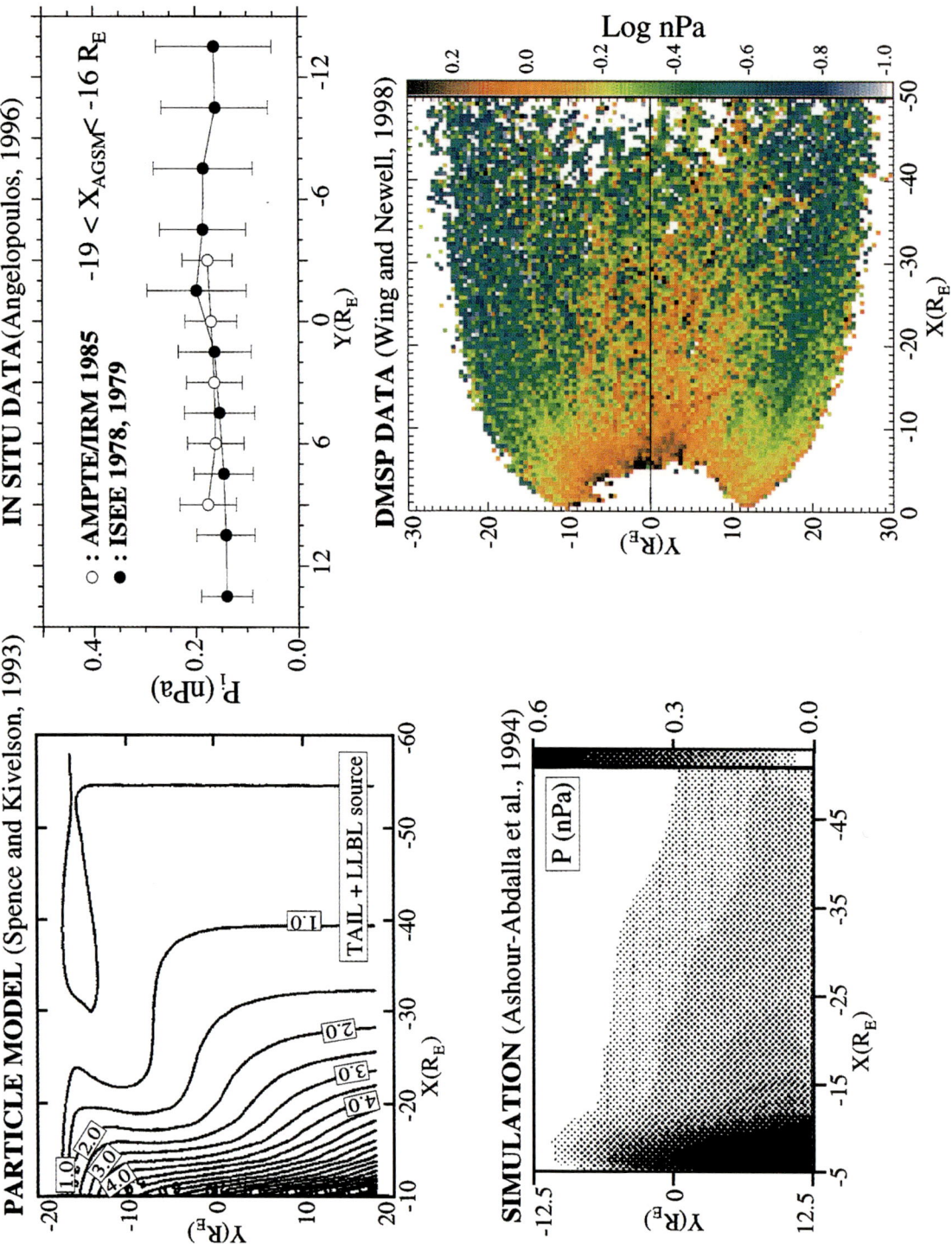

Plate 4. Model equatorial pressure distribution (left) disagrees with ensemble averages of measurements (right).

et al., 1997]. These satellites have low mass (<1 kg) and average orbit power (0.5 W) and are radiation tolerant to achieve longevity critical for overlapping with ensuing Constellation implementations. A classical design and higher accumulation data rates (16Hz) result in mass increase to 5 kg with a ~2W orbit average power. The latter was the approach validated by the SDT on geospace multiprobes.

Such low mass and power satellites have only recently been considered again [*Fleeter*, 1997] after a prolonged period since the early Vanguard satellites (Plate 5). The difference is that today modern electronics permit highly sophisticated on-board operations with extremely modest resources. The miniature satellite can be equipped with a three-axis fluxgate magnetometer at 1nT sensitivity, a central processing unit and commercial off-the-shelf components for attitude determination, power and communication subsystems. Using a store-and-dump scheme all accumulated data can be downlinked with very low transmission power.

The satellites can be placed by a re-ignitable engine into a moderate inclination transfer orbit. Aided by orbit precession and lunar perturbations the probe positions will cover a magnetotail region with the required coverage. After accounting for structure, propulsion system and hydrazine, 60 or more such satellites can be deployed using a single Pegasus launch. Studies of orbital parameters that optimize coverage are underway at several institutions (e.g., *Stern*, 1997; *Fazakerley et al.*, 1997, *Rayburn et al.*, 1998).

4.1.2 Energy and flux circulation. The addition of a plasma instrument on a magnetometer-bearing microsat increases data rates to a point that a classical (~5 kg, ~3W orbit average power) satellite design is the only option available. A single mother-ship is sufficient for a dense orbit deployment at high apogee, highly elliptical orbits sampling the equatorial plasma sheet and magnetopause. Cluster-like formations at strategic locations are also possible (apogees at 12, 20 and 40 R_E). By running the probes through an MHD simulation it is possible to reconstruct the equatorial distribution of pressure, flows and energy flux density (Plate 6). Such equatorial maps accumulated every few seconds will provide an unprecedented look at the energy flux density, pressure distribution and flow vorticity in the tail. The location of first substorm signature in the magnetosphere will be established as well as the unique pre-substorm particles and fields conditions there. The k-spectrum of the inferred electric field fluctuations in the magnetosphere will be established and that will help determine if plasma transport is convective or turbulent.

The above mission implementation can therefore address the following science objectives: i) Where does magnetotail particle acceleration start, why does it start there and how does it progress across the magnetosphere? ii) What pressure system in the magnetosphere energizes the quiet time aurora and powers its expansion at active times? iii) How do reconnection and eddy diffusion control the efficiency of solar wind magnetospheric coupling? iv) How does the solar wind energy density flux that couples into the magnetosphere ultimately drive plasma transport in near-Earth environment and the ionosphere? v) Is plasma transport throughout the magnetosphere convective or turbulent?

Although the magnetic field in the high-beta plasma sheet is highly variable and magnetometer measurements there cannot help towards global current reconstruction, the total plasma pressure computed as the sum of plasma and magnetic field pressure may produce the total lobe field accurately enough to warrant topological reconstruction. This is currently being tested on Geotail data by N. Tsyganenko. Detector inter-calibration and stability must also be evaluated before this mission implementation technique can permit reconstruction of the global magnetospheric topology. Once this approach is validated, mapping of plasma pressure and density along model flux tubes can provide three-dimensional images of these plasma parameters.

4.2 Imaging

Several pertinent imaging techniques are at different stages of development. ENA imaging demonstrated through analysis of data from previous missions (POLAR), tested on Astrid, flown on Cassini and to be implemented as a powerful research tool on IMAGE and TWINS will provide for the first time a global image of the plasma parameters from the inner magnetosphere. Extending our imaging capability beyond the inner magnetosphere is imperative so as to unravel the electrodynamic processes that control global circulation. Stereoscopic energetic particle imaging of the tenuous magnetotail is currently under consideration as a new mission concept.

Resonance scattering is another imaging technique implemented for the plasmasphere on the HeII (30.4 nm line) on missions Planet-B, IMAGE, and AKKA, and will be tested for the outer magnetosphere on Selene. It has been proposed for the magnetotail on the OII (83.4 nm) line. Ionospheric Oxygen is abundant in the tail lobes especially during active times. A hydrogen absorption cell to suppress the Lyman alpha (121.6 nm) contamination of the OII line is being considered. The technique can provide ion density, and line of sight speed [*Nakamura et al.*, 1997].

Remote sensing combined with in situ measurements and modeling leads to far more comprehensive information than can be obtained from images alone. Such is also the case for magnetospheric radio tomography [*Ergun et al.*, 1997] described herein for a meridional plane in the magnetotail. It employs ~20 satellites to obtain tomographically inverted images of electron density, N_e, at finer spatial resolution than in-situ instrumentation would allow. Each satellite emits two nearby frequency signals (1 MHz and 3 MHz) with the same phase for 0.1 seconds, while the others receive. Within 2 seconds, all 20 satellites have emitted once and received signals 19 times from all other satellites. Both differential Doppler delay and phase shift between the two signals upon detection are proportional to the total electron

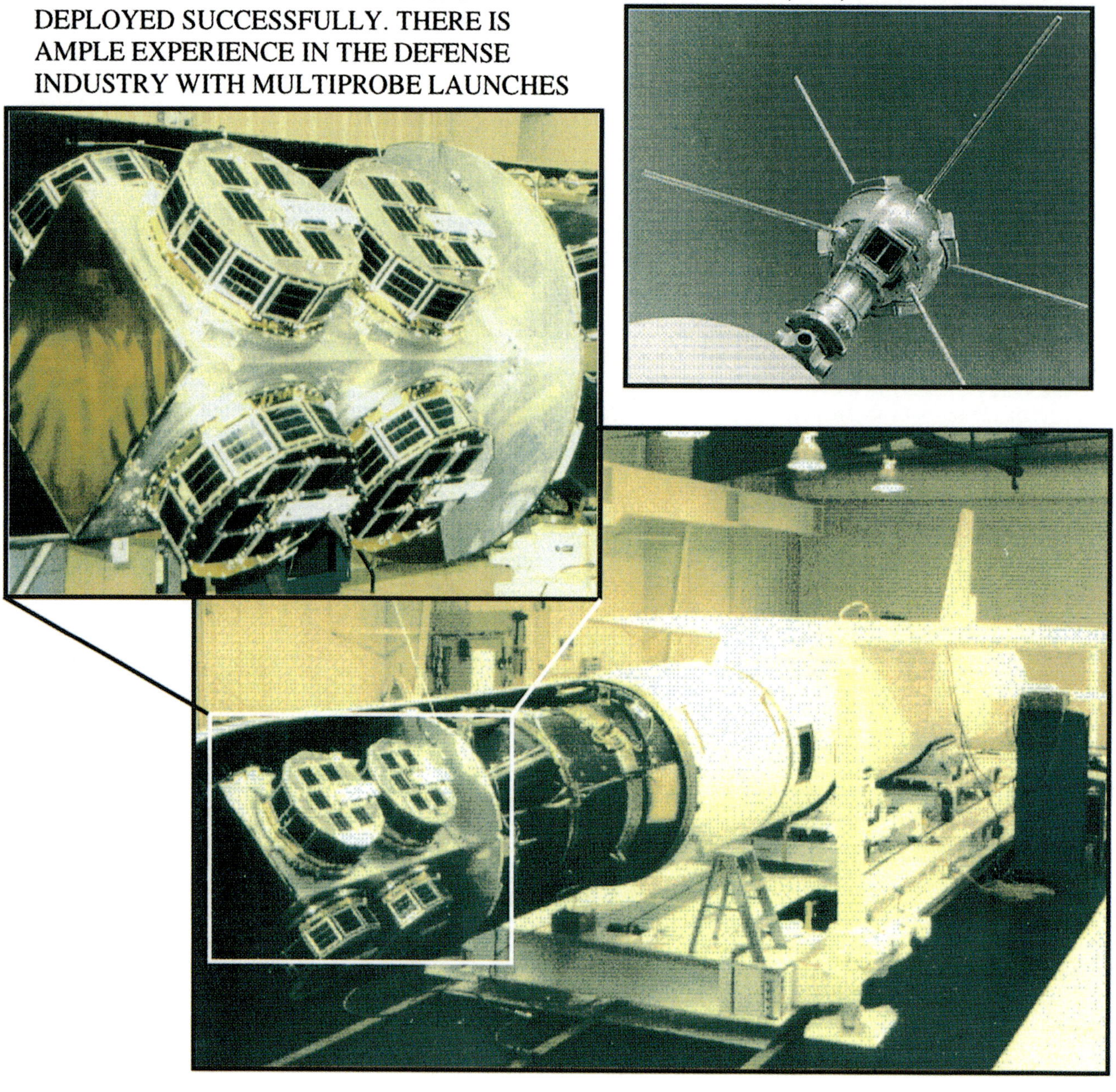

Plate 5. Microsatellites have been built by NASA and launched in significant numbers by the defense industry.

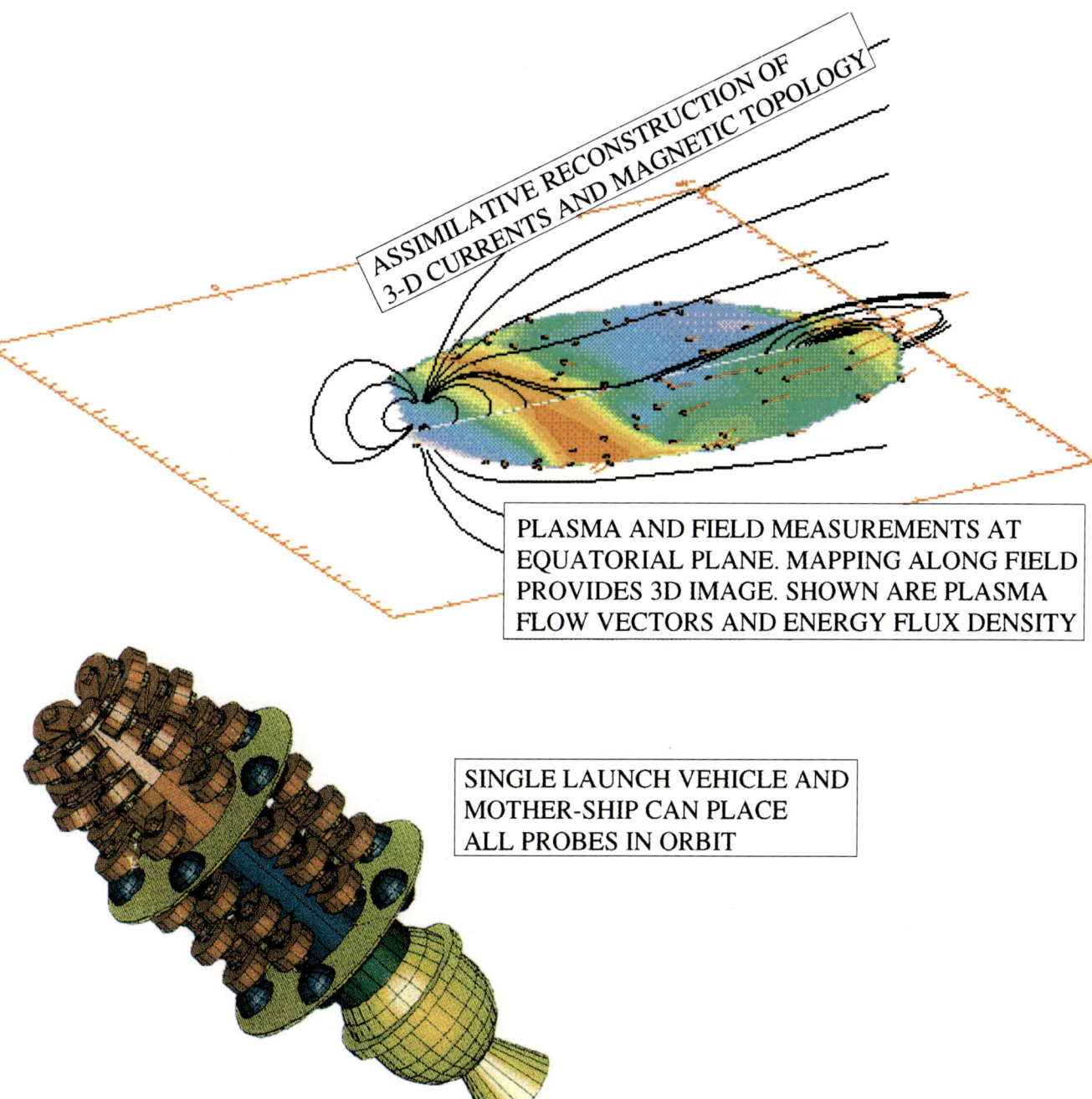

Plate 6. Equatorial Constellation: Deployment and assimilative reconstruction from runs through global MHD.

content (TEC) along the ray path. The TEC data when deconvolved with standard (e.g., least-squares fit) techniques produce a planar image of N_e. Such images (Plate 7) allow the study of magnetospheric plasma loading and topological evolution. At the dayside magnetosphere, the boundary shape and motion as well as the magnetosheath and bow shock will be imaged instantaneously over a large area. Their evolution in time and their deformations in response to the solar wind conditions along with the available in situ measurements will permit intelligent guesses on the location of reconnection sites and their evolution with time.

The application of the technique on a meridional plane allows the study of the following outstanding questions in magnetospheric physics: (i) What is the effect of solar wind disturbances, such as coronal mass ejections and interplanetary shocks, on the bow shock, magnetosheath and magnetopause? (ii) What is the inter-dependence of localized reconnection at the magnetopause boundary and the boundary's large scale evolution? (iii) What is the instantaneous average current density across the plasma sheet and its radial profile along the magnetotail? (iv) What is the inter-dependence and evolution of current disruption, fast flows, and plasma sheet thickness? (v) How does plasma circulate into, and throughout the magnetosphere? In situ measurements and modeling will be used to train scientists with a new way of observing known phenomena, such as solar wind shocks and hot diamagnetic cavities in the dayside or bursty bulk flows in the tail. The experience gained from the early observations of these phenomena will increase familiarity and provide confidence in the method, thereby allowing scientists to interpret the resulting images at all times.

5. ENHANCING TECHNOLOGIES

5.1 Detector Miniaturization

Efforts in miniature magnetometers led by JPL and APL are already bearing fruit and await flight opportunities for testing [*Miller et al.*, 1997; *Potemra et al.*, 1997]. Immediate benefit from such efforts will be the routine usage of sensitive magnetometers for dual, operational and scientific, purpose on commercial Constellations. Quite promising mission implementations have also been proposed [*Boehm*, 1997].

Plasma detectors specifically designed for low weight and low power consumption while retaining the capabilities of traditional designs are being developed at LANL and SWRI [*Lawrence et al.*, 1997; *Nordholt et al.*, 1997; *Funsten and Elphic*, 1997]. Several of these are slated for flight within the next 2 years [*Winningham et al.*, 1997; *Norberg et al.*, 1997].

Through many years of evolution on LANL/DOD and NASA missions, energetic particle detectors have now become extremely efficient and can accommodate both medium and high energies [*Reeves et al.*, 1997; *McNutt et al.*, 1997] under 1 kg and 1 W. Further reduction in the energy threshold of particle detection is an area of active research with cross-disciplinary interest, promising more comprehensive, compact and sensitive particle instruments [*Ritzau et al.*, 1997].

5.2 Solar Electric Propulsion

This efficient method for orbit control, to be validated in October 1998 on the DS1 mission, has been proposed on a new space physics mission concept study [*Kluever et al.*, 1997] and is applicable to Constellation. The NSTAR thruster at a total 110 kg, utilizing a single, 1.25 kW input power solar panel can achieve orbit raise from GTO to a high apogee, high inclination orbit within 3 months, for an initial payload of ~480 kg. The latter is the launch capability of a Taurus on a GTO. A comparable ascent duration with same initial mass is achieved utilizing a lighter, Hall thruster system, whose NASA validation is pending. Given the lift capacity of a Taurus, and assuming a 53 kg of Hall thruster engine, power processing unit, solar panel for satellite delivery from GTO to final orbits, and 20% mass for structure we obtain orbit placement for 70 satellites of ~5 kg each. At two instruments per satellite (magnetometer plus energetic particles, or plasma detector) this mass allocation is possible with current technology.

5.3 Microelectronics Technologies

Techniques for chip design, fabrication and testing borrowed from defense and commercial sector promise weight and power savings for Constellation's DPU and instruments. As an added benefit they improve radiation hardening, which is key to prolonged lifetime. For example a Time-of-Flight chip developed at APL with 3gr, 25mW, 1MRad specifications has substituted a 250 gr, 1W, 100kRad card on Cassini [*Paschalidis*, 1997].

5.4 Micropropulsion Systems

Conventional, hydrazine propulsion systems (with an overhead of a few kg minimum) are adequate for satellites in the range of 50-100 kg, but are too taxing for autonomous Constellation microsats, typically less than 10 kg. Technological development in this area is rapid and applications are cross-disciplinary [*Panetta*, 1997].

6. CONSTELLATION PATHFINDERS

The importance of developing data ingestion, assimilation and analysis tools that will demonstrate Magnetospheric Constellation's science closure cannot be overemphasized. Such science tools along with cost-effective operations mission planning tools will undoubtedly draw upon the industry's and the scientific community's experience from

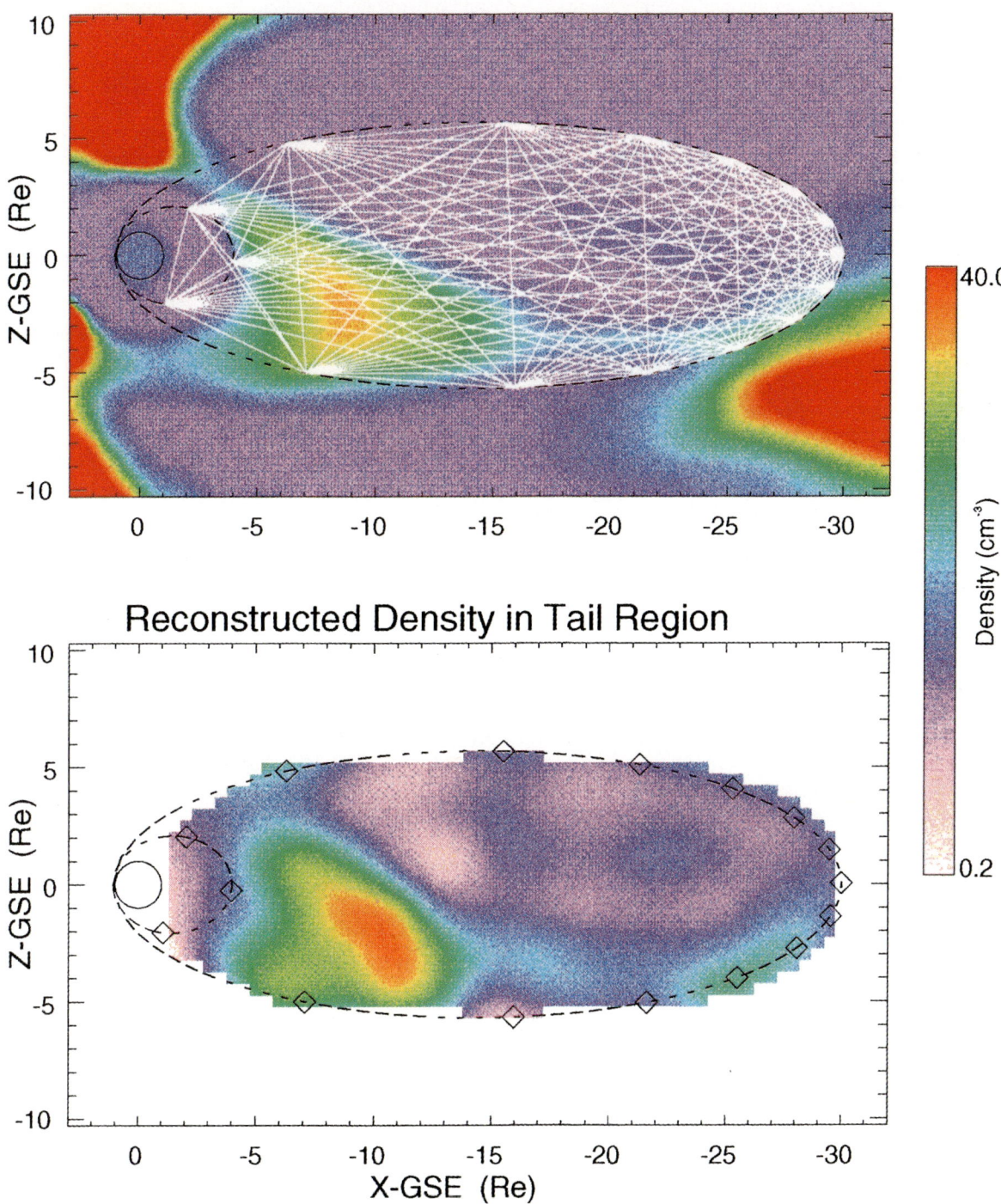

Plate 7. Meridional tomographic density reconstruction in the magnetotail from runs through global MHD.

existing Constellations. For example, operational LANL, GPS, and military (Molniya-orbit) satellites obtain dosimetry data that can also be used to assemble a first, albeit crude map of energetic particle behavior in the inner magnetosphere. These data and a combination of particle tracing and other models (e.g., Rice Convection Model) can help develop the first Constellation analysis methods for the near-Earth environment.

Attitude magnetometers on-board the Iridium Constellation (70 satellites fully deployed by the end of 1998) can be used for routine detection of field-aligned current location and magnitude. Preliminary, NSF-funded analysis of the above dataset is quite encouraging. Data analysis from these satellites can also help built the first Constellation data ingestion algorithms. Moreover, pursuing advance agreements with other telecommunications Constellations (e.g. Globalstar) for dual magnetometer usage should be a top priority for the Magnetospheric Constellation program.

With the advent of commercial Constellations significant experience has developed in the industry regarding tracking and commanding fleets of satellites. Low cost software packages can perform orbit-predicts, pass-schedules, data-dumps and antenna control [*Bester et al.*, 1997]. Experience with command sequences, automated contact planning, data dump prioritization, on-orbit data retention or memory-overwrite, data dissemination and visualization can be gained by such early Constellation implementations at minimal cost.

Validation of technical innovation or demonstration of proof-of-principle can occur through inexpensive, well focused, scientifically rewarding missions. For example [*Harvey et al.*, 1997], a small number of probes in strategic magnetospheric orbits can resolve some of the spatio-temporal ambiguities that riddle substorm research while serving as a technological pathfinder for Constellation. The community can benefit directly from the STEDI program's proven low-cost practices [*Chakrabarti et al.*, 1997]. The new millennium program (NMP) is yet another funding opportunity for Constellation to obtain its first low-cost demonstration flight.

Piggyback rides on civilian, commercial or military launches [*Rademacher and Leschly*, 1997] are ideal for low weight satellites. Such is the launch of the 5 kg Swedish scientific satellite MUNIN (Figure 3). Arianespace has developed the "ASAP" platform for routine secondary launches (four slots of 50 kg each on Ariane 4 and eight slots of 100 kg each on Ariane 5). To date, 32 secondary spacecraft have been launched by Arianespace. Opportunities on other, US or foreign launchers also exist. Piggyback rides can result in very inexpensive ($1-3 M per slot) Magnetospheric Constellation protoflights.

7. PHASED DEPLOYMENT

The Constellation mission objectives cannot be fully addressed by any single proposed approach. However, when the approaches discussed are implemented as a series of overlapping missions, each with well defined, strategically focused objective, they can synergistically bring about a revolution in our understanding and our predictive capability of magnetospheric processes. Incremental implementation distributes and reduces risk. It spurs technical innovation, justified by the economy of scale, but allows it to mature through multiple incarnations. Miniaturization technologies improve the mission's total science return per dollar by enabling a choice of either more comprehensive experiments to partake in the deployment, or by reducing launch costs. Thus, the declared goals of Magnetospheric Constellation become less daunting.

Magnetometer satellites deployed as the first means towards the final goal of topology reconstruction will permit the first Global Current Monitor mission (GCM). Orbit optimization for coverage, resolution, and prolonged survival in the expected radiation and thermal environment (shadows) is critical for overlapping with future implementations. The benefits from learning the practices that render this maiden voyage of Constellation successful far outweigh the importance of achieving a dense enough network with a single deployment. Despite the GCM mission's exploratory nature and invaluable scientific returns its primary achievement will be undoubtedly be the technical demonstration of fleet deployment from a single, low-cost (Pegasus) launch.

High apogee, elliptical, equatorial orbits can measure in situ the thermal particles and fields of the plasma sheet and produce the input necessary to understand and model magnetospheric circulation, energy throughput and energy transformation within the magnetosphere. With its more comprehensive measurements it will enhance GCM's ability to topologically reconstruct the instantaneous magnetosphere. Energetic particle measurements (10 keV - 2 MeV) are necessary on the probes that reside close to the inner edge of the plasma sheet and outer electron belt. Measurements near the equatorial plane are essential because high latitude satellites miss a significant part of the particle phase space, while the flux-tube mapping uncertainties complicate data interpretation.

A relatively small number of probes on highly elliptical orbits is sufficient to reconstruct the ordered field and adiabatic particles of the inner magnetosphere populations with the help of particle orbit integration and global models [*Reeves*, 1997]. This represents the Inner Magnetosphere Monitor (IMM). A larger number of probes is necessary in the outer magnetosphere as described in Section 4.1.2 and is termed the Outer Magnetosphere Monitor (OMM). A single launch vehicle (Taurus) can accommodate both missions since the orbits are similar.

A Meridional Tomographic Imager (MTI) of the plasma sheet and the magnetopause, as explained in Section 4.2 will provide instantaneous pictures of the density and when combined with available in situ diagnostics will lead to validation of general circulation models with unprecedented

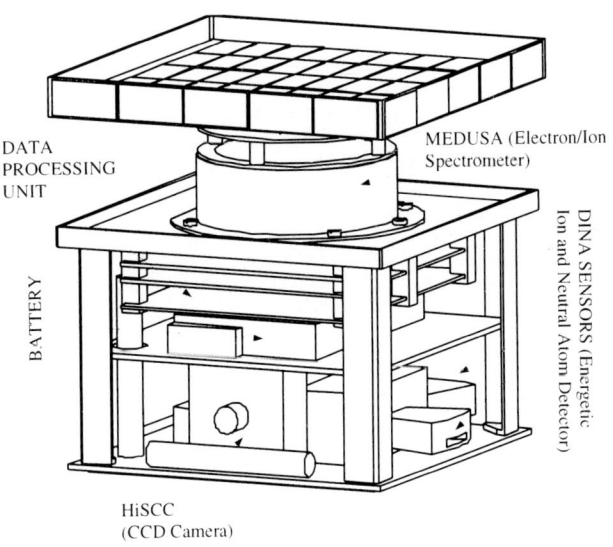

Figure 3. Following the success of Freja and Astrid, the Swedish Institute for Space Physics is building MUNIN, a 5 kg satellite to measure auroral particles and image the aurora in visible wavelengths [*Winningham et al.*, 1997]. It is slated for flight as piggyback on a DeltaII launch on May 27, 1999.

resolution and accuracy. The additional in situ measurements will provide a boost to the capabilities of Constellation to reconstruct the magnetospheric topology. A single Taurus launch vehicle may be possible for MTI deployment.

It is evident that the level of quantitative spatial information on solar wind-magnetosphere coupling sought after by Constellation will require far more detailed solar wind monitoring than is possible from the L1 point. The solar wind streamlines that determine the efficiency of this coupling are the ones that pass through the subsolar region. A small number of satellites can achieve continuous monitoring of the solar wind near the subsolar region but the required orbit necessitates additional launch. This can be either a series of piggybacks on large launchers or a single Pegasus launch.

Imaging of the auroral oval in UV and visible light has been shown to provide global information on the magnetospheric system's energy dissipation, accurate timing of phenomena such as substorm onset and cusp precipitation and estimates of the amount of energy storage in the magnetotail lobes. No full understanding of the magnetospheric system would be complete without validation of a predictive capability of the ionospheric energy dissipation as a function of geographic location and time. Continuous UV/Visible imaging of the auroral oval, hereby termed the Energy Dissipation Monitor (EDM), is an essential ingredient of Constellation and, if limited to that objective, feasible for a low cost.

A tiered development of Magnetospheric Constellation calls for three Armadas each composed of a number of fleets. Armada I could involve a GCM fleet of 20-60 microsats (depending on data accumulation rates) launched on a Pegasus, an EDM satellite (piggyback) and an ECM fleet of 4 microsats (Pegasus, or piggybacks). Armada II involves the IMM fleet of 10 microsats, and an OMM fleet of 60 microsats (all launched on a single, Taurus launch vehicle). Integrated instrument and bus design is important for achieving the mission weight targets outlined. Armada III involves an MTI fleet of 20 satellites (on a Taurus). Here station-keeping is important and each satellite should be equipped with its own lightweight propulsion system. The fully deployed Constellation will employ >100 microsats for magnetospheric reconstruction but each fleet will perform region-specific tasks. Launch costs are kept significantly below $100M (Taurus is $28M, Pegasus is $18M and each piggyback slot to GTO is assumed to be $2M). Multiple mission PIs, one for each fleet, can render the program manageable, distribute the expertise to the appropriate institutions, and take full advantage of the successes of the SMEX program in employing the "single PI mode" that is synonymous with fiscal scrutiny and schedule adherence.

8. AFTERWORD

Large, diverse datasets have been collected from previous magnetospheric missions, yet major questions, defined as the field's priorities from as early as the 1970s remain unanswered. Gaps in our understanding arise from the apparent observational void on particles and fields measurements in the wavenumber spectrum continuum. Moving into the next century, magnetospheric physics, a mature but still an experimentally-driven discipline, must generate datasets capable of improving and validating effective global circulation models, and theoretical constructs that stand the test of time. It must also adhere to NASA's mission statement and within a spirit of fiscal responsibility it must explore new territories while stimulating development of new technologies that can have broad usage in industry and other NASA disciplines. The Constellation mission can meet these challenges by involving NASA Centers of excellence, Industry and Academia and utilize the strength of each to achieve an optimal program implementation. The proposed phased deployment of a series of Constellation Armadas has the potential of securing wide community support, and the advantage that it distributes both its challenges and its rewards equally amongst its constituents.

Acknowledgments. We thank the members of the SDT team on Geospace Multiprobes, and in particular R. Heelis for useful discussions. We also thank the contributors to the 1997 AGU session on Constellation for useful discussions, and in particular R. Ergun, A. Fazakerley, R. Heelis, N. Paschalidis, J. Raeder, G. Reeves, and K. Tsyganenko who provided figures for the Yosemite 1998 talk and/or this paper.

REFERENCES

Angelopoulos, V. A., The role of impulsive particle acceleration in magnetotail circulation, in *Proceedings of the Third International Conference on Substorms, ESA SP-389*, 17, 1996.

Angelopoulos, V., et al., Magnetotail flow bursts: association to global magnetospheric circulation, relationship to ionospheric activity and direct evidence for localization, *Geophys. Res. Lett.*, 24, 2271, 1997.

Ashour-Abdalla et al., Structure of plasma flows in the Earth's magnetotail, in *Proceedings of the Second International Conference on Substorms, University of Alaska, Fairbanks*, 111, 1994.

Berchem, J., Interactive visualization of numerical simulation results: a tool for mission planning and data analysis, in Visualization techniques in space and atmospheric sciences, edited by E.P. Szuszczewics and J.H. Bredekamp, NASA SP-519, 1995.

Bester, M., et al., Tracking, data acquisition and data dissemination from multiprobe missions, *EOS Trans.*, 78, F571, 1997.

Boehm, M.H., A free-flying magnetometer sensor for the outer magnetosphere, *EOS Trans.*, 78, F573, 1997.

Chakrabarti, S., D. Cotton, University based space physics constellations: An extrapolation of TERRIERS experience to multiprobe missions, *EOS Trans.*, 78, F573, 1997.

Dungey, J., Inaugural Lecture as Professor of Physics at Imperial College, 1966.

Ergun, R. E., et al., Radio tomography investigation of the Earth's magnetosphere, *EOS Trans.*, 78, F567, 1997.

Fazakerley, A., et al., A study of multispacecraft magnetospheric magnetospheric monitoring mission concept, *EOS Trans.*, 78, F570, 1997.

Fleeter, R., Very low cost microspacecraft for magnetosphere mapping and other multiple spacecraft missions, *EOS Trans.*, 78, F574, 1997.

Funsten, H.O. and R.C. Elphic, Advanced miniature plasma spectrometer (AMPS): miniaturized instrumentation for plasma analysis, *EOS Trans.*, 78, F574, 1997.

Ghil, M. and P. Malanotte-Rizzoli, Data assimilation in meteorology and oceanography, *Adv. Geophys.*, 33, 141, 1991.

Harvey, P. et al., Stepwise approach towards a magnetospheric laboratory *EOS Trans.*, 78, F573, 1997.

Jacquey, C., et al., Tailward propagating cross-tail current disruption and dynamics of near-Earth tail: a multi-point measurement analysis, *Geophys. Res. Lett.*, 20, 983, 1993.

Kaufmann, R.L., D. J. Larson, Electric field mapping and auroral Birkeland currents, *J. Geophys. Res.*, 94, 15307, 1989.

Kennel, C. F., Convection and Substorms, Oxford University Press, p. 174, 1995.

Kluever, C.A. et al., The global magnetospheric dynamics mission: A optimized mission to study meso and micro physical processes in the magnetosphere, *EOS Trans.*, 78, F570, 1997.

Lawrence, D.J. et al., Initial calibration results for the plasma experiment for planetary exploration (PEPE), *EOS Trans.*, 78, F571, 1997.

McIlwain, C. E., Comments and speculations concerning the radiation belts, in *Annals of IQSY, MIT Press*, Cambridge, MA, 4, 1968.

McNutt, R.L. et al., A compact particle detector for low-energy particle measurements, *EOS Trans.*, 78, F574, 1997.

Miller, L.M. et al., Silicon micromachined micro-magnetometers for free flying magnetometer mission, *EOS Trans.*, 78, F571, 1997.

National Academy of Sciences, National Research Council Report on "Space Science in the 21st Century", 1988.

Nagai, T., Observed Magnetic Substorm Signatures at Sychronous Altitude, *J. Geophys. Res.*, 87, 4405, 1982.

Nakamura, M., et al., Feasibility study of magnetotail imaging and its contribution to the magnetospheric constellation mission, *EOS Trans.*, 78, F567, 1997.

Norberg, O. et al., The MUNIN nanosatellite--a precursor to magnetospheric constellations, *EOS Trans.*, 78, F574, 1997.

Nordholt, J.E. et al., Plasma experiment for planetary exploration (PEPE'), *EOS Trans.*, 78, F571, 1997.

Ohtani, S., et al., Tail current disruption in the geosynchronous region, in *Magnetospheric Substorms, AGU Geophys. Monogr. Ser.* 64, 131, 1991.

Panetta, P.V., Microsatellite technology development for the STP constellation missions, *EOS Trans.*, 78, F574, 1997.

Paschalidis, N., Microelectronics tecnologies enabling new generation spacecraft and instrumentation, *EOS Trans.*, 78, F574, 1997.

Petscheck, H.E. et al., Systems study of magnetospheric mapping mission, *EOS Trans.*, 78, F573, 1997.

Potemra, T.A. et al., Miniature magnetometers designed on xylophone resonators, *EOS Trans.*, 78, F571, 1997.

Rademacher, J. and K. Leschly, Microspacecraft secondary payload launch capabilities and mission possibilities, *11th AIAA/USU conference on small satellites*, SSC97-IX-3, 1997.

Raeder, J., Data assimilation from multipoint measurements, *EOS Trans.*, 78, F568, 1997.

Rayburn, C. et al., The dynamic evolution of a multi-satellite constellation, *EOS Trans.*, 79, in press, 1998.

Reeves, G.D., et al., Energetic particle contributions to a magnetospheric constellation mission, *EOS Trans.*, 78, F567, 1997.

Ritzau, S.M. et al., Solid state detection of low energy (1-20 keV) ions and electrons, *EOS Trans.*, 78, F571, 1997.

Sergeev, V. A. et al., Comparison of UV optical signatures with the substorm current wedge as predicted by an inversion algorithm, *J. Geophys. Res.*, 101, 2615, 1996.

Siscoe, G.L. Major outstanding questions and implementation principles of a constellation mission, *EOS Trans.*, 78, F567, 1997a.

Siscoe, G.L., Magnetospheric physics: Big storms make little storms, in *News and views, Nature*, 390, 448, 1997b.

Slavin, J.A., et al., WIND, GEOTAIL, and GOES 9 observations of magnetic field dipolarization and bursty bulk flows in the near-tail, *Geophysical Research Letters*, 24, 971, 1997.

Spence, H. E., and M. G. Kivelson, Contributions of the low-latitude boundary layer to the finite width magnetotail convection model, *J. Geophys. Res.*, 98, 15487, 1993.

Stern, D.P. and N.A. Tsyganenko, Uses and limitations of the Tsyganenko magnetic field models, *EOS Trans. AGU*, 73, 489, 1992.

Stern, D.P., "Science closure" of the "Profile" multiprobe mission, *EOS Trans.*, 78, F569, 1997.

Tsyganenko, N.A. and A.V. Usmanov, Determination of the magnetospheric current system parameters and development of experimental geomagnetic field models based on data from IMP and HEOS satellites, *Planetary and Space Science*, 30, 985-98, 1982.

Tsyganenko, N.A., Toward real-time magnetospheric mapping based on multi-probe space magnetometer data, *EOS Trans.*, 78, F568, 1997.

Winningham, J.D.,et al., MUNIN nanosatellite science goals, *EOS Trans.*, 78, F571, 1997.

Wing, S., and P. T. Newell, Central plasma sheet ion properties as inferred from ionospheric observations, *J. Geophys. Res.* 98, 6785, 1998.

V. Angelopoulos, Space Sciences Laboratory, University of California, Berkeley, CA 94720-7450, USA (E-mail: vassilis@ssl.berkeley.edu)

H. E. Spence, Boston University, Center for Space Physics, 725 Commonwealth Avenue, Boston, MA 02215, USA (E-mail: spence@buasta.bu.edu)

A Mercury Orbiter Mission

D.N. Baker

Laboratory for Atmospheric and Space Physics, University of Colorado at Boulder

By exploring and characterizing other planetary environments, we extend and generalize the knowledge gained from terrestrial studies. Mercury provides unique particles and fields features which are unobtainable at other planets due to the constraints of sampling times and the large dimensions of other magnetospheres relative to their planetary bodies. The highly variable interaction of the solar wind and the planetary surface with the Hermean magnetosphere is discussed in this paper. Of particular interest are substorm-like events occurring on very rapid time scales (tens of seconds) at Mercury. These bursts of activity can produce intensely energetic particle impacts upon the Hermean surface. Hermean magnetospheric and exospheric observations are considered which could provide essential data necessary to formulate the next generation of theories and models for terrestrial-type magnetospheric structure and dynamics. Critical issues necessary for the understanding of the surface history, atmospheric structure, and global magnetospheric dynamics of Mercury are presented and the elements of a Mercury Orbiter mission that can attain this understanding are described.

1. INTRODUCTION

The Mariner 10 flybys of Mercury in 1974 and 1975 resulted in the discovery of a strong planetary magnetic field and an active magnetosphere similar in many ways to that of Earth. Based upon the small size of the planet, Mercury's interior was expected to have cooled and solidified long ago. The presence of an intrinsic magnetic field, however, implied an internal dynamo in a fluid core, posing numerous, unresolved questions concerning the origin, composition, and thermal history of Mercury. The Mariner 10 spacecraft also detected intense particle bursts and magnetic field disturbances, indicating that phenomena analogous to terrestrial magnetospheric substorms occur at Mercury. The Mariner 10 images revealed a number of surface features unique to Mercury, including large-scale thrust faults apparently associated with crustal compression as the planet cooled and contracted.

The magnetospheric and planetary physics rationale for a Mercury orbiter mission has been reported upon previously in several NASA [*JPL*, 1977, 1990] and National Academy of Sciences [*NAS*, 1978, 1985, 1988] reports. The primary space physics science objectives identified in these reports are: 1) to map in three dimensions the magnetic structure and plasma environment of the "miniature" Mercury magnetosphere; 2) to study in detail the principal physical processes taking place during Hermean magnetospheric substorms with an emphasis on differences from Earth due to Mercury's lack of a highly conducting ionosphere; 3) to assess the role of interplanetary conditions in determining the rate at which the Hermean magnetosphere draws energy from the solar wind and the manner in which it is later dissipated; 4) to investigate heliospheric structure and dynamics inside of 0.5 AU; and 5) to utilize the proximity of Mercury to the Sun to achieve fundamental solar physics objectives by measuring neutrons and charged particles emanating from solar active regions. The primary planetology science objectives are: 1) to complete the global surface mapping initiated by Mariner 10; 2) to obtain global geochemical terrain maps of the occurrence of such elements as Fe, Th, K, Ti, Al, Mg, and Si; 3) to measure the intrinsic magnetic field in sufficient detail to allow for the detection of magnetic anomalies; and 4) to map Mercury's gravitational field and associated anomalies.

Previously, Mercury missions have been studied with the goal of addressing both planetary science and space physics goals. Such missions were major NASA programs, inconsistent with the present focus on cheaper and faster missions. As NASA reconsiders missions to Mercury, very critical space physics goals could be lost in favor of more limited or focused goals of studying planetary materials. It is therefore critical to define new missions which demonstrate that multidisciplinary goals, including space physics goals, can be achieved at Mercury in the present political climate of cheaper, faster, and better.

Recent advances in spacecraft miniaturization and innovative propulsion systems suggest the value of re-examining a mission to Mercury. Such a mission would provide unique particles and fields measurements which are unobtainable at other planets due to the constraints of orbital mechanics and the large dimensions of other magnetospheres relative to their planetary bodies. A Mercury Orbiter would be a mission of exploration and discovery that would provide the essential data necessary to formulate the next generation of theories and models for terrestrial-type magnetospheric structure and dynamics. Such a mission would also return critical measurements necessary for the understanding of not just the surface history and internal structure of Mercury, but the formation and chemical differentiation of the Solar System as a whole.

2. MARINER 10 PARTICLE AND FIELD OBSERVATIONS

In all, Mariner 10 performed three active close flybys of Mercury. The first occurred on March 29, 1974, with a periapsis distance of 1.3 R_M on the nightside of Mercury (1 R_M = 2439 km). The second encounter was on September 21, 1974, with a closest approach of 20 R_M far upstream on the dayside of the planet; this encounter produced no relevant in situ magnetospheric data. The third flyby occurred on March 16, 1975, with a closest approach of 327 km from the nightside surface at a high Hermean latitude (68° north). Thus encounters I and III consisted of flybys through the Mercury magnetosphere and constituted useful passages for magnetospheric studies. Figure 1 shows these flyby trajectories both in cylindrical coordinates (left) and as viewed in Y-Z coordinates from the sun.

The principal results concerning magnetic fields, plasmas, and energetic particle bursts in Mercury's magnetosphere came from Flyby I on March 29, 1974. Figure 2 shows data for the period 2030-2100 UT. The top panel shows counting rates from a sensor designed to measure electrons with E≥170 keV [*Simpson et al.*, 1974], while the middle and bottom panels show concurrently measured plasma number density (ρ) and electron temperature (T). The bottom three panels show magnetic field data including field magnitude (B), azimuth (φ), and inclination (θ). The field data are presented in a right-handed coordinate system in which X is toward the sun and Z is perpendicular to Mercury's orbital plane. Mariner crossed the magnetopause (MP) at 2037 UT inbound, reached closest approach at 2046:40 UT, and subsequently crossed the magnetopause again at ~2054:30 UT outbound. Several bow shock crossings were seen between 2057 and 2059 UT. The top panel of the figure shows several enhancements of the energetic electrons both in the magnetotail (events A, B, B', and C) and in the magnetosheath (events D and D').

Siscoe et al. [1975] and *Ogilvie et al.* [1977] showed that the energetic particle burst periods in Figure 2 tended to be times of substorm-like behavior. Mariner 10 entered the near-tail plasma sheet on the duskside and the sheath field was northward on tail entry while the field inside the magnetopause was very tail-like and relatively quiet. |B| then increased with time in approaching the planet, and according to *Siscoe et al.*, the higher-energy plasma electrons decreased in intensity. Shortly after closest approach, |B| decreased rapidly, and the field inclination increased markedly. This indicated a transition from a tail-like to a dipole-like field orientation and in the terrestrial case [*Baker et al.*, 1984, 1997] this would be a classic signature of substorm expansive phase onset. In fact, between 2047 and 2054 UT there were numerous large changes in **B** and these occurred in the same time period as did the large energetic particle bursts B, B', and C. The magnetic field was strongly southward upon Mariner's outbound exit.

Siscoe et al. [1975] focused on the fact that the interplanetary magnetic field (IMF) switched from northward to southward while Mariner was in the Hermean magnetosphere. They suggested, in analogy with Earth's case, that this initiated strong sunward plasma sheet convection and, presumably, enhanced magnetotail energy storage: When Mariner was about halfway through the tail the southward IMF initiated a series of substorms. *Siscoe et al.* [1975] showed by scaling arguments that substorm timescales should be of order 1-2 min in Mercury's case compared with 30-60 min in Earth's case [see also *Slavin and Holzer*, 1979]. Hence several substorms in a 20-min period is not unreasonable for Mercury. *Eraker and Simpson* [1986] and *Baker et al.*, [1986] developed this scenario further and suggested that the substorms in Mercury's magnetotail resulted from magnetic reconnection (neutral line formation) in the range of 3-6 R_M on the nightside. They suggested in further analogy with Earth that this substorm reconnection resulted in the impulsive acceleration of energetic particles.

3. SPACE PLASMA SCIENCE AT MERCURY

Mercury has a tenuous, neutral atmosphere whose constituents are poorly known. It is more properly termed an exosphere because the atmosphere is collisionless and the exobase is at the surface; i.e., an atmospheric neutral will typically fall back to the surface of Mercury before colliding with another neutral. The five known species in Mercury's exosphere – H, He, O, Na, and K – are also thought to be important constituents of the lunar atmosphere. Mariner 10 ultraviolet spectrometer observations detected H, He, and O at Mercury, while Na and K were later discovered by ground-based optical spectrophotometry (see Table 1 for surface density estimates). The mechanisms responsible for maintaining an atmosphere at Mercury, despite its high dayside surface temperature and low surface gravity, are not well understood. Atmospheric neutrals must continually fall to the surface and be re-ejected from it [*Morgan and Killen*, 1997]. Surface interactions are, therefore, critical in determining the atmospheric temperature, composition, and geographic distribution. Magnetospheric processes, including ion precipitation onto Mercury's surface and the pickup of photo-ions, may be extremely important for both atmospheric sources and losses (see Plate 1).

The absence of a collisional ionosphere has important consequences for global electric currents and plasma circulation patterns at Mercury. At Earth, high-latitude magnetospheric current systems close by flowing through the ionosphere. At Mercury, these currents cannot close through a collisional ionosphere (since none is present) or through the surface (because it is expected to be a good insulator). Closure of magnetospheric current systems through a resistive regolith or partially through an ionized exosphere rather than a collisional ionosphere would have important implications, both for the global current systems and magnetospheric convection as well as for dynamical processes such as substorms and flux transfer events. The heated and energized substorm particles could impact directly onto the cold regolith surface of Mercury, thereby creating a heated "auroral" band (see Figure 3) at the surface [*Baker et al.*, 1987]. Some theories hold that the timescale for the substorm growth phase at Earth is determined by ionospheric line-tying which limits the rate of magnetic flux return to the dayside following enhanced reconnection at the dayside magnetopause. Also, there are theories of the substorm expansion phase at Earth that consider active feedback between the magnetosphere and ionosphere (specifically enhanced conductivities in the auroral zones) as the essential ingredient for substorms. Such feedback is presumably absent at Mercury. Observations in Mercury's magnetosphere may determine whether ionospheric coupling, in the form of current closure through a resistive medium, is a necessary and central element in substorm-like energy conversion processes.

It is essential at Mercury to learn how magnetospheric structure evolves at relatively large distances and how the magnetotail responds to changes in the interplanetary medium. Does the magnetotail have a coherent structure; i.e., an identifiable plasma sheet and lobes, which extends to large distances? Structure and motion of the tail should be related to solar wind and near-planet magnetospheric changes. Many believe that during substorms in Earth's magnetosphere, the plasma sheet is severed by magnetic reconnection quite close to Earth and flows rapidly down the tail as a magnetically confined structure (a plasmoid). Some theories predict that this is the primary way that solar wind plasma and energy, earlier acquired by the magnetosphere, is dissipated and a portion returned to the solar wind. Thus, plasmoids may be of fundamental importance to magnetospheric physics. Observations of plasmoids in Mercury's magnetotail would provide important confirmation that magnetic reconnection and plasmoid formation are basic features of the process by which stored energy is released within planetary magnetospheres.

Solar wind energy coupling into Earth's magnetosphere is known to be strongly influenced by the polarity of the IMF. Southward IMF leads to strong coupling, through reconnection with the northward geomagnetic field at the surface of the magnetosphere. The occurrence of substorms, the basic mechanism for stored energy release and dissipation, clearly relates on a statistical basis to the occurrence of southward IMF. However, IMF direction typically varies on a time scale of a few minutes, i.e., much shorter than the time scale of energy storage and substorm occurrence at the Earth which is about an hour. Accordingly, detailed cause and effect relationships are very difficult to discern. In the case of Mercury, where the magnetospheric response time is believed, on the basis of Mariner 10 data, to be only a minute or so, relations between the IMF and internal magnetospheric processes could be studied with great

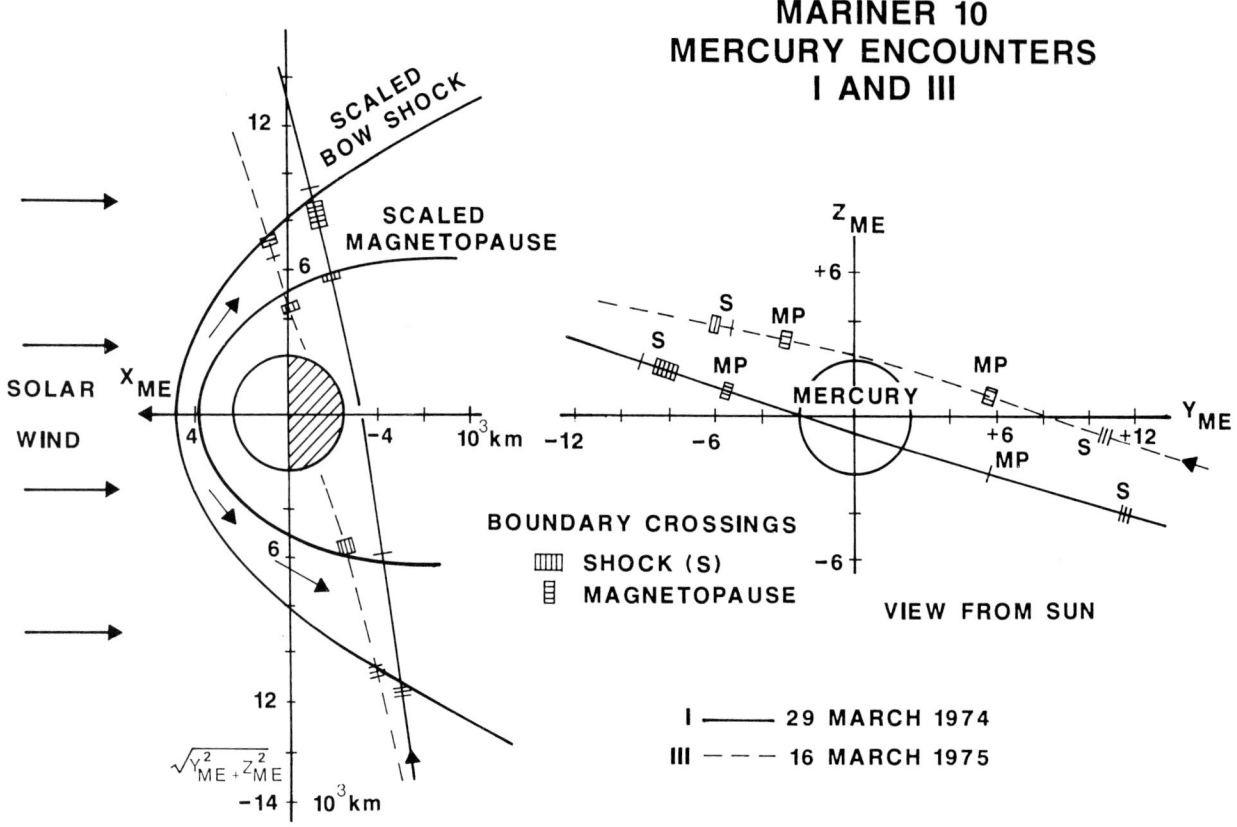

Figure 1. Mariner 10 flyby geometries at Mercury in 1974 and 1975 [adapted from *Ness*, 1979].

benefit. For example, it is not uncommon for the IMF to remain southward and constant for ten minutes. At Mercury this time span is long compared to the substorm cycle time and it would be possible to see whether the magnetosphere responded to this situation by repeated substorms and plasmoid releases as some substorm theories predict (see Table 2 for estimates of energy coupling strengths at Mercury).

4. COMPARATIVE MAGNETOSPHERIC STUDIES AT MERCURY

Mercury is the best place to test and extend the understanding of magnetospheric physics acquired by studying the Earth's magnetosphere. The major difference between Mercury and Earth, viz., the former's lack of an ionosphere, is highly valuable in that it will allow us to ascertain the degree to which theories developed at Earth can be extended to general magnetospheric systems in the case when one of the critical features of the system is radically altered. The slow rotation of Mercury causes solar wind driven convection to dominate throughout the magnetosphere: It is the magnetosphere among all the planets where this dominance is most extreme. Mercury's small magnetosphere may also solve the space-time ambiguity problem that has confounded efforts to perform synoptic studies of Earth's magnetosphere. Approximately once per hour the solar wind conditions change significantly, and magnetospheres must change to accommodate these new conditions. An Earth satellite takes many hours to a day to traverse each of the magnetosphere's structural units, which in the meantime is changing its shape and behavior. At Earth a satellite virtually never samples a complete structural unit before it changes its state. Hence, a statistical approach is necessary for synoptic studies of Earth's magnetosphere. A satellite at Mercury crosses the entire magnetosphere in one-third of an hour or less. The solar wind typically will not change during this time. Thus, the changes a satellite records in a magnetospheric structure at Mercury characterize that structure while the magnetosphere is in a fixed state.

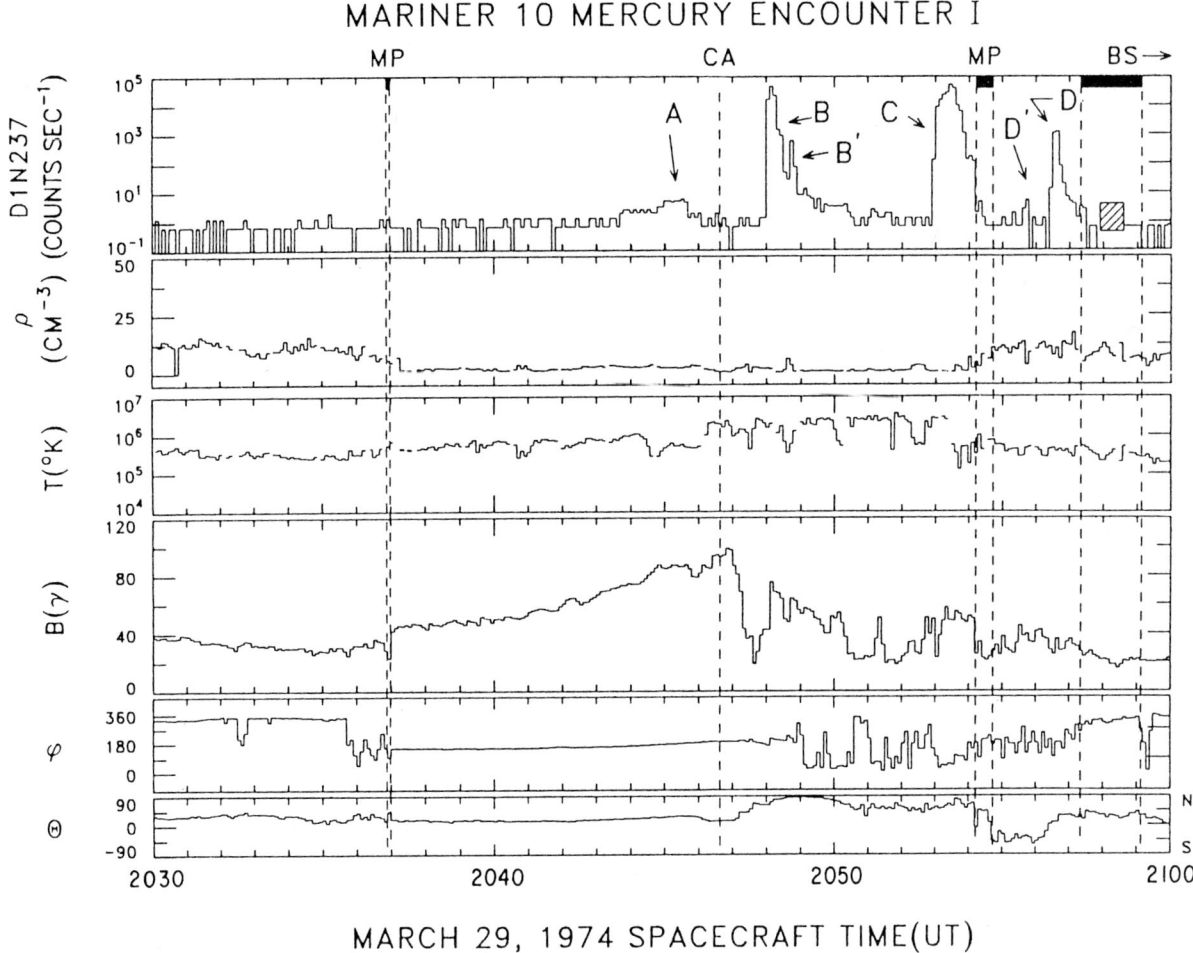

Figure 2. Mariner 10 energetic particle, plasma, and magnetic field data on 29 March 1974 [from *Eraker and Simpson*, 1986].

Table 1. Mercury exospheric number densities

Species	Surface Density (x10^3 cm^{-3})
H	0.02-0.2
He	6
O	40
Na	100
K	0.6
Ar	<7000

5. COMPARATIVE PLANETOLOGY AT MERCURY

Mercury represents an end member planet in Solar System origin and evolution in that it formed in the hottest part of the solar nebula, closer to the sun than any other planet. Outwardly, Mercury resembles the moon; however, a number of striking characteristics point to a very different beginning and geologic history for Mercury compared to the moon and other terrestrial planets: 1) Mercury's bulk density is 5.4 g/cm^3, indicating the highest iron to silicate mass ratio of the terrestrial planets; 2) Reflectance spectra of other terrestrial planets exhibit strong signatures of ferrous iron. Intensive Earth-based searches for these features on Mercury have been inconclusive. It is quite possible that iron does not play as significant a role in the mineralogy of Mercury's crust as it plays in other terrestrial planets; 3) The Mariner 10 discovery that Mercury has an intrinsic magnetic field (and therefore a molten core) is in direct conflict with conventional models which predict that Mercury's core should have solidified eons ago; and 4) Mercury exhibits a significant atmosphere which must be constantly replenished to offset a variety of active loss processes. Atmospheric composition and temporal behavior

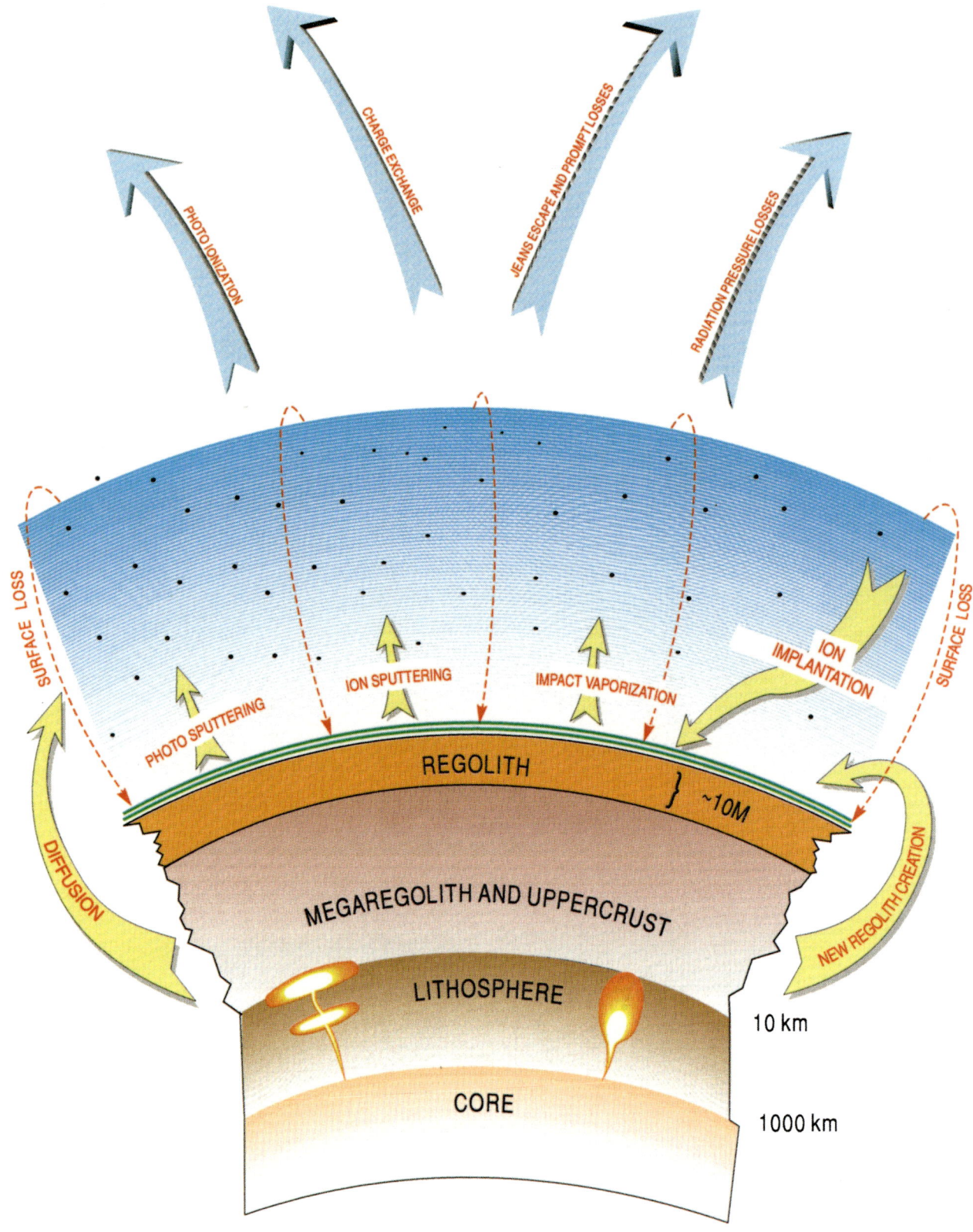

Plate 1. A diagram showing processes possibly acting between the surface and atmosphere of Mercury [from *Morgan and Killen*, 1997].

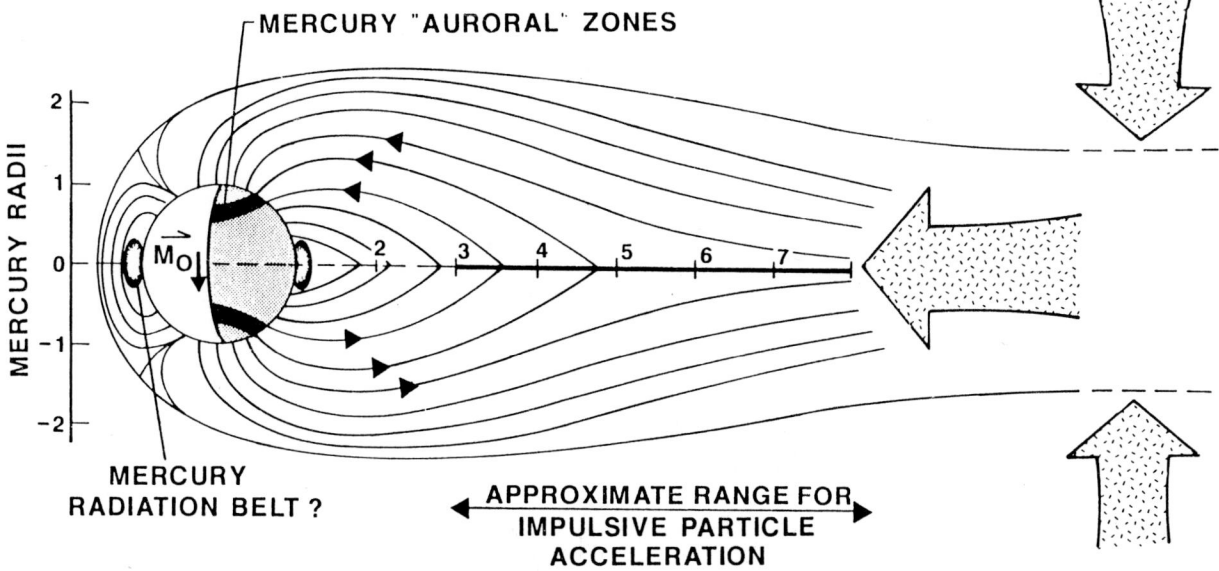

Figure 3. A schematic of the Mercury magnetosphere showing likely region of substorm particle acceleration and possible auroral zones due to particle impacts [from *Baker et al.*, 1987].

are controlled by its interaction with the magnetosphere and the surface, and are driven by the highly variable solar wind.

The composition, structure, and temporal behavior of the atmosphere must be determined. Because at least some of the neutral atmospheric species are derived from surface sputtering [*Morgan and Killen*, 1997], knowledge of atmospheric composition will be useful for inferring surface composition and investigating the long term weathering effects of the solar wind on the surface. In addition, detection and measurement of the amount of cold-trapped polar water ice is extremely important.

6. A MERCURY MISSION SCENARIO

Because of the large velocity changes required, it is a difficult task to place a spacecraft in orbit around Mercury under any circumstances, and particularly with a science payload capable of satisfying multidisciplinary aims. Thus, in recent times propulsion technologies alternative to chemical systems have been promoted as possibly providing much greater flexibility for Mercury mission scenarios. With the development of a new propulsion system for the "New Millennium" series of spacecraft, and with similar development under way in both Japan and Europe, Solar Electric Propulsion (SEP), utilizing electrostatic ion thrusters (or arc jets), appears to be a particularly promising alternative. By taking advantage of the extremely high specific impulses (thrust/mass) of the ion engines (2000-4000 s versus ~300 s for chemical systems), it is possible that much lower launch masses may be required relative to the instrument payload mass. However, at least at the present time, Solar Electric Propulsion cannot be considered a panacea. While the astronautics literature describes many electric propulsion scenarios for reaching several different planetary bodies, typically the true engineering complexities and hardware overheads are underestimated.

Another aspect of the chemical versus SEP trade evaluations is the mission trajectory. Figure 4, for example, provides an illustration of an SEP trajectory evaluation that was conducted for "Roadmap" studies. In this case, the spacecraft, launched by a Delta launch vehicle in December 2000, obtains a gravity assist from Venus, and with the ~5 kW solar electric propulsion system achieves orbit around Mercury with a total flight time of about 2.8 years. Such studies support the advertised capabilities of SEP, since for

Table 2. Magnetospheric Energy Budgets

Energy source	Power level
SW Input	
Kinetic energy flux	$10^{11} - 10^{13}$ W
Electromagnetic coupling	$10^{9} - 10^{12}$ W
Substorm Dissipation	
Earth-Scaling [*Siscoe*, 1975]	$10^{9} - 10^{10}$ W
Particle Bursts [*Eraker*, 1986]	$10^{11} - 10^{14}$ W
Auroral Bands [*Baker*, 1987]	$10^{-5} - 10^{-3}$ W/cm^2

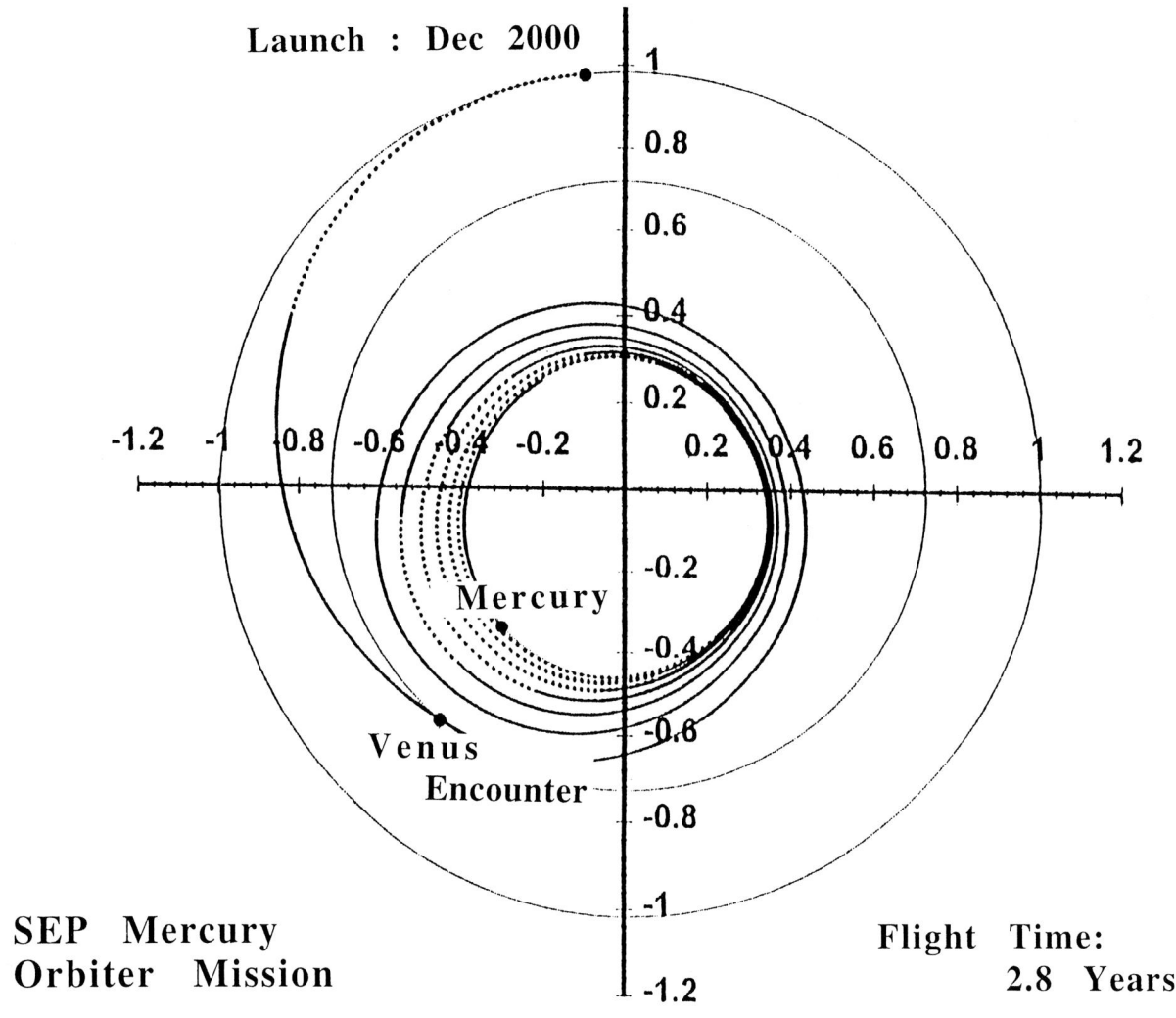

Figure 4. Possible Solar Electric (SEP) trajectory for Mercury mission launching in December 2000.

this approximate time frame there has not been found a reasonable chemical propulsion solution for getting to Mercury orbit (where "reasonable" means with fuel mass fractions that are consistent with standard engineering assumptions). Also, chemical solutions typically require substantially longer flight times. However, the inefficiencies are also quite significant. While chemical solutions require velocity changes of the order of ~3 km/s, SEP solutions require velocity changes of order ~11 km/s due to the inefficiencies of thrusting at non-optimum times (i.e., the very high specific impulses of SEP is squandered to a certain extent). These inefficiencies must be folded together with the substantial costs of including an SEP system on a proposed mission spacecraft.

6.1. Spacecraft Stabilization

Spinning a Mercury orbiter spacecraft effectively distributes the solar heat flux (and, in most instances, the planetary heat flux) around the perimeter. This reduces the peak surface temperatures of sensors, solar cells, and the sides of the spacecraft and helps distribute heat in the interior. Isothermalizing the interior of the spacecraft is important to reduce differences between the fuel and oxidizer temperatures and temperature gradients in each propellant. Thus, the "rotisserie" mode offers several potential operational benefits. In general, a spinning spacecraft also provides a simple, robust operational configuration and it aids in the complete and effective measurement of particles and fields.

On the other hand, a spinning spacecraft is not as desirable as a 3-axis stabilized vehicle for imaging and other planetary science objectives. Thus, tradeoff studies between these options have been a crucial part of recent studies.

6.2. Thermal Design Options

The hostile thermal environment in the vicinity of Mercury is one of the key drivers of any spacecraft design. The solar irradiance at Mercury varies from 4.6 to 10.6 times that at Earth, and the normal planetary albedo is extremely low. As a result, Mercury's surface absorbs 94% of the incident solar flux and re-emits this energy in the infrared waveband. Mercury rotates about its axis so slowly that the dayside surface temperature reaches a near-steady-state temperature of 700K at perihelion. Conversely the dark side temperature reaches as low as 100K. The direct solar flux on the spacecraft ranges from zero-solar constants (SC) during occultations to 1 SC at Earth to 10.6 SC at Mercury perihelion. Total solar flux depends on solar range, phase angle, altitude, and surface albedo in the vicinity of a spacecraft. Heat flux on a spacecraft reradiated from Mercury (in terms of equivalent solar constants) ranges from 0 SC when the spacecraft is far away or in occultation to $\approx$ 8.5 SC when a spacecraft is at periapsis above the subsolar point and Mercury is at perihelion. Other important considerations are that while the sun is small in the sky, Mercury reaches a large angular extent, and the energy from Mercury is predominantly at long wavelengths whereas that emitted by the sun is predominantly at short wavelengths.

Study of possible spacecraft thermal design is primarily based on controlling how environmental heat is distributed on the spacecraft surfaces, minimizing undesired heat input and transfer, controlling heat rejection, maximizing effective heat capacities, and isothermalizing interiors. Low solar-absorptance/high-emissivity, electrically-conductive exteriors may be used to reduce temperatures of surfaces that are exposed to direct or indirect solar flux. In areas used for cooling, direct solar heat input can be eliminated by keeping the sun direction parallel to the radiator surfaces and recessing the radiator assemblies. Indirect solar heating from reflections and reradiation from the antennas must be minimized by limiting the maximum temperatures of these assemblies, limiting their area which is in view of the radiators, and using low α/ϵ surfaces on the radiators.

7. SCIENTIFIC INSTRUMENTS

7.1. Instrument Miniaturization

Recent spaceflight programs such as Clementine have demonstrated the remarkable potential of "miniaturized" space instrumentation [Rustan, 1994]. CCD cameras and particle detectors [e.g., Baker et al., 1995] weighing hundreds of grams have been flown. The pressures are now on to reduce scientific payload masses even more while maintaining good instrument performance and overall sensitivity. Small, capable instruments are critical to the success of any future Mercury mission. This is especially true if both space physics and planetary science goals are to be accommodated.

7.2. Magnetometer

Spaceborne magnetometry is a very mature area with at least two technologies, ring core flux gates and optical pumping of helium, capable of delivering state-of-the-art measurements for only 1-2 kg and 1.5-2.5 watts. Magnetic cleanliness requirements and the probable need for a short boom upon which to mount the sensors have been included in recent studies.

7.3. Energetic Particles

Using innovative configurations along with the development of specialized VLSI circuitry, very compact, light, and low power energetic particle detectors are available that maintain very high performance characteristics. The sensors measure electrons and the time-of-flight and total energy of ions. The compact sensor can obtain the energy spectra of the ions, discriminated according to mass species (H, He, O, Na, K) for energies between about 10 keV/nuc and 10 MeV/nuc. Also, several view directions are accommodated over a total angle of $\leq$ 180°. High sensitivity is maintained, despite a very compact configuration (geometric factor/view direction = 0.1 to 0.2 cm^2-sr). Energetic particle instruments can be defined that are suitable for characterizing the intense, spectrally-soft Hermean energetic particle environment. Such a sensor's role includes detection of possible environmental background (solar and magnetospheric) signals in other onboard sensors.

7.4. Plasma Composition Measurements

To understand the origin, role, and fate of solar wind and planetary plasma species in the Hermean magnetosphere, high sensitivity, high time resolution, 3-D hot plasma composition measurements must be made. Such measurements provide information on the energy distribution, flow and density of species with drastically different sources and histories. For example, sputtered secondary ions or atmospheric photo-ions will be observed as pickup rings or shells, similar to those observed at comets. They provide information on electric fields far from the measurement points. Such measurements also provide information that is crucial to understanding global magnetospheric convection, entry of solar wind plasma (and exit of magnetospheric plasma) at the magnetopause and cusps, and heating, transport and loss of plasma sheet plasma. Because of the rapid timescales characteristic of the Hermean magnetosphere, high sensitivity, high time resolution measurements are crucial. The many species involved require high mass resolution and the relevant species range from photoions up to hot plasma sheet particles so an energy range from 1 eV to at least 30 keV is needed.

7.5. Electric Field Investigation

Electric field measurements are desirable to determine the global electric circuit for the unique plasma environment of Mercury. What is the electric potential drop across the planet and its magnetosphere (tens of kilovolts? more?) and how does this affect the resulting plasma convection and energetic particle acceleration? The understanding of how the electrical conductivity at the foot of magnetospheric field lines controls convection processes is central to all of magnetospheric physics. An electric field instrument flying on a orbiter mission at low altitudes across the polar cap of Mercury is the only way to directly answer this critical question.

7.6. Plasma Wave Investigation

Plasma wave instrumentation can detect and monitor emissions at Mercury that are associated with substorm activity (including local electron and ion acceleration and instability phenomena), as well as plasma waves that will be used as signatures of magnetospheric boundaries (such as the bow shock, the magnetopause, the cusp, auroral oval, etc.) that the spacecraft will encounter. Such instrumentation could also provide accurate plasma density measurements without the complication of spacecraft potential offsets suffered by thermal plasma instruments. The required instrument would measure the radio and plasma wave spectrum from 1 Hz to 300 kHz.

7.7. Combined Gamma Ray/Neutron Spectrometer

One of the most important planetary parameters is global elemental composition. By measuring the surface composition of a planet, and tying that composition to geologic features such as impact basins, highland deposits, ejecta blankets and volcanic features, one gains a more three-dimensional understanding of the planet's make-up. This is why gamma ray spectrometers (GRS) are being built for flight to Mars and the Moon. A GRS also doubles as a solar gamma ray detector for flare studies. Finally, a GRS is also sensitive to very high energy electrons (>1 MeV) arising from either magnetospheric or solar processes. This triple-duty nature of a GRS makes it a very valuable instrument for a Mercury mission.

For geochemical purposes the GRS detects gamma rays arising from the interaction of cosmic rays with surface planetary materials down to a depth of a few tens of cm. Gamma ray line emissions from radioactive decay, inelastic neutron scattering and neutron capture provide a detailed fingerprint of regional geochemistry for most mineralogically-important elements: Si, Mg, Fe, Al, Ti, K, Th, U and Ca all can be discerned with varying degrees of accuracy.

Neutron spectrometer (NS) measurements provide information on soil constituents with high neutron absorption cross-sections (such as Fe, Ti, Sm and Gd), assisting in the geochemical assay. But the most dramatic and important use of the NS is in the detection of water ice, the presence of which moderates the neutrons and very much changes the neutron energy spectrum. In particular, one finds that epithermal neutrons (energies between 0.3 eV and 500 keV) are very quickly moderated down to thermal energies in the presence of water. By measuring the ratios of fast (>500 keV) to epithermal to thermal fluxes, one can obtain an estimate of the water content in the relevant piece of planetary real estate.

7.8. Integrated Imaging and Spectroscopy Experiment

Global imaging and spectroscopic observations of the surface and of the atmosphere are key measurements required to address the planetology and space plasma science goals of a Mercury orbiter mission. The primary imaging measurement objective would be to map the entire surface of Mercury with a resolution of 1 km and up to 25% of the surface with a resolution of 100 m to support geologic investigations. A secondary objective would be to obtain multispectral images of the surface for composition studies. This would determine atmospheric composition, structure, and temporal behavior. These measurements rely on limb scan observations with a vertical resolution of about 50 km (approximately 1 atmospheric scale height for Na). Spectral coverage in the range 0.115 to 0.6 microns with a resolution of 0.1 nm is sufficient to measure prominent emissions of known and candidate species including H, O, Na, K, S, S+, Si, Al, Ca, Ca+, Mg, Mg+, Fe, and OH [*Morgan and Killen* 1997]. These measurements would determine the atmosphere's major sources and sinks (atmospheric processes) and investigate the interactions between the atmosphere, magnetosphere, and the solar wind. Measurement of atmospheric composition can also be used to infer surface composition. Except for the noble gases, hydrogen, and a few volatile species, atmospheric species are derived from the surface. By understanding the sources, sinks, and the gas surface interactions, and by measuring regolith derived elements (Ca, Mg, Na, K, and Fe) the relative ratios of these species in the surface rock can be determined. Recent advances in detector technology have led to the development of a new generation of low-mass, low-power remote sensing instruments (e.g., the visible imaging system for the Clementine mission and Ultraviolet Imaging Spectrographs [*McClintock*, 1996]).

8. SUMMARY

Planetary exploration programs have revealed the benefits of comparative magnetospheric and planetological studies. The known intrinsic planetary magnetospheres of Mercury, Earth, Jupiter, Saturn, Uranus, and Neptune all have similarities of structure which allow the development of analogies between them. However, as perhaps an even more important test of present theoretical understanding, each planetary magnetosphere has significant differences from the other systems. This causes a substantial contrast from one planet to the next. Mercury's most Earth-like of magnetospheres shows many familiar features such as energetic particle bursts and globally coherent dynamics. Thus, a pervasive feature of cosmic plasmas generally, and magnetospheres in particular, appears to be the rapid and efficient conversion of magnetic field free energy into the kinetic energy of suprathermal particle populations [e.g., *Rosner et al.*, 1984]. A Mercury orbiter mission could revolutionize our understanding of Earth and the entire inner Solar System. New spacecraft designs and miniaturized instruments now place a comprehensive Mercury mission into our fiscal grasp.

Acknowledgments. This paper represents the thinking of the Space Physics "New Concepts" team as well as the Sun-Earth Connections "Roadmap" group. The author thanks the entire membership of these teams for useful discussions and valuable inputs. This work was supported by NASA.

REFERENCES

An Implementation Plan for Priorities in Solar System Space Physics, National Academy of Sciences, Washington, DC, 1985.

Baker, D.N., S.I. Akasofu, W. Baumjohann, J.W. Bieber, D.H. Fairfield, E.W. Hones Jr., B.H. Mauk, R.L. McPherron, and T.E. Moore, Substorms in the magnetosphere, in *Solar Terrestrial Physics – Present and Future*, NASA, Washington, DC, 1984.

Baker, D.N., J.A. Simpson, and J.H. Eraker, A model of impulsive acceleration and transport of energetic particles in Mercury's magnetosphere, *J. Geophys. Res., 91,* 8742, 1986.

Baker, D.N., J.E. Borovsky, J.O. Burns, G.R. Gisler, and M. Zeilik, Possible calorimetric effects at Mercury due to solar wind-magnetosphere interactions, *J. Geophys. Res., 92,* no. A5, 4707-4712, 1987.

Baker, D.N., S. Kanekal, J.B. Blake, and J.H. Adams Jr., Charged-particle telescope experiment on Clementine, *J. Spacecraft and Rockets, 32,* no. 6, 1995.

Baker, D.N., A.J. Klimas, D. Vassiliadis, T.I. Pulkkinen, and R.L. McPherron, Re-examination of driven and unloading aspects of magnetospheric substorms, *J. Geophys. Res., 102,* 7169, 1997.

Eraker, J.H., and J.A. Simpson, Acceleration of charged particles in Mercury's magnetosphere, *J. Geophys. Res., 91,* 9875, 1986.

McClintock, W.E., and G.M. Lawrence, Low mass, low power ultraviolet telescope imaging spectrograph for planetary atmospheric remote sensing, *SPIE Proceedings, 2807,* 256-266, 1996.

Mercury Dual Orbiter: Mission and Flight System Definition, JPL D-7443, Pasadena, CA, 1990.

Morgan, T.W., and R. Killen, A non-stoichiometric model of the composition of the atmospheres of Mercury and the Moon, *Planet. Space Sci., 45,* 81-94, 1997.

Ness, N.F., The magnetosphere of Mercury, in *Solar System Plasma Physics*, vol. 2, edited by C.F. Kennel, L.J. Lanzerotti, and E.N. Parker, North-Holland, Amsterdam, 1979.

Ogilvie, K.W., J.D. Scudder, V.M. Vasyliunas, R.E. Hartle, and G.L. Siscoe, Observations at the planet Mercury by the plasma electron experiment Mariner 10, *J. Geophys. Res., 82,* 13, 1977.

Report of the Terrestrial Bodies Science Working Group, Volume II. Mercury, JPL Pub. 77-51, Pasadena, CA, September 15, 1977.

Rosner, R., E.L. Chupp, G. Gloeckler, D.J. Gorney, S.M. Krimigis, Y. Mok, R. Ramaty, D.W. Swift, L. Vlahos, and E.G. Zweibel, Particle acceleration, in *Solar Terrestrial Physics – Present and Future*, NASA, Washington, DC, 1984.

Rustan, P.L., Flight qualifying space technology with the Clementine mission, *EOS, 75,* 1994, pp. 161-165.

Simpson, J.A., J.H. Eraker, J.E. Lamport, and P.H. Walpole, Electrons and protons accelerated in Mercury's magnetic field, *Science, 185,* 160, 1974.

Siscoe, G.L., N.F. Ness, and C.M. Yeates, Substorms on Mercury?, *J. Geophys. Res., 80,* 4359, 1975.

Slavin, J.A., and R.E. Holzer, The effect of erosion on the solar wind stand-off distance at Mercury, *J. Geophys. Res., 84,* 2076, 1979.

Space Science in the Twenty-First Century: Planetary and Lunar Exploration, National Academy of Sciences, 1988.

Strategy for Exploration of the Inner Planets: 1977-1987, National Academy of Sciences, Washington, DC, 1978.

D.N. Baker, Laboratory for Atmospheric and Space Physics, University of Colorado, Boulder, CO 80309-0590.